Springer Theses

Recognizing Outstanding Ph.D. Research

Aims and Scope

The series "Springer Theses" brings together a selection of the very best Ph.D. theses from around the world and across the physical sciences. Nominated and endorsed by two recognized specialists, each published volume has been selected for its scientific excellence and the high impact of its contents for the pertinent field of research. For greater accessibility to non-specialists, the published versions include an extended introduction, as well as a foreword by the student's supervisor explaining the special relevance of the work for the field. As a whole, the series will provide a valuable resource both for newcomers to the research fields described, and for other scientists seeking detailed background information on special questions. Finally, it provides an accredited documentation of the valuable contributions made by today's younger generation of scientists.

Theses are accepted into the series by invited nomination only and must fulfill all of the following criteria

- They must be written in good English.
- The topic should fall within the confines of Chemistry, Physics, Earth Sciences, Engineering and related interdisciplinary fields such as Materials, Nanoscience, Chemical Engineering, Complex Systems and Biophysics.
- The work reported in the thesis must represent a significant scientific advance.
- If the thesis includes previously published material, permission to reproduce this must be gained from the respective copyright holder.
- They must have been examined and passed during the 12 months prior to nomination.
- Each thesis should include a foreword by the supervisor outlining the significance of its content.
- The theses should have a clearly defined structure including an introduction accessible to scientists not expert in that particular field.

More information about this series at http://www.springer.com/series/8790

Thomas James Whittles

Electronic Characterisation of Earth-Abundant Sulphides for Solar Photovoltaics

Doctoral Thesis accepted by
the University of Liverpool, Liverpool, UK

Author
Dr. Thomas James Whittles
Department of Physics and Stephenson Institute for Renewable Energy
University of Liverpool
Liverpool
UK

Supervisor
Dr. V. R. Dhanak
University of Liverpool
Liverpool
UK

ISSN 2190-5053 ISSN 2190-5061 (electronic)
Springer Theses
ISBN 978-3-030-06276-7 ISBN 978-3-319-91665-1 (eBook)
https://doi.org/10.1007/978-3-319-91665-1

© Springer Nature Switzerland AG 2018, corrected publication 2018
Softcover re-print of the Hardcover 1st edition 2018
This work is subject to copyright. All rights are reserved by the Publisher, whether the whole or part of the material is concerned, specifically the rights of translation, reprinting, reuse of illustrations, recitation, broadcasting, reproduction on microfilms or in any other physical way, and transmission or information storage and retrieval, electronic adaptation, computer software, or by similar or dissimilar methodology now known or hereafter developed.
The use of general descriptive names, registered names, trademarks, service marks, etc. in this publication does not imply, even in the absence of a specific statement, that such names are exempt from the relevant protective laws and regulations and therefore free for general use.
The publisher, the authors and the editors are safe to assume that the advice and information in this book are believed to be true and accurate at the date of publication. Neither the publisher nor the authors or the editors give a warranty, express or implied, with respect to the material contained herein or for any errors or omissions that may have been made. The publisher remains neutral with regard to jurisdictional claims in published maps and institutional affiliations.

This Springer imprint is published by the registered company Springer Nature Switzerland AG
The registered company address is: Gewerbestrasse 11, 6330 Cham, Switzerland

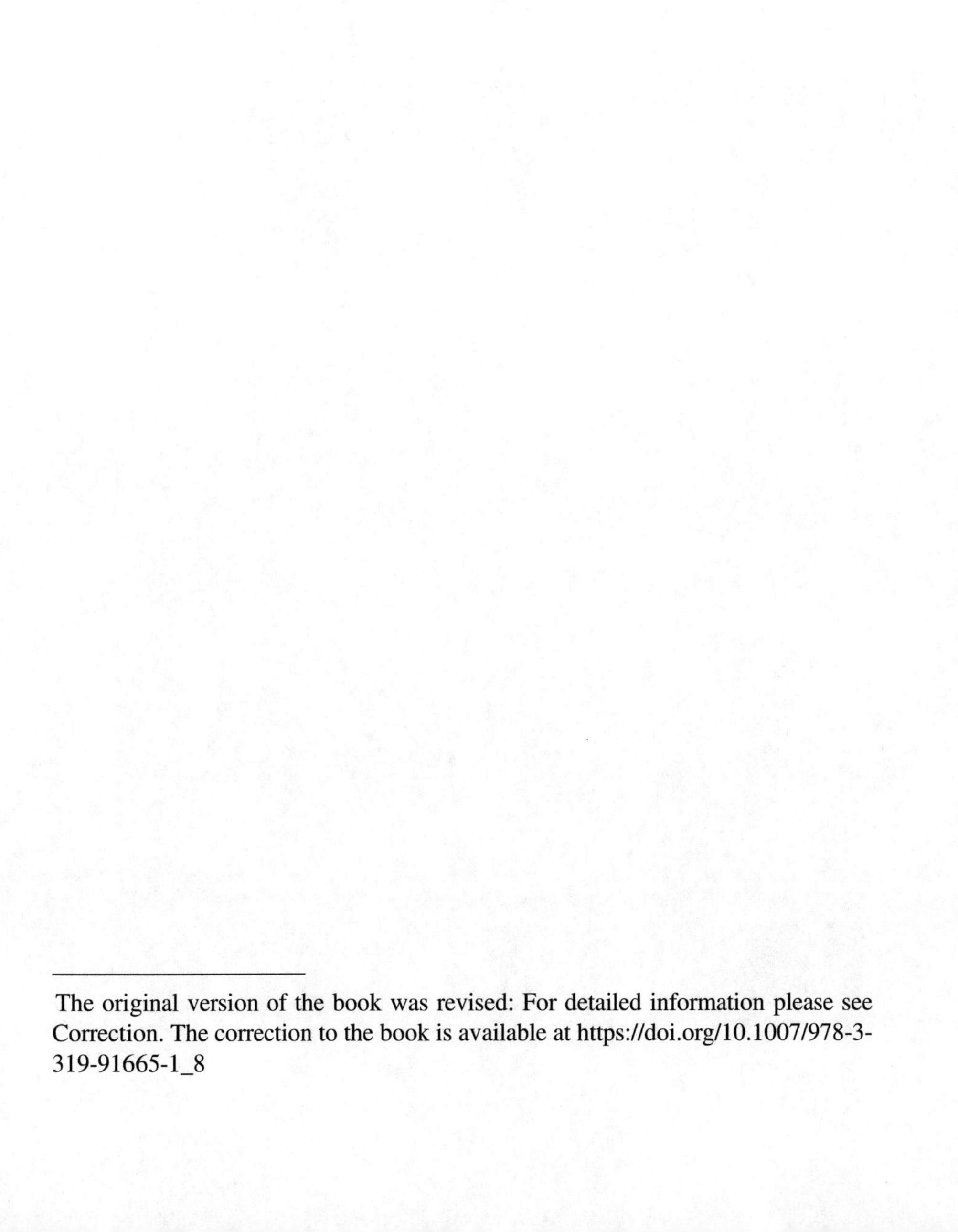

The original version of the book was revised: For detailed information please see Correction. The correction to the book is available at https://doi.org/10.1007/978-3-319-91665-1_8

Supervisor's Foreword

The generation and use of energy continue to increase because of human population increase and technological advancement, and by the year 2050, the world's energy need is projected to be about 30 TW per year. More than 80% of this currently comes from burning fossil fuels, which has led to adverse environmental and climate changes. The case for alternative renewable energy sources is therefore clear and while all renewable resources (solar, wind, hydroelectric, biomass, and geothermal) play an important role, only solar can meet this level of demand, since the sun continuously provides more energy to the earth in one hour than the world that consumes in an entire year. Currently, commercial solar cell technologies are based on silicon, cadmium telluride (CdTe), or copper indium gallium diselenide (CIGS) as the photovoltaic (PV) solar absorbers. While the latter two are thin-film technologies that are cheaper than silicon, there are serious concerns about the long-term sustainability and toxicity of some of the elements used in these thin films. There is thus justifiable search for new solar cell materials using earth-abundant, cheap, and environmentally friendly elements. The search for a suitable solar absorber combines sophisticated computational screening together with experimental characterisation of the electronic structure of the material. Efficiency losses in solar cells can occur because of an inappropriate bandgap or because of band misalignments with the other layers within the cell; therefore, studying band alignments in heterojunctions is of vital importance. The bandgap, position of the band levels, and optical absorption characteristics are important properties for solar absorbers. Determining the band structures of materials can allow relationships between the energy levels to be drawn, leading to an understanding of how electronic orbitals of the different atoms hybridise and give rise to the electronic band positions. This also allows the bandgap and band alignment to be determined. Also, from a solar cell device perspective it is important to determine the effect that imperfections have on the absorber material. These imperfections can manifest as unwanted material made of the constituent elements of the absorber, but in a different phase, or with some form of external contaminant such as oxygen or carbon; thus, electronic characterisation is extremely important in this search for new PV materials. The work of Thomas Whittles presented in this book demonstrates the use of electronic characterisation, combined with theoretical insights into the origin of electronic levels in some of these earth-abundant PV materials, and crucially identifies the important parameters required in the search for new materials. The methods and procedures that are shown are not confined to PV materials, but are general to the electronic characterisation of any materials for devices.

Liverpool, UK
May 2018

Dr. V. R. Dhanak

Abstract

This thesis explores the electronic characterisation of materials for use as absorber layers within photovoltaic solar cells: an attractive solution to the energy crisis. Primarily, XPS was used to characterise $CuSbS_2$, Cu_3BiS_3, SnS, and Cu_2ZnSnS_4. All of these materials can be classified as earth-abundant: an important factor when considering materials that are both readily available and environmentally friendly.

To varying degrees, all of these materials are established as potential absorber layers, with reports of successful, albeit low-efficiency devices. Also, the literature is often scant with regards to more fundamental characterisation of these materials, specifically in terms of how the underlying electronic structure affects properties that are pertinent to solar cells. Where XPS is utilised, it often lacks rigour and is obfuscated by the complexities of the spectra.

Work is then presented with two aims. First, to use a combination of high-quality XPS measurements and density of states calculations, in order to give insight into the electronic structure of the materials. Second, to present the full potential of XPS, as applied to solar absorbers, and how this technique complements others that are used for characterisation.

Without exception, each material demonstrated natural band positions that make them unsuitable for use with established solar cell technologies. That is, a low IP (4.71–5.28 eV), resulting in a large CBO (0.5–0.85 eV) with CdS, for example. This offers some explanation to the poor efficiencies. These properties were found to be a consequence of the bonding nature of the valence and conduction bands, differing to the conventional absorber materials because of the influence of lone-pair electrons or other second-cation states.

The power of XPS was demonstrated for application to absorber materials. The presence, effects, and formation pathways of contamination of the materials were elucidated using XPS, showing how the presence of such can act adversely to the performance of cells, especially with regards to oxidation, and also how these factors have been overlooked in the past. When coupled with other complementary techniques, it has the ability to aid in the phase identification of a grown material and also to help determine the presence of unwanted phases, which cause detriment to the device. For these applications, the methods, fitting procedures, and analysis considerations detailed in this thesis should be followed.

Research interest in these materials should be maintained based on the findings, with new approaches to cell design being aided by the characterisation methods developed here.

Publications and Presentations

Publications Directly Associated with this Thesis

- Peccerillo, E.; Major, J.; Phillips, L.; Treharne, R.; Whittles, T. J.; Dhanak, V.; Halliday, D.; Durose, K. Characterization of Sulfurized $CuSbS_2$ Thin Films for PV Applications. In *2014 IEEE 40th Photovoltaic Specialist Conference (PVSC)*; IEEE, 2014; pp 0266–0269.
- Burton, L. A.; Whittles, T. J.; Hesp, D.; Linhart, W. M.; Skelton, J. M.; Hou, B.; Webster, R. F.; O'Dowd, G.; Reece, C.; Cherns, D.; Fermin, D. J.; Veal, T. D.; Dhanak, V. R.; Walsh, A. Electronic and Optical Properties of Single Crystal SnS_2 : An Earth-Abundant Disulfide Photocatalyst. *J. Mater. Chem. A* **2016**, *4* (4), 1312–1318.
- Whittles, T. J.; Burton, L. A.; Skelton, J. M.; Walsh, A.; Veal, T. D.; Dhanak, V. R. Band Alignments, Valence Bands, and Core Levels in the Tin Sulfides SnS, SnS_2, and Sn_2S_3 : Experiment and Theory. *Chem. Mater.* **2016**, *28* (11), 3718–3726.
- Gupta, S.; Whittles, T. J.; Batra, Y.; Satsangi, V.; Krishnamurthy, S.; Dhanak, V. R.; Mehta, B. R. A Low-Cost, Sulfurization Free Approach to Control Optical and Electronic Properties of Cu_2ZnSnS_4 via Precursor Variation. *Sol. Energy Mater. Sol. Cells* **2016**, *157*, 820–830.
- Whittles, T. J.; Veal, T. D.; Savory, C. N.; Welch, A. W.; de Souza Lucas, F. W.; Gibbon, J. T.; Birkett, M.; Potter, R. J.; Scanlon, D. O.; Zakutayev, A.; Dhanak, V. R. Core-Levels, Band Alignments and Valence Band States in $CuSbS_2$. *Manuscript in Preparation* **2017**.
- Whittles, T. J.; Veal, T. D.; Yates, P.; Savory, C. N.; Scanlon, D. O.; Major, J. D.; Durose, K.; Dhanak, V. R. Core-levels, Band Alignments and Valence Band States in Cu_3BiS_3 for its use as a PV Absorber. *Manuscript in Preparation* **2017**.

Presentations Directly Associated with this Thesis

- EUPVSEC 2014, Amsterdam NL
  Poster: Tin Sulfide for Low-Cost Solar Cells
- Postgraduate Poster Day 2015, University of Liverpool UK
  Poster: *Earth Abundant Materials for Solar Cells*

- EMRS Spring 2015, Lille FR
  Talk: Tin Sulfide for PV: A Reconfirmation of Potential
- EUPVSEC 2015, Hamburg DE
  Poster: Earth Abundant Absorbers $CuSbS_2$ & Cu_3N: Photoemission and Optical Studies
- EUPVSEC 2015, Hamburg DE
  Talk: Tin Sulfide for PV: A Reconfirmation of Potential

Publications Arising from Work Conducted during the Course of this Thesis but not Directly Related to it

- Li, W.; O'Dowd, G.; Whittles, T. J.; Hesp, D.; Gründer, Y.; Dhanak, V. R.; Jäckel, F. Colloidal Dual-Band Gap Cell for Photocatalytic Hydrogen Generation. *Nanoscale* **2015**, *7* (40), 16606–16610.
- Shaw, A.; Whittles, T. J.; Mitrovic, I. Z.; Jin, J. D.; Wrench, J. S.; Hesp, D.; Dhanak, V. R.; Chalker, P. R.; Hall, S. Physical and Electrical Characterization of Mg-Doped ZnO Thin-Film Transistors. In *2015 45th European Solid State Device Research Conference (ESSDERC)*; IEEE, 2015; pp 206–209.
- Ahmed, A.; Robertson, C. M.; Steiner, A.; Whittles, T.; Ho, A.; Dhanak, V.; Zhang, H. Cu(I)Cu(II)BTC, a Microporous Mixed-Valence MOF via Reduction of HKUST-1. *RSC Adv.* **2016**, *6* (11), 8902–8905.
- Speckbacher, M.; Treu, J.; Whittles, T. J.; Linhart, W. M.; Xu, X.; Saller, K.; Dhanak, V. R.; Abstreiter, G.; Finley, J. J.; Veal, T. D.; Koblmüller, G. Direct Measurements of Fermi Level Pinning at the Surface of Intrinsically N-Type InGaAs Nanowires. *Nano Lett.* **2016**, *16* (8), 5135–5142.
- Neri, G.; Forster, M.; Walsh, J. J.; Robertson, C. M.; Whittles, T. J.; Farràs, P.; Cowan, A. J. Photochemical CO 2 Reduction in Water Using a Co-Immobilised Nickel Catalyst and a Visible Light Sensitiser. *Chem. Commun.* **2016**, *52* (99), 14200–14203.
- Shaw, A.; Wrench, J. S.; Jin, J. D.; Whittles, T. J.; Mitrovic, I. Z.; Raja, M.; Dhanak, V. R.; Chalker, P. R.; Hall, S. Atomic Layer Deposition of Nb-Doped ZnO for Thin Film Transistors. *Appl. Phys. Lett.* **2016**, *109* (22), 222103.
- Sansom, H. C.; Whitehead, G. F. S.; Dyer, M. S.; Zanella, M.; Manning, T. D.; Pitcher, M. J.; Whittles, T. J.; Dhanak, V. R.; Alaria, J.; Claridge, J. B.; Rosseinsky, M. J. $AgBiI_4$ as a Lead-Free Solar Absorber with Potential Application in Photovoltaics. *Chem. Mater.* **2017**, *29* (4), 1538–1549.
- Major, J. D.; Phillips, L. J.; Al Turkestani, M.; Bowen, L.; Whittles, T. J.; Dhanak, V. R.; Durose, K. P3HT as a Pinhole Blocking Back Contact for CdTe Thin Film Solar Cells. *Sol. Energy Mater. Sol. Cells* **2017**, *172*, 1–10.

Work Done in Conjunction with Others

With the exceptions listed below, this thesis contains an account of my research carried out at the Stephenson Institute for Renewable Energy and Department of Physics, University of Liverpool, between October 2013 and July 2017 under the supervision of Dr. V. R. Dhanak and funded through the EPSRC.[1] The research here has not been previously submitted, wholly or in part, for admission to a higher degree at this, or any other academic institution. Where collaborators have provided data, all analysis and interpretation are the work of the author.

- The **$CuSbS_2$** films measured in Chap. 3 were provided by **Francisco Willian de Souza Lucas**,[2] **Adam W. Welch** (see Footnote 2), and **Andriy Zakutayev** (see Footnote 2).
- The **Cu_3BiS_3** film measured in Chap. 4 was provided by **Peter Yates**.[3]
- The **Tin Sulphides** crystals measured in Chap. 5 were provided by **Lee A. Burton**.[4]
- The **CZTS** crystals measured in Chap. 6 were provided by **Akira Nagaoka**.[5]
- The **RS** measurements in Chaps. 3 and 4 were performed by **James T. Gibbon** (see Footnote 3).
- The **XRD** measurements in Chap. 4 were performed by **Peter Yates** (see Footnote 3).
- The **Optical Absorption Spectroscopy** measurements in Chaps. 3 and 5 were performed by **Max Birkett** (see Footnote 3) and **Wojciech M. Linhart** (see Footnote 3).

[1] UK Engineering and Physical Sciences Research Council (Grant # EP/L505018/1 and EP/K503095/1).

[2] National Renewable Energy Laboratory, Golden CO, USA.

[3] Stephenson Institute for Renewable Energy, University of Liverpool, Liverpool, UK.

[4] Centre for Sustainable Chemical Technologies, University of Bath, Bath, UK.

[5] Department of Applied Physics and Electronic Engineering, University of Miyazaki, Miyazaki, Japan.

- The **EDX** measurements in Chap. 6 were performed by **Tobias Heil**.[6]
- The **XRD** measurements in Chap. 6 were performed by **Christopher A. Muryn**.[7]
- The **DFT** calculations in Chaps. 3, 4, and 6 were performed by **Christopher N. Savory**[8] and **David O. Scanlon** (see Footnote 8).
- The **DFT** calculations in Chap. 5 were performed by **Lee A. Burton** (see Footnote 4) and **Aron Walsh** (see Footnote 4).

[6] Nanoinvestigation Centre at Liverpool, Liverpool, UK.

[7] School of Chemistry, University of Manchester, Manchester, UK.

[8] Department of Chemistry, University College London, London, UK.

Acknowledgements

Doing a Ph.D. is hard. There is no doubt about that. I feel this must be especially the case for the experimental sciences. The world felt against me, and progress was slow, non-existent, or even retrograde at times. Some days, all I would have to show for my toil would be a UHV system that leaked only slightly less than it did the day before and an immaculately organised literature folder or tool cabinet: resultant from the time waiting for the excess helium to be pumped away after the joys of leak chasing. Despite this, it would appear that a fair amount of results emanated from this time, are presented within this thesis, and will hopefully add to the scientific knowledge in this promising subject area.

It is saddening that the quality of some of the results that make it through the peer-reviewed process is extraordinarily poor, even in reputable journals. However, at times, some of the associated text, interpretations, and figures provided the comic relief necessary to undertake a Ph.D. I would therefore encourage anyone struggling or doubting themselves along this journey to read the literature, find something that makes you smile—or despair—print it out, stick it on the wall, and remember; you can do better than this.

In my time doing this Ph.D., thanks must go out to those who kept me sane and helped make it so.

First, thanks to Dr. Vin Dhanak who, as a supervisor, has allowed me to develop my own research style, but also has been there with guidance and advice, for which I could not ask for better. I am also extremely grateful for the amount of Indian restaurants I have visited with him and his family, where he has always been most generous. Ketki Dhanak must also be mentioned here, for without her, conversations about physics during these meetups would have continued long into the night, but thanks also for the kind gift of Philip Hofmann's excellent book [1].

Throughout these four years, many people deserve thanks for their help with work and life at the University of Liverpool: David Hesp, for teaching me all he knew and telling how it is; Dr. Tim Veal, for invaluable discussions; Jack Swallow, Tom Baines, Graeme O'Dowd, Peter Yates, and of course Vince Vasey for never being one to turn down a beer; everyone in the Stephenson Institute for Renewable

Energy who has helped with measurements, granted me co-authorship on papers, provided stimulating conversation at lunch, or challenged me to a game of chess—and lost; and everyone who has passed through the laboratory either as Ph.D. students, visitors, master's students or anything else: hopefully I taught you something and you enjoyed the experience, thank you for allowing me that pleasure. There are many others whom I can't possibly name here, so read into that what you will!

Extra thanks must go to all of my family, especially my fantastically supportive parents. I don't think they have much clue of what I do, but I know they are proud and will support me in any way they can. I'm pretty sure my aunt still thinks I'm going to be a physician.

Finally, my utmost gratitude must go to Franzi. Without her unwavering support, in all the ways she knows how, I don't think I'd be in the position that I am today. Therefore, it is to her that I dedicate this thesis; hopefully, this will somewhat make up for the time I couldn't spend at home whilst writing it. She also has the best suggestions, so here you go:

Liverpool, UK
July 2017

Dr. Thomas James Whittles

Reference

1. Hofmann P. Surface physics: An introduction. Philip Hofmann; 2013.

Contents

Abbreviations

Adv.	Adventitious
ALD	Atomic layer deposition
Ann.	Anneal
Arb.	Arbitrary
a–Si	Amorphous silicon
ASTM	American Society for Testing and Materials
AZO	Aluminium-doped zinc oxide
BD	Berndtite
BE	Binding energy
CAS	$CuSbS_2$
CASe	$CuSbSe_2$
CB	Conduction band
CBD	Chemical bath deposition
CBM	Conduction band maximum
CBO	Conduction band offset
CBS	Cu_3BiS_3
CE	Configuration energy
CGS	$CuGaSe_2$
CHA	Concentric hemispherical analyser
ChC	Chalcocite
ChP	Chalcopyrite
CIGS	$Cu(In,Ga)Se_2$
CIS	$CuInSe_2$
CL	Core level
CTS	Copper tin sulphide
CV	Covellite or capacitance–voltage profiling
CVD	Chemical vapour deposition
CVT	Chemical vapour transport
CZS	Copper zinc sulphide
CZTS	Cu_2ZnSnS_4

CZTSe	$Cu_2ZnSnSe_4$
CZTSSe	$Cu_2ZnSn(S,Se)_4$
DFT	Density functional theory
DIN	Deutsches Institut für Normung
DoS	Density of states
DSSC	Dye-sensitised solar cell
EA	Electron affinity
EBPVD	Electron beam physical vapour deposition
EDX	Energy-dispersive x-ray spectroscopy
Env.	Envelope
EPMA	Electron probe microanalysis
EPSRC	Engineering and Physical Sciences Research Council
EQE	External quantum efficiency
ESCA	Electron spectroscopy for chemical analysis
EU	European Union
Exp.	Experimental
FCC	Face-centred cubic
FEG	Free electron gas
FET	Field-effect transistor
FF	Fill factor
FG	Flood gun
fs-UPS	Femtosecond ultraviolet photoemission spectroscopy
FTIR	Fourier transform infrared spectroscopy
FTO	Fluorine-doped tin oxide
FWHM	Full width at half maximum
GGA	Generalised gradient approximation
HC	Hydrocarbon
HSE	Heyd–Scuseria–Ernzerhof
HV	High vacuum
HZ	Herzenbergite
ICDD	International Centre for Diffraction Data
ICP-MS	Inductively coupled plasma mass spectrometry
$IMFP_e$	Inelastic mean free path of an electron
IP	Ionisation potential
IPES	Inverse photoemission spectroscopy
IQE	Internal quantum efficiency
ISE	Institute for Solar Energy Systems
ITO	Indium tin oxide
IUPAC	International Union of Pure and Applied Chemistry
JQA	Japan Quality Assurance Organisation
KE	Kinetic energy
KPFM	Kelvin probe force microscopy
KS	Kesterite
LEED	Low-energy electron diffraction
NEXAFS	Near-edge x-ray absorption fine structure

NiCaL	Nanoinvestigation Centre at Liverpool
NREL	National Renewable Energy Laboratory
OoM	Order of magnitude
OT	Ottemannite
PAW	Projector augmented wave method
PBE	Perdew–Burke–Ernzerhof
PDF	Powder diffraction file
pDoS	Partial density of states
PES	Photoemission spectroscopy
PL	Photoluminescence
PLD	Pulsed laser deposition
PPMS	Physical property measurement system
PV	Photovoltaic
PVD	Physical vapour deposition
PYS	Photon yield spectroscopy
RGA	Residual gas analyser
RoEU	Rest of European Union
RS	Raman spectroscopy
SCF	Stoichiometry correction factor
SEC	Secondary electron cutoff
SEM	Scanning electron microscope
SILAR	Successive ionic layer adsorption and reaction
SLG	Soda–lime glass
SNR	Signal-to-noise ratio
SP	Sphalerite
Sp.	Sputter
SQL	Shockley-Queisser limit
SRPES	Synchrotron radiation photoemission spectroscopy
ST	Stannite
STM	Scanning tunnelling microscope
TDL	Thermodynamic limit
TEM	Transmission electron microscopy
TFSC	Thin-film solar cell
TPE	Tunable photon energy
TSP	Titanium sublimation pump
TT	Thermal treatment
UHV	Ultrahigh vacuum
UPS	Ultraviolet photoemission spectroscopy
UV	Ultraviolet
VASP	Vienna ab initio simulation package
VB	Valence band
VBM	Valence band maximum
VBO	Valence band offset
VdW	Van der Waals
WF	Work function

XANES	X-ray absorption near edge structure
XAS	X-ray absorption spectroscopy
XES	X-ray emission spectroscopy
XPS	X-ray photoemission spectroscopy
XRD	X-ray diffraction
XRF	X-ray fluorescence
ZTS	Zinc tin sulphide

Symbols

a	Lattice constant
a_B	Bose–Einstein parameter
A	Anneal or proportionality constant
A_{eff}	Effective working area
B	Background
c	Speed of light
C, C_{Tot}	Conductance, total conductance
Cu_{ex}	Extra copper species
$\text{Cu}_{\text{Tet/Tri}}$	Tetrahedrally/trigonally coordinated copper
d	Interplanar spacing or thickness
e	Doubly degenerate orbital
$E_{B/K}$	Binding/kinetic energy
E_0	Bandgap at 0 K
$E_{\text{CBM/VBM}}$	CBM/VBM energy
E_{CL}	Core-level energy
E_F	Fermi-level energy
$E_g^{d/i}$	Bandgap (direct/indirect)
$E_{P/R}$	Pass/retardation energy
$E_{\text{SEC/Vac}}$	SEC/vacuum-level energy
f	Atomic scattering factor
F_{hkl}	Structure factor
G	Outgassing rate
h	Planck constant
$h\nu$	Photon energy
I	Current or intensity
$I_{D/Ph/Sh}$	Current (diode reverse saturation/photo/shunt)
I_{flux}	Flux ion current
I_{sol}	Power density of the sun
j	Total angular momentum quantum number
$J_{\text{Max/SC}}$	Current density (maximum/short circuit)

k	Wavevector
k_S	Shirley factor
l	Azimuthal quantum number
m	Mass
M	Molar mass
n	Principle quantum number or diffraction order or number of moles or electron concentration
N	Number of molecules
p	Pressure or hole concentration
$P_{\text{Light/Max}}$	Power (light/maximum)
R	Reflectance
$R_{c/i/o}$	Radius (central/inner/outer)
R_{eff}	Effective resolution
$R_{S/Sh}$	Resistance (series/shunt)
S	Sputter or pumping speed
S_{eff}	Effective pumping speed
t_2	Triply degenerate orbitals
T	Transmittance
V	Volume
$V_{\text{OC/Max}}$	Voltage (open circuit/maximum)
w	Slit width
X	Atomic percentage
α	Absorption coefficient or Auger parameter or Varshni parameter
α'	Modified Auger parameter
β	Varshni parameter
$\Delta\alpha$	Angular acceptance
ΔE	FWHM of PES feature
ΔE_A	Analyser resolution
$\Delta E_{\text{CL}/X}$	Linewidth (core level/x-ray)
$\Delta\omega$	Raman shift
η	Efficiency
θ, φ	Angle
Θ_E	Einstein temperature
$\lambda_{I/S}$	Wavelength (incident/source)
ρ	Density
$\phi_{A/G/S}$	Work function (analyser/electron gun/sample)
ω	Screening parameter

Chapter 1
Introduction

One thing I feel sure of... is that the human race must finally utilise direct sun power or revert to barbarism.

Frank Schuman, Solar Energy Pioneer, 1913

We, as humans, are at a unique point in history. Throughout the ages, mankind has predicted and prophesised its own demise, but never yet has this come to fruition. At present, we have reached such a level of technological advancement that we are able to predict how this will happen. On a universal scale, the heat death of the universe [1] is a theory with substantial support, but on a more local scale, a deficit of humanly extractable energy is likely to terminate the human race, and our continued reliance and exponentially increasing need for energy, in various forms, could be our downfall [2, 3].

The original version of this chapter was revised: Table was reformatted and the references to the table were cited. The correction to this chapter is available at https://doi.org/10.1007/978-3-319-91665-1_8

© Springer International Publishing AG, part of Springer Nature 2018

T. J. Whittles, *Electronic Characterisation of Earth-Abundant Sulphides for Solar Photovoltaics*, Springer Theses, https://doi.org/10.1007/978-3-319-91665-1_1

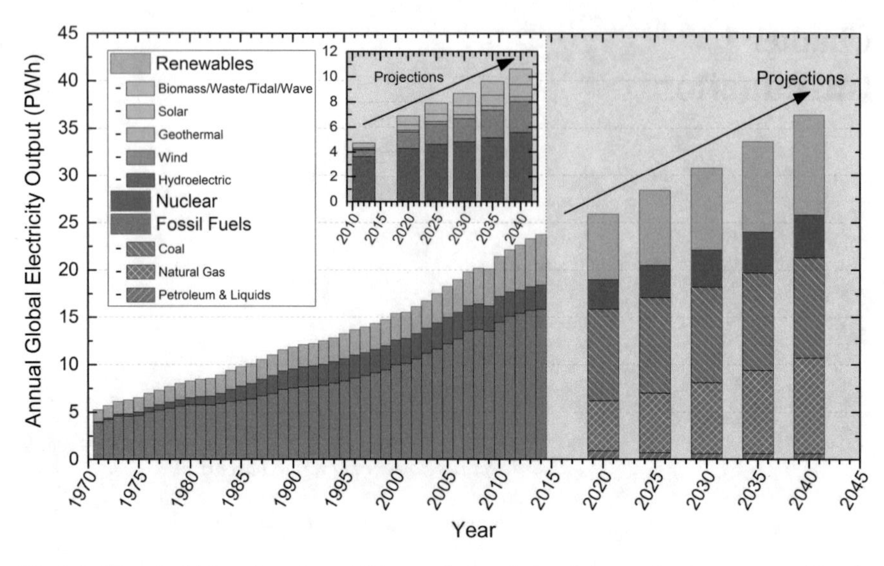

Fig. 1.1 Historical [5] and projected [6] annual global electricity production, separated by the generation method. Inset shows the projected production data from renewable sources

1.1 Energy

A widely encompassing term, energy is generally used outside of the physicists' vernacular to mean a measure of the ability to exert power in order to produce an effect, and can manifest in many forms. Humans need energy in this sense in order to perform the tasks elemental to survival. This could be electricity for heating, lighting, computers, and the multitude of other devices upon which we luxuriate, but also now rely. Else, it could be fuels to power machines and engines which seem somewhat archaic in this age of electricity. Finally, it could be human energy, which we acquire from our fuel of food; ultimately produced using some of the other forms of energy mentioned above.

In the long run, it would seem that the generation and use of electricity is of primary concern when considering energy production; however, it is 'primary fuel sources',[1] from which the world gathers most of its energy [4]. Despite this, the use and production of electricity is the focus of this thesis, as new technologies at the forefront of research are now almost wholly dependent on electricity as the power source.

As time progresses, the requirement for electricity increases, as a result of population growth and technological advancement. Historical global electricity production [5] is shown in Fig. 1.1, and over the past 40 years, the worldwide electricity output has increased almost fivefold, surpassing the 20 PWh mark in 2008.

[1] An apologetic term used for non-renewable resources when they are not used to generate electricity.

1.1.1 The Problem

As can be seen in Fig. 1.1, world electricity use is increasing. In order to meet this demand, new methods of electricity generation were developed and implemented, shown by the emergence and growth of the use of nuclear and renewable sources in Fig. 1.1. However, the use of fossil fuels for electricity production has also seen a continued increase.

Fossil fuels, by their very nature, must be extracted, transported, processed, and then combusted in order to produce thermal energy in the form of steam, which then is converted to electricity through turbines and generators. This multistage process consumes energy throughout and results in waste products which must either be disposed of in an environmentally friendly way,[2] or unsustainably.[3]

These waste products contributing towards climate change [7] is one of the main driving factors that is changing the way that electricity is generated [8]. Also important are the reserves of fossil fuels that remain available on Earth. Whilst there is much debate about the remaining amount of coal, oil, and gas, regardless of when the supplies will run out, solutions must be found before this happens. A proactive approach is favourable, especially given that the current projections of electricity generation expect the use of fossil fuels to continue rising as well[4] [6].

1.1.2 The Solutions

The increased use of renewable sources for electricity generation is promising and generally accepted to be the answer to the problems of fossil fuels, or non-renewable sources, but the use of the term renewable is often misleading. Given enough time, new fossil fuels will be naturally generated, but on a timeframe which is impractical for human endeavours. This, and the definition of 'renewable',[5] led to the connotations that renewable sources of energy must be able to be depleted. This is generally not the case, and instead, renewable sources of energy exploit natural processes which are already in motion.

These processes are many and varied, but the major contenders are: hydroelectric power, which involves the control of a body of water, whose kinetic energy is used to spin turbines; geothermal power, which exploits geothermal activity to produce steam; wind power, which uses the wind to drive turbines; wave and tidal power, which convert the kinetic energy of the oceans, seas, and estuaries into electricity; biomass and waste power, which involve the combustion of organic material that is more readily available than fossil fuels, or that would otherwise go to waste;

[2] Which costs further energy.

[3] Which is environmentally damaging.

[4] See Fig. 1.1.

[5] "Of a natural resource or source of energy: capable of being replenished, not depleted by its utilization" [9].

and solar power, a broad term which encompasses technologies that convert power directly from the sun into electricity.

All of these technologies are established and deployed in some form, and in 2012, contributed 22% of the world's total electricity production [6]. A summary of the methods used to generate electricity in 2012 is given in Fig. 1.2a, and the projections for 2040 in Fig. 1.2b. The overall share of fossil fuels is expected to decrease, and the level of nuclear power generation is expected to remain steady, probably due to the worldwide apprehensions associated with this method [10–12]. Renewable sources are seeing the fastest growth in the energy sector [13], and within the renewable category, the favoured technologies are also shifting. Hydroelectric power is by far the major contribution to the renewables, probably because of the large scale of such installations and the large amount of electricity they can generate. The other technologies are expected to increase significantly, with the wind and geothermal share more than doubling, and solar power taking more than four times the current share.

Although this increase in the use of renewable sources is promising, and by their nature, these sources do not suffer from the problems of fossil fuels, there are other problems with each of these methods which are restricting the deployment viabilities. Hydroelectric stations have a very large initial cost, and also have the environmental

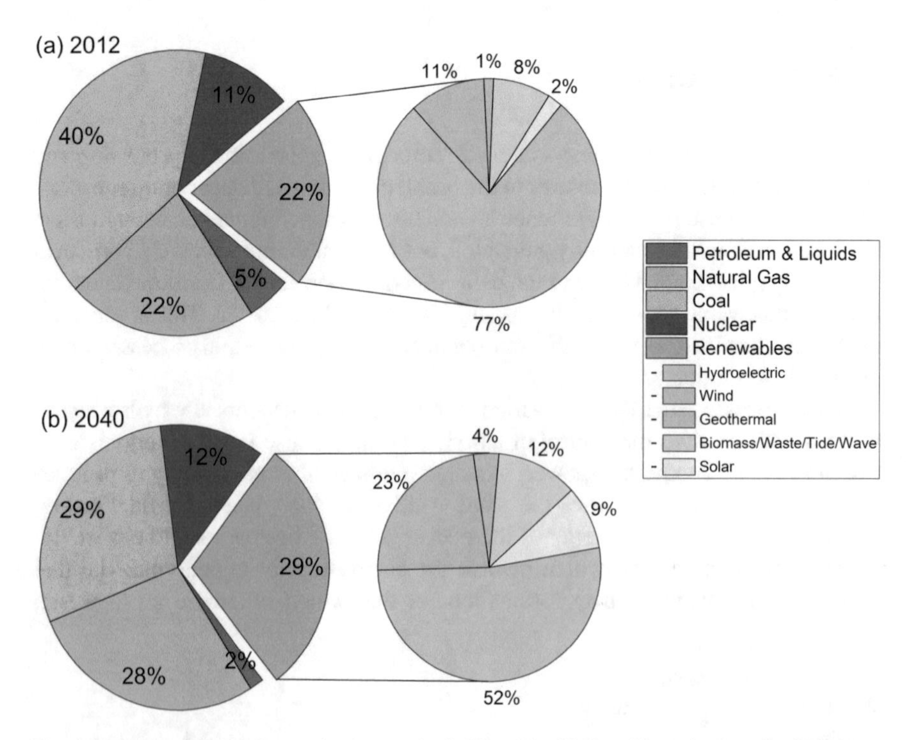

Fig. 1.2 Share of electricity production methods [6]. **a** For 2012 and **b** projections for 2040

and ethical impacts associated with the requirement to flood the surrounding land and the displacement of peoples. Wind and wave power are dependent upon prevailing conditions, and offshore installations have associated high installation costs. Geothermal installations are limited to geothermally active areas of the world, which are generally in remote locations. The use of biomass is controversial because it still produces waste, and also consumes and depletes a resource, rather than exploiting a natural process like the other methods. This leaves solar power.

1.1.3 The Case for Solar Energy

Exploitation of the power delivered by the sun to the earth is arguably the only source which is truly inexhaustible. The fossil fuels will one day run out, geothermal activity may cease, rivers may dry up, or the wind may stop being effective, and human life will continue, but when the sun stops delivering power to Earth, the human race will experience direr problems than the cessation of solar power.

The greatest benefit of the sun is the sheer scale of power that is delivered to Earth. More than 150 PW is delivered to Earth's upper atmosphere [14], which when considering the effect of the atmosphere, obliqueness, seasonal variation, and diurnal variation, results in ~24 PW reaching the surface of the earth [15]. If this energy was completely extractable and usable, it is 4 orders of magnitude greater than the current global electricity demand.[6] Obviously, not all of this energy is available for conversion to electricity, but the amount of surplus is attractive.

1.2 Solar Photovoltaics

Although solar power is utilised widely to directly heat water through the use of solar panels, and there is a growing sector using concentrated solar power to heat water and drive conventional generators, the largest sector of solar electricity generation is through the use of solar photovoltaics (PV). This technology converts the light energy from the sun directly into electricity through the photovoltaic effect, which will be explained in Sects. 1.2.1 and 1.2.2.

There are great benefits of using this method of electricity generation, beyond the large amount of available power from the sun. These technologies directly convert light energy into electrical energy and therefore do not rely upon intermediary conversion methods such as turbines, which have their own efficiency limitations. They therefore contain no mechanically moving parts, relying upon electronic phenomena, and this results in them requiring almost zero maintenance, and benefitting from a long lifespan. Also, they produce no emissions and can be installed discreetly and widely because of the various form factors that they can take. With regards to

[6] Calculations for this can be found in Appendix A, Sect. A.1.

deployment, they do not require a centralised power station because they are modular, and therefore the power output is independent of the scale of the system, increasing proportionately to the number of cells, in theory ad infinitum [16].

Compared with other deployed electricity generation systems, PV technologies are relatively fledgling; one of the reasons why the deployment of solar PV is only 2% of the 22% of renewables as of 2012.[7] Nevertheless, this relates to 90 GW of solar PV currently installed worldwide, which with current targets is set to rise to 350 GW by 2020, and 60 GW per year is being manufactured [6].

1.2.1 Semiconductor Theory

PV devices are reliant upon semiconductors, and it is now prudent to introduce these materials. Quantum mechanics dictates that the electrons within an atom are confined to orbitals of discrete energies [17]. As individual atoms are brought into proximity, the wave functions of the individual electrons overlap, and form bands by the Pauli exclusion principle [18]. In the solid state, materials can be categorised into three classifications: metals, insulators or semiconductors. The band structures of these three materials are shown in Fig. 1.3.

Electrons in a solid in the ground state fill energy levels until the Fermi-level (E_F). Depending upon the band structure and density of states, determined by the periodicity and nature of the lattice, the Fermi-level can reside within a band of electron states, or within a region devoid of electron states. In the first case, this is a metal, with no forbidden band between the filled and empty states and therefore the electrons can move freely within the material with good conductivity. In the latter cases, there exists a forbidden band (bandgap, E_g) between the valence band (filled states) and conduction band (empty states), which an electron must overcome in order to conduct. An insulator has a bandgap greater than ~4 eV, and a semiconductor has a bandgap less than this, so that thermal excitations can raise electrons to the

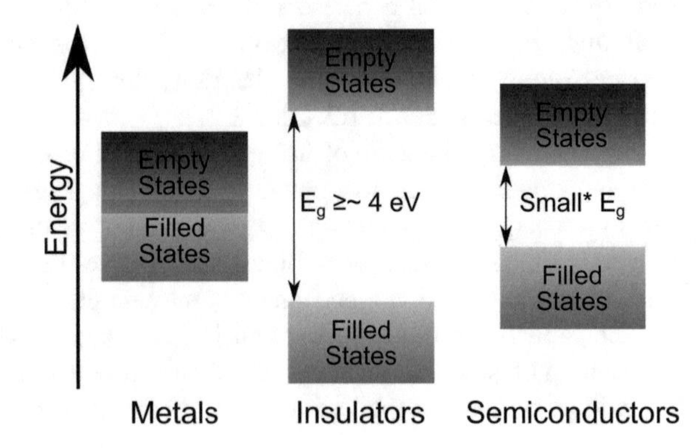

Fig. 1.3 Relative band positions for metals, insulators and semiconductors. *Small bandgap of semiconductors is also dependent upon their ability to be doped and the thermal excitation of carriers across the bandgap

[7] See Fig. 1.2a.

conduction band, allowing (semi)conduction. Such materials also have the ability to be doped, which affects the conductivity [8, 18].

In an intrinsic semiconductor crystal at equilibrium, the concentration of electrons in the conduction band (n) is equal to the concentration of holes in the valence band (p). The alteration of the conductivity of a semiconductor via the introduction of impurities is called doping [8]. The addition of an impurity with an extra valence electron than that of the host material[8] introduces an extra electron to the conduction band through the creation of a donor state, in which the valence electrons are much closer to the conduction band of the host and allows easy transfer into the conduction band, increasing the concentration of electrons. These extra states within the bandgap result in a shift of the Fermi-level to a position between this extra state and the bottom of the conduction band. This is then termed an n-type semiconductor, as the majority carriers are negatively charged. Conversely, the addition of an impurity with one fewer valence electron[9] produces the opposite effect, introducing extra holes to the valence band through the creation of an acceptor state and lowering the position of the Fermi-level within the bandgap. This is then termed a p-type semiconductor, as the majority carriers are positively charged. These phenomena are shown in Fig. 1.4.

Such effects are easily exploited for perfect single crystal elemental semiconductors such as silicon or germanium, whose covalent bonding lends to the accurate prediction of conductivity changes through impurity doping. In other semiconductor materials, where ionic bonding plays some role, the inadvertent introduction of defects in the crystal structure is the cause of conductivity changes. Fortunately, most compound semiconductors can be assigned an ionic formula, with the constituents taking integer oxidation states, based upon full electron shells [19]. For example, in SnS, where the oxidation states are Sn^{2+} and S^{2-}, it is the V_{Sn} defect which is responsible for the intrinsic p-type conductivity observed in this material [20], because it is this defect which has been shown to have the lowest enthalpy of formation. Sn^{2+} ions contribute two electrons to bonding, so where the vacancy is present, these two electrons are missing, and this site effectively acts as a dopant with too few electrons,

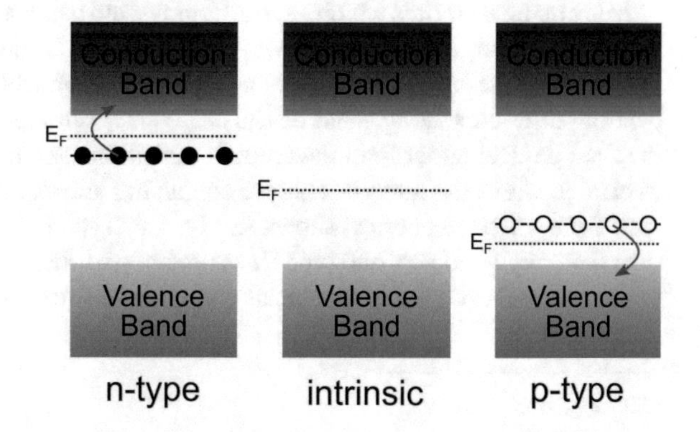

Fig. 1.4 Fermi-level positions for n-type, intrinsic, and p-type semiconductors. Donor states in the n-type allow electrons (closed circles) to transfer into the conduction band and acceptor states in the p-type allow holes (closed circles) to transfer into the valence band

[8]E.g. P in Si.

[9]E.g. B in Si.

resulting in p-type conductivity because these states lie close to the valence band and act as acceptors [21].

The properties of semiconductors can be predicted to a certain degree through the patterns observed in the chemical trends of materials. This is useful when considering new materials for use in applications, and was found with the development of solar cells, which will be discussed in Sect. 1.3. Generally, if a material has the same crystal structure, oxidation states of the elements, and stoichiometry as the parent semiconductor, then this will also be a semiconductor, where the bandgaps will be related [19].

Whilst semiconductors have applications when used on their own, it is the effects produced when two (or more) semiconductors are contacted together that must be considered for solar cells. The most useful of these junctions, the pn junction, is of interest here. This involves the connecting of a p- and n-type semiconductor, and is illustrated in Fig. 1.5. Before the materials are contacted, the n-type material has free electrons, and the p-type material has free holes. Just after contact, at the interface of the two materials, some of the free carriers will diffuse to the other side of the junction, where they recombine with the charge-opposed carriers in the other material, forming a depletion region where there exist no free carriers. This recombination causes excess charge at the interface due to the now-absent charge carriers, positive in the n-type, and negative in the p-type, which forms an electric field in this region, halting further diffusion.[10] The pn junction acts as a diode, limiting current flow to one direction upon the application of an external voltage, allowing current flow via diffusion.[11]

In terms of the band levels at the pn junction, when brought into contact, the Fermi-levels of the two materials align, and the valence and conduction bands bend within the depletion region in order to overcome the band discontinuities formed by the alignment of the Fermi-levels.[12]

The case considered above concerns the formation of a homojunction, that is, the formation of a pn junction between two instances of a material which have the same bandgap, but opposite conductivity types[13]; however, pn junctions are also possible using materials which have different sized bandgaps, which when contacted, redistribute charge in the same way, forming a heterojunction. The difference in bandgaps of the materials in this case form band offsets, which because the band bending after contacting must be the same size, can result in band discontinues in the final device, rather than the smooth transition seen in Fig. 1.5b. The difference in band levels of the materials before contacting allows these types of junction to be sorted into three categories, shown in Fig. 1.6. Type I are known as straddling gap, type II as staggered gap, and type III as broken gap. Such junctions are more relevant to the materials studied in this thesis and shall be further explored in Sect. 1.2.2.

---

[10]See Fig. 1.5a.

[11]See Fig. 1.5c.

[12]See Fig. 1.5b.

[13]E.g. silicon with n- and p-type doping.

1.2.2 Photovoltaic Devices

The Photovoltaic Effect

The pn junction is the basis for many electronic components, including diodes and transistors and is also used as the source of many parts of integrated circuits, but these exploit the rectifying behaviour of the junction with the application of external biases. Solar cells use the properties of a pn junction, coupled with the photovoltaic effect.

First discovered in 1839 by Edmond Becquerel [22], this phenomenon is related to the photoelectric effect in that interactions with light can excite electrons in a

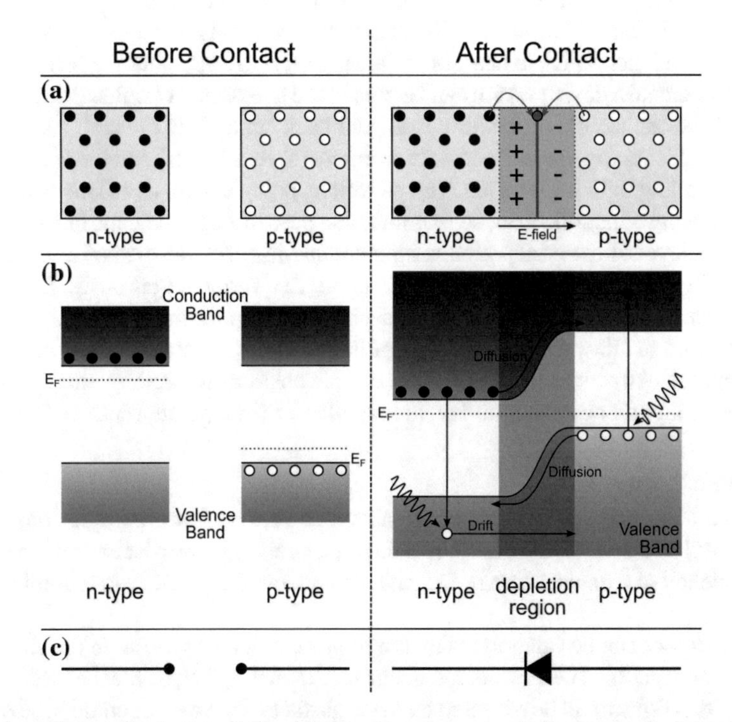

Fig. 1.5 **a** Charge distributions, **b** band level schematic, **c** circuit diagram equivalents, for a pn junction before and just after contacting the constituent n- and p-type semiconductors. Filled circles represent electrons, open circles represent holes, and arrows demonstrate the movement of charge carriers

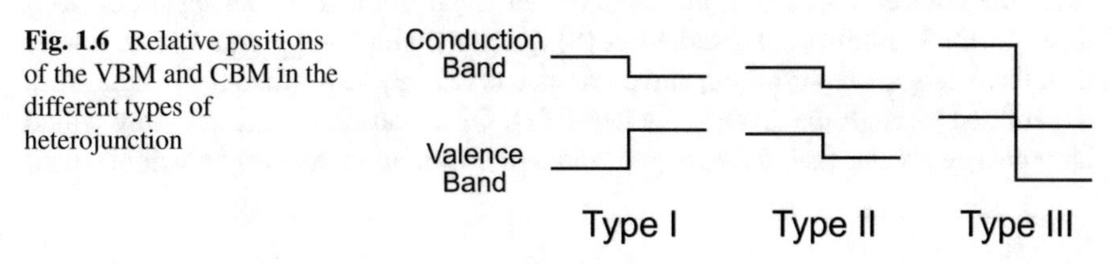

Fig. 1.6 Relative positions of the VBM and CBM in the different types of heterojunction

material. If light is absorbed by a semiconductor whose bandgap is less than the energy of the photons, then this absorption can excite an electron into the conduction band of the semiconductor. If this effect occurs in a semiconductor which is forming a pn junction, then the electron-hole pair created can be separated: either by the acceleration of the charge carriers by the electric field in the junction, or by the diffusion of the charge carrier into the depletion region, where it is accelerated by the field.[14] This creates a photovoltage across the junction, and allows a photocurrent to flow if an external load is connected.

Cell Architecture

Solar cells are usually designed so that the majority of the charge carriers are generated in the semiconductor which has the largest minority carrier diffusion length, often resulting in a thicker p-type layer, with the thinner n-type layer on top. The p-type layer is then referred to as the absorber layer, and the n-type as the window layer. The pn junction is the heart of the solar cell; however, other parts are required in order to create a device which can be connected to external circuits. Primarily, this involves the connection of metallic front and back contacts to the cell, which act as electrodes. Generally, the back contact requires a sheet of metal, but the front contact must also allow light into the junction as well as being conductive. This is achieved by either using a metallic grid, so that light can pass through the gaps in this grid, or by using a layer of material which is both conducting and transparent to light.

Other layers are often incorporated which serve a variety of purposes. Depending on the form factor for the device, some mechanical support is required which may or may not need to allow the transmission of light. Layers are sometimes used in order to aid charge extraction, or improve the effectiveness of the cell. A variety of light management and encapsulation layers may also be incorporated into cells in order to improve performance.

Solar Cell Losses

Ideally, a solar cell converts all of the absorbed energy from the light into usable electricity, but this is not the case in real devices, as energy losses occur, which reduce the efficiency of a device; that is, the ratio of generated electricity to incoming light energy.

These losses can be categorised by the stage at which they occur in the electricity generation process. The first losses occur as the incident light reaches the device. Some of this light could be reflected from the surface of the cell, or could be absorbed by the contacts before reaching the junction region. In any case, these photons are then not available for charge generation.

The subsequent losses which occur within the cell are summarised in Fig. 1.7. Once the photon reaches the junction, the bandgap of the absorber layer acts as a filter for the incoming solar radiation [8]: photons with energy equal to or higher than the bandgap are absorbed, and photons with energy lower than the bandgap are transmitted through the absorber material (1). Of the absorbed photons, any which have energy greater than the bandgap lead to the formation of charge carriers which

---

[14] See Fig. 1.5b.

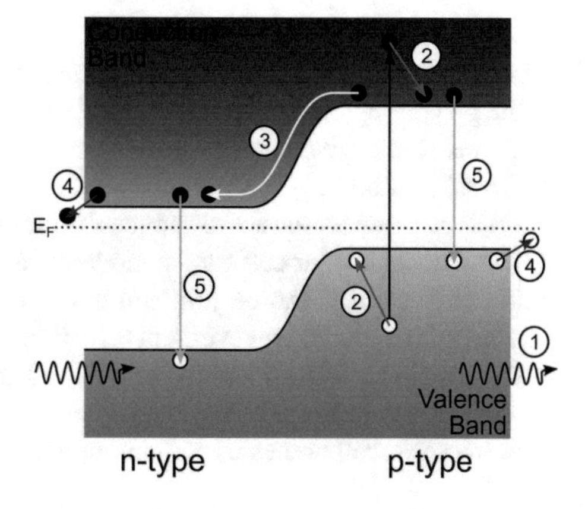

Fig. 1.7 Loss mechanisms that can occur within the junction of a solar cell. Electrons are represented by filled circles, holes by open circles. Numbered mechanisms are described in the text

are not located at the band extrema, and will relax to the band edge, transmitting the extra energy to lattice phonons (2). This thermalisation results in energy which does not contribute to the electricity generated. Following the generation of charge carriers in the absorber material, they need to be separated in order to contribute to the photovoltage. Before separation, charges can undergo losses at the junction, where they diffuse to lower potentials, before being accelerated by the electric field, resulting in a lower photovoltage (3) [23]. After separation, energy can be lost through discontinuities at the front and back contacts of the cell (4). Depending on the size of the discontinuity, further photovoltage losses will occur in the transfer of the electron or hole to the Fermi-levels of these contacts. Furthermore, if the carriers are not separated effectively, they will recombine (5). This can occur for several reasons, but results in the annihilation of the charge carriers, which are then not available for electricity generation. If the charge carriers are generated outside of the region over which the electric field acts, they must diffuse to this region before being accelerated by the field, and if their diffusion length is not long enough to achieve this, then they will recombine before this happens. Recombination can also happen because of states generated within the bandgap, which allow the charge carriers to recombine via this extra state, making this event more likely. These states can result as an effect of defects and grain boundaries in the material [8].

Beyond the phenomena which can occur within the junction as described above, other losses can occur as a result of the band alignments of heterojunctions, as discussed in Sect. 1.2.1. In terms of the conduction band offsets, which relate to the electron transport, two situations can exist, shown in Fig. 1.8. There can either be a negative CBO, where the CB of the p-type layer is above that of the n-type, or a positive CBO, the reverse. Following the joining of the two materials,[15] the redistribution of charge results in band bending which causes the band discontinuities to form a 'cliff-like' structure for the negative CBO, and a 'spike-like' structure for the positive CBO. For electron transfer, the CBO should be near zero, as is the case in

[15] See the bottom panels of Fig. 1.8.

homojunctions. For the cliff-like structure, electrons generated in the n-type material experience a barrier against diffusion, resulting in recombination, and those created in the p-type region can suffer voltage losses through the offset. For the spike-like structure, if the spike is too large, this acts as a barrier to electron transport, leading to recombination [24, 25].

Each of these losses act to reduce the amount of extractable electricity produced by the cell and a solar cell can be modelled as an electrical circuit of the type shown in Fig. 1.9. Ideally, the pn junction under illumination acts as a current source in parallel with a diode; however, imperfections also mean that there exists some value of shunt resistance (R_{Sh}) within the cell, and also there is a resistance associated with the contacts and external circuitry, the series resistance (R_S). The extractable current from the solar cell comes as a combination of these components, so that

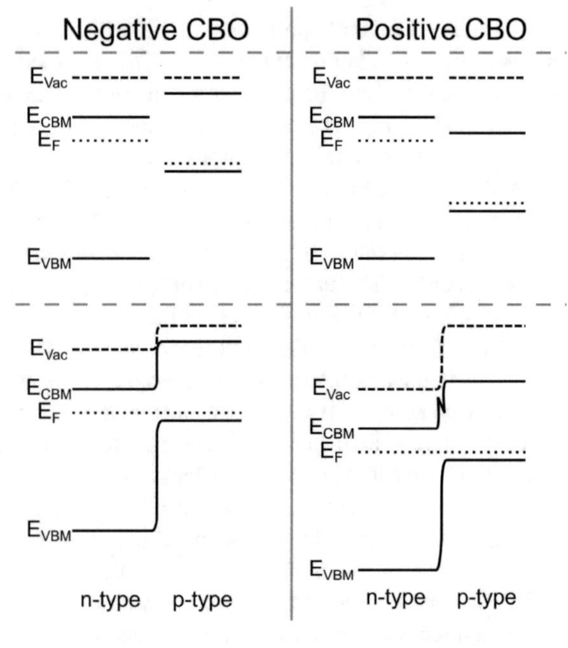

Fig. 1.8 Schematic band alignment diagrams for conduction band offsets (CBO): positive (left) and negative (right). Vacuum aligned diagrams (top) represent before contact, and Fermi-level aligned diagrams (bottom) represent after contact

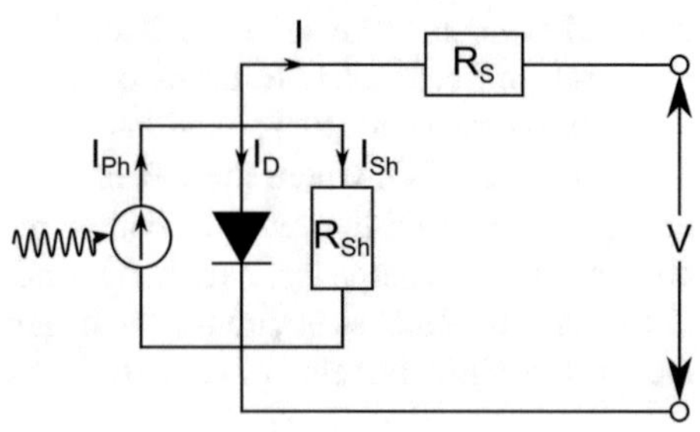

Fig. 1.9 Equivalent circuit of a solar cell device

$$I = I_{Ph} - I_D - I_{Sh}, \quad (1.1)$$

where I_{Ph} is the generated photocurrent; I_D is the current through the diode, dependent upon the reverse saturation current of the diode itself and also the voltage across it, which in turn is dependent upon the series resistance; and I_{Sh} is the current through the shunt resistor. Therefore, to optimise the performance of the solar cell, the photocurrent and shunt resistance should be maximised, whilst minimising the series resistance and reverse saturation current of the diode [8, 16].

Absorber Material Requirements

It is important that all aspects of the solar cell are designed in order to maximise the efficiency, but in this thesis, the absorber layer is the main consideration. There are several requirements which a material must fulfil for it to be considered for use as an effective solar absorber.

As discussed above, the band alignments can affect the overall efficiency of a cell, and from a technological design point of view, it is better to start with a material that has favourable band positions which are intrinsic, rather than the need to tune the band structure at a later point.

Defects in the material can lead to recombination if the produced states are found deep within the bandgap, therefore it is more favourable to use materials which are either less susceptible to defects, or have ones which form shallow defect states, leading to relative benignity upon their formation [26, 27]. In order to give charge carriers more chance to be produced and separated, materials should have strong absorption coefficients in the visible range, long carrier lifetimes and high mobilities, which come as a result of small effective masses and greater band dispersion.

The generation of charge carriers as depicted in Fig. 1.5 suggests that the excitation of an electron to the conduction band occurs only through a change in energy, and such is the case for direct bandgap materials, that is when the VBM and CBM occur at the same point in k-space. However, for materials which do not have the VBM and CBM located at the same point, a phonon is required to produce the momentum change required as well, because photon absorption does not produce a momentum change [8], which is depicted in Fig. 1.10. This is not necessarily a problem, especially if there exists a direct transition which is of suitable energy [28], but it will still lead to photons which cannot be absorbed, and so a direct bandgap material, which will have a higher absorption coefficient, is preferable.

Whilst recombination is a problem for solar cells, arguably the greater problems are those which come from inefficient photon absorption[16] [23]. In order to achieve the highest possible photovoltage from a cell, the bandgap needs to be large as the photovoltage cannot exceed the bandgap. However, due to the nature of the solar spectrum, to absorb as much light as possible, and to reduce the number of sub-bandgap photons which are transmitted, the absorber needs a small bandgap. These conflicting statements result in an ideal bandgap for solar absorbers, which absorbs as much light as possible but maintains high photovoltages, and various studies have placed this value in the region of 1.1–1.6 eV [16, 27, 29].

[16] See (1) and (2) in Fig. 1.7.

The optimal bandgap value is based on the work of Shockley and Queisser [30], who calculated the maximum efficiency possible as a function of absorber bandgap, finding a peak value of 30% efficiency with an absorber bandgap of 1.1 eV, known as the Shockley–Queisser Limit (SQL). The range of workable bandgaps quoted above comes as a result of assumptions made in their study which do not necessarily hold in real devices, and also that the function experiences an approximate plateau in this bandgap range. Despite these assumptions, the SQL is still often quoted throughout the literature [31], and overcoming it is a goal that has now been achieved by various methods, which are introduced in Sect. 1.3.4. The maximum efficiency as a function of absorber bandgap found by Shockley and Queisser is shown in Fig. 1.11, along with the contributions of the other loss mechanisms.

Device Metrics

The defining metric for the performance of a solar cell is the efficiency; the ratio of the extracted electrical energy to the incident light energy. However, because of the interplay of the various aspects of a solar cell discussed above, it is far from simple to determine the cause for an observed lower efficiency than is expected.

The efficiency of a device can be determined by measuring the current–voltage (J–V) characteristics of the device under illumination of a source with known flux.

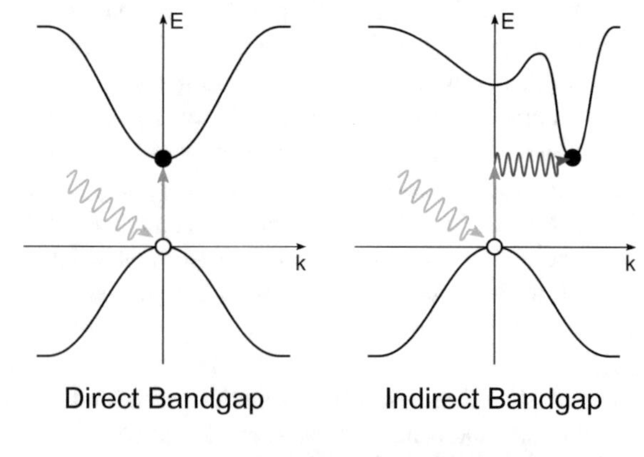

Fig. 1.10 Photon (orange arrow) absorption for direct and indirect bandgap semiconductors. Note the required phonon (blue arrow) to facilitate momentum transfer in the indirect bandgap material

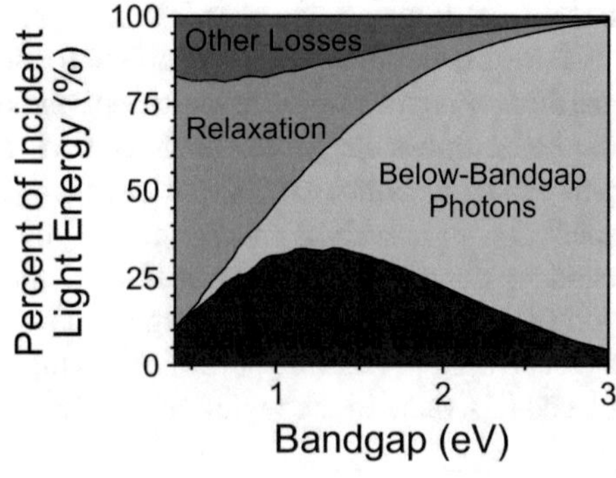

Fig. 1.11 Percentage of solar cell loss mechanisms as a function of bandgap. Maximum cell efficiency is resultant of the calculations made in calculating the SQL [30]. Other losses include those shown in Fig. 1.7. Adapted from a figure courtesy of Steven J. Byrnes

A typical J–V curve for a solar cell under illumination is shown in Fig. 1.12 (red curve), and resembles the J–V characteristics of a diode, displaced on the current axis because of the constant current source aspect of the cell.

From this curve, several features can be determined. With infinite load connected to the cell, the open circuit voltage (V_{OC}) is the maximum voltage produced, and its difference from the bandgap can be a measure of the amount of recombination present within the device. With a zero load connected across the cell, the short circuit current (J_{SC}) is the maximum current produced, and can be a determiner for the carrier lifetimes within the cell. Under ideal conditions, the change in current value with voltage will be an abrupt transition; however, in a real cell, the resistances present will deviate this 'squareness' of the J–V curve. The power of the solar cell,

$$P = J \cdot V, \tag{1.2}$$

is also shown in Fig. 1.12 (blue curve), and the current and voltage values at which the power is maximum,

$$P_{\text{Max}} = J_{\text{Max}} \cdot V_{\text{Max}}, \tag{1.3}$$

are the points taken in order to calculate a measure of the ideality factor for the cell. The Fill Factor (FF) is defined as the ratio of the maximum power of the cell to the ideal power of the cell:

$$FF = \frac{J_{\text{Max}} \cdot V_{\text{Max}}}{J_{SC} \cdot V_{OC}}, \tag{1.4}$$

and acts as a determiner of the loss from ideality. The efficiency of the cell is then calculated as,

$$\eta = \frac{J_{SC} \cdot V_{OC} \cdot FF \cdot A_{\text{eff}}}{P_{\text{Light}}}, \tag{1.5}$$

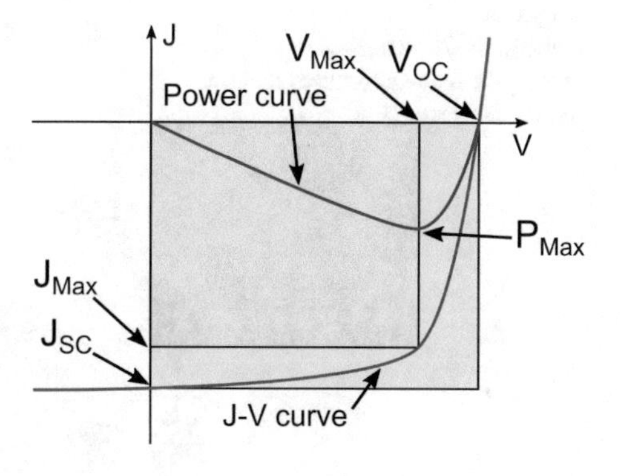

Fig. 1.12 Typical solar cell J–V characteristic curve under illumination. The fill factor is the ratio of the areas of the two shaded boxes

where A_{eff} is the effective working area of the cell, and P_{Light} is the power of the incoming light [8].

1.3 A Brief History of Solar Cells

Different types of solar cell will be introduced in this section, following the history of their development. Although all aspects of the cell are important, to reduce confusion, descriptions of cells here shall be in reference to their absorber layers unless otherwise specified.

Throughout the history of solar cells, the absorber layer has changed and evolved, but they have remained in the same broad material families. Silicon, ubiquitous with semiconductors, has played a large role in solar cells and acts as the parent material for many of those which followed it. This evolution of semiconductors based on the silicon structure is demonstrated in Fig. 1.13 [32, 33]. The compounds shown here all take similar, tetrahedral crystal structures with related properties, and shall be introduced further in the subsequent sections; however, chemical trends apply, with an increasing bandgap following increasing atomic radii, shorter bond lengths, and larger differences in electronegativity [19].

The efficiency of solar cells is the main technological motivating factor for their development; however, because they will also be deployed into power generation grids, there are economic impacts which must also be addressed. Therefore, a balance between the efficiency of devices and the power generation cost is used as a general arbiter for the development of solar cell technologies [15]. This development is grouped into three categories, or 'generations' of solar cells, and their trends in terms of cost and efficiency are shown in Fig. 1.14 [23]. Generation A followed from the historical invention of the solar cell and resulted in cells of modest efficiency but which were expensive to produce. This led to the development of new technologies

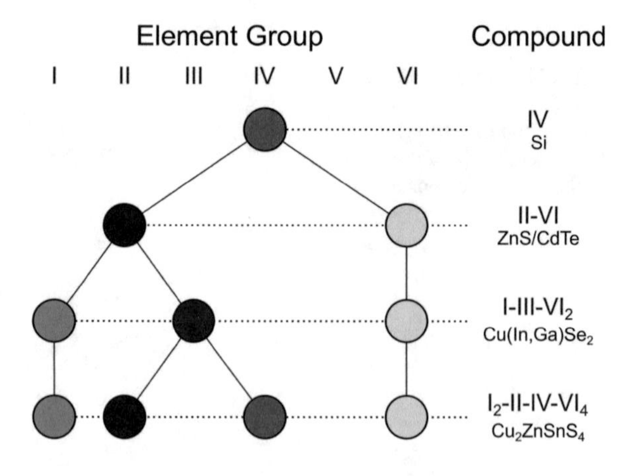

Fig. 1.13 Generational progression of semiconductor materials sharing the same structures and related properties

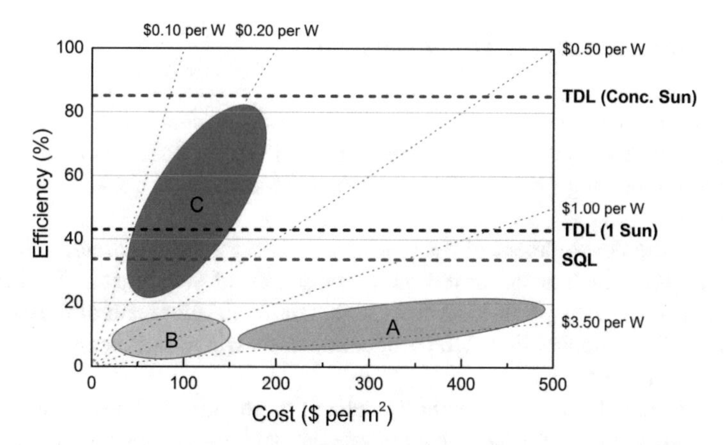

Fig. 1.14 Three generations of solar cell technologies in terms of their cost and efficiency. Diagonal tie lines show regions of equal cost per power. Horizontal lines show the Shockley Queisser Limit (SQL, red dash), thermodynamic limit (TDL) for a cell illuminated by the power density of 1 sun (blue dash), and TDL for a cell illuminated by the concentrated power density of the sun (green dash)

for generation B, resulting in cheaper cells, but at the slight expense of efficiency. Now, in the current research climate, the goal is to maintain the cheap cost of cells, but to increase their efficiencies, using hitherto unknown technologies which are able to exceed the SQL and also the thermodynamic limit (TDL) of a solar cell illuminated by 1 sun [34]. It is noted that the borders of generation C are not well-defined, and that many new technologies are incorporated into it which set out to study many different aspects.

1.3.1 0th Generation: The Early Years

The physical phenomenon that is key to solar cells; the photovoltaic effect, was first discovered in 1839, where it was observed in an electrochemical cell [22]. Moving to the solid state, the effect was observed in a selenium rod in 1876 by Adams and Day [35], and the development of a solid state cell by Fritts, utilising selenium covered with gold plate, demonstrated a conversion efficiency of less than 1% in 1883 [36–38]. Whilst somewhat primitive, and not reminiscent of modern cells, this early research continued and the photovoltaic effect was also observed in compound semiconductors; notably CdS by Reynolds et al. [39] in 1954.

The development of the first 'modern' cell[17] was created using high purity silicon, and was demonstrated by Bell Labs in 1954 to give an efficiency of 6% [40]. This led to the first generation of solar cells, based on high-purity crystalline silicon.

[17]Utilising a pn junction.

1.3.2 1st Generation: Wafer-Based PV

PV technologies based on crystalline silicon (c-Si) wafers are the most widely employed and have seen the most development. They are based on homojunctions of n- and p-doped high-purity silicon wafers and took 87% of the PV market share in 2011 [8].

Initially, the development of devices was limited by the requirement for single crystal silicon, which at the time was the only way of ensuring the high purities required. This kept the production costs high, but the development of the casting of multicrystalline silicon reduced the production costs whilst sacrificing only a small amount of efficiency.

The high-abundance, reproducibility, reliability, stability and non-toxic nature of silicon are the reasons why these technologies are still dominating the market [8, 15] and cells are achieving efficiencies over 20% [41].

Silicon is not the only wafer technology used as a solar absorber, with III-V semiconductors also used, and traditionally, it is these cells which produced the highest efficiencies [8]. However, they have not seen such a high market share because of the complex processing methods required and the much higher materials costs because of the use of rare elements [8, 15].

Despite the wide availability of c–Si solar cells and the advantages thereof, there are several reasons why some focus has moved away from this technology towards other materials. First, there is the problem that silicon is an indirect bandgap material, and even though the required phonon can be provided thermally at room temperature [8], silicon still has a low absorption coefficient, requiring a 200 μm thick layer of material in order to absorb 82% of the light. This thickness requirement, coupled with the high purity required, drives up the production costs, which are further exacerbated considering that 40% of the material is lost to kerf waste during wafer slicing [42]. These problems provided the motivation for investigating technologies which require much less material.

1.3.3 2nd Generation: Commercial Thin-Film

The costs involved with using wafer technologies can be largely overcome by using deposited thin-films, rather than wafers, because of the amount of material that is necessary. Materials with much higher absorption coefficients are required to ensure that enough light is absorbed to generate high photocurrents. c-Si is not suited to this, but thin-films of silicon in the amorphous phase (a-Si) increase light scattering within the material, which in turn enhances the absorption.

As a technology, a-Si was developed throughout the late 1960s [43, 44], with the first cells being reported in 1976 at 2.4% efficiency [45]. This technology has not gained much hold in the market, due to the limited efficiencies [15]. Alongside it therefore, other thin-film technologies were developed. These materials are usually ones with direct bandgaps that have high absorption coefficients so that the thickness can be minimal and still absorb most of the incoming light, but with smaller film thicknesses, the absorption of longer wavelengths of light is sacrificed [8]. As mentioned in Sect. 1.2.1, these materials are usually compound semiconductors which have intrinsic p-type conductivity due to the presence of defects and so do not require doping. The additive manufacturing techniques used in thin-film production lends to the use of minimal feedstock material and cheaper production methods [8], and the varied deposition techniques allow much variation for integration and form factors of the final devices, as they are not limited by the wafer dimensions [15].

The first thin-film technology used as a solar absorber that was not based on silicon was arguably Cu_2S [46–48]; however, degradation problems suffered because of the mobility of the copper ions caused interest to be lost. Throughout the years, various semiconductors have been tested for their use as thin-film PV absorbers, such as Cu_2O [49, 50] and FeS_2 [51]; however, two technologies have emerged, which alongside a–Si make up almost all of the thin-film solar cells in use today.

Cadmium telluride (CdTe) as an absorber material goes back to 1963 where it was investigated in bulk [52], and from there using it in thin-film form developed. It has a bandgap of 1.45 eV, high optical absorption, and is generally grown in a superstrate configuration with CdS as the window layer [8].

Copper indium gallium diselenide (CIGS) developed from the Cu_2S cells, where it was found that the inclusion of indium prevented the diffusion of the copper [53], leading to interest in $CuInSe_2$ cells in both single crystal [54] and thin-film form [55] in 1974 and 1976, respectively. Although the bandgap of $CuInSe_2$ is too low[18] for effective solar absorption, the alloying of the indium with gallium can tune the bandgap to a better suited value,[19] and hence, CIGS has developed as a successful solar absorber.

Compared with a-Si, CdTe and CIGS both demonstrated higher efficiencies and greater stabilities, which led to the favouring of these two technologies over the former [8]. The record efficiencies have passed between the technologies, but currently CIGS holds the record at 22.6%, with CdTe at 22.1% [41]. The historical market share of thin-film technologies, shown in Fig. 1.15, reflects the development of the various technologies, with the increase in CdTe and CIGS deployment, but over the past few years, the market has seen a relative decrease in overall thin-film deployment, caused by the renewed deployment of silicon cells in this time [57].

Despite the increasing deployment of thin-film technologies, CdTe and CIGS both suffer from their own problems. First, the efficiencies of the deployed modules are lower than the records because the production methods used for the record devices are not suitable for industrial scale-up [58–60]. CdTe performance is very dependent upon grain size, because deep level traps occur here, which act as recombination centres, and the very high work function leads to problems with band alignments at the back contact [8]. CIGS cells, because of their multi-element nature, can suffer from a high variability of film stoichiometry and resultant properties [15]. The main

[18] ~1.0 eV [56].

[19] ~1.1 to 1.2 eV.

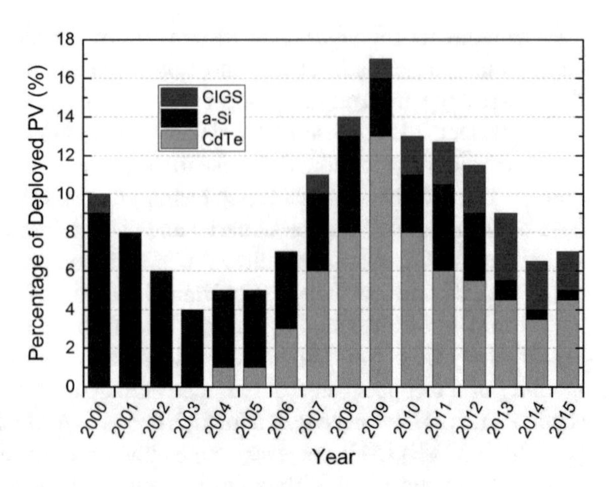

Fig. 1.15 Historical market share of deployed thin-film technologies from total PV deployment [57]

problems faced by both CdTe and CIGS however, are associated with the environmental and economic impact of the constituent elements of the materials.

The toxicity of CdTe is the subject of much debate, and while cadmium as a metal is indeed toxic [61], proponents of CdTe argue that the compound itself is not toxic, that the stability of the compound reduces the chance of leakage, and that in thin-films it is used in such little quantity that it is not of concern [62]. It is also cited that cadmium is extracted as a by-product of zinc refining [63], regardless of cadmium demand, and that it is better stabilised in compounds, rather than stockpiled or used in other applications which do have toxic effects [64]. Nevertheless, whilst toxicology data on CdTe itself is not available [62], it is also possible that exposure to the metal could come as a result of the growth, or that the necessary 'activation' step in the production of cells could lead to exposure to CdCl, which is toxic [8]. Tellurium is also a potential problem for CdTe solar cells. A very rare element, tellurium finds little use outside of the PV sector [65, 66] and as such, demand is not high, with studies citing tellurium scarcity as a limiting factor for the deployment of CdTe [67], which will be further discussed in Sects. 1.4 and 1.5.3.

Similar problems are faced with CIGS. Although selenium itself is toxic only in high concentrations [68], during the growth of CIGS it is common to find the use of H_2Se [62], which is highly toxic [69]. Whilst this factor is not usually addressed, the problems associated with the use of indium often are. Another rare element [15], indium use has expanded due to the surge in demand for consumer electronics [65], where indium is widely used. World reserve estimates of indium vary [70], but even if more indium is discovered, the changing economic climate could result in a price rise, raising a potential barrier against the deployment of CIGS [67]. Indium reserves will be further discussed in Sects. 1.4 and 1.5.3.

Motivation to develop the next generation of absorber materials for use in PV is driven in part by these toxicity and scarcity concerns associated with CdTe and CIGS [71] and it is hoped that these problems can be overcome by using materials which do not face these problems.

1.3.4 3rd Generation: Everything Else, Dependent upon Research Interests

This wide ranging generation encompasses all technologies that are now working to overcome problems faced by the previous two generations. From Fig. 1.14, it would appear that the main goal of these technologies is to drive the efficiency up; however, other technologies are also being developed which bring novelty to the PV sector, and can include new form factors, changes in production methods, or external additions to cells that can improve performance and reliability. Materials that are more sustainable and do not have the associated problems of CdTe or CIGS are also being developed, which is the motivation for this work and shall be introduced fully in Sect. 1.4, but it is prudent to acknowledge the other areas here.

As of 2015, deployed PV is shared amongst the technologies as shown in Fig. 1.16, with c-Si taking the overwhelming majority share, and thin-film constituting only 7% (exploded segments). It is clear from this deployment that other technologies, which make up the entirety of generation C, have not yet reached market viability, although this is the ultimate goal.

Many of the technologies being studied that aim to reach efficiencies above the SQL work to overcome the losses experienced by photons that have too little energy to overcome the bandgap, or by those that have too high energy, resulting in losses through thermalisation. This is commonly achieved by using multi-junction solar cells, which use several pn junctions connected together in a single device, designed in such a way that absorbers with different bandgaps allow photons to be absorbed by the most relevant absorber [72]. Such technologies have not yet seen deployment because there are losses associated with connecting the junctions which need to be overcome, the costs of production are high, and the long term reliability and scale up

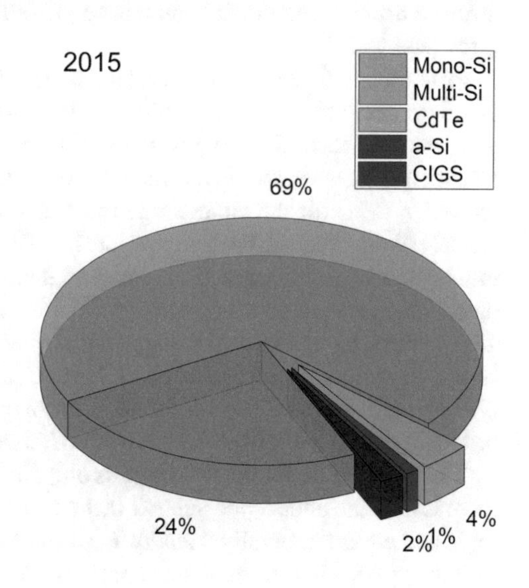

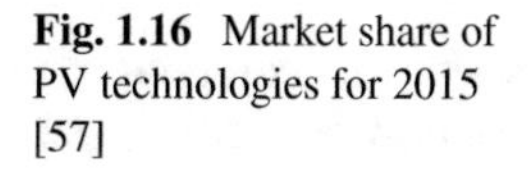
Fig. 1.16 Market share of PV technologies for 2015 [57]

uniformity are poor [15]. Other techniques are also available, which narrow the range of the solar spectrum to better suit the device in question, and these can include [8]: up- and down-converters which change the nature of the spectrum before it enters the device; the use of light capture devices; and the incorporation of materials that give more chance of the absorption of a photon, such as anti-reflection coatings at the front of the device, or reflective coatings at the back of the device.

New absorber materials different to the type studied in this thesis are also being considered for their use in PV. These include organic, dye-sensitised, and perovskite solar cells, which all benefit from low cost, simplistic manufacturing processes but are limited by their stability [15]. There are also materials which exploit quantum confinement in order to fine-tune the absorption characteristics [73], or else change the transport properties within the cell through the creation of quantum wells [74].

The materials studied in this thesis do not strictly fit into generation C as shown in Fig. 1.14, and this grouping has been criticized given the current atmosphere of PV research and development [15]. More accurately, it can be said that the new materials studied in this thesis bridge the gap between generation B and C, and even that they bring improvements along another axis; that is sustainability.

1.4 Earth-Abundant Thin-Film Solar Cells

The main motivation to replace both CdTe and CIGS with other materials that demonstrate similar properties, are the toxicity and scarcity issues that were discussed in Sect. 1.3.3. One commonly used metric for determining the suitability of an element is crustal abundance; the concentration of a certain element within the earth's crust. This is a good starting point for selecting suitable elements and can be used as an ultimate arbiter. The crustal abundance [75] of the elements are shown in Fig. 1.17 (grey data and lines).

Problems with this metric are demonstrated in these data as well, as it does not take into account the extractability or demand for each element, which can affect how much of each is accessible and extracted. Perhaps a better metric to use when considering new materials are the reserve estimates, which are shown in Fig. 1.17 (coloured data) for the elements of the materials studied in this thesis, and the historical PV absorbers discussed in Sect. 1.3. This is better as these estimates also take into account current extraction rates and include only known reserves. From these data, it can be seen that the scarcity problems associated with indium and tellurium are apparent, but it would also suggest that gallium and selenium could also be a concern, and there is much associated literature in whether these concerns are founded [76–79]. However, if these potentially problematic elements are removed from the production of solar cells, then there is no longer a problem; potential or not.

The term 'earth-abundant' suggests only that an element be plentiful in nature, but there is also the underlying subtext that an earth-abundant absorber material be non-toxic and environmentally friendly. So, even though cadmium is in plentiful supply, the concerns with its toxicity are a problem, as there will be a stigma associated with

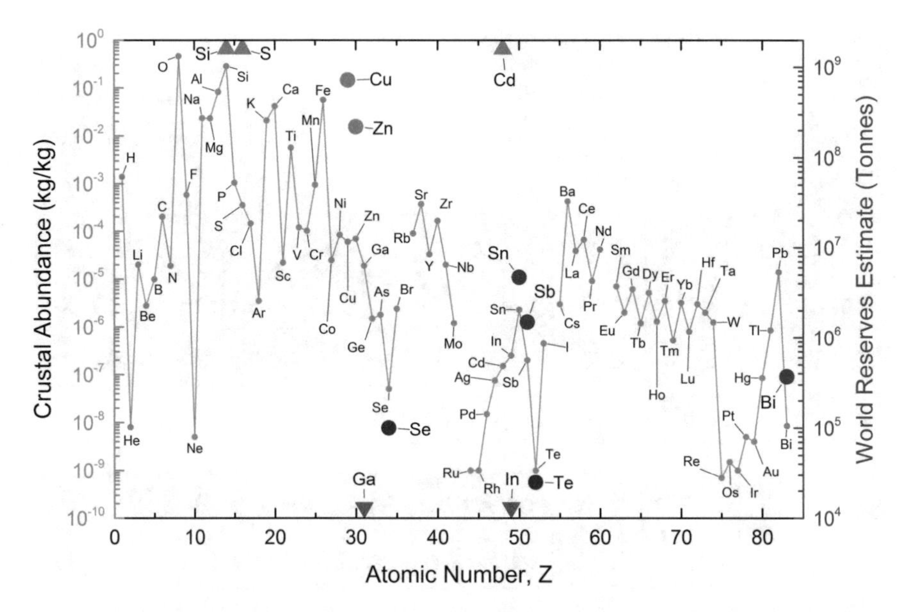

Fig. 1.17 Crustal abundances [75] (grey data) and world reserve estimates [65] (coloured data) for the elements used in thin-film PV technologies and the materials studied in this thesis. Green triangles have no data because they are considered inexhaustible. Red triangles have no data because their reserves are estimated to be very low

it; as was the case with nuclear energy [10]. These problems are of no concern if the use of cadmium is removed entirely, providing the motivation to move away from CdTe deployment.

To generate materials that do not use the elements discussed above, but which also show similar properties to the parents, as was discussed in Sect. 1.2.1, the elements should be isoelectronically substituted into the structures, and as such this limits the particular choice of elements to certain areas within the periodic table; in order to preserve the chemical trends. Figure 1.18 shows the periodic table with the values of crustal abundance displayed using the overlaid spectrum, and also shows the key areas to which the anions (purple) and cations (blue) of a new material should be limited in order to better predict their properties. Also, throughout the history of PV, which will be shown in Sect. 1.5, large improvements in efficiency over a short timeframe are very rare, and it is continued research that improves the efficiencies of the current technologies. For this reason, a drastic shift away from the known material behaviours is risky, and as such, the new materials should have similar structures, comprised of elements restricted to the highlighted areas of the periodic table shown in Fig. 1.18.

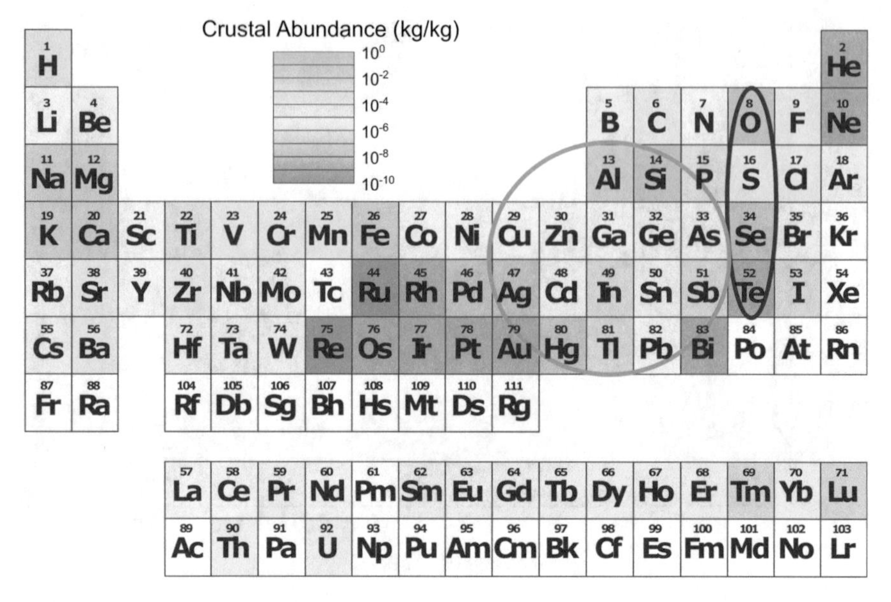

Fig. 1.18 Periodic table coloured by crustal abundance [75]. Key areas highlighted for study in this sector with further discussion in the text. Uncoloured elements have no abundance data available

1.4.1 Why Sulphides?

In CIGS and CdTe, the anions of selenium and tellurium are unsuitable for earth-abundant materials. These two elements belong to the same group; that is the chalcogens. From Fig. 1.18, the abundance of the chalcogens increases with decreasing atomic number. This would suggest that the use of either sulphur or oxygen as an anion would be more beneficial for an earth-abundant semiconductor. Oxygen, which has a much higher electronegativity than the other group VI elements, is sometimes not included in the chalcogen group because the properties between oxides and sulphides are less predictable than between the other chalcogens, and this leads to oxide materials with large bandgaps which are unsuitable for absorbers, and as such they shall not be considered [80]. Non-toxic and abundant in many forms [81], sulphur would seem an ideal choice for the anion in these materials.

Beyond the attractiveness of the use of sulphide materials because of the abundancy of sulphur, there are several other advantages. The low melting point of sulphur gives sulphides low melting points as well, and as such, processing temperatures can be kept low, reducing production costs [82]. Sulphur occurs in various natural forms with various properties. Therefore, if synthetic materials take one of these forms, then there is promise for the stability of that material [17]. Finally, there is a wealth of research already available for natural minerals, which can provide a starting platform for new research.

Table 1.1 Possible elements and oxidation states for the constituents of sulphosalts, which take the form $A_xB_yC_z$

Oxidation State	A	B	C
+	Cu, Ag, Tl, Alkalis	—	—
2+	Pb, Sn, Fe, Mn, Hg, Alkaline Earths	—	—
3+	Lanthanides	As, Sb, Bi	—
4+	Sn	—	—
2-	—	—	S, Se, Te

Underlined elements meet the criteria in the text for being used in an earth-abundant absorber material

1.4.2 Cu-Based Earth-Abundant Post-transition Metal Sulphides

For CIGS, the use of copper is not a problem in abundance or environmental terms [83, 84] and the low-lying *d* orbitals in copper have been shown to lower the bandgap in copper-containing materials, encourage the adoption of the tetrahedral crystal structure, and also improve the stability [85]. For these reasons, the materials of interest in this thesis are copper-based sulphides, but as was found with the evolution of Cu_2S into CIGS, there is a need for another cation in order to inhibit the copper mobility.

Isoelectronic substitution of the problematic indium and gallium in CIGS[20] results in the quaternary compound Cu_2ZnSnS_4 (CZTS), which uses abundant and safe zinc [86] and tin [87], and is explored in Chap. 6. With a similar crystal structure and properties to CIGS, this material has already gained substantial grounding in the research community and efficiencies continue to improve. A large problem of CZTS is the formation of unwanted secondary phases; however, some of these are interesting materials in their own right, the most promising of which for their use in PV are the tin sulphides, explored in Chap. 5.

The other materials that are of interest here are members of the family of sulphosalts. The definition of a sulphosalt is a mineral of the form $A_xB_yC_z$, where A, B, and C can be atoms in the oxidation states shown in Table 1.1 [88]. Although single crystals of these minerals have been grown since 1957 [89], only recently have studies extended beyond structural determinations [88].

The first solar cells based on a sulphosalt absorber were designed in 2007 and 2009, achieving 1% efficiency in the Sn–Sb–S [88] and Pb–Sb–S [82] families, respectively. However, when considering the criteria mentioned earlier, the elements shown in Table 1.1 are whittled down to only those underlined because the others are either rare, toxic, or do not fit the chemical trends to produce attractive properties. Also, the inclusion of copper in such materials is beneficial, and so one is left with

[20]See Fig. 1.13.

the Cu–Sb–S and Cu–Bi–S systems, from which to pool materials. From Fig. 1.17, it can be seen that antimony has a similar estimated reserve to tin and so should not be a problem. Bismuth has a much lower estimated reserve, higher than selenium, but still low. Nevertheless, increased interest in the extraction of bismuth for other uses,[21] means that there is a drive for more to be extracted. Both antimony and bismuth also have the benefit of being non-toxic metals, so their use is warranted [90, 91]. Reasons for the selection of $CuSbS_2$ and Cu_3BiS_3 for study from the Cu–Sb–S and Cu–Bi–S families shall be discussed in their respective chapters.[22]

In summary, the earth-abundant PV absorber materials chosen for study in this thesis are: $CuSbS_2$ (CAS), Cu_3BiS_3 (CBS), SnS, and Cu_2ZnSnS_4 (CZTS), each of which shall be introduced fully in their relevant chapter.

1.5 Efficiency: The Quintessential Goal of Solar Cell Research?

The definition of efficiency is the ratio of the power extracted from a system to power delivered to the system; however, this value can also be dependent upon other external factors. Given the range of locations that solar cells are deployed in, coupled with the time of day and prevailing conditions, and the type of light received, it is clear that a solar cell will not always run at its quoted efficiency.

1.5.1 Standards

In order to overcome these variations and allow the efficiency of a solar cell to be used as a metric that acts as the common denominator for all types of PV, test condition standards have been agreed upon in order to calculate device efficiencies. In Sect. 1.1.3, factors which affect the solar spectrum were mentioned, and in its simplest approximation, the sun can be modelled as a black body point source at a fixed distance from the earth, emitting spectral radiation that follows Planck's Law [93]. Studies that modelled better approximations to the true solar spectrum resulted in a standard solar spectrum,[23] which is used for PV applications.

Deviations from a true black body spectrum occur because of the finite size of the sun and the inhomogeneity of it, which causes some absorption. Then, the radiation which is delivered at the top of Earth's atmosphere, zenith to an observer, takes the form shown in Fig. 1.19 (black curve) and is termed AM0; that is air mass 0 [8]. Then, because the majority of the world does not receive solar radiation at zenith, and due to further absorption and scattering experienced through the atmosphere, the global

[21] Especially its use as a replacement for lead in solder [92].

[22] Chapters 3 and 4, respectively.

[23] ASTM International standard (G173–03) [94].

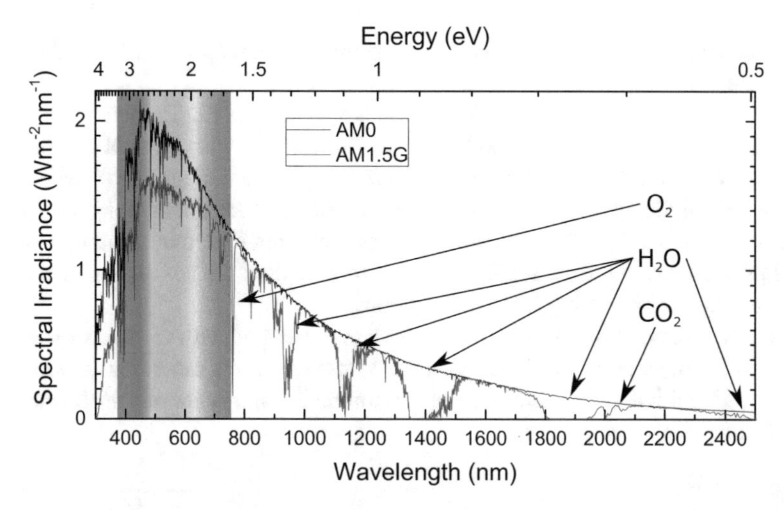

Fig. 1.19 Solar irradiance spectra according to the ASTM G173–03 standard [94]. Dips in the spectra and the deviation from a black body spectrum occur partly as a result of the gas absorption due to the materials highlighted

standard for solar radiation that is used to test solar cells is termed AM1.5G; that is air mass 1.5 globally, which equates to an average power density of 1000 Wm^{-2}. This is also shown in Fig. 1.19 (red curve), labelled with the dips in the intensity due to atmospheric gaseous absorption. Also shown is the visible spectrum of light, showing that the most intense parts of the solar radiation to reach the earth's surface are from visible, and infrared light.

Therefore, when testing a solar cell device, the illuminating light should be one which is calibrated to produce the AM1.5G spectrum, in order that true comparisons can be made between cells. Historically, this standard did not exist and as such, device efficiencies from historical reports must be cautiously assessed, if the illumination source is not stated. Whilst the light source is a major factor which can affect the measured efficiencies, other factors can also affect it between measurements, and for this reason, standard laboratories exist throughout the world which can provide a standard test service to certify the efficiency of devices. These include NREL and Sandia in USA, Fraunhofer ISE in Germany, and JQA in Japan [16].

1.5.2 Record Efficiencies

There is no defining threshold of efficiency which a technology must surpass in order to be granted market viability, but the higher the better. NREL maintains a chart of efficiency advancements for the materials receiving the most research attention, and this is shown in Fig. 1.20. Most of the deployed single junction technologies have record efficiencies just above 20%, the main exception being a–Si at only 14%, so reaching 20% efficiency is a good starting point on the road to market viability for new materials.

The known record efficiencies for the four PV absorbers studied here are shown in Table 1.2, as they do not appear in Fig. 1.20.[24] Compared with some of the materials shown in were developed with their, these are fledgling materials, and therefore have not reached higher efficiencies perhaps only because of a lack of cumulative research intensity. It is therefore the motivation of this thesis not to demonstrate increases to these efficiencies, but to provide reasons as to why they are not higher, and also to suggest ways of improving them in the future.

All of the materials shown in Table 1.2 were developed with their earth-abundant nature in mind, and this raises the question: what level of efficiency can be sacrificed if these technologies can overcome the environmental and economic problems faced by the current market leaders?

1.5.3 A Few Words on Deployment

There is no definitive answer to the question posed at the end of Sect. 1.5.2. Nevertheless, the level of current deployment of solar cells should be introduced, to give some insight as to how the market currently stands. Once a PV technology has proven itself to be reliable, stable and able to deliver good efficiencies that are reproducible, it must face the challenges which restrict its deployment throughout the world. As of 2013, the EU contained 60% of the world's solar capacity [95], and the irradiance received by Europe is shown in Fig. 1.21a. As can be seen, the economically strongest areas of Europe receive relatively little sunlight in comparison with even the northern tip of Africa, and indeed the rest of the world [15], yet it is here that most of the world's solar capacity is deployed. The share of deployment between the countries of the EU is shown in Fig. 1.21b, showing that solar capacity in Europe is dominated by Germany, despite this country receiving little sunlight compared with the countries of southern Europe. This is due to many factors, including the introduction of sophisticated feed-in tariffs [96], and the successful closures of nuclear plants [97].

Table 1.2 Record efficiencies for solar cells utilising the materials studied in this thesis

Material	Record Efficiency	Reference
$CuSbS_2$	3.22%	Banu et al. [98]
Cu_3BiS_3	No tested devices	—
SnS	4.36%	Sinsermsuksakul et al. [99]
CZTS	9.2%	Sun et al. [100]

[24]CZTSSe contains selenium and is not of interest here.

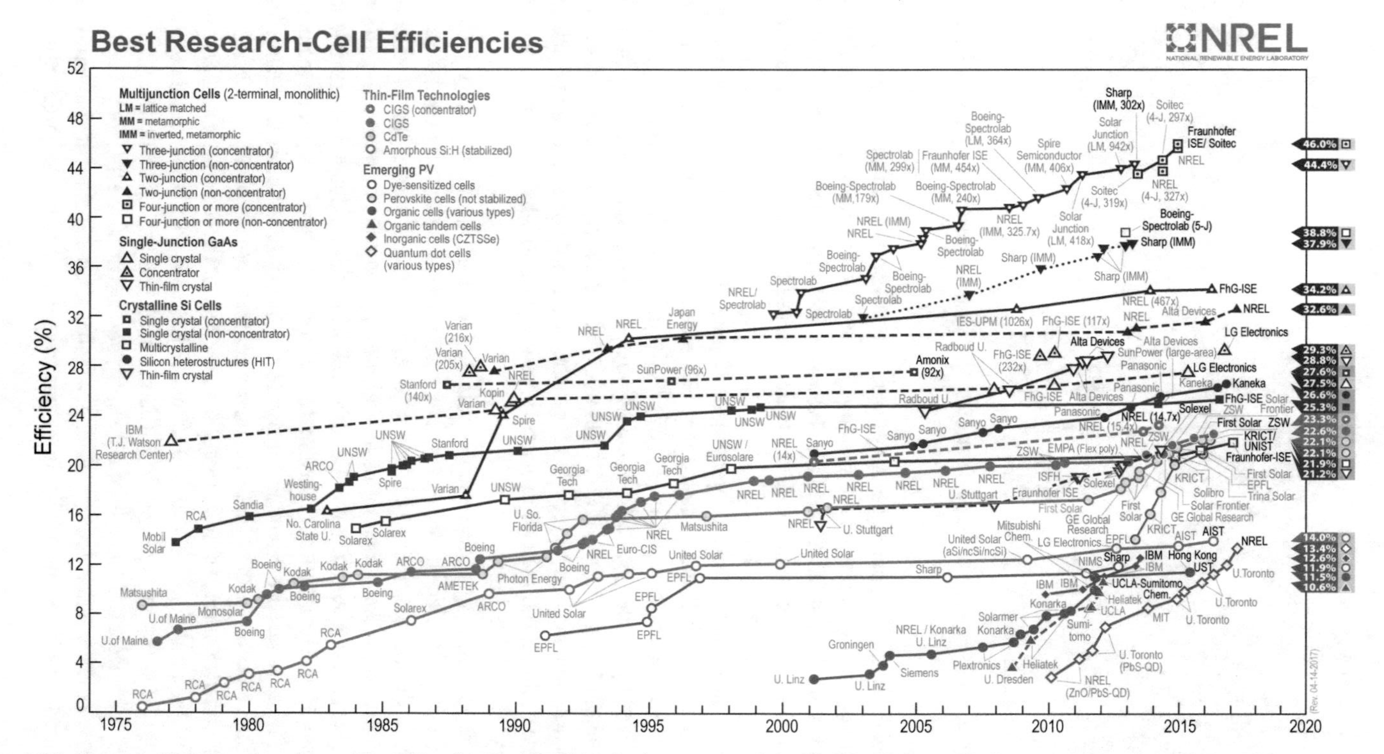

Fig. 1.20 Record efficiency chart for various PV technologies. This plot is courtesy of the National Renewable Energy Laboratory, Golden, CO

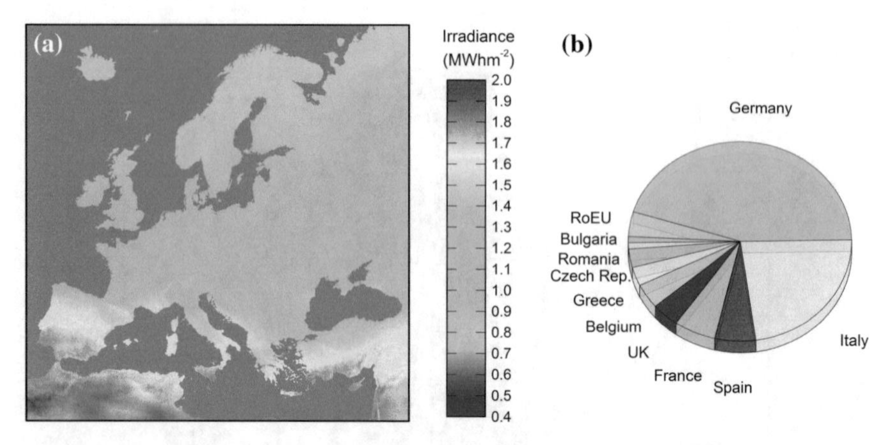

Fig. 1.21 **a** Solar irradiance of Europe [101, 102]. **b** EU nation share of deployed PV capacity [95]

So, given that PV technologies may be employed in areas that are not the most suitable for generating electricity from sunlight, it would be beneficial for the deployed modules to be as efficient as possible. However, it is also important that the production processes and constituent elements that are used in the cells are sustainable, in order to meet the energy needs.

Due to the current level of deployment of solar cells, and the increasing requirements for energy, it is often quoted that PV technologies must be capable of being deployed with terawatt capacity [15, 67, 79, 103]. For a certain technology, this can be achieved by greater production and deployment; however, this can be limited by the sustainability and availability of the materials. Because the use of a certain element is not monopolised by the PV sector, the availability of an element is therefore not the defining limit, but can provide some insight into whether it poses a serious barrier to that technology.

Figure 1.22 shows estimates of the potential limits to the deployed PV technologies, and those studied in this thesis. The assumptions made and calculations used to produce these are explained in Appendix A, Sect. A.2. It shows the absorber layer raw materials cost for a module of that technology capable of producing 1 GW of power (bars), along with the number of possible 1 GW modules that are able to be produced, limited by the reserve estimate for the rarest element within a technology (line). These values are intended for use as comparators between the technologies rather than absolute values, and as such do not take into account: processing costs, the cost and availability of other materials required, or the need for elements outside of the PV sector. In Fig. 1.22a, calculations are made based on the current record efficiencies, and are contrasted with Fig. 1.22b, which shows the calculations assuming that each technology is capable of delivering devices at 20% efficiency.

At the current efficiencies,[25] silicon is more than capable of delivering TW-deployment based on the availability; however, because of the larger amount of

[25]See Fig. 1.22a.

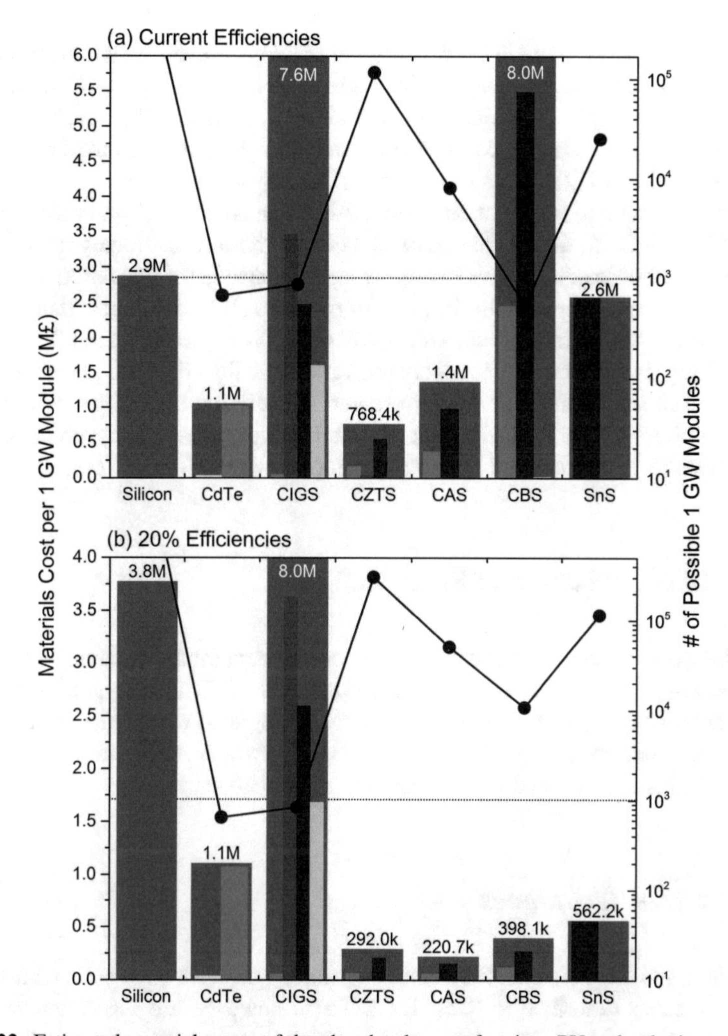

Fig. 1.22 Estimated materials costs of the absorber layers of various PV technologies, separated into the individual elements of each technology (bars). Number of possible 1 GW modules based on the world reserve estimate of the most limiting element (data points). The terawatt capability threshold is also marked. Estimates were calculated using the assumptions detailed in Appendix A, Sect. A.2 for **a** current record efficiencies for each technology (CBS assumed at 1%), **b** assuming that each technology can produce cells with 20% conversion efficiency

material required, this drives up the cost of the modules. The cost of the absorber layer for CdTe is much cheaper due to the smaller amount of material necessary, but the limiting element of tellurium means that this technology will fall short of TW-deployment. The same is true for CIGS, limited by the current estimated reserves and price of indium. CZTS, CAS and SnS all contain cheaper materials than CIGS or silicon, but the current efficiencies of CAS, CBS, and SnS still make GW modules more expensive than CdTe. What is promising however, is that even with the relatively poor efficiencies demonstrated so far, neither CZTS, CAS, nor SnS are TW-scale limited by the elemental reserves of tin or antimony. Bismuth is rarer than antimony or tin, and as such, coupled with the extremely poor current efficiencies of CBS cells,[26] limits CBS cells from reaching the TW-scale.

Comparing with the calculations from 20% efficient devices,[27] the values for silicon, CdTe, and CIGS do not change much because these technologies are already at this level of efficiency, but a drastic improvement is seen in the technologies utilising the materials studied here, with the price of modules all falling below that of CdTe, and the availability of the materials allow not only TW deployment, but PW capabilities, making these technologies much more attractive than the current technologies. It is therefore important to continue research into driving the efficiencies of these devices higher, but perhaps, from an availability and sustainability point of view, 20% efficiency is not required in order to compete with CdTe or CIGS.

1.6 Characterisation of Solar Cells

Given the development of solar cells and the requirement for the discovery and implementation of new materials for use as absorber layers, it is thus the goal of research to first acknowledge, and then develop the technologies necessary to achieving the above-mentioned goals of producing solar absorbers which can be classified as earth-abundant. This is facilitated by rigorous materials characterisation.

1.6.1 Usual Techniques

A standard set of characterisation techniques have arisen for solar cells which measure the metrics described in Sect. 1.2.2. These however, are associated with the characterisation of complete cells, and the scope here is of the characterisation of the absorber materials in their own right.

In reports of solar cells, there is usually some form of materials characterisation, at least for the absorber layer, and these typically take the form of structural and compositional characterisations; in order to ensure that the correct material has been produced. Whilst this is important for the growth of well-established materials, the same cannot be said for new materials, where properties may not be fully established. Therefore, in order to better progress development, other characterisation techniques should also be used, with the goal of better understanding the fundamental material properties. Such studies do exist throughout the literature, but they

[26]None have yet been reported, but it is assumed to be 1% for these calculations.

[27]See Fig. 1.22b.

are usually included supplementary to other characterisations and do not receive the spotlight that they deserve.

1.6.2 The Importance of Electronic Characterisation

As was shown in Sect. 1.2, the properties of a solar absorber are dependent upon the electronic structure and this can be determined through the electronic characterisation of the material.

Before a material is grown for its use as a solar absorber, there now exist sophisticated theoretical screening techniques which allow the selection of certain materials with suitable properties from a vast pool, without the need to grow the materials. Such approaches can be useful given the large variety of materials which are being suggested; however, these techniques are relatively new and not widely implemented. Once a material has been selected using one of these approaches, it is necessary to grow quality samples of the material in order that it can be studied further to corroborate the theoretical findings, before it is incorporated into devices. This bottom-up approach to materials selection and characterisation for solar absorbers should reduce the wasting of resources investigating materials which are unsuitable.

There are many various techniques which can be used to study different aspects of the electronic structure of materials, but several are more suited to PV absorbers because of their relation to the properties which are important for this application. The individual techniques themselves shall be introduced in Chap. 2. The bandgap, position of the band levels, and optical absorption characteristics are important properties for solar absorbers, along with structural and chemical stability, which are important for materials in general. Studies of the band structures of materials can allow relationships between the energy levels to be drawn, determining how the orbitals hybridise and give rise to the electronic band positions. This also allows the bandgap to be determined. The nature of the states in the conduction and valence band also allows the transport properties and optical behaviour of a material to be determined. With relation to the vacuum level, the band positions allow the extraction of the values of electron affinity (EA), work function (WF), ionisation potential (IP), and bandgap (E_g); which should be unique to a material. The definitions of these values are shown in Fig. 1.23.

Losses occur in a solar cell mainly because of the inappropriate bandgap, or because of band misalignments within the cell, therefore studying band alignments in heterojunctions is important. As discussed in Sect. 1.2, the contact of two semiconductors causes charge to equilibrate, the Fermi-levels to align, and the bands to bend. However, before this happens, the natural band positions of a material can help to predict this and so measurements of the values shown in Fig. 1.23 are important for new solar absorber materials.

Beyond measuring the properties of ideal materials, from a device design perspective it is also important to determine the formation mechanism and effect that imperfections have on the absorber material and by extension, on devices. These

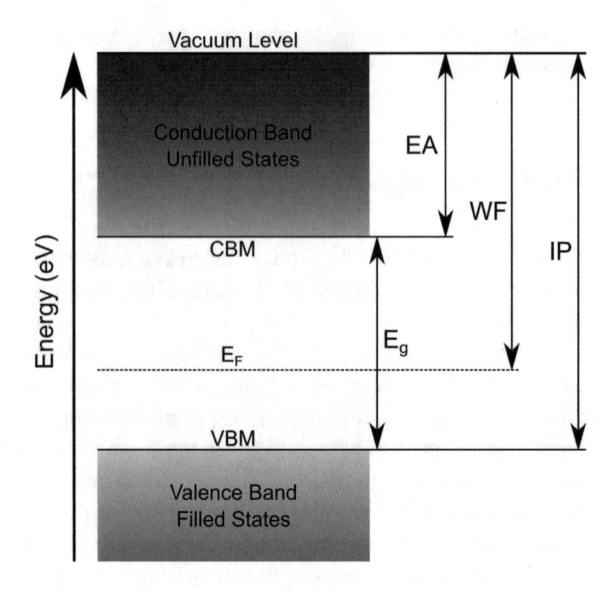

Fig. 1.23 Schematic definitions of energy level differences in a semiconductor material. Vacuum level to conduction band minimum (electron affinity, EA). Vacuum level to valence band maximum (ionisation potential, IP). Vacuum level to Fermi-level (work function, WF). Conduction band minimum to valence band maximum (bandgap, E_g)

imperfections can manifest as unwanted material made of the constituent elements of the absorber, but in a different phase, or as these elements mixed with some form of external contaminant, such as oxides or hydrocarbons. Both of these can result during, or after the growth of the absorber material, and can act detrimentally to the cell. The electronic characterisation of these unwanted materials can therefore help determine the reasons that they formed, but also the effect that they have.

Therefore, through the rigorous electronic characterisation of absorber materials, standards can be set, and create a more complete understanding of the nature of these materials and why they should, or indeed should not be used in solar cells as the absorber layer.

1.7 Overview of the Thesis

1.7.1 Structure

The work presented in this thesis consists of the electronic characterisation of the materials introduced in Sect. 1.4.2, mainly through photoemission spectroscopy techniques, and how this applies to the use of these materials as PV absorbers, developed to address the problems presented in this chapter. As similar techniques are used throughout this thesis to study each material, there is much scope for direct comparison between the results. However, the results are structured so that each chapter is dedicated to each of the materials individually. This is because whilst the materials share common backgrounds and goals,[28] they still retain individual characteristics, which are worth introducing and exploring separately.

Chapter 2 concerns the experimental techniques used throughout this thesis. It introduces them from a physical standpoint, and explains their methods of implementation. Furthermore, it introduces and describes the necessity for, and commissioning of an experimental UHV system that was designed in order to implement the majority of the experimental measurements made for this project.

Chapters 3 and 4 present the electronic characterisation of $CuSbS_2$ and Cu_3BiS_3. They introduce these materials in the context of PV, and present the core-level electronic structures, band levels, and valence bands of these materials, measured using XPS and compared with DFT calculations, as well as structural analyses by XRD and RS, and measurements of the bandgap using absorption spectroscopy. They also explore the required cleaning of this material for measurement, and analyses of the contaminants. The discussions then analyse these findings with regards to PV applications.

Chapter 5 presents the electronic characterisation of SnS, SnS_2, and Sn_2S_3. It introduces these materials in the context of PV, and presents comparisons of the core-level electronic structures, band levels, and valence bands of the materials, measured using PES and absorption spectroscopy, compared with DFT calculations. It also explores the required cleaning of these materials for measurement, and the analysis of the contaminants. The discussion then analyses these findings with regards to PV applications.

Chapter 6 presents the electronic characterisation of CZTS and its possible binary secondary phases. It introduces CZTS in the context of PV, highlighting the difficulties associated with its characterisation, and then introduces the possible binary secondary phases of CZTS and explains the importance of determining the presence of these phases. It then presents full XPS characterisations of these secondary phases, which can be used as standards in the future, followed by the core-level electronic structure, band levels, and valence band of single crystal CZTS, measured using PES and compared with DFT calculations, as well as structural analysis by XRD, RS, and EDX, which can also be used as standard comparators. These characterisation techniques are then compared with regards to their efficacy in the identification of secondary phases within CZTS. It also explores the required cleaning of this material for measurement, and the analysis of the contaminants. The discussion then analyses all of these findings with regards to PV applications.

Chapter 7 concludes the thesis, summarising all of the findings of the individual chapters, and also drawing comparisons between them. It assesses the use of the analysis techniques presented throughout, and presents an overview comparison of the materials studied, assessing their prospects for use as PV absorbers in light of the findings.

Following these chapters are a set of appendices, which give extra details, information, or results, related to different parts of the thesis. If presented within the

[28]See Sect. 1.4.

individual chapters, these would prove excessive, but they are included in the appendices for the sake of completeness and interest.

1.7.2 Aims

The main aim of this thesis is to determine whether the materials presented in Sect. 1.4.2 have the potential to become market viable absorber materials for thin-film solar cells, and if they do, what the reasons are that have led to the poor efficiencies so far and what needs to be done in order to further their advancement. This is hoped to be achieved by studying the underlying electronic structure of the materials using a combined experimental and theoretical approach.

Furthermore, XPS is a technique that is widely employed throughout the literature, but unfortunately suffers from complex analyses, which sometimes leads to misconstrued conclusions in the literature. It is therefore another aim of this thesis to demonstrate the full potential of this analysis technique when it is properly applied, and how it can be an important tool in the characterisation of PV absorber materials.

References

1. Dyson L, Kleban M, Susskind L. Disturbing implications of a cosmological constant. J High Energy Phys. 2002;2002(10):011.
2. Matheny JG. Reducing the risk of human extinction. Risk Anal. 2007;27(5):1335–44.
3. Sagan C. Nuclear war and climatic catastrophe: some policy implications. Foreign Aff. 1983;62(2):257.
4. International Energy Agency (IEA). World Energy Outlook 2012; 2012.
5. International Energy Agency (IEA). Key World Energy Statistics 2016; 2016.
6. U.S. Energy Information Administration (EIA). International Energy Outlook 2016; 2016.
7. Petit JR, Jouzel J, Raynaud D, Barkov NI, Barnola J-M, Basile I, Bender M, Chappellaz J, Davis M, Delaygue G, Delmotte M, Kotlyakov VM, Legrand M, Lipenkov VY, Lorius C, Pépin L, Ritz C, Saltzman E, Stievenard M. Climate and atmospheric history of the past 420,000 years from the Vostok ice core, Antarctica. Nature. 1999;399(6735):429–36.
8. Irvine SJC. Materials challenges: inorganic photovoltaic solar energy. In: Irvine SJC, editor. RSC energy and environment series. Cambridge: Royal Society of Chemistry; 2014.
9. "renewable, Adj. and N.". OED Online; Oxford University Press, 2017.
10. Horlick-Jones T, Prades A, Espluga J. Investigating the degree of "stigma" associated with nuclear energy technologies: a cross-cultural examination of the case of fusion power. Public Underst Sci. 2012;21(5):514–33.
11. Taylor JJ. Improved and safer nuclear power. Science. 1989;244(4902):318–25.
12. Johnstone P, Sovacool BK, MacKerron G, Stirling A. Nuclear power: serious risks. Science. 2016;354(6316):1112.
13. International Energy Agency (IEA). World Energy Outlook 2016; 2016.
14. Hermann WA. Quantifying global exergy resources. Energy. 2006;31(12):1685–702.
15. Jean J, Brown PR, Jaffe RL, Buonassisi T, Bulović V. Pathways for solar photovoltaics. Energy Environ Sci. 2015;8(4):1200–19.
16. Pfisterer, F. Photovoltaic cells. In: Ullmann's encyclopedia of industrial chemistry. Weinheim: Wiley-VCH Verlag GmbH & Co. KGaA; 2000. p. 135–54.

17. Pickart SJ. Physical properties of sulfide materials. Mineral Soc Am Spec Pap. 1970;3:145–53.
18. Mizutori M, Yamada R. Semiconductors. In: Ullmann's encyclopedia of industrial chemistry, vol. 9. Weinheim: Wiley-VCH Verlag GmbH & Co. KGaA; 2000. p. 245–60.
19. Goodman CHL. The prediction of semiconducting properties in inorganic compounds. J Phys Chem Solids. 1958;6(4):305–14.
20. Vidal J, Lany S, D'Avezac M, Zunger A, Zakutayev A, Francis J, Tate J. Band-structure, optical properties, and defect physics of the photovoltaic semiconductor SnS. Appl Phys Lett. 2012;100(3):32104.
21. Burton LA, Colombara D, Abellon RD, Grozema FC, Peter LM, Savenije TJ, Dennler G, Walsh A. Synthesis, characterization, and electronic structure of single-crystal SnS, Sn_2S_3, and SnS_2. Chem Mater. 2013;25(24):4908–16.
22. Becquerel E. Memoire Sur Les Effets Electriques Produits Sous L'influence Des Rayons Solaires. C R Hebd Seances Acad Sci. 1839;9:561–7.
23. Conibeer G. Third-generation photovoltaics. Mater Today. 2007;10(11):42–50.
24. Minemoto T, Matsui T, Takakura H, Hamakawa Y, Negami T, Hashimoto Y, Uenoyama T, Kitagawa M. Theoretical analysis of the effect of conduction band offset of window/CIS layers on performance of CIS solar cells using device simulation. Sol Energy Mater Sol Cells. 2001;67(1–4):83–8.
25. Bär M, Weinhardt L, Heske C. Soft X-ray and electron spectroscopy: a unique "tool chest" to characterize the chemical and electronic properties of surfaces and interfaces. In: Advanced characterization techniques for thin film solar cells. Weinheim: Wiley-VCH Verlag GmbH & Co. KGaA; 2011. p. 387–409.
26. Zakutayev A, Caskey CM, Fioretti AN, Ginley DS, Vidal J, Stevanovic V, Tea E, Lany S. Defect tolerant semiconductors for solar energy conversion. J Phys Chem Lett. 2014;5(7):1117–25.
27. Ganose AM, Savory CN, Scanlon DO. Beyond methylammonium lead iodide: prospects for the emergent field of ns^2 containing solar absorbers. Chem Commun. 2017;53(1):20–44.
28. Yu L, Kokenyesi RS, Keszler DA, Zunger A. Inverse design of high absorption thin-film photovoltaic materials. Adv Energy Mater. 2013;3(1):43–8.
29. Loferski JJ. Theoretical considerations governing the choice of the optimum semiconductor for photovoltaic solar energy conversion. J Appl Phys. 1956;27(7):777–84.
30. Shockley W, Queisser HJ. Detailed balance limit of efficiency of P-N junction solar cells. J Appl Phys. 1961;32(3):510–9.
31. Rühle S. Tabulated values of the Shockley-Queisser limit for single junction solar cells. Sol Energy. 2016;130:139–47.
32. Walsh A, Chen S, Wei S-H, Gong X-G. Kesterite thin-film solar cells: advances in materials modelling of Cu_2ZnSnS_4. Adv Energy Mater. 2012;2(4):400–9.
33. Schorr S. Structural aspects of adamantine like multinary chalcogenides. Thin Solid Films. 2007;515(15):5985–91.
34. Brendel R, Werner JH, Queisser HJ. Thermodynamic efficiency limits for semiconductor solar cells with carrier multiplication. Sol Energy Mater Sol Cells. 1996;41–42:419–25.
35. Adams WG, Day RE. The action of light on selenium. Proc R Soc London. 1876;25(171–178):113–7.
36. Fritts CE. On the Fritts selenium cells and batteries. J Franklin Inst. 1885;119(3):221–32.
37. Siemens W. On the electro motive action of illuminated selenium, discovered by Mr. Fritts, of New York. J Franklin Inst. 1885;119(6):453–IN6.
38. Fritts CE. On a new form of selenium cell, and some electrical discoveries made by its use. Am J Sci. 1883;s3-26(156):465–72.
39. Reynolds DC, Leies G, Antes LL, Marburger RE. Photovoltaic effect in cadmium sulfide. Phys Rev. 1954;96(2):533–4.
40. Chapin DM, Fuller CS, Pearson GL. A new silicon P-n junction photocell for converting solar radiation into electrical power. J Appl Phys. 1954;25(5):676–7.
41. Green MA, Emery K, Hishikawa Y, Warta W, Dunlop ED, Levi DH, Ho-Baillie AWY. Solar cell efficiency tables (version 49). Prog Photovoltaics Res Appl. 2017;25(1):3–13.

42. Wang TY, Lin YC, Tai CY, Fei CC, Tseng MY, Lan CW. Recovery of silicon from kerf loss slurry waste for photovoltaic applications. Prog Photovoltaics Res Appl. 2009;17(3):155–63.
43. Chittick RC, Alexander JH, Sterling HF. The preparation and properties of amorphous silicon. J Electrochem Soc. 1969;116(1):77.
44. Spear WE, Le Comber PG. Investigation of the localised state distribution in amorphous Si films. J Non Cryst Solids. 1972;10:727–38.
45. Carlson DE, Wronski CR. Amorphous silicon solar cell. Appl Phys Lett. 1976;28(11):671–3.
46. Hall RB, Birkmire RW, Phillips JE, Meakin JD. Thin-film polycrystalline $Cu_2S/Cd_{1-x}Zn_xS$ solar cells of 10% efficiency. Appl Phys Lett. 1981;38(11):925–6.
47. Vanhoecke E, Burgelman M. Reactive sputtering of Thin Cu_2S films for application in solar cells. Thin Solid Films. 1984;112:97–106.
48. Partain LD, Schneider RA, Donaghey LF, McLeod PS. Surface chemistry of Cu_xS and Cu_xS/CdS determined from x-ray photoelectron spectroscopy. J Appl Phys. 1985;57(11):5056.
49. Rakhshani AE. Preparation, characteristics and photovoltaic properties of cuprous oxide—a review. Solid State Electron. 1986;29(1):7–17.
50. Rai BP. Cu_2O solar cells: a review. Sol Cells. 1988;25(3):265–72.
51. Ennaoui A, Fiechter S, Pettenkofer C, Alonso-Vante N, Büker K, Bronold M, Höpfner C, Tributsch H. Iron disulfide for solar energy conversion. Sol Energy Mater Sol Cells. 1993;29(4):289–370.
52. Cusano DA. CdTe solar cells and photovoltaic heterojunctions in II–VI compounds. Solid State Electron. 1963;6(3):217–32.
53. Welch AW, Zawadzki PP, Lany S, Wolden CA, Zakutayev A. Self-regulated growth and tunable properties of $CuSbS_2$ solar absorbers. Sol Energy Mater Sol Cells. 2015;132:499–506.
54. Wagner S, Shay JL, Migliorato P, Kasper HM. $CuInSe_2/CdS$ heterojunction photovoltaic detectors. Appl Phys Lett. 1974;25(8):434–5.
55. Kazmerski LL, White FR, Morgan GK. Thin-film $CuInSe_2/CdS$ heterojunction solar cells. Appl Phys Lett. 1976;29(4):268–70.
56. Tinoco T, Rincón C, Quintero M, Pérez GS. Phase diagram and optical energy gaps for $CuIn_yGa_{1-y}Se_2$ alloys. Phys Status Solidi. 1991;124(2):427–34.
57. Fraunhofer ISE. Photovoltaics Report; 2016.
58. Ferekides CS, Balasubramanian U, Mamazza R, Viswanathan V, Zhao H, Morel DL. CdTe thin film solar cells: device and technology issues. Sol Energy. 2004;77(6):823–30.
59. Wu X. High-efficiency polycrystalline CdTe thin-film solar cells. Sol Energy. 2004;77(6):803–14.
60. Bosio A, Menossi D, Mazzamuto S, Romeo N. Manufacturing of CdTe thin film photovoltaic modules. Thin Solid Films. 2011;519(21):7522–5.
61. Schulte-Schrepping K-H, Piscator M. Cadmium and cadmium compounds. In: Ullmann's encyclopedia of industrial chemistry, vol. 100 C. Weinheim: Wiley-VCH Verlag GmbH & Co. KGaA; 2000. p. 121–45.
62. Fthenakis VM, Moskowitz PD. Photovoltaics: environmental, health and safety issues and perspectives. Prog Photovoltaics Res Appl. 2000;8(1):27–38.
63. Fthenakis V. Sustainability of photovoltaics: the case for thin-film solar cells. Renew Sustain Energy Rev. 2009;13(9):2746–50.
64. Fthenakis VM. Life cycle impact analysis of cadmium in CdTe PV production. Renew Sustain Energy Rev. 2004;8(4):303–34.
65. U.S. Geological Survey. Mineral Commodity Summaries 2017; 2017.
66. Knockaert G. Tellurium and tellurium compounds. In: Ullmann's encyclopedia of industrial chemistry, vol. 115. Weinheim: Wiley-VCH Verlag GmbH & Co. KGaA; 2011. p. 5–6.
67. Candelise C, Speirs JF, Gross RJK. Materials availability for thin film (TF) PV technologies development: a real concern? Renew Sustain Energy Rev. 2011;15(9):4972–81.
68. Langner BE. Selenium and selenium compounds. In: Ullmann's encyclopedia of industrial chemistry, vol. 9. Weinheim: Wiley-VCH Verlag GmbH & Co. KGaA; 2000. p. 245–60.

69. Cerwenka EA, Cooper WC. Toxicology of selenium and tellurium and their compounds. Arch Environ Heal An Int J. 1961;3(2):189–200.
70. Phipps G, Mikolajczak C, Guckes T. Indium and gallium: long-term supply. Renew Energy Focus. 2008;9(4):56–9.
71. Delbos S. Kësterite thin films for photovoltaics: a review. EPJ Photovoltaics. 2012;3:35004.
72. Henry CH. Limiting efficiencies of ideal single and multiple energy gap terrestrial solar cells. J Appl Phys. 1980;51(8):4494–500.
73. Mazzer M, Barnham KWJ, Ballard IM, Bessiere A, Ioannides A, Johnson DC, Lynch MC, Tibbits TND, Roberts JS, Hill G, Calder C. Progress in quantum well solar cells. Thin Solid Films. 2006;511–512:76–83.
74. Ross RT, Nozik AJ. Efficiency of hot-carrier solar energy converters. J Appl Phys. 1982;53(5):3813–8.
75. Lide DR. CRC handbook of chemistry and physics. 85th ed. Boca Raton, FL: CRC Press; 2004.
76. Wadia C, Alivisatos AP, Kammen DM. Materials availability expands the opportunity for large-scale photovoltaics deployment. Environ Sci Technol. 2009;43(6):2072–7.
77. Andersson BA. Materials availability for large-scale thin-film photovoltaics. Prog Photovoltaics Res Appl. 2000;8(1):61–76.
78. Andersson B, Azar C, Holmberg J, Karlsson S. Material constraints for thin-film solar cells. Energy. 1998;23(5):407–11.
79. Feltrin A, Freundlich A. Material considerations for terawatt level deployment of photovoltaics. Renew Energy. 2008;33(2):180–5.
80. Makovicky E. Crystal structures of sulfides and other chalcogenides. Rev Mineral Geochemistry. 2006;61(1):7–125.
81. Nehb W, Vydra K. Sulfur. In: Ullmann's encyclopedia of industrial chemistry, vol. 1. Weinheim: Wiley-VCH Verlag GmbH & Co. KGaA; 2006. p. 1–32.
82. Dittrich H, Stadler A, Topa D, Schimper H-J, Basch A. Progress in sulfosalt research. Phys status solidi. 2009;206(5):1034–41.
83. Lossin A. Copper. In: Ullmann's encyclopedia of industrial chemistry. Weinheim: Wiley-VCH Verlag GmbH & Co. KGaA; 2001. p. 467–97.
84. Zhang J, Richardson HW. Copper compounds. In: Ullmann's encyclopedia of industrial chemistry. Weinheim: Wiley-VCH Verlag GmbH & Co. KGaA; 2016. p. 1–31.
85. Dufton JTR, Walsh A, Panchmatia PM, Peter LM, Colombara D, Islam MS. Structural and electronic properties of $CuSbS_2$ and $CuBiS_2$: potential absorber materials for thin-film solar cells. Phys Chem Chem Phys. 2012;14(20):7229.
86. Schwab B, Ruh A, Manthey J, Drosik M. Zinc. In: Ullmann's encyclopedia of industrial chemistry. Weinheim: Wiley-VCH Verlag GmbH & Co. KGaA; 2015. p. 1–25.
87. Graf GG. Tin, tin alloys, and tin compounds. In: Ullmann's encyclopedia of industrial chemistry, vol. 37. Weinheim: Wiley-VCH Verlag GmbH & Co. KGaA; 2000. p. 119–26.
88. Dittrich H, Bieniok A, Brendel U, Grodzicki M, Topa D. Sulfosalts—a new class of compound semiconductors for photovoltaic applications. Thin Solid Films. 2007;515(15):5745–50.
89. Wernick JH, Benson KE. New semiconducting ternary compounds. J Phys Chem Solids. 1957;3(1–2):157–9.
90. Grund SC, Hanusch K, Breunig HJ, Wolf HU. Antimony and antimony compounds. In: Ullmann's encyclopedia of industrial chemistry, vol. 100 C. Wiley-VCH Verlag GmbH & Co. KGaA: Germany; 2006. p. 41–93.
91. Krüger J, Winkler P, Lüderitz E, Lück M, Wolf HU. Bismuth, bismuth alloys, and bismuth compounds. In: Ullmann's encyclopedia of industrial chemistry, vol. 100 C. Weinheim: Wiley-VCH Verlag GmbH & Co. KGaA; 2003. p. 121–45.
92. Gerein NJ, Haber JA. One-step synthesis and optical and electrical properties of thin film Cu_3BiS_3 for use as a solar absorber in photovoltaic devices. Chem Mater. 2006;18(26):6297–302.
93. Chen CJ. Physics of solar energy. Hoboken: Wiley; 2011.

94. ASTM International. Standard Tables for Reference Solar Spectral Irradiances: Direct Normal and Hemispherical on 37° Tilted Surface. ASTM G173-03(2012). West Conshohocken 2012.
95. International Energy Agency (IEA). World Energy Outlook 2014; 2014.
96. Jacobsson S, Lauber V. The politics and policy of energy system transformation—explaining the German diffusion of renewable energy technology. Energy Policy. 2006;34(3):256–76.
97. Jorant C. The implications of Fukushima. Bull At Sci. 2011;67(4):14–7.
98. Banu S, Ahn SJ, Ahn SK, Yoon K, Cho A. Fabrication and characterization of cost-efficient $CuSbS_2$ thin film solar cells using hybrid inks. Sol Energy Mater Sol Cells. 2016;151:14–23.
99. Sinsermsuksakul P, Sun L, Lee SW, Park HH, Kim SB, Yang C, Gordon RG. Overcoming efficiency limitations of SnS-based solar cells. Adv Energy Mater. 2014;4(15):1400496.
100. Sun K, Yan C, Liu F, Huang J, Zhou F, Stride JA, Green M, Hao X. Over 9% efficient Kesterite Cu_2ZnSnS_4 solar cell fabricated by using $Zn_{1-X}Cd_XS$ buffer layer. Adv Energy Mater. 2016;6(12):1600046.
101. Šúri M, Huld TA, Dunlop ED, Ossenbrink HA. Potential of solar electricity generation in the European Union member states and candidate countries. Sol. Energy. 2007;81(10):1295–305.
102. Huld T, Müller R, Gambardella A. A new solar radiation database for estimating PV performance in Europe and Africa. Sol Energy. 2012;86(6):1803–15.
103. Polman A, Knight M, Garnett EC, Ehrler B, Sinke WC. Photovoltaic materials—present efficiencies and future challenges. Science. 2016;352(6283):307.

Chapter 2
Experimental Methods

We have this handy fusion reactor in the sky called the sun, you don't have to do anything, it just works. It shows up every day.

Elon Musk, CEO of Tesla Motors, 2015

This chapter introduces the experimental techniques used throughout this thesis, including the methods used to produce the studied materials, and also an introduction to the practices used to perform the characterisations.

A large part of the characterisation work in this thesis is involved with the electronic characterisation of PV absorber materials using various implementations of photoemission spectroscopy. For this reason, a large part of this chapter is dedicated to these techniques, their implementation, and also the need for the specialist environment of ultrahigh vacuum.

Where a specific experimental system is described, it is the one which was used in the work presented in further chapters. The minutiae of each technique pertinent

---

The original version of this chapter was revised: Table was reformatted. The correction to this chapter is available at https://doi.org/10.1007/978-3-319-91665-1_8

© Springer International Publishing AG, part of Springer Nature 2018

T. J. Whittles, *Electronic Characterisation of Earth-Abundant Sulphides for Solar Photovoltaics*, Springer Theses, https://doi.org/10.1007/978-3-319-91665-1_2

to the different experiments can be found in the individual chapters, with reference to additional specifics contained in Appendix C.

2.1 Material Growth Methods

The materials that were studied throughout this thesis were grown by various collaborators using the methods outlined in the following sections.

For PV applications, an important area of study is the growth of the materials to be used as absorber layers. In the final device, the absorber will be present as a thin-film, and therefore thin-film growth techniques are relevant. Many of the studies that focus on the growth of a thin-film of a particular material, are concerned with the ability to grow phase-pure material. Consequently, often reported is the presence of unwanted secondary phases, contamination, impurities, or other imperfections within a film such as grain boundaries, defects, or inhomogeneities. These aspects of thin-film growth are important to consider, but for new materials, where the properties have perhaps still not been fully disseminated, it is prudent to determine the properties of a form of the material which is free from these imperfections. This is achieved by the growth of single crystals.

2.1.1 Single Crystal Growth Methods

Whilst single crystal growth methods are not within the scope of this thesis, because they are well established [1, 2], it is practical to introduce the methods used to produce the crystals that were studied here.

Chemical Vapour Transport

The single crystals of SnS, SnS_2, and Sn_2S_3, which are presented in Chap. 5 were grown by the chemical vapour transport (CVT) method. This method functions by containing precursor elements of the crystal with another element to act as an intermediary medium which facilitates the vaporisation and subsequent crystallisation of the desired material [3], and has shown to be successful for the growth of various halides, oxides and chalcogenides [4].

For the tin sulphides, stoichiometric amounts of sulphur and tin were placed in a silica ampoule with iodine as the carrier agent. The evacuation and sealing of this ampoule then ensured that no external impurities could affect the crystal growth and they were placed in a zone-controlled furnace. A temperature gradient was then applied across the ampoule, which was greater than the boiling point of the intermediary phase, but lower than the boiling point of the crystal phase in order to ensure the vaporisation and crystallisation of each phase, respectively. This temperature gradient then drove the reaction,

$$\mathrm{Sn} + \mathrm{S} + \mathrm{I}_2 \rightarrow \mathrm{SnI}_2 \rightarrow \mathrm{SnS} + \mathrm{I}_2. \tag{2.1}$$

The high temperature caused the iodine to react with tin, and tin iodide to be transported to the cooler side where the tin is released, reacts with the sulphur, and the iodine diffuses back toward the tin, forming a cycle.

Travelling Heater Method

The single crystal CZTS which is presented in Chap. 6 was grown by the travelling heater method, which has proved successful for the growth of CdTe [5] and HgSe [6]. A polycrystalline CZTS ingot was first synthesised by heating stoichiometric amounts of copper, zinc, tin, and sulphur in an evacuated ampoule above the melting point of CZTS. This ingot and tin were then sealed in another evacuated ampoule, which was lowered extremely slowly through a furnace, encouraging the single crystal growth of CZTS as the ampoule passed out of the heated region.

2.1.2 Thin-Film Growth Methods

The various methods used for the growth of thin-films are beyond the scope of this thesis, but a general introduction to their implementation along with the techniques used to grow the materials studied here shall be given. The method chosen for the growth of the absorber layer can impact the cell both technologically but also economically and environmentally, because even though the materials used may be considered earth-abundant, if the processing methods are not sustainable also, then this goal is jeopardised.

Vacuum Versus Non-vacuum

Thin-film deposition techniques are often categorised into vacuum based and non-vacuum based, especially with regards to CZTS [7], and both have significant advantages and disadvantages.

Vacuum based techniques involve the successive deposition of the atoms of a source material, from which they are transported to a substrate for deposition. The most commonly used techniques which require the use of vacuum are sputtering, evaporation, and pulsed-laser deposition (PLD), the processes for which are shown in Fig. 2.1a–c, respectively.

Sputtering ejects atoms of a solid target material, depositing them onto a substrate by means of bombarding the target with energetic ions of a gas [8]. For the thin-films of $CuSbS_2$ measured in Chap. 3, this method was employed, using targets of Cu_2S and Sb_2S_3. Further details of this procedure can be found in Sect. 3.2.1 and the appropriate literature [9–11].

In evaporation, a source of material is heated, by means of an electrical filament or an electron beam, which causes the material to evaporate and travel to a substrate where again it condenses [12]. For the thin-films of Cu_3BiS_3 measured in Chap. 4, this method was used. Further details of this procedure can be found in Sect. 4.2.1.

PLD uses a high power laser beam to vaporise material from a target in pulses which is then deposited onto a substrate [13].

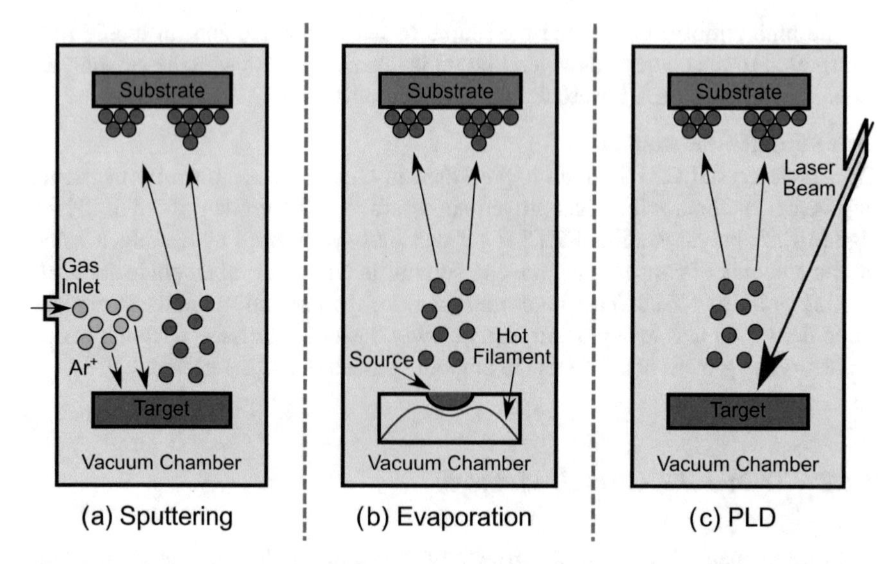

Fig. 2.1 Physical vapour deposition (PVD) methods for thin-film growth. **a** Sputtering by means of argon ion bombardment of a source material. **b** Thermal evaporation of a source material using a heated filament. **c** Pulsed-laser deposition (PLD) of a source material by energetic pulses from a laser. All methods require a level of high vacuum in order to keep the atmosphere free from contaminants and to allow the unimpeded transport of the deposition material to the substrate

As all three of these techniques cause the depositing material to undergo a phase change to the vapour phase, they are collectively known as physical vapour deposition (PVD) [14]. The conditions of the deposition because of the vacuum afford these techniques the ability to produce uniform films with good reproducibility. Unfortunately, the necessary equipment is rather expensive both in terms of cost and energy, and they also do not scale-up well, with low throughput and low efficiency of source material use [14–16].

Non-vacuum based techniques involve the production of a precursor solution which is then applied to a substrate by various means. These can be direct and physical as is the case with inkjet printing [17] or spin coating [18], or require the use of reaction techniques such as spray pyrolysis [19], electrodeposition [20], or the SILAR technique [21]. Such methods utilise more of the materials that they are supplied with and are therefore cheaper, also because of the lack of expensive equipment. However, these benefits come with a sacrifice in film reproducibility and overall quality because of the lack of precise control during the growth, in contrast with the highly controllable vacuum techniques [14–16].

Variations within all of these techniques are possible, with the most common being the order and method by which the precursors are incorporated. Whilst it has been shown to be possible to deposit and crystallise CZTS in a one-stage process, via sputtering [22], evaporation [23], or electrochemical means [24], it is much more common to utilise a multi-stage process that involves the deposition of precursors, which are then incorporated into the final material by annealing in a sulphur atmosphere.

In the case of CZTS, the most widely studied material of interest in this thesis, it was the vacuum techniques that historically produced the highest efficiencies and received the most research attention [15], as a result of the high controllability and desire to study the new material. However, now CZTS has established itself as a viable technology, at least in the research community, trends are shifting toward non-vacuum based growth with the motivation for achieving high efficiencies with techniques that are better suited to low-cost industrial scale-up [15]. The non-vacuum based techniques described above are all suitable to meet this end; however, one technique excels somewhat in its simplicity and adaptability.

Chemical Bath Deposition

Some of the results from the single crystal CZTS, presented in Chap. 6 were compared with measurements of thin-films of CZTS grown by chemical bath deposition (CBD) [25]. This process dissolves precursor salts in a solution, which are then applied to a substrate by immersing it in the chemical bath and annealing. This technique is more controllable than some other non-vacuum processes as the choice of salts and solvent, composition of the solution, bath temperature, immersion time, annealing temperature and atmosphere can all be easily controlled [26, 27]. The speed and simplicity of the process allows many variables to be considered at the same time, improving research throughput [15].

Despite the promise of this technique, there are several drawbacks which limit its effectiveness. The quality, homogeneity, uniformity, and reproducibility of the produced films are generally lacking when compared to other techniques. This is because of the ineffective dissolving of the precursors [28], differences in volume expansion rates of the various precursors during annealing [29], or the presence of remnant precursor or solution residue in the finished films [30, 31], all of which can affect the performance of devices. Nevertheless, the cost-effectiveness and simplicity of this technique show promise for future research.

2.2 Ultrahigh Vacuum

From a surface science point of view, the use of vacuum is almost ubiquitous. However, the techniques primarily used in surface science are now also finding application for technical materials as well.

2.2.1 The Necessity for Vacuum

There are two primary reasons why vacuum is a necessity for certain studies [32]. First, the characterisation techniques involve the generation or measurement of elec-

Table 2.1 Categories, or levels of vacuum classified by lowest pressure

Level of Vacuum	Pressure (mbar)	Mean Free Path (m)
Atmosphere	1000	1×10^{-7}
Rough	1	1×10^{-4}
Medium	1×10^{-3}	0.1
High (HV)	1×10^{-6}	100
Ultrahigh (UHV)	1×10^{-10}	1×10^{6}

Also shown is the mean free path of a nitrogen molecule at that pressure [32]

trons (or other particles) and their interaction with matter. This requires that the electrons are able to move from source, through the sample, to the detector, without losing energy. Also, the necessary equipment can require a level of vacuum, for example the production of electrons from a hot filament without oxidation. Second, because these techniques are surface sensitive, it is important that the surfaces being measured remain somewhat constant over the time period in which the measurement takes place.

There are different classifications of the quality of vacuum, dependent upon the pressure. These are categorised in Table 2.1, along with the mean free path of a nitrogen molecule at that pressure. Given that a vacuum system must be able to be sited within a laboratory, then unimpeded electron travel is satisfied by high vacuum, where an electron is expected to collide with the chamber walls before encountering a gas molecule.

The second condition may seem somewhat inconsequential for technical samples as they will already contain a layer of contamination present from atmospheric handling. However, further contamination may influence the signal from a measurement, or cause the sample to change in some other way. Furthermore, if a sample is required to be cleaned in vacuo, then it is important to maintain this cleanliness throughout measurements, which may take several hours. From the kinetic theory of gases, this level of cleanliness can be maintained for several hours with pressures of 1×10^{-9} mbar and below (UHV), where it takes 1 h or longer for the adsorption of one monolayer of contaminant [33].

It is noted that this level of cleanliness is of vital importance for surface science studies where the first few atomic layers and/or adlayers are under scrutiny [33, 34]; however, this can be relaxed somewhat due to the technical nature of the samples under study throughout this thesis. Nevertheless, care was ensured to achieve and maintain the best level of vacuum possible in order to ensure the non-dynamicity of all measurements performed here.

2.2.2 Achieving UHV

This section outlines the requirements, techniques, and equipment used in providing the UHV environment necessary for undertaking the measurements presented throughout this thesis. Further details of the specifics of the equipment are given in Appendix B, Sect. B.1.

Primarily, a modern UHV system comprises a stainless steel chamber with several ports to which different components can be attached, including: pumps, characterisation probes, means of manipulating and positioning samples around the chamber, features to prepare or grow samples, and even other complete systems that can be interconnected.

The ability to achieve UHV conditions is one of the main factors to be considered in system design, because it restricts the size, shape, and materials used in construction. If one assumes that a vacuum system has constant volume, then from the ideal gas law, the rate of change of number of molecules (N) is proportional to the rate of change in pressure (p) [34–36], so that

$$\frac{dN}{dt} \propto V\left(\frac{dp}{dt}\right) = G - pS, \tag{2.2}$$

where G is the outgassing rate in litres millibar per second, and S is the combined pumping speed in litres per second. At higher pressures, the outgassing rate is small and the solution to Eq. (2.2) takes exponential form [32]. This results in the rapid removal of most of the gas molecules from the volume of the chamber, known as the viscous flow regime, where the gas in the chamber behaves as a massive fluid. At lower pressures, outgassing, the release of adsorbed gases upon a surface, affects the ability to remove further molecules from the chamber and therefore the ultimate pressure that a system can reach [32], in the steady state is

$$p = \frac{G}{S}. \tag{2.3}$$

At these lower pressures, gases no longer behave as a fluid because of the low number of molecules within the system, acting independently, hence the term molecular flow regime.

In order to lower the ultimate pressure of a system, either the pumping speed should be increased, or the outgassing rate should be minimised [34]. The pumping speed is dictated by the efficiency of the pumps used in the system. Because the flow regime changes from viscous to molecular some point between atmospheric and UHV conditions, no single pump can cover this entire pressure range, because there are different pumping principles. For this reason, several pumps are attached to a system, with different uses for different gases under different conditions [34, 36], and the operating principles of the main types are given in Appendix B, Sect. B.1. In the systems that will be described in Sects. 2.2.3 and 2.2.4, a level of medium vacuum was first achieved using scroll pumps, and then turbomolecular pumps were used to achieve UHV. UHV conditions were then maintained using a combination of ion pumps, titanium sublimation pumps (TSP) and hydrogen getters.

The other way to improve the overall pressure of the system, and sometimes the most important factor, is to decrease the outgassing rate. At low pressures, molecules adsorbed to any surface within the chamber will begin to desorb, effectively raising the pressure [34]. The majority of the adsorbates are from water [37], adsorbed to the chamber walls at atmospheric pressure. Eventually, this outgassing rate will decrease as the adsorbates desorb, but this will take a long time. Instead, the rate of desorption can be temporarily increased by raising the temperature of the entire system to around 150 °C for a few days (bakeout) [32–34]. This causes the adsorbates to desorb much more quickly, so that after cooling down the system again to room temperature, the adsorbates should now be of such a low quantity that the outgassing rate is negligible, and the ultimate achievable pressure is much lower.

After bakeout, the limiting outgassing rates for a UHV system are from gases which either continue to desorb from inner surfaces, diffuse from the materials that the chamber itself is made from, or permeate through the chamber walls from the atmosphere [32]. Further desorption and permeation can be controlled through careful chamber design and materials use [33]. Adsorbates can only occur at surfaces, so the surface-to-volume ratio should be minimised by using spherical chambers and cylindrical connectors, and by minimising the use of corners and corrugations. Materials for use in UHV conditions are also restricted to those with low vapour pressures, with most components made from stainless steel, refractory metals, or various ceramics [33]. As such, the use of many plastics, adhesives, brass, rubber, solder, and various other volatile materials are unsuitable for UHV use [33, 34].

So that an entire system need not undergo bakeout each time a sample is to be introduced, it is usually separated into at least two parts: the main chamber of the system, which is kept at UHV conditions except during maintenance; and a secondary part, termed the load lock [32]. This is a small chamber which can be isolated and pumped separately from the main chamber. The reduced size allows much quicker pump down times, and UHV conditions are not necessarily required, as the sample is transferred to the main chamber through a low conductance port.

By observing these considerations, UHV pressures of $\sim 1 \times 10^{-10}$ mbar are achievable and maintainable. The pressure in an UHV system is generally measured using an ion gauge, the operating principles of which are given in Appendix B, Sect. B.1.

2.2.3 A 3-Chamber Surface Science System

An UHV system that was available to use for the measurements required in this thesis is shown in Fig. 2.2. The vacuum schematic views in Fig. 2.2a, b show the arrangement of peripherals attached to the chambers, as well as the pumping and sample manipulation availabilities.[1] Photographs of this system are shown in Fig. 2.2c.

[1] The key to vacuum symbols is given in Appendix B, Sect. B.2.

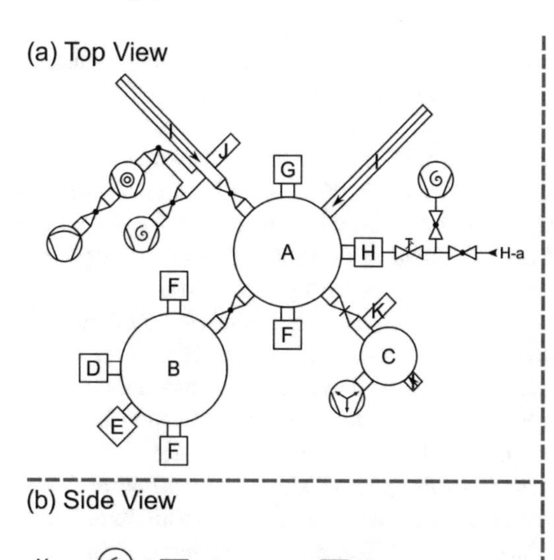

A - Preparation Chamber
B - Analysis Chamber
C - STM Chamber
D - Electron Energy Analyser
E - X-ray Gun
F - Evaporator
G - LEED
H - Ion Gun
  a - Argon Gas
I - Transfer Arm
J - Load Lock
K - Sample Storage
L - RGA
M - UV Gun
  a - Helium Gas
N - 4-Axis Manipulator
  a - Sample Heating Stage
  b - Sample Cooling Stage
  c - Sample Bias Connection
  d - Sample Heating Connection
  e - Sample Cooling Connection

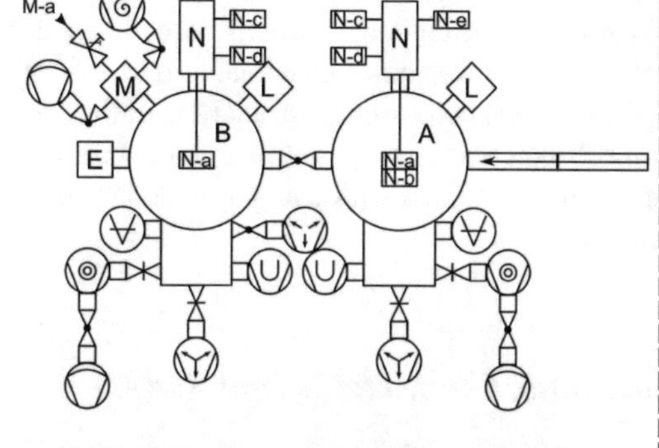

(c) Photographs

Fig. 2.2 Schematic diagrams of the 3-chamber surface science UHV system available in the laboratory where some of the work for this thesis was performed. **a** Top view, **b** side view, and **c** photographs

It is capable of maintaining UHV pressures and performing the photoemission measurements needed. It consists of three chambers, designated analysis chamber,[2] preparation chamber,[3] and STM chamber.[4] In addition to these main features, various evaporators can be added in order to perform in vacuo depositions, and the manipulators are equipped with means of annealing and cooling samples. In addition, the chambers and electron energy analyser are made from mu-metal; a nickel-iron alloy which has high permeability and acts to shield the chamber from stray magnetic fields [38], which could affect the trajectory of electrons inside the vacuum.

These features are very useful concerning surface science studies, given that this system was designed as a synchrotron beamline endstation. However, they are somewhat superfluous to the electronic characterisation of the technical materials of interest here in the laboratory. Furthermore, the size of this system allows it to be easily customised, but the increased amount of features also lends to greater difficulty in achieving UHV pressures. The load lock of this system is situated in a different chamber to the analysis tools required here and therefore several transfers are required to bring a sample from entry to the point of analysis. Therefore, there was motivation to develop a new system that was better suited to the characterisations needed here, but the main limitation of this system is that the x-ray source is non-monochromated. This results in the production of satellite features that can complicate analysis of the spectra and also results in much poorer resolution when compared to a monochromated x-ray source. Discussion of the improvements gained with using a monochromated x-ray source is given in Sect. 2.3.2.

2.2.4 A Single-Chamber Electronic Characterisation System

An UHV system was therefore designed and built with the requirements of:

1. Performing photoemission measurements required for the work presented here.
2. Incorporating a monochromated x-ray source.
3. Having a compact design with good conductance and minimal sample transfers.
4. Being constructed from mu-metal in order to shield from stray magnetic fields.
5. Having the ability to clean the surfaces of samples.
6. Having the ability to store samples within vacuum.

The conductance of a system causes detriment to the effective pumping speed of the pumps (S_{eff}), such that [35]

$$\frac{1}{S_{\mathrm{eff}}} = \frac{1}{S} + \frac{1}{C_{\mathrm{Tot}}}, \tag{2.4}$$

where S is the total pumping speed and C_{Tot} is the total conductance of the components in question, given through the combination of the individual conductances either connected in series,

[2] Equipped with x-ray and UV guns, and an electron energy analyser for photoemission measurements.

[3] Equipped with an ion gun, sample heating and cooling, and a LEED optic.

[4] For STM measurements and long term sample storage.

$$\frac{1}{C_{\mathrm{Tot}}} = \sum_i \frac{1}{C_i}, \tag{2.5}$$

or parallel,

$$C_{\mathrm{Tot}} = \sum_i C_i. \tag{2.6}$$

From Eq. (2.4), the effective pumping speed equals the given pumping speed if $C_{\mathrm{Tot}} = \infty$, but otherwise the conductance should be maximised by either connecting components together in parallel, or by using components with high conductances [39]. The conductance of a component is dependent upon the geometry and is reduced by length, corners, obstructions, abrupt changes in cross-sectional area, blind voids, and other increases in surface-to-volume ratios [36].

For these reasons, spherical components were used where possible, otherwise short, wide cylindrical pipework and tubing was used, and extra pipework was included which connected low conductance areas of the system to the pumps in parallel, which increased the conductance in places.

The system in its current state is shown in Fig. 2.3, and at various stages throughout assembly in Appendix B, Sect. B.3. The vacuum schematic views in Fig. 2.3a, b show the arrangement of peripherals attached to the chamber, as well as the pumping and sample manipulation availabilities.[5] Photographs of this system are shown in Fig. 2.3c.

The system has a single main chamber, in which all of the characterisation peripherals[6] are confocal to the centre, where the sample is positioned. Sample entry and transfers are facilitated through a two stage entry system, with the first stage capable of loading four samples simultaneously and the second stage containing means of cleaning the sample, akin to the preparation chamber in the system described in Sect. 2.2.3, but on a much smaller scale and without the need for an extra manipulator. All three of these sections of the system can be pumped individually, allowing maintenance or modification to be performed modularly.

In order to achieve a representative measurement using photoemission techniques, it is sometimes necessary to remove contaminants from the surface of the samples, which have formed as a result of atmospheric handling, or during the growth [33]. Photoemission techniques are very surface sensitive in nature, and as such, contamination at the surface can attenuate signal arising from the underlying material of interest. Cleaning therefore must be performed in vacuo, because any ex situ cleaning will be somewhat reversed during the time spent between cleaning and entry to the UHV system. Contaminants present at the surface of samples are usually atmospheric in nature and can vary from chemisorbed species, such as surface oxidation,

[5]The key to vacuum symbols is given in Appendix B, Sect. B.2.

[6]X-ray gun with monochromator, UV gun, electron energy analyser, electron gun, and IPES detector.

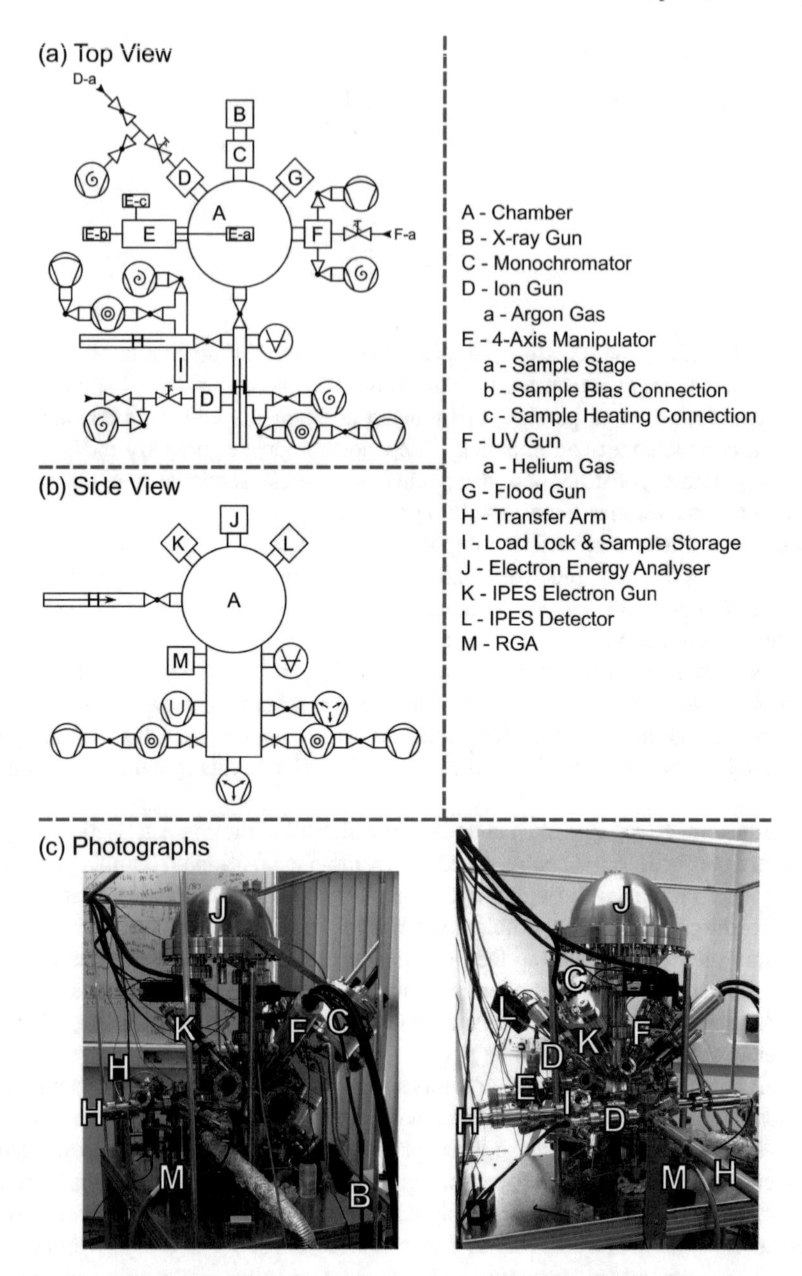

Fig. 2.3 Schematic diagrams of the single-chamber electronic characterisation UHV system designed and built by the author in order to perform the PES measurements required for this thesis. **a** Top view, **b** side view, and **c** photographs

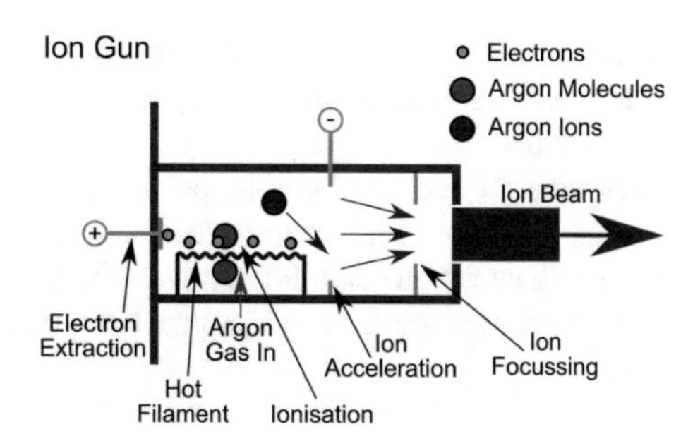

Fig. 2.4 Schematic diagram of an ion gun used for surface cleaning, showing how argon gas is ionised and then formed into a beam which is accelerated towards a sample

to physisorbed species, which are bonded by van der Waals forces such as water vapour or hydrocarbons [33].

Several methods can be used for this cleaning, including crystal cleavage, or the introduction of reactive gases [33, 34]. Neither of these techniques was suitable for the technical samples and single crystals studied in this work, as they are more suited for large metallic single crystals or refractory metals. Instead, sputtering and annealing was employed, which is also used for the cleaning of bulk metal single crystals for surface science studies [40], whereby the annealing also segregates bulk contaminants at the surface, which are then etched away.

Bombardment of a sample surface with energetic argon ions results in the removal of surface layers by sputtering. This can be used for any sample [34], but will result in some level of disorder at the surface [33], which can then be remedied by annealing, allowing the atoms to relax into their original positions in the crystal structure [33]. Thermal annealing can also result in the direct desorption of surface contamination, further aiding the cleaning process [34]. In the system shown in Fig. 2.3, thermal annealing was possible in the chamber due to the radiative heating element mounted beneath the sample in the manipulator. Argon ions could be directed at a sample using one of the two ion guns[7] present in the system. One gun was located in the chamber, allowing surface cleaning and measurement without the need for sample transfer, and another directed at the preparation transfer arm, allowing the cleaning of particularly contaminated samples, where the removal of the contaminants could compromise the vacuum in the chamber. In both cases, argon ions were produced using the same technique, and the gun is shown in Fig. 2.4. A constant stream of clean argon gas[8] was allowed into the vacuum at the back of the gun via a leak valve. The gas molecules then pass over a hot filament which thermionically produces electrons, ionising the gas molecules. The ions are then accelerated through a potential, focussed, and emitted toward the sample as a beam [32].

The cleanliness of the samples is important in order to achieve a representative measurement, but a high level of rigour in the employed cleaning methods is necessary only for surface science characterisation techniques that demand that the sample be completely free of contamination [40]. For photoemission spectroscopy, measure-

[7]PSP Vacuum Technology ISIS-3000.

[8]The argon feed line was cleaned by means of the scroll pump attached to the gas line (see Fig. 2.3a).

ments can still take place with a small amount of remnant contamination, and the effects can be addressed. Indeed, the nature of the single crystals and technical samples studied here[9] suggests that preparation to a surface science standard would be excessive, or even impossible, not least because when cleaning binary sulphides in vacuo, there is a high chance that prolonged ion beam exposure or high temperatures will cause preferential removal of the sulphur [41–45], inducing stoichiometric changes in the samples and jeopardising the goal of achieving a representative surface.

2.3 Photoemission Spectroscopy

The electronic properties, electronic structure, and chemical environments present in a material are commonly studied using different forms of photoemission spectroscopy (PES). These are based on the measurement of the energy of electrons emitted from a sample via the photoelectric effect. Electron spectroscopies are attractive measurement techniques because the generation, manipulation, and detection of electrons, is simple and easily implemented.

2.3.1 Principles and Theory

Different types of PES are described primarily by the type of excitation used, and then secondarily by any other distinguishing features, which leads to the various techniques and their associated, often voiced acronyms [46, 47]. In this thesis, x-ray photoemission spectroscopy (XPS), ultraviolet photoemission spectroscopy (UPS), and inverse photoemission spectroscopy (IPES) are of interest, and because much of the theory is shared amongst them, it will be described here in terms of the more commonly used XPS, with the specifics for UPS and IPES discussed later.

XPS is used to probe the occupied electronic states of a material by measuring the kinetic energy of electrons that have been emitted from the sample due to the transfer of energy from incoming x-ray photons, which have energy ~1000 eV.

Photoemission

The photoemission process is based on the photoelectric effect, separated into three stages [32]:

1. x-ray photons of known energy interact with electrons in the atomic orbitals of a material, creating photoelectrons.
2. These electrons then move from the point of generation to the surface of the sample, undergoing various scattering processes.

[9] Small, irregular morphology, unpolished, rough surface.

3. At the surface of the sample, the electrons overcome the work function, and are emitted into the vacuum.

In the simplest description of photoemission, the electron undergoes no energy losses as it moves through the sample, resulting in a photoelectron kinetic energy that is related to the photon energy, its binding energy, and the work function of the sample. In reality, many other effects can come into play, which will be introduced later, but the main limiting factor of the photoemission process after photon absorption is the inelastic mean free path of the electrons ($IMFP_e$) [48]. The incoming photons penetrate the sample to a depth ~1 μm [39], and so photoelectrons are produced also to this depth. The $IMFP_e$ is energy dependant, but generally independent of material, giving rise to the 'universal curve' [49], which shows that over the electron kinetic energy range 10–1000 eV,[10] the $IMFP_e$ is ~1 nm [34, 38], meaning that the majority of electrons contributing to the XPS signal, having undergone no lossy collisions, originate from the top few atomic layers of a sample. This is why XPS is considered a surface sensitive technique [50]. The rest of the photoelectrons produced either escape the sample having lost some kinetic energy, or do not have the energy to overcome the work function once reaching the surface of the sample.

At the third stage of the photoemission process, the electrons undergo a loss of energy overcoming the work function of the sample. This loss is not included with those from the second stage because it is a necessary one in order to perform the measurements. This loss mechanism is therefore integral to the photoemission process and is included in the characteristic PES equations. It is therefore prudent to introduce the concept of work function and how the PES measurements are affected by it.

In Sect. 1.6.2, the definition of the work function of a material was introduced as the energy difference between the Fermi-level and the vacuum level. In this description, the work function can be imagined as an energy barrier that an electron must overcome in order to pass from the surface of a material, to a point at rest just outside the surface of a material, in vacuum. Such a barrier exists due to the contribution of two factors [35, 51]. The bulk contribution is dominant and arises because of the inner potential of the crystal structure. In addition to this is the surface contribution from the dipole formed by the electron density overlapping slightly into the vacuum at the boundary of the solid, which is dependent upon surface features and can be different for dissimilar surfaces of the same material.[11] Consequently, the concept of vacuum level is only well-defined in a region proximal to the surface so that the effect of the surface dipole is experienced. The vacuum level is therefore material dependent and is different to the concept of the vacuum level at infinity which is invariant [52]. It would appear that reference to the vacuum level at infinity would be more suitable, but because any measurement of an electron energy must involve the use of an analyser, which has its own work function and vacuum level, then the vacuum level at infinity is immeasurable and not of consideration here. The resulting difference in vacuum levels between two materials is not of concern because it is

[10]Most of XPS electron energies fall in this range.

[11]For example, different crystallographic facets.

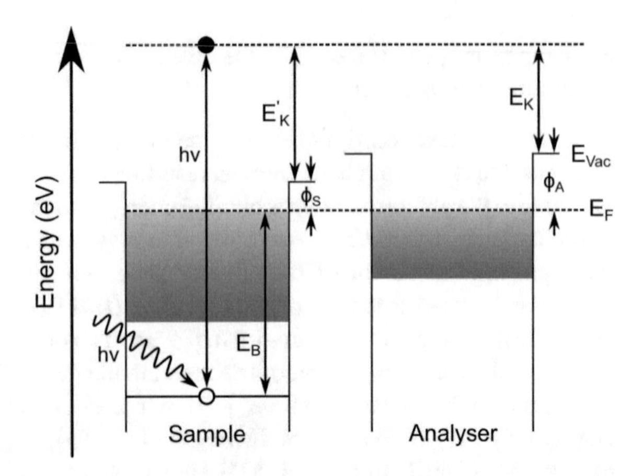

Fig. 2.5 Relative band levels and Fermi-level alignment between sample and analyser connected electronically for use in a PES measurement. Symbols are as-defined in the text

compensated for by the Fermi-level alignment of two materials in electrical contact: a necessity for PES measurements.

In connecting a sample electrically to the analyser, the Fermi-levels of the two systems align, resulting in the situation shown in Fig. 2.5. Electrons in the sample have a binding energy (E_B) with respect to the Fermi-level. Incident photons transfer their energy (hν) to the electrons, producing photoelectrons which escape the sample after overcoming the BE and the work function of the sample (ϕ_S), so that they have a vacuum kinetic energy (E'_K) of

$$E'_K = h\nu - E_B - \phi_S. \quad (2.7)$$

Then, because the analyser is connected to the sample and because it has its own work function (ϕ_A), which must be yielded by a measured electron, the measured KE is

$$E_K = E'_K - (\phi_A - \phi_S) = h\nu - E_B - \phi_S - (\phi_A - \phi_S) = h\nu - E_B - \phi_A, \quad (2.8)$$

which is more commonly written as

$$E_B = h\nu - E_K - \phi_A. \quad (2.9)$$

Equation (2.9) is used in order to convert the measured kinetic energies into more meaningful binding energies and is independent of sample work function. This means that different samples can be measured, given that they are in electrical contact with the analyser. The practical method of implementing this relationship is described in Sect. 2.3.3.

Chemical Shifts

The atomic binding energies vary for different elements, and as such, allow one to determine which elements are present in a sample. However, it is the information

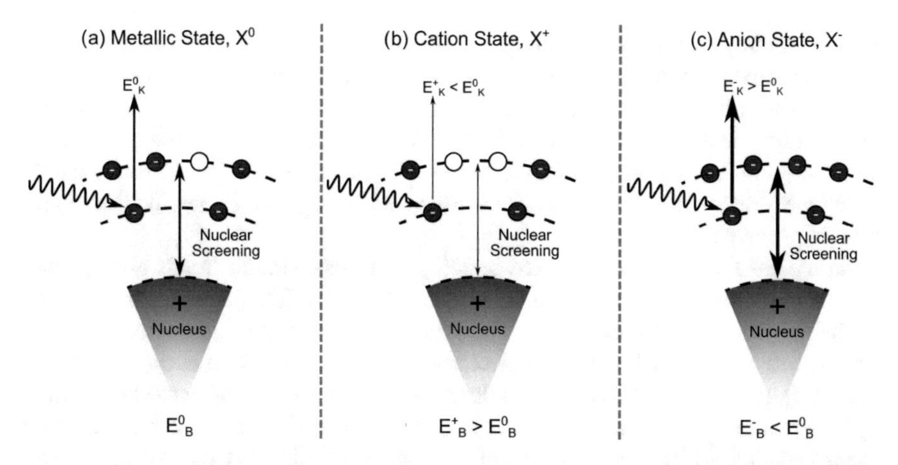

Fig. 2.6 Demonstration of the differences in nuclear screening and therefore binding energies of an electron in **a** a metallic species with zero oxidation number, **b** a cationic species with positive oxidation number, and **c** an anionic species with negative oxidation number. Such changes in the energies of photoemitted electrons result in the chemical shifts observed in spectra. Symbols are as-defined in the text

about the chemical environments within a material that XPS can provide that makes it such a powerful technique. Indeed, an early pioneer of XPS, Kai Siegbahn coined the term electron spectroscopy for chemical analysis (ESCA) instead of XPS in order to demonstrate this [53, 54]. This term has now fallen out of favour because of the need to distinguish between the many other types of electron spectroscopy available [48].

The BE of an electron arises from the electrostatic interaction between itself and the positive nuclear charge of an atom, but is also dependent upon the presence of other electrons around the atom, which can cause screening of the nuclear potential, and results in a lowering of the BE of the electron. With relation to photoemission, nuclear screening results in the photoelectron having greater KE than it otherwise would from overcoming the nuclear potential alone. Such is the case for the single atomic picture.

In the case of solids, one can imagine that the nuclear screening is affected also by interatomic interactions, both nuclear and electronic. However, for metallic elements, that is elements with an oxidation state of 0, analogies can be drawn to the single atomic picture and act as a baseline BE, shown in Fig. 2.6a, neglecting effects of the free electron gas (FEG) for metallic solids, which will be addressed regarding photoemission in the next section.

Changes in electron density surrounding an atom can cause the amount of nuclear screening to increase or decrease, which in turn produces a lower or higher BE for an electron, respectively. This can be induced in many different ways, but a simple interpretation of the concept comes from ionic bonding. For a cation,[12] there

[12]See Fig. 2.6b.

is less electron density surrounding the atom because some electrons have been transferred to the anion, which results in less nuclear screening, and a lower measured photoelectron KE, relating to a higher BE. The opposite is true in the case of anions, where more electron density surrounds the nucleus.[13] Generally, when considering an ionic picture, a higher[14] oxidation state results in a larger BE shift from the case of the metallic species, such that $E_B^{2+} > E_B^{+} > E_B^{0} > E_B^{-} > E_B^{2-}$, but there are other factors which can affect this.

In the case of covalent or mixed bonding materials, similar trends apply, given that most inorganic materials can be assigned formal oxidation states, but the size of the BE shifts are also dependent upon the level of covalency, the strength of the bonds, and also the differences in electronegativity between two atoms. These are again responsible for a shift in electronic density either away from, or toward an atom by different amounts, in turn affecting the nuclear screening, but generally, a larger electronegativity difference will result in a larger BE shift because of the greater shift in electron density [55], and a longer bond length results in smaller BE shifts because of the lessened action at a distance to transfer electron density [56, 57].

Described above is the simplest, somewhat naïve description of chemical BE shifts in XPS, which is good as an initial starting point and shall be the model used to account for the majority of the shifts seen in the spectra presented here. However, there are many other effects which can alter binding energies. Some of these occur after the production of the photoelectron, are termed final-state effects and will be discussed in the next section. Others, which are present regardless of the photoemission process and depend on the chemical environment, are deemed initial-state effects [32], include the shifts described above, but can also depend on the number of bonds present [39], crystal field effects [54], or ionic potentials [58, 59]. Combinations of these effects within a material can make predictions of absolute binding energies difficult.

Spectral Features

Ideally, photoemission will produce a series of peaks with binding energies representative of each of the orbitals of each of the elements within a sample, with shifts between peaks arising as a result of differing chemical environments. However, other processes can occur during photoemission which means that other features are sometimes present within a spectrum.

Because it is the kinetic energies of the electrons that are measured in an XPS experiment, but the binding energies have greater use, it is customary to display an XPS spectrum of intensity as a function of decreasing BE [32].

If photoemission results in an electron that has not lost any energy and originates from a non-valence orbital, then the peak is described as a core-level (CL) peak. These are identified using spectroscopic notation of the form nl_j, representing the principle, azimuthal, and total angular momentum quantum numbers [32]. Core-levels with BE up to the photon energy are accessible, and an example of an XPS survey spectrum is shown in Fig. 2.7, for a CdTe sample, with the CL features marked in blue. For $l = 0$,

[13]See Fig. 2.6c.

[14]That is, a greater difference from 0 in either direction.

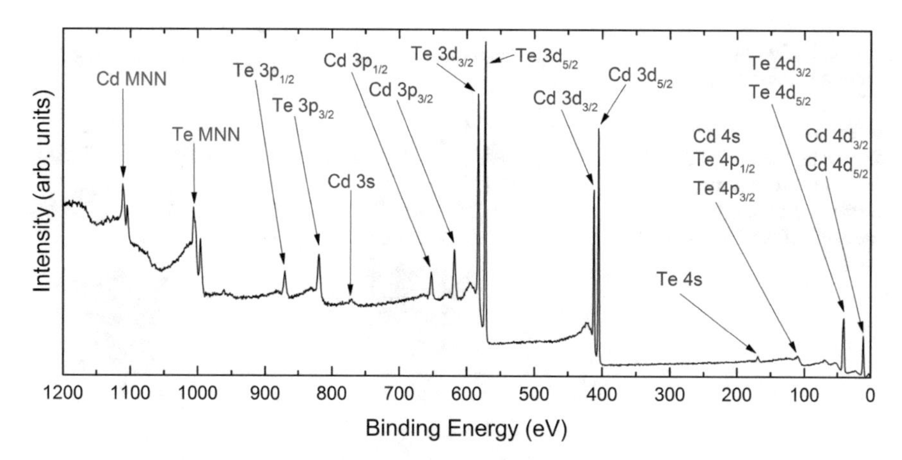

Fig. 2.7 Example XPS survey spectrum from a sample of CdTe measured by the author [60]. Core-level features, including spin-orbit split components are marked in blue, and Auger features are marked in red

Table 2.2 XPS spin-orbit doublet intensity ratios due to the electronic occupancy of the orbitals [32]

ℓ	j	Occupancy $(2j+1)$	Doublet Intensity Ratio
0, s	1/2	2	—
1, p	1/2, 3/2	2, 4	1:2
2, d	3/2, 5/2	4, 6	2:3
3, f	5/2, 7/2	6, 8	3:4

s orbitals demonstrate a singlet peak as only one j-value is possible

$j = 1/2$ is the only possible value, and so *s* orbitals result in a singlet. For $l > 0, j$ can take two values, and so results in a doublet, separated by an energy known as the spin-orbit splitting. Due to the occupancy of each of the doublet orbitals, this results in the intensity ratios between the doublet peaks as shown in Table 2.2.

If the photon is absorbed by a valence electron, the photoemission process can still occur, but because the valence orbitals are involved in bonding, the resulting spectrum is a series of broad features, reflecting the DoS in the VB.

During the second stage of the photoemission process, there is a finite probability that the photoelectron will lose energy as a result of inelastic scattering as it moves through the sample. This manifests as a rising background of secondary electrons, increasing with BE, onto which the photoemission peaks are superimposed, as seen in the example in Fig. 2.7.

Features due to the Auger effect are also present in PES spectra[15] because the photoemission process can also induce the Auger effect, which is shown in Fig. 2.8.

[15]Shown in red in Fig. 2.7.

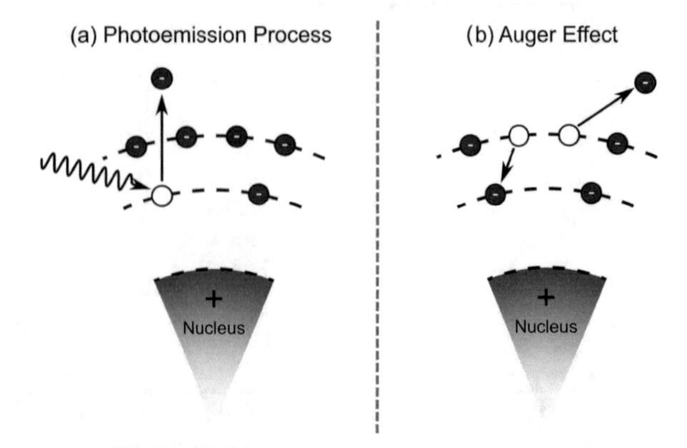

Fig. 2.8 Electron emission from **a** the photoelectric effect, and **b** the Auger effect, where an electron from a higher orbital fills the core-hole, transferring this energy difference to another electron which is emitted as well

After photoemission, the core-hole left by the photoemitted electron can be filled by an electron from a higher orbital relaxing into it, which in turn causes the emission of a second, Auger electron, whose KE is related to the energy level differences and is independent of the photon energy [61]. Because of the multitude of orbitals which can be involved in the Auger process, these features manifest in PES spectra as broad, energy-expansive features with multiple peaks and are identified using x-ray notation of the form [32]:

$$n^{(1)}_{l+j+\frac{1}{2}} n^{(2)}_{l+j+\frac{1}{2}} n^{(3)}_{l+j+\frac{1}{2}}, \tag{2.10}$$

with $n^{(1)}$, $n^{(2)}$, and $n^{(3)}$ representing the orbital in which the photoemission process took place, the orbital from which the relaxing electron originated, and the orbital from which the Auger electron was emitted, respectively.

Final-state effects produce quantised loss features in XPS spectra, manifesting as extra peaks or features to the higher BE side of the main peak.

In materials with unpaired electrons in the outer shell, photoelectrons can couple with these via exchange interactions, resulting in extra peaks for the parallel and anti-parallel spin couplings in these interactions [32, 39, 55]. Such features are named multiplet splittings.

As a photoelectron is emitted, there is a chance that it can interact with a valence electron, promoting it to an empty conduction state and hence reducing the KE of the photoelectron by the energy difference between the excited state and the ground state, thereby producing features at higher BE in the spectrum. These are called shake-up features and are more likely for certain elements in certain oxidation states, especially where unpaired electrons exist in the VB [32, 39]. An example of this is shown in Fig. 2.9, which shows the Cu 2*p* region for a material that contains a mixture of the Cu^{+} and Cu^{2+} oxidation states [62]. For the Cu^{+} oxidation state, the valence 3*d* orbital is full, and no shake-up features are observed; however, for the Cu^{2+} oxidation state, the 3*d* orbital contains nine electrons and the photoemitted 2*p* electrons can excite the unpaired 3*d* electron, resulting in strong features at higher BE (marked in blue).

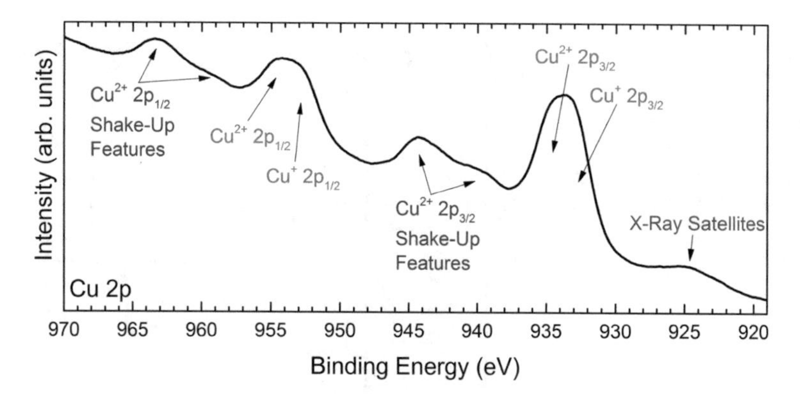

Fig. 2.9 Example XPS Cu 2*p* region [62] from a sample containing a mixed amount of Cu^{+} and Cu^{2+}, demonstrating the satellite features from the shake-up process (blue) and using a non-monochromatic x-ray source (red)

A photoelectron can also excite valence electrons out of the material, producing shake-off features, which often have low intensity and are masked by the background [39]. For metals, the electron density is high at the Fermi-level, with a continuum of states above the Fermi-level with no forbidden gap, resulting in a continuum of shake-up possibilities. Instead of producing discrete features in the spectrum, this manifests as an asymmetric tail of states to the higher BE side of the main peak [32].

Metals also contribute another mechanism that can result in photoelectron energy losses. It can occur because of the coupling of the photoelectron with plasmon oscillations in the FEG and results in a series of loss features separated from the main peak by multiples of the plasma frequency, reducing in intensity. Both bulk and surface plasmon oscillations can contribute to these loss features, with the latter occurring at smaller shifts, related to the surface plasma frequency. Such loss features can also occur in non-metals, but with reduced intensity [32, 39].

Ultraviolet Photoemission Spectroscopy

The photoemission principle is the same for UPS as it is for XPS, with the difference being the energy of the incoming photons. For XPS, x-ray photons with energy ~1000 eV are used, but for UPS, the UV photons have energy ~10 eV. At this level, only electrons originating from the VB of materials are probed [48]. The much narrower linewidth of the excitation sources used here also allow much higher resolutions to be achieved with UPS and this shall be further discussed in Sect. 2.3.4. UPS finds greater use in surface science studies [38], but can also be used in elucidating the electronic structure of technical samples.

Inverse Photoemission Spectroscopy

Where PES investigates the occupied states of a material, there exists a complementary technique which allows the unoccupied states above the Fermi-level to be investigated by equivocally the opposite process, that is, inverse photoemission spectroscopy (IPES). Directing a beam of energetic electrons at a sample causes them to

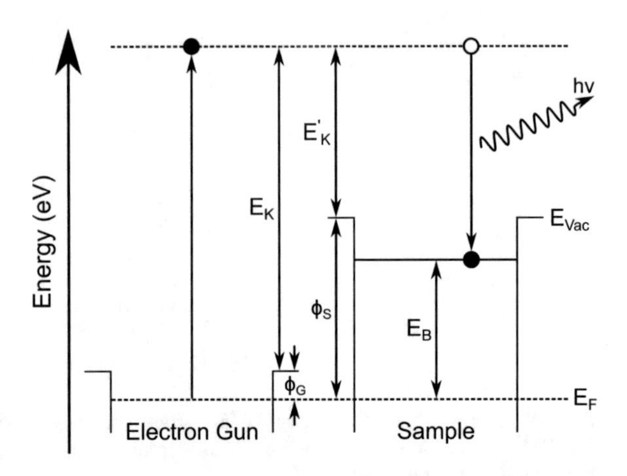

Fig. 2.10 Relative band levels and Fermi-level alignment between electron gun and sample connected electrically for use in an IPES measurement. Symbols are as-defined in the text

relax into any of the unoccupied states within a material that have energy less than the KE of the electron. This relaxation process then emits a photon whose energy is related to the KE of the electron and the energy of the unoccupied state [48]. In this way, the process is very similar to the photoemission process, with the difference that the electron levels involved are above the Fermi-level, and have negative BE. Equations similar to those for PES also hold, and are derived from the energy level diagram shown in Fig. 2.10.

Electrons with kinetic energy (E'_K) relative to the vacuum level of the sample, relax to an occupied state with binding energy (E_B), and release a photon with energy ($h\nu$), such that

$$h\nu = E'_K + \phi_S - E_B, \tag{2.11}$$

where ϕ_S is the work function of the sample. Because the electrons were emitted from an electron gun that is electrically connected to the sample, a similar condition to that in PES holds, allowing samples of different work functions to be measured by accelerating the electrons to a kinetic energy relative to the vacuum level of the electron gun (E_K), so that

$$E_K = E'_K + \phi_S - \phi_G, \tag{2.12}$$

then,

$$h\nu = E_K + \phi_G - \phi_S + \phi_S - E_B = E_K - E_B + \phi_G, \tag{2.13}$$

which is more commonly written as

$$E_B = E_K + \phi_G - h\nu. \tag{2.14}$$

So, the binding energies of the unoccupied states can be determined from Eq. (2.14) by measuring the emitted photon energies and the accelerating kinetic energy of the incoming electrons. Such it is that Eq. (2.14) is the negative of that for PES, Eq. (2.9).

2.3.2 Operation and Implementation

In this section, the experimental implementation of the PES techniques will be described, including the necessary equipment. Common to all of the techniques is the necessity of using UHV for the reasons discussed in Sect. 2.2.1 and to connect the sample electrically to the rest of the system[16] in order to align the Fermi-levels so that Eqs. (2.9) and (2.14) hold true. This can be achieved by attaching part of the surface of the sample and also supporting the back of the sample with conducting materials, and then connecting these to the remainder of the UHV system, so that the Fermi-levels of all of these parts align [53]. However, for reasons that will be discussed in Sect. 2.3.4, the sample is usually connected to a floating ground, which passes through a power supply, before both this and the UHV system are connected to a common ground.

X-Ray Photoemission Spectroscopy

X-rays are produced by bombarding a target material with high energy electrons [55]. The schematic of a typical x-ray gun is shown in Fig. 2.11. It consists of an isolated rod, the end of which is coated with the target material. It is customary to coat the end of the rod with two different materials in order that different photon energies can be produced. In close proximity to the target material is a filament,[17] which can thermionically produce electrons if a current is passed through it. By applying a high positive voltage[18] to the target-capped rod (anode), the electrons are accelerated and bombard the target material. This causes the emission of x-rays with various energies, which are then directed towards a sample. The inside of the anode is supplied with circulating water, which counteracts the electron impact-induced heating of the anode.

Two physical processes are involved in the production of x-rays by electron bombardment and are shown in Fig. 2.12a. The first is the production of bremsstrahlung, caused by the deceleration of electrons by the positively charged nucleus of the target material and produces x-rays with a continuum of energies, shown by the background of the red spectrum in Fig. 2.12c. The peaks shown in this spectrum are results of characteristic x-rays of the target material, which are produced when an impinging electron transfers some of its KE to a core electron in the target material, ejecting it. This then results in a secondary inelastically scattered electron, the ejected electron, and a core-hole from where the ejected electron originated. An electron in a higher energy level then relaxes into this core-hole, emitting a photon with energy equal to

[16]Specifically the analyser and electron gun.

[17]One for each target material.

[18]$\sim$12 kV.

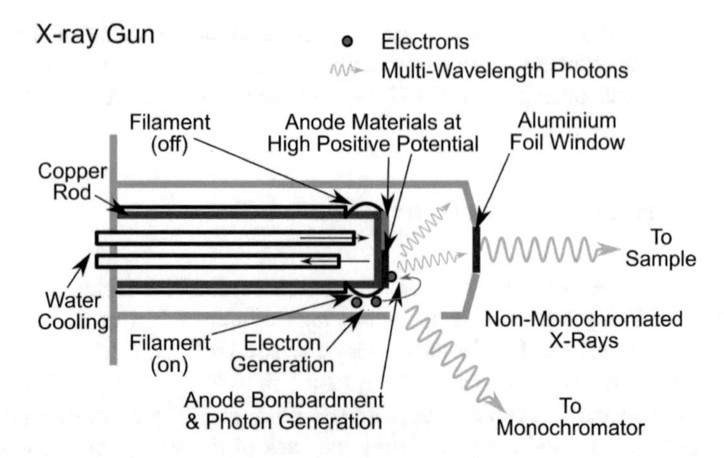

Fig. 2.11 Schematic diagram for a laboratory x-ray source for use in XPS measurements. For non-monochromated sources, x-rays are passed through an aluminium foil window, but for monochromated sources, they are passed directly to the monochromator

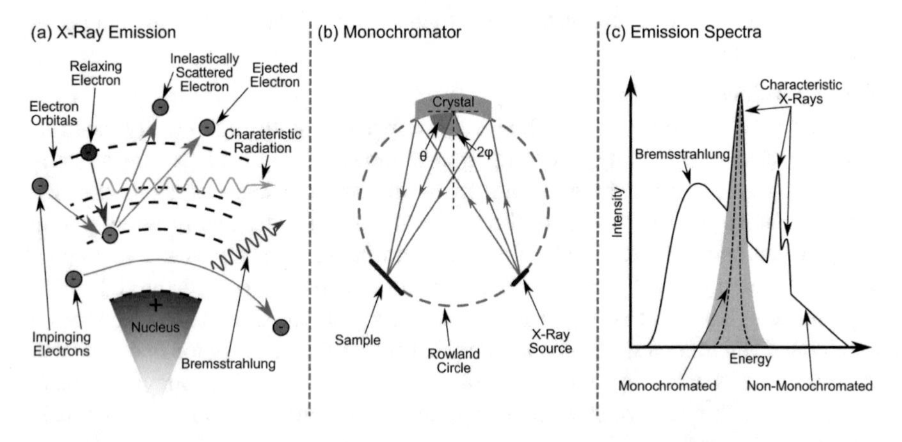

Fig. 2.12 X-ray generation. **a** Mechanisms for x-ray production using the bombardment of high-energy electrons into a target material. **b** Monochromation of x-rays using diffraction in a crystal. **c** Resultant x-ray emission spectra for: non-monochromated x-rays (red line) with the primary characteristic x-ray line highlighted (pink shading); and monochromated x-rays (black dash line) showing the exclusion of the Bremsstrahlung background and secondary characteristic x-ray lines, and also the narrower linewidth of the emission line. These spectra are normalised to the intensity of the main characteristic peak, but in reality, this is much lower for the monochromated source

the difference between the two energy levels involved. These then appear as a series of peaks in the emission spectrum, with energies and intensities ratios dependent upon the target material [48].

In order to build a meaningful spectrum, Eq. (2.9) assumes that the photons that are incident on the sample are of a single energy; however, the production of secondary characteristic x-rays and also the bremsstrahlung means that photons of many energies reach the sample. This is known as a non-monochromated x-ray source and results in extra features in the spectra [53]. The bremsstrahlung adds to the background of spectra because they are a continuum, and can be suppressed somewhat by the aluminium window through which the photons are directed in non-monochromated sources, shown in Fig. 2.11. This window also serves to disperse some of the anode heating and acts as a stray electron extractor [32]. On the other hand, the secondary characteristic x-rays produce features which mirror the photoemission processes but have lower intensities and appear shifted on the BE scale of a spectrum because it is calibrated to the main photon energy using Eq. (2.9). Such features are shown in Fig. 2.9 (x-ray satellites) and can complicate the analysis of spectra [32]. The primary emission lines used in non-monochromatic XPS are usually Al Kα_1 (1486.6 eV) and Mg K$\alpha_{1,2}$ (1253.6 eV), both of which have a natural linewidth < 1 eV, with magnesium generally favoured because of its slightly narrower linewidth [63]. These lines are represented by the red shading in Fig. 2.12c.

The selection of one specific energy of photon from the emission spectrum can be achieved through diffraction [32, 38, 39, 48] in a process called monochromation, which is usually designed to select the Al Kα_1 line by using a quartz $\left(10\bar{1}0\right)$ crystal. The conditions used to produce this are shown in Fig. 2.12b. From the source, x-rays are incident on the crystal at an angle (θ) to the plane of the crystal. From the Bragg condition for diffraction,[19] constructive interference occurs between crystal planes if

$$2d\ \sin(\theta) = n\lambda, \tag{2.15}$$

where d is the interplanar distance, n is an integer order of diffraction, and λ is the wavelength of the radiation. For Al Kα_1 radiation,

$$\lambda = \frac{hc}{E} = \frac{hc}{1486.6\,\text{eV}} \approx 8.34\,\text{Å}, \tag{2.16}$$

and for quartz $\left(10\bar{1}0\right)$ [64],

$$d = a\ \cos(30°) \approx 4.25\,\text{Å}. \tag{2.17}$$

This results in a Bragg angle of

$$\theta = \sin^{-1}\left(\frac{\lambda}{2d}\right) \approx 78.87^{\circ}, \tag{2.18}$$

and an angular separation between source and sample[20] of

$$2\varphi = 2(90 - \theta) \approx 22.27° \tag{2.19}$$

Then, by satisfying the reflection conditions and using the angular separation found here, a focussed, monochromatic x-ray beam will be incident on the sample. This is

[19]Further explored in Sect. 2.6.1

[20]As shown in Fig. 2.12.

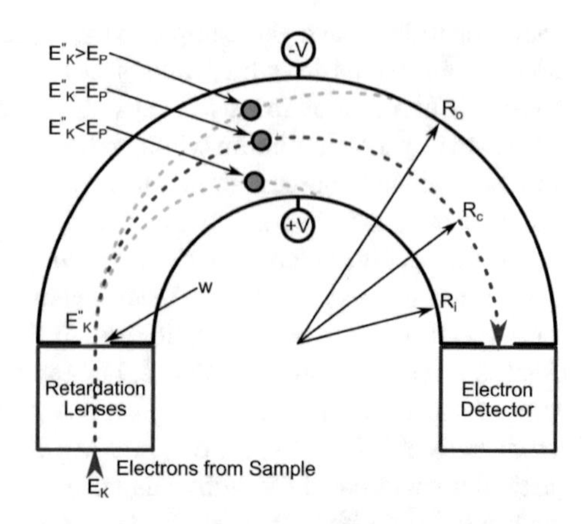

Fig. 2.13 Concentric hemispherical analyser (CHA) for measuring the kinetic energies of photoemitted electrons in a PES experiment. Symbols and explanation to the principles are given in the text

fulfilled by placing the source, sample, and crystal tangent to the Rowland circle and equating the incidence and reflection angles.

At a sacrifice of some intensity, the monochromated x-ray source benefits from no satellite features, no contribution to the background by bremsstrahlung, and because after monochromation, the linewidth is independent of the natural linewidth from emission and is dependent on the errors associated with the properties of the crystal, a much higher resolution of spectra is possible, represented by the black dashed line in Fig. 2.12c.

In the system described in Sect. 2.2.3, a twin anode, non-monochromated x-ray source was available,[21] but the system described in Sect. 2.2.4 benefitted from a monochromated x-ray source.[22]

The spectrometer used in XPS can be divided into two parts, the analyser, which only allows electrons of a certain energy to pass through, and the detector, which counts the number of electrons that were allowed to pass through the analyser. Then, by sweeping across the possible kinetic energies, and counting at each point, a spectrum can be generated.

There are many ways of analysing the kinetic energies of electrons, but the most used in PES is the concentric hemispherical analyser (CHA), shown in Fig. 2.13. This consists of two concentric, hemispherical plates, between which, a potential is applied [32, 34]. An incident electron will traverse around if its kinetic energy (E''_K) is such that the deflecting potentials keep it on a concentric course with respect to the plates, this is termed the pass energy (E_P) of the analyser [34]. Electrons with energy greater than the pass energy will collide with the outer plate, and those with lower energy will collide with the inner plate. By changing the potential applied across the plates, one can alter the pass energy and theoretically sweep through the range of kinetic energies required, counting the intensity of electrons at each dwell point.

[21] PSP Vacuum Technology TX400.

[22] SPECS XRC-1000M.

From the physical size of the analyser and the slits, a small range of kinetic energies can pass through the analyser taking different trajectories. This uncertainty contributes to the widths of the peaks observed in a spectrum and is given by [32, 34, 38]

$$\Delta E_A = E_P\left(\frac{w}{2R_c} + (\Delta\alpha)^2\right), \tag{2.20}$$

where w is the average width of the slits used in the analyser, $\Delta\alpha$ is the acceptance angle of the analyser, and R_c is the radius of the trajectory for an electron at exactly the pass energy, calculated as the average radius of the inner (R_i) and outer (R_o) hemispheres [32, 39]. In order to pass an electron of maximum KE from an XPS experiment,[23] assuming a slit width of 2 mm, and neglecting angular divergence,[24] then a resolution of 1 eV requires an analyser of radius,

$$R_c = \frac{1500\,\text{eV}}{1\,\text{eV}} \cdot \frac{2\,\text{mm}}{2} = 1.5\,\text{m}, \tag{2.21}$$

resulting in an analyser so large that it is impractical. It is also noted that because different KE electrons require different pass energies, this results in different resolutions throughout the spectrum [38]. In order to maintain a constant resolution throughout the spectrum, and also to ensure good resolutions with an analyser of manageable size, the analyser is operated with a constant pass energy and a series of electrostatic lenses at the entrance serve to retard the electrons to the appropriate energies [32], so that

$$E_K'' = E_K - E_R, \tag{2.22}$$

where E_R is the retardation energy resulting from the lenses. For example, if a pass energy of 20 eV is chosen, then the hemispherical voltages are set to maintain this. Then, if the retarding lenses are set to retard all of the entering electrons by 200 eV, only electrons which had KE from the sample of 220 eV will be able to pass through the analyser and be counted. Then, by sweeping the retarding voltage across the range of energies required, the spectrum can be generated with a constant resolution throughout.

Counting the electrons that are allowed to pass through the analyser is achieved by using electron multipliers. These consist of a tube coated in a material that allows secondary electron generation after being struck by an electron emerging from the analyser. Then, by applying a positive potential to the back of the tube, a current can be measured that is related to the amount of electrons entering the multiplier.

[23] $E_K \approx 1500\,\text{eV}$.

[24] $\Delta\alpha = 0$.

In all cases throughout this thesis, the CHA analysers used were the same,[25] had an average radius of 120 mm, a variety of slit widths were available for selection, and the detector consisted of five multipliers placed in the exit plane of the analyser.

Ultraviolet Photoemission Spectroscopy
The generation of UV photons is achieved by using gas discharge lamps, which generate photons through the partial ionisation of the gas. A high voltage strikes the gas molecules, and the ions then collide into neutral molecules, gaining electrons from them in the process. The subsequent relaxation of these electrons then emits photons of characteristic energies. For UPS in this thesis, He(I) discharge was chosen.[26] A problem with this type of UV source is that secondary emission lines are produced as well as the main one, similar to non-monochromated x-ray sources, and as such, these cause satellites in the spectra. A great advantage of using a UV source over an x-ray source is the much narrower natural linewidth of the emission line.[27]

The same CHA was used for UPS measurements, but with a much lower pass energy to benefit from the improved resolutions afforded by the narrower linewidths.

Inverse Photoemission Spectroscopy
An IPES experiment requires a source of electrons and a way of counting and measuring the energies of the emitted photons.

Using electrons of high enough KE results in the emission of photons with a spectrum of energies below the KE of the electrons, the measurement of which is used to produce the spectrum. This method, shown in Fig. 2.14a, is called tunable photon energy (TPE), and whilst possible, requires expensive and complex equipment to perform, because the energy analysis of photons is difficult. Instead, IPES experiments were performed using the isochromat method, which exploits the fact that the KE of the electrons acts as a cutoff for the energy of the emitted photons. By using a photon detector coupled with a photon bandpass filter, this analyser can then count photons of a single, fixed energy. From Eq. (2.14), it can be seen that if the photon energy remains constant, then the different binding energies can be investigated by sweeping through the electron kinetic energies instead; a more easily achieved process, shown in Fig. 2.14b.

In this thesis, the electron gun[28] consisted of a BaO electron dispenser and accelerating and focussing electrodes to produce a collimated beam of electrons with controllable kinetic energies. The photon detector[29] consisted of a SrF_2 window acting as a low-pass filter for the photons, which then produce photoelectrons on striking a NaCl-coated tantalum cone, where the photoionisation limit of NaCl acts as a high-pass filter. These electrons are then detected using an electron multiplier and related to photon intensity by the same method as in the CHA.

[25] PSP Vacuum Technology RESOLVE120MCD5.

[26] $h\nu = 21.22\,\text{eV}$.

[27] 1.2 meV for He(I) [65].

[28] PSP Vacuum Technology IPES Electron Gun.

[29] PSP Vacuum Technology IPES Detector.

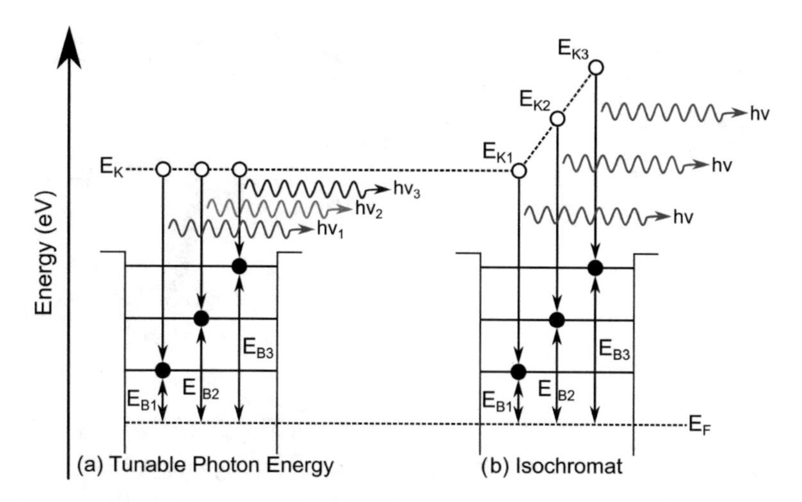

Fig. 2.14 Modes of operation for an IPES experiment. **a** Tunable photon energy mode, where electrons of a single kinetic energy produce photons with various energies relating to the unoccupied band structure of a sample. Measurement of the energies of these photons then produces a spectrum. **b** Isochromat mode, where electrons of various kinetic energies produce photons with energies less than the electron energy, but only photons of a single energy are detected, resulting in a spectrum that is dependent upon the electron kinetic energy. Symbols are as-given in the text

2.3.3 Spectrometer Calibrations

Equations (2.9) and (2.14) suggest that in order to reproduce the binding energies of a sample, the work function of the analyser or electron gun must be known. In practice, this determination is difficult, and instead is compensated for through the benefits afforded by the alignment of the Fermi-levels when the sample is electrically connected to the system.

Before a PES or IPES experiment can be performed, an energy-scale calibration must be implemented by taking measurements of a pure metal sample. Silver foil is commonly used for this purpose because of its relative inertness and ease to clean, using high energy Ar^+ bombardment, which removes the surface contamination.

Because metals have occupied density of states up to the Fermi-level, in PES there is a cutoff, above which no states are occupied and therefore no photoelectrons can be produced. By definition, the binding energy is referenced to the Fermi-level, and so this cutoff occurs at $E_B = 0\,\text{eV}$. Measurement of the KE of electrons arising from the Fermi-edge therefore allow the BE scale to be set, placing the origin at the Fermi-edge, and Eq. (2.9) becomes

$$E_B = E_K(\text{Fermi}) - E_K. \tag{2.23}$$

In IPES, the same cutoff exists, but for the unoccupied density of states, and so at this point Eq. (2.14) becomes

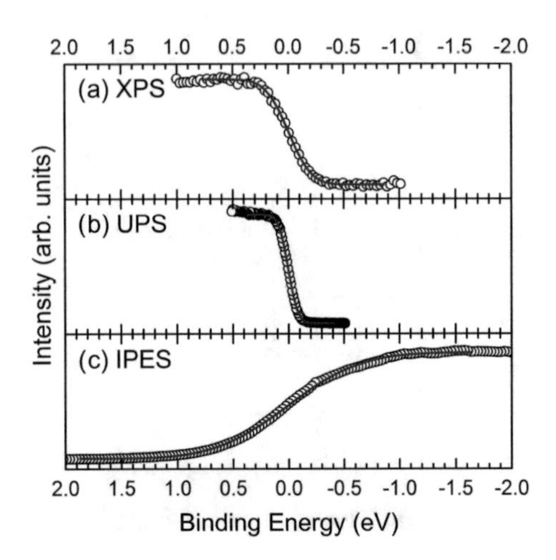

Fig. 2.15 Fermi-edge spectra of a cleaned silver foil measured using **a** XPS, **b** UPS, and **c** IPES. Fitted Fermi-Dirac distribution functions shown in red

$$E_B = E_K - E_K(\text{Fermi}). \tag{2.24}$$

After performing this measurement, subsequent measurements of any sample will result in the calculation of accurate binding energies as long as the sample is electrically connected to the system, that is, if the binding energies are always referenced from the coincidental Fermi-edge.

The Fermi-edges for a silver foil measured by XPS, UPS, and IPES are shown in Fig. 2.15a–c, respectively, after the BE scale has been corrected. Theoretically, the Fermi-level results in a sharp transition between the (un)occupied density of states to zero probability; however, it is physically modelled by the Fermi-Dirac distribution function, the fitting of which to the spectrum allows the determination of the Fermi-level position. These fittings are also shown in Fig. 2.15.

XPS measurements of the silver foil also allow the linearity of the spectrometer to be checked, that is, whether the electrons with a range of KE are measured with the same accuracy by the spectrometer. This is done by measuring various CL regions and Auger features of the silver and checking their BE, which are well known and standardised for the pure, clean metal [66].

The measurement of the cleaned metal foil also allows one to estimate some of the factors contributing to the overall resolution of the spectrometer. The FWHM of a feature in PES (ΔE) is limited by a combination of factors that add in quadrature (to first approximation) [32] to give

$$\Delta E = \sqrt{\Delta E_{CL}^2 + \Delta E_X^2 + \Delta E_A^2}, \tag{2.25}$$

where ΔE_{CL} is the linewidth of the core-level being measured and is element, orbital, and chemical environment dependant, mainly due to the lifetime of the core-hole formed in the photoemission process [39], ΔE_X is the natural linewidth of the excitation emission line, and ΔE_A is the analyser resolution, calculated from Eq. (2.20).

Table 2.3 Measured Fermi-edge broadenings of a cleaned silver foil from XPS, UPS, and IPES, with the estimated contributions to the broadenings from the excitation source linewidth (ΔE_X), and the analyser resolution (ΔE_A)

Technique	Measured E_F Broadening	ΔE_X	ΔE_A
XPS	0.41	~0.17[a] (~0.85[b])	~0.16[d] (~0.42[e])
UPS	0.16	~0.001[c]	~0.03[f]
IPES	1.27	~0.25[a]	~0.7[a]

Values for the non-monochromated x-ray source shown in parentheses. The analyser broadening is calculated using Eq. (2.20), and the parameters given in the footnotes
[a]From manufacturer's guide
[b]Citrin et al. [63]
[c]Zhang [65]
[d]$E_P = 10\,\text{eV}, w = 3.75\,\text{mm}$
[e]$E_P = 20\,\text{eV}, w = 5\,\text{mm}$
[f]$E_P = 2\,\text{eV}, w = 3.75\,\text{mm}$

There are several other contributions to the FWHM of a feature, but they are either constant throughout experiments,[30] or give contributions which are negligible. By assessing the broadening (12–88%) [65] of the Fermi-edge measured by each technique, the spectrometer resolution can be estimated, without lifetime effects. Estimates of the different contributions to the resolution for each technique are detailed in Table 2.3. It can be seen that the resolution of IPES is much poorer than direct PES techniques, which is a result of the energy spread of the electron beam and the acceptance window of the bandpass photon analyser. With a very narrow emission linewidth for He(I), the UPS used here was limited in resolution by the analyser radius. For XPS, before monochromation, the resolution was limited by the emission linewidth, meaning that some analyser resolution could be sacrificed in favour of an increased count rate without affecting the overall resolution. After monochromation, the improvement in the emission linewidth caused the overall resolution to be analyser-limited, which then benefited from a lower pass energy with a sacrifice in count rate.

2.3.4 Spectral Analysis

In order to extract meaningful interpretations or properties from spectra, at least some level of analysis is required.

Sample Charging

In Sects. 2.3.1 and 2.3.3, it was shown that in order for Eq. (2.9) to hold and result in a spectrum that is properly calibrated, then the sample must be connected electrically to the spectrometer. This also has the added benefit of supplying electrons to conducting

[30]Such as thermal broadening [65].

materials which can then replace those lost to the photoemission process. If these electrons were not available to replenish the emitted ones, then this would cause a region of positive charge to build up at the surface of the sample [32]. This can happen in semiconducting and insulating samples where the conductivity prevents other electrons from replenishing the emitted ones, leading to an accumulation of positive charge at the surface. In terms of the spectra, this causes the emitted electrons to lose some KE as a result of their interaction with this positive charge region, causing the spectrum to shift to higher biding energies [67]. In some samples, notably layered structures or very wide-bandgap materials, the charge region can vary as a function of depth into the sample, and the emitted electrons also experience an origin-depth dependent level of retardation, resulting in spectra which are not only shifted, but also skewed, leading to difficult or impossible interpretation of the spectra [67, 68].

Charging in XPS can be accounted for by two methods, that of charge elimination or charge referencing. The first method acts to replenish the electrons lost from the surface and neutralise the surface charge [53]. This can be achieved by making more electrons available at the surface of the sample by adding direct metallic connections. Other techniques involve creating a flood of electrons which are then able to neutralise any charge. In non-monochromatic x-ray sources, the wide x-ray beam consists of low energy photons from the bremsstrahlung, which collide with anything inside the chamber, producing many low energy photoelectrons, which are then able to neutralise the charge or, in monochromated systems, where the bremsstrahlung is eliminated, the electrons can be produced in a flood gun (FG) [69], which then fires a disperse beam of low energy electrons toward the sample, resulting in a similar effect [53].

The problem with these methods is that whilst they serve to eliminate the charging effects, it is always necessary to have a reference point to check that the charge has indeed been eliminated, and not just reduced [39]. For this reason, charge referencing is usually the first analysis procedure to be applied to any spectrum.

Assuming that the charging is constant, a spectrum can be charge-corrected by applying an extra term to Eq. (2.9), shifting the BE scale using some known reference energy. The application of inert reference materials to a sample can be used to achieve this, either by evaporating a small amount of metal to the surface, or by implanting Ar^+ ions into the sample using the ion gun. Then, by knowledge of the BE of standard elemental peaks, one can correct the entire spectrum to these standards. The main problem with these methods is that they can physically alter the sample under consideration [53].

Instead, it is adventitious carbon that is commonly used to reference a spectrum in order to correct for charging effects [53, 67]. Carbon will be present on any sample that has been exposed to atmosphere or poor vacuum before being introduced to UHV and takes the form of different types of hydrocarbons. The C 1*s* photoelectrons for these species can then be used as the reference BE, usually set to 284.8 eV. The main problem with using this method is that studies have shown this value to sometimes be unreliable and also that the BE can be dependent upon the underlying material [70]. Therefore, the method of charge referencing should always be explicitly stated, so that future considerations can be made if needed. Throughout this thesis, different methods of charge referencing were employed and these, along with the justifications for their choice can be found in the individual chapters.

Core-Levels: Qualitative

The most widely used aspect of XPS measurements is that of CL analysis, which provides information about the composition and chemistry of a sample via elemental identification and chemical analysis from the BE shifts described in Sect. 2.3.1[48, 55]. CL peaks in XPS have finite width for the reasons described in Sect. 2.3.2, and are superimposed on top of the secondary electron background. Therefore, to be meaningfully analysed, peak fitting is necessary, especially if there is possibility of the peaks overlapping, demonstrated by the many spectral features in Fig. 2.9.

Before peak models can be applied, the background must be modelled in the region of interest as it cannot be measured directly [32]. The simplest background is a straight line between the start and end points of a region of interest, but it is clear that this is not physically sound. Instead, two mathematical backgrounds were developed for application to CL XPS spectra [71]. The number of electrons that contribute to the background not only depends on the BE, but also upon the intensity of any photoemission features, because these are the source of the secondary electrons. The Shirley background is iterative, splitting a region into many BE steps and calculating the background of the next step (B_{i+1}) depending on the background and intensity values of the previous steps [72], given by

$$B_{i+1} = k_S \sum_{i=0}^{i_{\mathrm{Max}}} (I_i - B_i), \tag{2.26}$$

where I is the intensity of the signal, and k_S is the Shirley factor, derived from the condition that $I = B$ at the region edges. The implementation of this background is simple; however, it lacks a fully physical basis as it does not take into account the fact that features have finite widths and therefore can contribute intensity at higher BE. Instead, the Tougaard background was developed which takes into account the intensity across the entire region of interest in its calculation of the background [73]. The form of this background means that in order to achieve a good fit, a large range at the high BE side is required, which is sometimes not possible. Compromise between the physical rigour but complicated implementation of the Tougaard background, and the less rigour but easier implementation and better results achieved with the Shirley background has led to the wide adoption of the latter when considering CL peaks. It is this background type which was therefore used throughout this thesis unless otherwise stated.

The contributions to the finite widths of XPS peaks in Eq. (2.25) are Lorentzian in nature for the lifetime broadening, and Gaussian in nature for the others [74]. A mixed Gaussian–Lorentzian (Voigt) peak profile therefore models the majority of CL XPS peaks well [32] and is the type used here. The main exception is for metallic species which have a high BE asymmetric tail,[31] and are modelled well using the peak profile developed by Doniach and Sunjic [75].

The goal of peak fitting is to extract the intensity and BE of the different chemical species present in a sample by adding synthetic peaks with variable position, area and FWHM, to a model whose enveloping total matches the experimental data as closely as possible. Each chemical environment can be represented by an individual peak/doublet present for each orbital, and whilst a perfect fit to the experimental data can be achieved by adding an infinite number of peaks, constraints to the fitting must be imposed in order for a model to be physically representative. In the case of spin-orbit split doublets, the energy difference of the splitting and area ratio between the peaks is known, and generally, the FWHM between the peaks of a spin-orbit split doublet should be the same size for a single chemical environment; however, this is sometimes not the case. For transition metals, the lower-j peak of a doublet is wider than the higher-j peak, sometimes significantly [76]. This is due to an extra broadening process that these electrons can undergo, termed Coster–Kronig broadening [77], which occurs because of a short-lifetime Auger process in which the decaying electron arises from an orbital of the same quantum number, and occurs for the lower-j peak because it is the higher-j electron from the same orbital which must fill it [63, 74, 77, 78].

Having developed a fitting model for an XPS spectrum, assignments of the peaks develop as a result of pre-measurement assessment of the expected features, experience, and comparison of the spectra to either standard data banks, or to previous studies of similar materials, which should either demonstrate similar BE or at the least provide a starting point from which assessments can be made.

Many materials can be assigned an ionic formula, and because a change in oxidation state usually results in a change in BE,[32] the oxidation states present within a sample can be determined. This is especially important for the growth of materials which contain elements that can take multiple oxidation states.

There is a possibility that during the growth of a material, contamination could arise for several reasons. As the chemical environments of these contaminants are different to the underlying material, the BE will be different, resulting in extra peaks for the contaminants, which can then be determined through the peak fitting.

Differences from the expected BE for a material can give insights into chemical changes that have occurred which have resulted in charge transfer. This is important with regards to PV absorbers because it is the electronic properties that can affect important properties of the cell.

Core-Levels: Quantitative

To a first approximation, the intensity of an XPS peak is proportional to the amount of that species present in the analysis volume [32]. This allows stoichiometry determinations of samples, but also can be used to assess the amount of contamination present, albeit with some necessary considerations.

---

[31] See Sect. 2.3.1.

[32] See Sect. 2.3.1.

In the case of overlayers, effects from the $IMFP_e$ as a function of energy can affect analysis. For example, consider a sample of GaAs which has been exposed to atmosphere, and a thin layer of Ga_2O_3 has formed at the surface [79]. Using a monochromated Al Kα x-ray source, two CL features are available for analysis, Ga $2p$[33] or Ga $3d$,[34] which should, in terms of initial-state effects, demonstrate the same features. The results however, show that the peaks associated with GaAs are smaller than those for Ga_2O_3 in the Ga $2p$ spectrum, but that the opposite is true for the Ga $3d$ spectrum. This is because the lower KE of the Ga $2p$ electrons grants a shorter $IMFP_e$ to these electrons, resulting in an increased surface sensitivity, because where electrons with higher KE would be emitted from the sample, at this KE, they are scattered in the material.

Even for samples where the entire analysis volume over the full energy range is homogeneous, the peak areas are still not directly proportional to the amount of that species within a sample. This is because of the probabilities of photoemission (photoionisation cross sections), which differ between elements and orbitals. Calculations of these probabilities still do not necessarily yield accurate results as there are other factors that affect the photoemission process, including the transmission function of the analyser, which is energy dependent, and other matrix elements which affect the probability of transferring an electron into the final state.

In the system described in Sect. 2.2.4, the angle between the x-ray beam and the path to the analyser entrance was set to 54.74°. Under this condition, anisotropy of photoemission is removed, simplifying the quantitative analysis [32]. Then, if the same x-ray source and spectrometer are used throughout measurements, many other factors remain constant, and the analysis is affected mostly by $IMFP_e$ and cross-section effects, so that empirical stoichiometry correction factors (SCF) can be derived for different orbitals by measuring samples which have well-defined stoichiometries. Then, the atomic percentage of a chemical species (X_i) can be determined by

$$X_i = \frac{\frac{I_i}{\mathrm{SCF}_i}}{\sum_{j=1}^{n}\left(\frac{I_j}{\mathrm{SCF}_j}\right)}, \tag{2.27}$$

where I is the area of a peak, and the sum is over all of the species of interest.

Auger Features

The presence of Auger features in XPS spectra can also be beneficial to analyses [32]. In a broad set of Auger features, there is usually one transition which results in a peak that stands out from the rest and it is the position of this which is used to define the energy of the Auger feature, found by taking the zero-point of the differential of the spectrum [32].

Binding energy shifts in CL can often be small between different species, leading to ambiguities in the analysis. The Auger features can be used to help overcome these problems because they also experience chemical shifts, but often on a larger scale due to the three electron levels involved rather than one with CL [80, 81]. The

[33]BE $\approx$ 1120 eV.

[34]BE $\approx$ 20 eV.

Auger parameter, α is defined as the difference in KE between the most prominent Auger line and CL:

$$\alpha = E_K(\text{Auger}) - E_K(\text{CL}), \quad (2.28)$$

which results in a value that combines the shifts of each feature, producing much larger shifts that are discernible between species [38]. As stated at the beginning of this section, any charge referencing that a spectrum undergoes contains a potential source of error with the reliability of the chosen referencing method, and so, as a relative value, independent of any absolute binding energies [82] and charge referencing,[35] the use of the Auger parameter overcomes this error.

A problem with the Auger parameter shown in Eq. (2.28) is that it is photon energy dependent, because the measured CL kinetic energy is. To overcome this, the modified Auger parameter is used [80–82], defined by

$$\alpha' = \alpha + h\nu = E_K(\text{Auger}) - E_K(\text{CL}) + h\nu = E_K(\text{Auger}) + E_B(\text{CL}), \quad (2.29)$$

which takes the photon energy into account and therefore α′ can be used comparatively, given that the spectrometer is properly calibrated.

Valence Band Spectra

Electrons that are photoemitted from the VB of materials create broad features in the spectrum at low BE, introduced in Sect. 2.3.1. Ignoring any features of the spectrum which arise due to final-state effects, the VB spectrum from PES gives a representation of the VBDoS of a material. Early work on studies of metal surfaces showed that the measured valence bands, both from XPS and UPS, were able to confirm band structure calculations for these materials [84, 85], if the surface contamination was removed [86]. Further work then demonstrated that PES could also be used to measure the VBDoS of semiconducting materials [87–89], allowing much potential for PES to corroborate theoretically calculated DoS. In order to make a direct comparison between the spectra and theoretical DoS, experimental corrections can be applied to the latter, modelling the effects caused by the photoemission process. Because the KE of the electrons arising from the VB are sufficiently high and equal, there is little variation in the analysis depth in this region between the elements and therefore the intensity correction factors reduce to the photoionisation cross-sections [90]. After these have been applied to the partial DoS curves for each orbital, the DoS can be convoluted with a Gaussian function, accounting for the analyser resolution, linewidth of the excitation source and thermal effects, and then with a Lorentzian function, which accounts for the core-hole lifetimes [91].

For a more representative fit to the spectra, a background should also be subtracted from the data, and again in XPS VB spectra, the Shirley background has demonstrated good results because the rise in background at these energies is minimal [86]. For UPS, the background is not often discussed, but is important, because the entire spectrum consists of valence electrons, and can strongly vary. Different approaches

[35]To an extent [83].

have been implemented with varying degrees of success [92] and the one chosen for the UPS measurements presented in Chap. 6 shall be discussed there.

Band Alignments

For PV devices, the importance of band alignments was introduced in Sect. 1.2.2, and PES can be used to calculate these, which can be achieved by either an indirect, or direct method.

The indirect method, termed Kraut's method, involves the measurement of three samples in order to determine the valence band offset (VBO) of a junction [93–95]: the substrate (a), the thick overlayer[36] (b), and the interfacial sample[37] (i). It also exploits the facts that: the Fermi-level of any sample will align with that of the spectrometer; the energy separation between a CL and the VBM is constant for a given material, independent of any band bending which may occur; and the band bending is equal for all of the levels. The VBO of a junction is thus determined from the positions of a CL in each of the materials for each sample, and the positions of the VBM for the non-interfacial sample:

$$\text{VBO} = \left(E^b_{\text{CL}} - E^b_{\text{VBM}}\right) - \left(E^a_{\text{CL}} - E^a_{\text{VBM}}\right) - \left(E^b_{\text{CL}}(i) - E^a_{\text{CL}}(i)\right). \tag{2.30}$$

The CBO can then likewise be calculated using values of the bandgap of the two materials. Further details of this method can be found in the literature [96–99] as it is not implemented in this thesis because it would be impractical to grow the interfacial sample on the materials used.

It is also of interest to study the band positions of a material relative to the vacuum level, termed the natural band alignments of materials [100]. This results in measurements of the IP, WF, and EA, as shown in Fig. 1.23. Such a measurement does not take the band bending and charge redistribution upon junction formation into account, but can serve as a good starting point for predicting the behaviour of materials when formed into junctions, as the band offsets will be similar [101, 102]. The following method is beneficial for fledgling PV absorber materials that require further study, such as those studied in this thesis.

The secondary electrons produced in the photoemission process cannot have KE up to the photon energy because there is the hard threshold of the work function of the sample to overcome before they can be emitted [103]. This results in a very sharp, intense secondary electron cutoff (SEC) at a KE relative to the sample of zero.[38] This feature allows the WF of the sample to be determined using Eq. (2.7), giving

$$\phi_S = h\nu - E_{\text{SEC}}. \tag{2.31}$$

[36]Thickness > 10nm so that signal from the substrate is not detected.

[37]Thin enough so that signal from both the substrate and overlayer are detectable.

[38]$E'_K = 0$.

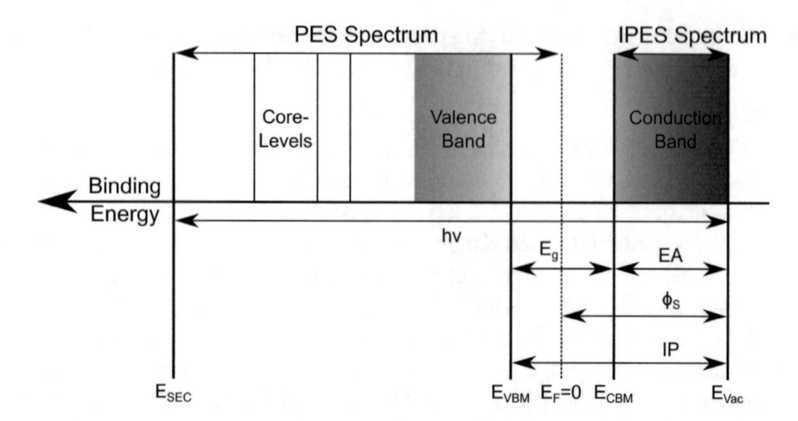

Fig. 2.16 Electronic energy level determination from a combination of PES and IPES spectra. Symbols are as-given in the text

Because of the Fermi-level alignment between samples and the spectrometers, PES and IPES spectra can be combined to represent the occupied and unoccupied DoS simultaneously [104], as is shown in Fig. 2.16. Then, by determining the binding energies of the VBM and CBM, the IP, EA and E_g of a sample can be determined by

$$\mathrm{IP} = \phi_S + E_{\mathrm{VBM}}, \tag{2.32}$$

$$\mathrm{EA} = \phi_s + E_{\mathrm{CBM}}, \tag{2.33}$$

and

$$E_g = \mathrm{IP} - \mathrm{EA}. \tag{2.34}$$

Within the spectrometer, secondary electrons can be generated by the collision of electrons with surfaces of it, causing the spectrum near the cutoff to be a combination of all of these electrons. In order to separate the sample SEC from the spectrometer SEC, a negative bias is applied to the sample, giving extra KE to any electrons originating from the sample, but not to those from the spectrometer, which allows the independent measurement of the sample SEC.

Difficulties arise in the determination of the position of the band edges because of the method used. Historically, VB measurements were limited to metals, or simple semiconductors such as Si and Ge. Because of this, accurate calculations for the DoS were available, which Kraut et al. [93] used to develop a method of determining the VBM position. This involved fitting the available theoretical DoS to the measured spectrum by the method described above and then working backwards to determine the energy of the DoS function onset before any experimental corrections were applied [105]. As the use of PES for VBO determination grew, accurate DoS calculations were not available for every material, and so the method of linearly extrapolating the leading edge of the VB spectrum to the baseline became the most

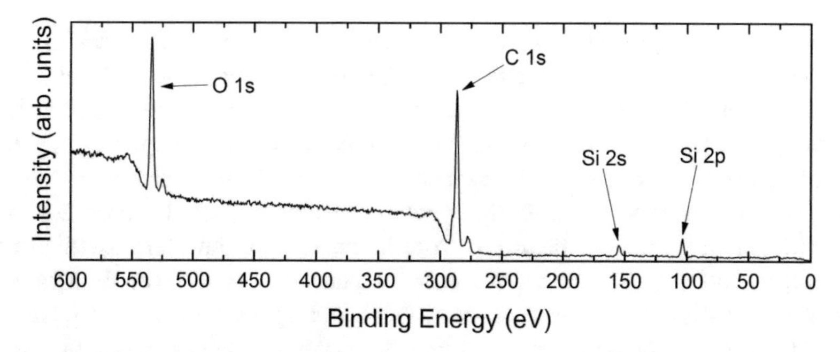

Fig. 2.17 XPS survey spectrum for the double-sided carbon tape used to affix samples to the sample plates. This spectrum is used to exclude spectral feature overlaps in the case that signal from the underlying tape could be detected

popular method of determining the VBM position, and yields accurate results in some cases [95, 106]. Advances in theoretical methods have given rise to accurate DoS calculations for many materials and Chambers et al. [106] compared the accuracy of these two VBM determination methods, and the application of which will be discussed in Sects. 6.7.3 and 7.2.2.

Sources of Misinterpretation

Different aspects of PES spectra can be the source of various types of misinterpretation. During measurement, especially VB studies, the presence of contamination can cause unprecedented differences in the spectra which cannot be accounted for, except by cleaning the surface. Indeed, the intensity of a signal originating from a material lying underneath a layer of contamination decreases exponentially as a function of contamination thickness [107].

There is also a possibility, for inhomogeneous or non-contiguous samples, that the exciting photons will produce photoelectrons from either the underlying substrate, or part of the mounting materials, which could either overlap with, or distort the spectrum arising from the material of interest. This was particularly the case for the crystals measured in Chap. 5, where imperfect mounting of the crystals could have led to the presence of signal arising from the double-sided carbon tape used for mounting the crystals to the sample plate. To check this, measurements were made of a piece of uncovered carbon tape, the survey XPS spectrum of which is shown in Fig. 2.17, to determine the elements present in this tape and check whether these may overlap with the signal from other samples.

With regards to measurements of PV absorbers using XPS, the most common use is the confirmation of the expected oxidation states of a grown material, and whilst this is possible using XPS, it requires rigorous analysis in order to exclude other interpretations of the observed spectra. Throughout the literature, this is sometimes lacking, and will be addressed with regards to the materials studied here in their respective chapters.

One final point that should be addressed with regards to misinterpretation of PES spectra is the errors associated with BE determinations. The sources of error in PES, like any experimental technique, are numerous, and differ on a case-by-case basis. In an ideal measurement of a single peak, the error associated with the BE is due only to the peak broadening, which results from the factors discussed earlier, but in reality, there is also error from the statistical noise, which is dependent upon the amount of times the counting experiment was repeated. In spectra which involve overlapping peaks, there are also fitting errors which arise from the accuracy of the fitted background, and also the fitting procedure and algorithms used. Furthermore, an incorrect interpretation of any spectra[39] negates any error estimate regardless. Whilst it is possible to account for all of these factors through error analysis, it is more prudent to estimate BE errors in PES spectra using an educated combination of the knowledge of the instrumental resolution and confidence in the fitting procedure. As a rule of thumb, for the instrumental broadenings shown in Table 2.3, a minimum starting value for the confidence in quoted binding energies throughout this thesis were estimated to be $\pm(0.05 - 0.1)$ eV. Of course, these values can increase for noisy or complex data sets. The values measured in this thesis are quoted to two decimal places. Whilst this is not always completely correct practice, it offers better transparency for comparisons to literature reports that also quote values to this level of precision.

2.4 Density Functional Theory

Density functional theory (DFT) is employed in many aspects of physics, chemistry and materials science to predict the electronic structures of materials. In this thesis, it is used to predict the DoS, band structure, and certain electronic properties of the materials in question.

2.4.1 Introduction

Whilst the full formalisms of DFT are beyond the scope of this thesis, its implementation as a corroborative technique to experimental results is valuable. The specific implementations of the DFT used throughout this thesis are described in the appropriate chapters, and the reader is directed to literature that introduces the methods employed [108–112].

Predictions of the electronic structure of materials stem from attempts to solve the Schrödinger equation. The solutions to the Schrödinger equation, the wavefunctions, contain all of the information about the energies, positions, and properties of the particles that comprise a system, and the equation describes how they interact with one another.

[39] Such as the addition of superfluous peaks, or use of an unphysical model.

Whilst the wavefunction solutions to the Schrödinger equation are well defined for the case of a single electron, once a many-electron system is taken into account, this cannot be solved exactly, and approximations must be used. Quantum mechanical wavefunctions are not observable, and so DFT was developed as a method which instead deals with a measurable quantity, electron density, by calculating the ground state electron density for a system and deriving other properties from there.

2.4.2 *Implementation*

Within solids, the periodic lattice allows the calculations to be non-infinite, and solutions to be found. From these solutions, various properties of a material can be predicted, and the ones used in this thesis shall be described below.

From DFT, the band structure of a material can be calculated and the band energies given as a function of k–space across the first Brillouin zone. The DoS can then be calculated by integrating the bands across k-space and the resulting DoS curve can be compared to experimental measurements by the method described in Sect. 2.3.4. The DoS can be further analysed by integrating bands arising from the different elemental orbitals, resulting in a series of partial DoS (pDoS) curves, which can be used to analyse the levels of hybridisation between the orbitals within a material, giving insight into the bonding nature.

From the DoS curves, the bandgap of the material can be predicted as the difference in the onsets of the VB states and the CB states, and by implementing a vacuum potential in the calculations, it is also possible to predict the positions of the band edges with respect to the vacuum level, giving a direct comparison to these values from experimental PES as well.

2.5 Fourier Transform Infrared Spectroscopy

Infrared spectroscopy is employed as a technique to study the vibrational properties of materials. In the case of PV materials, it is used in order to calculate the absorption spectrum of materials and from there, the light absorption properties can be determined, including a measurement of the bandgap.

2.5.1 *Principles and Theory*

Light, incident upon a material, can be either reflected, transmitted, or absorbed, and the relationship between the amounts that each of these happen is dependent upon the wavelength, or energy of the incident light. Historically, reflectance and transmittance spectra as a function of wavelength were generated by dispersive spectrometers, that

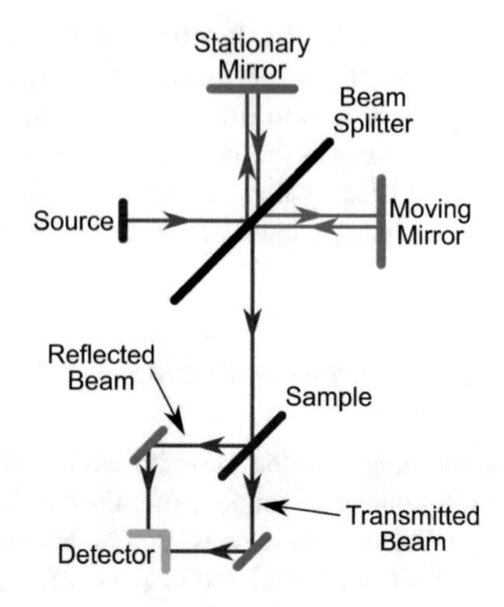

Fig. 2.18 Schematic of the ray diagram for a FTIR spectrometer using a Michelson interferometer

is, the wavelength of the incoming light was varied and the spectrum at the detector was generated from the intensity as a function of the wavelength of the incoming light [38]. However, dispersive monochromators are expensive and difficult to manage, and this, combined with the amount of time taken to generate a spectrum because of all the moving optics necessary, led to the development of a different technique to generate spectra.

Fourier transform infrared spectroscopy (FTIR), instead of using monochromatic light to generate a spectrum as a function of wavelength, uses an interferometer to generate an interferogram as a function of phase difference, utilising non-monochromatic light [113]. The experimental system of a FTIR spectrometer is shown in Fig. 2.18, utilising the common Michelson interferometer. A light source, consisting of a range of wavelengths, is split into two paths by a beamsplitter, directing one toward a stationary mirror, and the other to a movable mirror. After reflecting from the mirrors, the beams recombine again at the beamsplitter and the combined beam is then directed toward the sample. Depending upon the experimental system, after interaction with the sample, the reflected or transmitted beam is then collected by the detector and the intensity is recorded [114].

The path that the beam to the movable mirror takes differs in length by changing the position of the movable mirror. The intensity of the recombined beam directed to the sample is therefore dependent upon the interference caused by the path difference between the beams, and is a function of the position of the movable mirror. Therefore, by scanning the position of the movable mirror, an interferogram is generated as a function of phase difference [113]. Because of the non-monochromatic nature of the incident light, each wavelength contributes an individual interference pattern to the interferogram, which is a combination of all of these. Taking the Fourier transform of the interferogram then generates the spectrum as a function of wavelength [53].

This method of generating the spectrum is beneficial over the dispersive method because spectra are collected in a shorter time, experimental systems are simpler, and the resolution is better. The position of the mirror can be determined very accurately by including a monochromatic laser of known wavelength that follows the same optical path through the interferometer.

Throughout this thesis, measurements using FTIR were made in order to generate the absorption spectrum from a sample, which is calculated best by using a combination of reflectance and transmittance spectra, collected using a series of optics that can be altered to direct either the reflected or transmitted beam after interaction with the sample to the detector.[40]

2.5.2 Operation and Implementation

The absorption coefficient of a PV absorber material is dependent upon the energy of the incident light, as introduced in Chap. 1, and this can be calculated from the reflectance and transmittance spectra, as a function of photon energy. When considering materials measured for their use in PV, internal reflections within a thin-film can affect the amount of light detected in the reflected beam. To account for this, an extra term is included in the transmittance, which includes the probability of reflection at the first interface, second interface, and also the probability of internal reflections [115], so

$$T = \frac{(1-R)^2 \mathrm{e}^{-\alpha d}}{1 - R^2 \mathrm{e}^{-2\alpha d}}, \tag{2.35}$$

where T is the transmittance spectrum, R is the reflectance spectrum, and α is the absorption coefficient, all as a function of photon energy, and d is the thickness of the sample. Solving for the absorption coefficient gives

$$\alpha = \frac{1}{d} \ln\left(\frac{2TR^2}{\sqrt{(1-R)^4 + 4T^2R^2} - (1-R)^2}\right), \tag{2.36}$$

and one can calculate the absorption spectrum.

In semiconductors with strong absorption characteristics in the visible range, as is the case for PV absorbers, the bandgap acts as a filter, absorbing photons with energy equal to or greater than the bandgap. This manifests in the absorption spectrum as an onset of absorption at the bandgap energy. Physical and experimental considerations mean that this onset is not instantaneous and some level of fitting must be employed to determine the bandgap energy. The eponymous method that developed from the historical study of Tauc et al. [116, 117] has found widespread use throughout the literature, and has been generalised to different semiconductors [118]. It says that the absorption function just above the absorption onset of a semiconductor takes the form

$$(\alpha h\nu)^{\frac{1}{n}} = A\big(h\nu - E_g\big), \tag{2.37}$$

[40]See Fig. 2.18.

where α is the absorption coefficient, $h\nu$ is the photon energy, A is a proportionality constant, E_g is the bandgap of the semiconductor, and n is dependent upon the nature of the gap transition, taking $n = 1/2$ for direct, and $n = 2$ for indirect transitions. Thus, by plotting $(\alpha h\nu)^{1/n}$ against $h\nu$ (a Tauc plot), the trend of the function above the absorption onset will be linear and a linear extrapolation will yield the bandgap value where it intercepts the energy axis [119].

The bandgap of semiconductors is temperature dependent because of changes in lattice parameters and electron–phonon interactions [120]. Generally, for PV devices, the bandgap at operating temperature[41] is important. As theoretical determinations of the bandgap generally calculate the bandgap at absolute zero (E_0), it is prudent to study the temperature dependence of the bandgap at finite temperatures and to apply theoretical models in order to extrapolate and find the bandgap at absolute zero, so that comparisons can be made between theory and experiment. The use of a cryostat-cooled sample chamber in the FTIR spectrometer allows measurements to 4 K to be performed. Several models have been developed which attempt to describe the temperature dependence of the bandgap of semiconductors. Two of the more popular methods are the Varshni and Bose–Einstein equations, and these are used at various points throughout this thesis with discussion of their implementation found at these points. Here, the mathematical forms of these equations are introduced.

The Varshni relationship [121] is commonly used in the literature, and takes the form

$$E_g = E_0 - \frac{\alpha T^2}{T + \beta}, \tag{2.38}$$

where α and β are constants. This relationship has been used extensively and attempts to model the two mechanisms which are thought to account for the bandgap changes. First, the band positions alter as an effect of the expansion of the lattice, which is linear at high temperature, and non-linear at low temperature. Second, the band positions alter also because of the changing interaction between the electrons and the lattice, which is linear above the Debye temperature, and goes as T^2 below the Debye temperature.

The Bose–Einstein model [122] takes the form

$$E_g = E_0 - \frac{2a_B}{\exp\left(\frac{\Theta_E}{T}\right) - 1}, \tag{2.39}$$

where a_B is related to the exciton–phonon interactions, and Θ_E is the Einstein temperature. This model considers the interactions of the electrons in a crystal structure, modelling them as Einstein oscillators, following Bose–Einstein statistics [123].

In this thesis, measurements were performed on a Bruker Vertex 70 V FTIR spectrometer equipped with a reflectance and transmittance accessory which directs the beam to the sample at an incident angle of 11°. A continuous-flow helium cryostat was used to perform the low temperature measurements.

[41] Usually room temperature.

2.6 X-Ray Diffraction

X-Ray Diffraction (XRD) has become one of the defining techniques for materials characterisation, especially in the growth of semiconductors, and is key in structural analysis. Quick and easy to implement, first-order analysis is available to most researchers, where it is used as an identification tool. However, in-depth analyses allow much richer studies to be performed using this technique as well.

2.6.1 Principles and Theory

The principles of XRD have at their core, the Bragg condition, which can be explained through a classical treatment of the scattering of electromagnetic waves. When incident upon a solid, electromagnetic waves can be elastically scattered by the atoms in the crystal structure, emitting radiation of the same frequency from the atom [124]. Interference between these emitted waves produces constructive or destructive interference dependent upon the phase difference between them. Diffraction occurs in a solid crystal because of the periodicity of the lattice. Consider a series of parallel crystallographic planes within a crystal structure with interplanar spacing, d (Fig. 2.19). Incident electromagnetic waves at an angle (θ), are scattered by the atoms in the crystal structure in all directions, but are coherently reflected from the atoms at the same angle. Now, if two waves are incident upon the solid, one (ABC) which is reflected by the top plane, and another (DEF) which is reflected by the second plane, and if a detector is placed on the line (CF), then the two arriving waves will be in phase if, and only if the phase difference between them is a multiple of 2π, which equates to a path difference of an integer multiple of wavelengths, that is

$$DEF - ABC = n\lambda, \tag{2.40}$$

and due to the geometry of the crystallographic planes

$$\begin{aligned} DEF = DE + EF &= (AB + d\ \sin(\theta)) + (d\ \sin(\theta) + BC) \\ &= ABC + 2d\ \sin(\theta). \end{aligned} \tag{2.41}$$

Therefore

$$DEF - ABC = (ABC + 2d\ \sin(\theta)) - ABC = 2d\ \sin(\theta) = n\lambda, \tag{2.42}$$

the familiar Bragg condition for diffraction [124–126].

The case shown in Fig. 2.19 shows one set of diffraction planes, but in a powder or polycrystalline sample, the different crystallites are arranged randomly, causing all planes to be perpendicular to the normal incidence of the waves[42] at some point

[42] $\theta = 90°$.

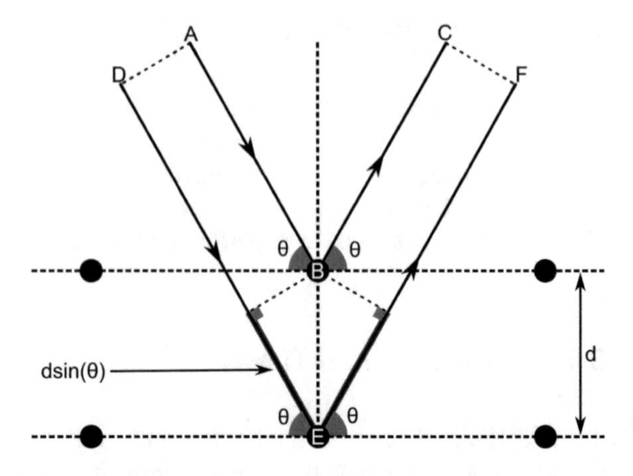

Fig. 2.19 Derivation of the Bragg condition for the diffraction of electromagnetic radiation incident upon a crystalline solid

within a sample and therefore by changing the angle of incidence and detection simultaneously, a diffraction pattern can be generated, as a function of 2θ.

Then, by calculating the interplanar spacing for the different families of planes, using the geometry and atomic positions within a crystal structure, one can determine the angles at which diffraction will occur for each series of planes [125]. These expected diffraction angles are sometimes restricted or forbidden by the presence of other atoms within the crystal structure, for example by interference between waves emitted from interplanar atoms, and if the geometry allows, this may result in destructive interference between some or all of the waves in that plane. The intensity of a Bragg peak, I_{hkl} in a diffraction pattern is given by the absolute square of the structure factor

$$I_{hkl} = |F_{hkl}|^2, \tag{2.43}$$

where *hkl* are the Miller indices of a certain series of planes.

The structure factor is calculated by summing the individual contributions of the diffracted waves from each non-equivalent atom in a crystal structure [124, 126], so

$$F_{hkl} = \sum_{n=0}^{N} f_n e^{2\pi i (hx_n + ky_n + lz_n)}, \tag{2.44}$$

where n is each non-equivalent atom in the crystal structure with coordinates (x, y, z), and f_n is the atomic scattering factor, dependent on the element. From Eq. (2.44), it can be seen that for certain arrangements of atoms, some families of planes result in a structure factor of zero, leading to zero intensity in the diffraction pattern; these are the forbidden reflections for that crystal structure.

2.6.2 Operation and Implementation

Given that the peak intensities produced for a material are dependent both on the crystal structure, and atoms present within the lattice, unique patterns are produced, allowing materials to be identified by their crystal structure. However, problems can arise when materials share the same crystal structure or when certain symmetries result in the same interplanar spacing, because this causes overlaps between the peaks in a diffraction pattern, complicating the analysis. Nevertheless, diffraction in crystals is still a powerful technique used to determine structures.

The description of the principles in Sect. 2.6.1 relates generally to the diffraction of electromagnetic waves in crystal structures, but here the technique of interest uses x-rays as the exciting radiation, hence x-ray diffraction. X-rays are used because they are easy to produce via the method outlined in Sect. 2.3.2, and also because the wavelength of x-rays is of a similar order of magnitude to the interplanar spacing in crystals, resulting in strong diffraction. Because photons scatter from atoms via interactions with the electrons, XRD is rather insensitive to differences between the nuclei of atoms. Neutrons with de Broglie wavelengths of the same order as the interplanar spacings can also be used for diffraction, and in this case are much more sensitive to nuclear differences because of the neutron scattering via the nuclei of atoms. This benefit would suggest that neutron diffraction is the superior technique [124]; however, the difficulties involved with producing the coherent neutrons required,[43] compared with x-ray production, means that the use of neutron diffraction is limited to advanced facilities, rather than used in laboratories. Throughout this thesis, different experimental systems have been used for performing XRD, which are described in the relevant chapters, but they consist of a source of x-rays,[44] a sample stage, and an x-ray detector, all mounted to a goniometer, which also allows focussing of the beams.

XRD is widely practiced, and results are often compiled into databases which can be searched and indexed for matching to experimentally produced patterns. This makes XRD very useful when trying to determine the structure of a new material, or an established one which is grown by a different method. Also, it is possible to calculate the theoretical diffraction pattern of a material whose crystal structure is known. By calculating the interplanar spacings and applying the selection rules of Eq. (2.44), the allowed reflection planes and their intensities can be determined, and by applying the Bragg condition, the angles at which these peaks appear can be found. In this thesis, theoretical XRD patterns were modelled using the reciprOgraph software[45] [127], and the published crystal structures.

Differences between the observed experimental peaks and the theoretical ones can be indicative of several details of the material, and there are various methods for extracting materials properties from the diffraction pattern. Beyond phase identifica-

---

[43] A spallation source [39].

[44] Usually Cu Kα, $\lambda = 1.541$ Å.

[45] Courtesy of Nicolas Schoeni and Gervais Chapuis of École Polytechnique Fédérale de Lausanne, Switzerland.

tion, one can determine the lattice parameters of a sample, changes in peak positions can be used to determine the effects of strain on the material, changes in peak intensities can show orientation effects, and differences in peak widths are indicative of the crystallite size [124].

2.7 Raman Spectroscopy

In the context of PV materials, Raman spectroscopy (RS) is used to complement XRD in structural identification. Developed to study the vibrational states of a material, it also relies on the interaction of light with matter in such a way that probes the vibrational modes of a material. As these modes are somewhat unique to a crystal structure [53], it can be used to phase-match a material in a similar way as XRD, but complementary to it.

2.7.1 Principles and Theory

Three of the ways in which light can interact with a solid are depicted in Fig. 2.20a. Incident light causes the electronic polarisation field in a material to oscillate at the same frequency as the exciting radiation. If these oscillations reradiate photons, then they will have the same frequency as the incident radiation. This is elastic (Rayleigh) scattering as there is no energy change. At finite temperatures, the atoms in a solid have vibrational modes, dependent upon the structure. These phonons can then interact with the electrons, changing the frequency of the polarisation field after photon interaction. The radiated photons can therefore have a shift in frequency, resultant from the difference between the energy of the ground state orbital and the vibrational mode [53, 128]. A decrease in energy is termed Stokes scattering, and an increase in energy is termed anti-Stokes scattering. Together, a shift of the same energy in both directions is Raman scattering.

The Raman shift in wavenumbers, the measurable quantity used to implement RS, is given as

$$\Delta\omega = \frac{1}{\lambda_I} - \frac{1}{\lambda_S}, \tag{2.45}$$

where λ_I is the wavelength of the incident radiation, and λ_S is the measured wavelength of the scattered radiation. Therefore, by measuring the energies of the scattered radiation, a spectrum can be generated as a function of Raman shift, with peaks appearing at frequencies equal to the energy of the vibrational modes of a material.

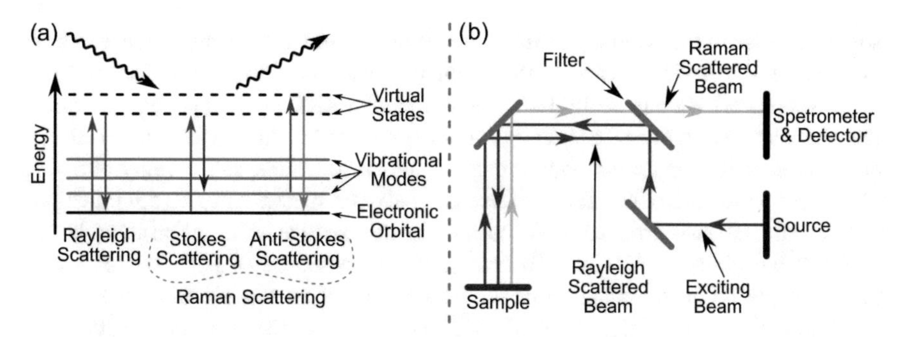

Fig. 2.20 **a** Light scattering in a solid. **b** Ray diagram for Raman spectroscopy

2.7.2 *Operation and Implementation*

The Raman spectrometers used throughout this thesis are described in their relevant chapters, but the general implementation of one is shown in Fig. 2.20b [53, 128]. Monochromatic laser light of known wavelength is directed at a sample, which undergoes the different scattering mechanisms. Some of this light is backscattered in the direction from whence it came and is a combination of Rayleigh and Raman scattered light. This then passes through a filter, which excludes the original wavelength of light because the Rayleigh scattering cross section is much higher than the Raman ones and can flood the Raman signal [128]. The Raman scattered light is then analysed using a dispersive spectrometer and detected in order to produce the spectrum. High-resolution spectrometers are required as the shifts to be detected can be very small.

A Raman spectrum will not directly represent the vibrational modes present in a material because there are selection rules, meaning that some modes are not Raman-active because they create phonon oscillations that cancel each other out, and also some modes are degenerate due to the crystalline symmetry. Predicting the phonon modes, calculating their energies, and determining the selection rules for a material in order to predict a Raman spectrum is not as easy as the method used for predicting XRD patterns, and for this reason, it is important to consult databases of standard spectra when performing RS measurements, if available. In the case of new materials, it is prudent to compare measured spectra with measurements from a high-quality sample of the material of interest, for example, a single crystal [128].

The scattering mechanisms are somewhat independent of the wavelength of the exciting radiation, but there are some considerations which can be made in order to exploit or suppress certain mechanisms [128]. A shorter wavelength increases the probability of scattering, but also increases the risk of fluorescence, and sample damage through heating. Especially for PV materials that have high absorption, the analysis volume can be altered by changing the excitation wavelength, because this will be affected by the amount of absorption. Also, where the energy of the excitation source is close to the bandgap of the material under study, resonant effects can alter the spectra. Further discussion of these considerations is addressed in Sect. 6.8.1.

As was stated, RS can be used in conjunction with XRD in order to help determine the phase and structure of a material, this is because the frequencies at which the characteristic Raman peaks appear can be unique to a material, thereby allowing identification. But also like XRD, differences between spectra can give other insights into certain materials properties. Shifts in the peaks can give information about lattice strain as the phonon energies will change, the dependence of peak intensities on the wavelength and polarisation of the excitation source can elucidate the crystal orientation and symmetry, and the width of the peaks can give information about the crystallinity [128, 129].

2.8 Energy-Dispersive X-Ray Spectroscopy

Whilst it is possible to perform stoichiometry determinations using PES, there are various sources of error by using this method, which are discussed in Sect. 2.3.4, and are sometimes detrimentally large. Instead, energy-dispersive x-ray spectroscopy (EDX) is used to determine the bulk stoichiometry of a sample.

Though not included in this thesis because it is not associated with the electronic characterisations, scanning electron microscopy (SEM) is a tool used in various aspects of materials science, including absorber material growth [130]. By including a photon spectrometer and detector into a SEM experimental system, EDX measurements can also be performed.

X-rays are generated from a sample in the SEM by the same processes used to generate x-rays for PES. High energy electrons are bombarded into a sample, and through the mechanisms described in Sect. 2.3.2 and shown in Fig. 2.12a, characteristic x-rays and bremsstrahlung are produced and emitted. The emission spectrum of the x-rays is then similar to that shown in Fig. 2.12c, but is a superposition of the characteristic x-ray peaks from each element present within a sample. The intensity of the x-rays is then related to the amount of that element present in the sample, and by correcting the intensities using factors derived from standard samples, to account for emission cross-sections, the stoichiometry of a sample can be determined [38, 130].

Historically, the energies of the x-rays were measured using a wavelength-dispersive spectrometer, which exploits Bragg diffraction through a crystal at certain angles, dependent upon the wavelength of the x-rays. Measurements using such a spectrometer were expensive, in terms of time, space, and money. For these reasons, and also because some resolution was able to be sacrificed, solid-state photon detectors have become the norm. These use a FET to collect the photons, which produce a photoelectron that loses its energy to the creation of excitons. In this way, the size of the voltage pulse, or number of excitons produced is proportional to the energy of the original x-ray, and the rate of pulses is related to the intensity of photons. This method allows spectra to be collected in a much shorter time and the physical size of the equipment is much smaller [38, 130].

Although this technique is regularly available with SEM operations, its main drawback is that the results only provide information on the amount of each element within a sample area. Whilst it can indeed map an area of a sample because of the scanning ability of the incident electron beam, allowing elemental segregation to be determined [131, 132], it cannot provide any information about the chemical environment of the elements [38]; a task reserved for PES.

2.9 Summary

The main experimental methods, growth techniques, and characterisation methods used throughout this thesis have been introduced. The principles and theoretical background to the methods have been explained and the specific implementations of these techniques pertinent to this thesis have been presented.

Greater attention in this chapter has been paid to PES techniques, including more in-depth details of their implementation, and also a comprehensive overview of the necessity and application of UHV science in this context. The majority of the work presented throughout this thesis was conceived and developed with this technique in mind, partly because of the widespread, and often misguided use of it throughout the literature for these materials. As such, it is important to understand the physical processes which govern and affect PES, including those which can affect the analysis and interpretation of the results, in order that its use is not treated as a 'black box'. This problem is not encountered with respect to the other techniques used throughout this thesis, as they are presented as somewhat ancillary to PES, and therefore do not require the same level of familiarity, at least here.

With regards to PES, a detailed understanding of both the mechanisms involved in its use, and the acknowledgement of the limitations of the technique allow not only the implementation of high-quality, meaningful analyses, but also the development of richer and more rigorous interpretations, through the exploitations afforded sometimes by the quirks and idiosyncrasies of these techniques.

References

1. Mizutori M, Yamada R. Semiconductors. In: Ullmann's encyclopedia of industrial chemistry, vol. 9. Weinheim: Wiley-VCH Verlag GmbH & Co. KGaA; 2000. p. 245–60.
2. Stockbarger DC. The production of large single crystals of Lithium Fluoride. Rev Sci Instrum. 1936;7(3):133–6.
3. Schäfer H, Jacob H, Etzel K. Chemische Transportreaktionen. II. Die Verwendung Der Zerfallsgleichgewichte Der Eisen(II)- Und Nickel(II)-Halogenide Zum Metalltransport Im Temperaturgefälle. Zeitschrift für Anorg. und Allg Chemie. 1956;286(1–2):42–55.

4. Binnewies M, Glaum R, Schmidt M, Schmidt P. Chemical vapor transport reactions—a historical review. Zeitschrift für Anorg und Allg Chemie. 2013;639(2):219–29.
5. Mochizuki K, Masumoto K. Growth of CdTe single crystals by THM (Travelling Heater Method) and its repetition effect. Mater Lett. 1986;4(5–7):298–300.
6. Reig C, Paranchych YS, Muñoz-Sanjosé V. Crystal growth of HgSe by the cold travelling heater method. Cryst Growth Des. 2002;2(2):91–2.
7. Das S, Mandal K. Low-cost Cu_2ZnSnS_4 thin films for large-area high-efficiency heterojunction solar cells. In: 2012 38th IEEE photovoltaic specialists conference, vol. 4; 2012; p. 2668–73.
8. Kelly PJ, Arnell RD. Magnetron sputtering: a review of recent developments and applications. Vacuum. 2000;56(3):159–72.
9. Welch AW, Zawadzki PP, Lany S, Wolden CA, Zakutayev A. Self-regulated growth and tunable properties of $CuSbS_2$ solar absorbers. Sol Energy Mater Sol Cells. 2015;132:499–506.
10. Willian de Souza Lucas F, Welch AW, Baranowski LL, Dippo PC, Mascaro LH, Zakutayev A. Thermal treatment improvement of $CuSbS_2$ absorbers. In: 2015 IEEE 42nd photovoltaic specialist conference (PVSC); 2015; IEEE. p. 1–5.
11. de Souza Lucas FW, Welch AW, Baranowski LL, Dippo PC, Hempel H, Unold T, Eichberger R, Blank B, Rau U, Mascaro LH, Zakutayev A. Effects of thermochemical treatment on $CuSbS_2$ photovoltaic absorber quality and solar cell reproducibility. J Phys Chem C. 2016;120(33):18377–85.
12. Rocket A. Physical vapor deposition. In: The materials science of semiconductors. Boston: Springer US; 2008. p. 505–72.
13. Schou J. Physical aspects of the pulsed laser deposition technique: the stoichiometric transfer of material from target to film. Appl Surf Sci. 2009;255(10):5191–8.
14. Irvine SJC. Materials challenges: inorganic photovoltaic solar energy. In: Irvine SJC, editor. RSC energy and environment series. Cambridge: Royal Society of Chemistry; 2014.
15. Abermann S. Non-vacuum processed next generation thin film photovoltaics: towards marketable efficiency and production of CZTS based solar cells. Sol Energy. 2013;94:37–70.
16. Suryawanshi MP, Agawane GL, Bhosale SM, Shin SW, Patil PS, Kim JH, Moholkar AV. CZTS based thin film solar cells: a status review. Mater Technol. 2013;28(1–2):98–109.
17. Jung S, Sou A, Banger K, Ko D-H, Chow PCY, McNeill CR, Sirringhaus H. All-inkjet-printed, all-air-processed solar cells. Adv Energy Mater. 2014;4(14):1400432.
18. Taylor JF. Spin coating: an overview. Met Finish. 2001;99(1):16–21.
19. Mooney JB, Radding SB. Spray pyrolysis processing. Annu Rev Mater Sci. 1982;12(1):81–101.
20. Fulop GF, Taylor RM. Electrodeposition of semiconductors. Annu Rev Mater Sci. 1985;15(1):197–210.
21. Pathan HM, Lokhande CD. Deposition of metal chalcogenide thin films by Successive Ionic Layer Adsorption and Reaction (SILAR) method. Bull Mater Sci. 2004;27(2):85–111.
22. Xie M, Zhuang D, Zhao M, Li B, Cao M, Song J. Fabrication of Cu_2ZnSnS_4 thin films using a ceramic quaternary target. Vacuum. 2014;101:146–50.
23. Shi C, Shi G, Chen Z, Yang P, Yao M. Deposition of Cu_2ZnSnS_4 thin films by vacuum thermal evaporation from single quaternary compound source. Mater Lett. 2012;73:89–91.
24. Vaccarello D, Tapley A, Ding Z. Optimization of the Cu_2ZnSnS_4 nanocrystal recipe by means of photoelectrochemical measurements. Rsc Adv. 2013;3(11):3512–5.
25. Gupta S, Whittles TJ, Batra Y, Satsangi V, Krishnamurthy S, Dhanak VR, Mehta BR. A low-cost, sulfurization free approach to control optical and electronic properties of Cu_2ZnSnS_4 via precursor variation. Sol Energy Mater Sol Cells. 2016;157:820–30.
26. Aono M, Yoshitake K, Miyazaki H. XPS depth profile study of CZTS thin films prepared by spray pyrolysis. Phys Status Solidi. 2013;10(7–8):1058–61.
27. Doona CJ, Stanbury DM. Equilibrium and Redox kinetics of Copper(II)-Thiourea complexes. Inorg Chem. 1996;35(11):3210–6.
28. Altamura G, Vidal J. Impact of minor phases on the performances of CZTSSe thin-film solar cells. Chem Mater. 2016;28(11):3540–63.

29. Ki W, Hillhouse HW. Earth-abundant element photovoltaics directly from soluble precursors with high yield using a non-toxic solvent. Adv Energy Mater. 2011;1(5):732–5.
30. Nguyen VT, Nam D, Gansukh M, Park S-N, Sung S-J, Kim D-H, Kang J-K, Sai CD, Tran TH, Cheong H. Influence of sulfate residue on Cu_2ZnSnS_4 thin films prepared by direct solution method. Sol Energy Mater Sol Cells. 2015;136:113–9.
31. Park S-N, Sung S-J, Son D-H, Kim D-H, Gansukh M, Cheong H, Kang J-K. Solution-processed Cu_2ZnSnS_4 absorbers prepared by appropriate inclusion and removal of Thiourea for thin film solar cells. Rsc Adv. 2014;4(18):9118–25.
32. Hofmann S. Auger- and X-ray photoelectron spectroscopy in materials science, vol. 49; Springer series in surface sciences. Berlin, Heidelberg: Springer Berlin Heidelberg; 2013.
33. Woodruff DP, Delchar TA. 1—Introduction. In: Modern techniques of surface science. Cambridge: Cambridge University Press; 1994. p. 1–14.
34. Chung Y-W. Fundamental concepts in ultrahigh vacuum, surface preparation, and electron spectroscopy. In: Practical guide to surface science and spectroscopy. Elsevier; 2001. p. 1–22.
35. Hofmann P. Surface physics: an introduction. Philip Hofmann; 2013.
36. Weston GF. Ultrahigh vacuum practice. Cambridge: Butterworth & Co. Ltd; 1985.
37. Berman A. Water vapor in vacuum systems. Vacuum. 1996;47(4):327–32.
38. Henning Bubert HJ. Surface and thin film analysis (Bubert H, Jenett H, editors), vol. 4. Weinheim, FRG: Wiley-VCH Verlag GmbH; 2002.
39. Vickerman JC, Gilmore IS. In: Vickerman JC, Gilmore IS, editors. Surface analysis– the principal techniques. Chichester: Wiley; 2009.
40. Soriaga MP. Ultra-high vacuum techniques in the study of single-crystal electrode surfaces. Prog Surf Sci. 1992;39(4):325–443.
41. Steichen M, Djemour R, Gütay L, Guillot J, Siebentritt S, Dale PJ. Direct synthesis of single-phase P-type SnS by electrodeposition from a dicyanamide ionic liquid at high temperature for thin film solar cells. J Phys Chem C. 2013;117(9):4383–93.
42. Sundberg J, Lindblad R, Gorgoi M, Rensmo H, Jansson U, Lindblad A. Understanding the effects of sputter damage in W-S thin films by HAXPES. Appl Surf Sci. 2014;305:203–13.
43. Velásquez P, Ramos-Barrado JR, Leinen D. The fractured, polished and Ar^+-sputtered surfaces of natural enargite: an XPS study. Surf Interface Anal. 2002;34(1):280–3.
44. Loeffler MJ, Dukes CA, Chang WY, McFadden LA, Baragiola RA. Laboratory simulations of sulfur depletion at Eros. Icarus. 2008;195(2):622–9.
45. Nossa A, Cavaleiro A. Chemical and physical characterization of C(N)-doped W-S sputtered films. J Mater Res. 2004;19(8):2356–65.
46. Woodruff DP, Delchar TA. Abbreviations. In: Modern techniques of surface science. Cambridge: Cambridge University Press; 1994. p. xiii–xiv.
47. Fadley CS. X-ray photoelectron spectroscopy: progress and perspectives. J Electron Spectros Relat Phenom. 2010;178–179:2–32.
48. Woodruff DP, Delchar TA. 3—Electron spectroscopies. In: Modern techniques of surface science. Cambridge: Cambridge University Press; 1994. p. 105–265.
49. Seah MP, Dench WA. Quantitative electron spectroscopy of surfaces: a standard data base for electron inelastic mean free paths in solids. Surf Interface Anal. 1979;1(1):2–11.
50. Rivière JC, Arlinghaus HF, Hutter H, Jenett H, Bauer P, Palmetshofer L. Surface and thin-film analysis, 2. Electron detection. In: Ullmann's encyclopedia of industrial chemistry. Weinheim: Wiley-VCH Verlag GmbH & Co. KGaA; 2011. p. 335–76.
51. Kahn A. Fermi level, work function and vacuum level. Mater Horiz. 2016;3(1):7–10.
52. Cahen D, Kahn A. Electron energetics at surfaces and interfaces: concepts and experiments. Adv Mater. 2003;15(4):271–7.
53. Vij DR. In: Vij DR, editor. Handbook of applied solid state spectroscopy. Boston: Springer US; 2006.
54. Lindgren I. Chemical shifts in X-ray and photo-electron spectroscopy: a historical review. J. Electron Spectros. Relat. Phenom. 2004; 137–140 (SPEC. ISS.):59–71.
55. Chung Y-W. Photoelectron spectroscopy. In: Practical guide to surface science and spectroscopy. Elsevier; 2001. p. 45–67.

56. Woicik JC. Hard X-ray photoelectron spectroscopy (HAXPES). In: Woicik JC, editors. Springer series in surface sciences. Cham: Springer International Publishing; 2016.
57. Wu J-B, Lin Y-F, Wang J, Chang P-J, Tasi C-P, Lu C-C, Chiu H-T, Yang Y-W. Correlation between N 1s XPS binding energy and bond distance in metal Amido, Imido, and Nitrido complexes. Inorg Chem. 2003;42(15):4516–8.
58. Bagus PS, Ilton ES, Nelin CJ. The interpretation of XPS spectra: insights into materials properties. Surf Sci Rep. 2013;68(2):273–304.
59. Gaarenstroom SW, Winograd N. Initial and final state effects in the ESCA spectra of Cadmium and Silver Oxides. J Chem Phys. 1977;67(8):3500–6.
60. Major JD, Phillips LJ, Al Turkestani M, Bowen L, Whittles TJ, Dhanak VR, Durose K. P3HT as a pinhole blocking back contact for CdTe thin film solar cells. Sol Energy Mater Sol Cells. 2017;172:1–10.
61. Chung Y-W. Auger electron spectroscopy. In: Practical guide to surface science and spectroscopy. Elsevier; 2001. p. 23–43.
62. Ahmed A, Robertson CM, Steiner A, Whittles T, Ho A, Dhanak V, Zhang H. Cu(i)Cu(i)BTC, a microporous mixed-valence MOF via reduction of HKUST-1. RSC Adv. 2016;6(11):8902–5.
63. Citrin PH, Eisenberger PM, Marra WC, Åberg T, Utriainen J, Källne E. Linewidths in X-ray photoemission and X-ray emission spectroscopies: what do they measure? Phys Rev B. 1974;10(4):1762–5.
64. Brice JC. The lattice constants of a-quartz. J Mater Sci. 1980;15(1):161–7.
65. Zhang W. Angle-resolved photoemission spectroscopy. In: Photoemission spectroscopy on high temperature superconductor. Springer Theses. Berlin, Heidelberg: Springer Berlin Heidelberg; 2013. p. 19–49.
66. Seah MP. Summary of ISO/TC 201 standard: VII ISO 15472:2001-surface chemical analysis-X-ray photoelectron spectrometers-calibration of energy scales. Surf Interface Anal. 2001;31(8):721–3.
67. Cros A. Charging effects in X-ray photoelectron spectroscopy. J Electron Spectros Relat Phenom. 1992;59(1):1–14.
68. Suzer S. Differential charging in X-ray photoelectron spectroscopy: a nuisance or a useful tool? Anal Chem. 2003;75(24):7026–9.
69. Baer DR, Engelhard MH, Gaspar DJ, Lea AS, Windisch CF. Use and limitations of electron flood gun control of surface potential during XPS: two non-homogeneous sample types. Surf Interface Anal. 2002;33(10–11):781–90.
70. Barr TL, Seal S. Nature of the use of adventitious carbon as a binding energy standard. J Vac Sci Technol A Vacuum Surf Film. 1995;13(3):1239.
71. Repoux M. Comparison of background removal methods for XPS. Surf Interface Anal. 1992;18(7):567–70.
72. Végh J. The Shirley Background Revised. J Electron Spectros Relat Phenom. 2006;151(3):159–64.
73. Tougaard S. Quantitative Analysis of the inelastic background in surface electron spectroscopy. Surf Interface Anal. 1988;11(9):453–72.
74. Weightman P. X-ray-excited auger and photoelectron spectroscopy. Reports Prog Phys. 2000;45(7):753–814.
75. Doniach S, Sunjic M. Many-electron singularity in X-ray photoemission and X-ray line spectra from metals. J Phys C: Solid State Phys. 1970;3(2):285–91.
76. Lebugle A, Axelsson U, Nyholm R, Mårtensson N. Experimental L and M core level binding energies for the metals ^{22}Ti to ^{30}Zn. Phys Scr. 1981;23(5A):825–7.
77. Coster D, Kronig L, De R. New type of Auger effect and its influence on the X-ray spectrum. Physica. 1935;2(1–12):13–24.
78. Nyholm R, Martensson N, Lebugle A, Axelsson U. Auger and Coster-Kronig broadening effects in the 2p and 3p photoelectron spectra from the metals ^{22}Ti-^{30}Zn. J Phys F: Met Phys. 1981;11(8):1727–33.
79. Speckbacher M, Treu J, Whittles TJ, Linhart WM, Xu X, Saller K, Dhanak VR, Abstreiter G, Finley JJ, Veal TD, Koblmüller G. Direct measurements of fermi level pinning at the surface of intrinsically N-type InGaAs nanowires. Nano Lett. 2016;16(8):5135–42.

80. Wagner CD. Chemical shifts of Auger Lines, and the Auger parameter. Faraday Discuss Chem Soc. 1975;60:291.
81. Wagner CD. Auger parameter in electron spectroscopy for the identification of chemical species. Anal Chem. 1975;47(7):1201–3.
82. Wagner CD, Joshi A. The Auger parameter, its utility and advantages: a review. J Electron Spectros Relat Phenom. 1988;47:283–313.
83. Oswald S, Gonzalez-Elipe AR, Reiche R, Espinos JP, Martin A. Are measured values of the Auger parameter always independent of charging effects? Surf Interface Anal. 2003;35(12):991–7.
84. Fadley CS, Shirley DA. X-ray photoelectron spectroscopic study of Iron, Cobalt, Nickel, Copper, and Platinum. Phys Rev Lett. 1968;21(14):980–3.
85. Shirley DA. High-resolution X-ray photoemission spectrum of the valence bands of Gold. Phys Rev B. 1972;5(12):4709–14.
86. Shirley DA, Fadley CS. X-ray photoelectron spectroscopy in North America—the early years. J Electron Spectros Relat Phenom. 2004;137–140:43–58.
87. Ley L, Kowalczyk S, Pollak R, Shirley DA. X-ray photoemission spectra of crystalline and amorphous Si and Ge valence bands. Phys Rev Lett. 1972;29(16):1088–92.
88. Ley L, Pollak RA, McFeely FR, Kowalczyk SP, Shirley DA. Total valence-band densities of states of III-V and II-VI compounds from X-ray photoemission spectroscopy. Phys Rev B. 1974;9(2):600–21.
89. Pollak RA, Ley L, Kowalczyk S, Shirley DA, Joannopoulos JD, Chadi DJ, Cohen ML. X-ray photoemission valence-band spectra and theoretical valence-band densities of states for Ge, GaAs, and ZnSe. Phys Rev Lett. 1972;29(16):1103–5.
90. Yeh JJ, Lindau I. Atomic subshell photoionization cross sections and asymmetry parameters: $1 \leqslant Z \leqslant 103$. At Data Nucl Data Tables. 1985;32(1):1–155.
91. Fadley CS, Shirley DA. Electronic densities of states from X-ray photoelectron spectroscopy. J Res Natl Bur Stand Sect A Phys Chem. 1970;74A(4):543.
92. Li X, Zhang Z, Henrich VE. Inelastic electron background function for ultraviolet photoelectron spectra. J Electron Spectros Relat Phenom. 1993;63(3):253–65.
93. Kraut EA, Grant RW, Waldrop JR, Kowalczyk SP. Precise determination of the valence-band edge in X-ray photoemission spectra: application to measurement of semiconductor interface potentials. Phys Rev Lett. 1980;44(24):1620–3.
94. Kraut EA, Grant RW, Waldrop JR, Kowalczyk SP. Semiconductor core-level to valence-band maximum binding-energy differences: precise determination by X-ray photoelectron spectroscopy. Phys Rev B. 1983;28(4):1965–77.
95. Eich D, Ortner K, Groh U, Chen ZH, Becker CR, Landwehr G, Fink R, Umbach E. Band discontinuity and band gap of MBE grown HgTe/CdTe (001) heterointerfaces studied by K-Resolved photoemission and inverse photoemission. Phys Status Solidi. 1999;261(1):261–8.
96. Waldrop JR, Kraut EA, Farley CW, Grant RW. Measurement of InP/$In_{0.53}Ga_{0.47}As$ and $In_{0.53}Ga_{0.47}As$/$In_{0.52}Al_{0.48}As$ heterojunction band offsets by X-ray photoemission spectroscopy. J Appl Phys. 1991;69(1):372.
97. Santoni A, Biccari F, Malerba C, Valentini M, Chierchia R, Mittiga A. Valence band offset at the CdS/Cu_2ZnSnS_4 interface probed by X-ray photoelectron spectroscopy. J Phys D Appl Phys. 2013;46(17):175101.
98. Waldrop JR, Grant RW. Measurement of AlN/GaN (0001) heterojunction band offsets by X-ray photoemission spectroscopy. Appl Phys Lett. 1996;68(20):2879.
99. List RS. Can photoemission measure valence-band discontinuities? J Vac Sci Technol B Microelectron Nanom Struct. 1988;6(4):1228.
100. Klein A. Energy band alignment in chalcogenide thin film solar cells from photoelectron spectroscopy. J Phys: Condens Matter. 2015;27(13):134201.
101. Klein A. Energy band alignment at interfaces of semiconducting oxides: a review of experimental determination using photoelectron spectroscopy and comparison with theoretical predictions by the electron affinity rule, charge neutrality levels, and the common anion. Thin Solid Films. 2012;520(10):3721–8.

102. Schlaf R, Lang O, Pettenkofer C, Jaegermann W. Band lineup of layered semiconductor heterointerfaces prepared by van Der Waals epitaxy: charge transfer correction term for the electron affinity rule. J Appl Phys. 1999;85(5):2732–53.
103. Helander MG, Greiner MT, Wang ZB, Lu ZH. Pitfalls in measuring work function using photoelectron spectroscopy. Appl Surf Sci. 2010;256(8):2602–5.
104. Bär M, Weinhardt L, Heske C. Soft X-ray and electron spectroscopy: a unique "Tool Chest" to characterize the chemical and electronic properties of surfaces and interfaces. In: Advanced characterization techniques for thin film solar cells. Weinheim: Wiley-VCH Verlag GmbH & Co. KGaA; 2011. p. 387–409.
105. Waldrop JR. Measurement of semiconductor heterojunction band discontinuities by X-ray photoemission spectroscopy. J. Vac Sci Technol A Vacuum Surf Film. 1985;3(3):835.
106. Chambers SA, Droubay T, Kaspar TC, Gutowski M. Experimental determination of valence band maxima for $SrTiO_3$, TiO_2, and SrO and the associated valence band offsets with Si(001). J Vac Sci Technol B Microelectron Nanom Struct. 2004;22(4):2205.
107. Smith GC. Evaluation of a simple correction for the hydrocarbon contamination layer in quantitative surface analysis by XPS. J Electron Spectros Relat Phenom. 2005;148(1):21–8.
108. von Barth U. Basic density-functional theory an overview. Phys Scr. 2004;T109:9.
109. Kryachko ES, Ludeña EV. Density functional theory: foundations reviewed. Phys Rep. 2014;544(2):123–239.
110. Cohen AJ, Mori-Sánchez P, Yang W. Challenges for density functional theory. Chem Rev. 2012;112(1):289–320.
111. Burke K, Werschnik J, Gross EKU. Time-dependent density functional theory: past, present, and future. J Chem Phys. 2005;123(6):62206.
112. Jones RO. Density functional theory: its origins, rise to prominence, and future. Rev Mod Phys. 2015;87(3):897–923.
113. Gremlich H-U. Infrared and Raman spectroscopy. In: Ullmann's encyclopedia of industrial chemistry, vol. 21. Weinheim: Wiley-VCH Verlag GmbH & Co. KGaA; 2000; p. 85–138.
114. Griffiths PR, de Haseth JA. Fourier transform infrared spectrometry. Hoboken: Wiley; 2007.
115. Ghezzi C, Magnanini R, Parisini A, Rotelli B, Tarricone L, Bosacchi A, Franchi S. Optical absorption near the fundamental absorption edge in GaSb. Phys Rev B. 1995;52(3):1463–6.
116. Tauc J, Grigorovici R, Vancu A. Optical properties and electronic structure of amorphous Germanium. Phys Status Solidi. 1966;15(2):627–37.
117. Tauc J. Optical properties and electronic structure of Amorphous Ge and Si. Mater Res Bull. 1968;3(1):37–46.
118. Davis EA, Mott NF. Conduction in non-crystalline systems V. Conductivity, optical absorption and photoconductivity in amorphous semiconductors. Philos Mag. 1970;22(179):0903–22.
119. Viezbicke BD, Patel S, Davis BE, Birnie DP. Evaluation of the Tauc method for optical absorption edge determination: ZnO thin films as a model system. Phys Status Solidi. 2015;252(8):1700–10.
120. Ünlü H. A thermodynamic model for determining pressure and temperature effects on the bandgap energies and other properties of some semiconductors. Solid State Electron. 1992;35(9):1343–52.
121. Varshni YP. Temperature dependence of energy gap in semiconductors. Physica. 1967;34(1):149.
122. Sarswat PK, Free ML. A study of energy band gap versus temperature for Cu_2ZnSnS_4 thin films. Phys B Condens Matter. 2012;407(1):108–11.
123. Lautenschlager P, Garriga M, Logothetidis S, Cardona M. Interband critical points of GaAs and their temperature dependence. Phys Rev B. 1987;35(17):9174–89.
124. Schorr S, Stephan C, Törndahl T, Mainz R. X-ray and neutron diffraction on materials for thin-film solar cells. In: Advanced characterization techniques for thin film solar cells. Weinheim: Wiley-VCH Verlag GmbH & Co. KGaA; 2011. p. 347–63.
125. Paulus EF, Gieren A. Structure analysis by diffraction. In: Ullmann's encyclopedia of industrial chemistry, vol. 21. Weinheim: Wiley-VCH Verlag GmbH & Co. KGaA; 2001. p. 85–138.

126. Cullity BD. Elements of X-ray diffraction. Reading: Addison-Wesley Publishing Company; 1956.
127. Chapuis G, Schoeni N. Towards a web-based interactive environment for the teaching of crystallography. Acta Crystallogr Sect A: Found Crystallogr. 2005;61(a1):c121–2.
128. Álvarez-García J, Izquierdo-Roca V, Pérez-Rodríguez A. Raman spectroscopy on thin films for solar cells. In: Advanced characterization techniques for thin film solar cells. Weinheim: Wiley-VCH Verlag GmbH & Co. KGaA; 2011. p. 365–86.
129. Lund EA, Du H, Hlaing OO WM, Teeter G, Scarpulla MA. Investigation of combinatorial coevaporated thin film Cu_2ZnSnS_4 (II): beneficial cation arrangement in Cu-rich growth. J Appl Phys. 2014; 115(17):173503.
130. Abou-Ras D, Nichterwitz M, Romero MJ, Schmidt SS. Electron microscopy on thin films for solar cells. In: Advanced characterization techniques for thin film solar cells. Weinheim: Wiley-VCH Verlag GmbH & Co. KGaA; 2011. p. 299–345.
131. Hoffmann V, Klemm D, Efimova V, Venzago C, Rockett Aa, Wirth T, Nunney T, Kaufmann Ca, Caballero R. Elemental distribution profiling of thin films for solar cells. In: Advanced characterization techniques for thin film solar cells. Weinheim: Wiley-VCH Verlag GmbH & Co. KGaA; 2011; p. 411–48.
132. Czanderna AW, Madey TE, Powell CJ. Beam effects, surface topography, and depth profiling in surface analysis. In: Czanderna AW, Madey TE, Powell CJ, editors. Methods of surface characterization. Boston: Kluwer Academic Publishers; 2002 Vol. 5.

Chapter 3
The Electronic Structure of $CuSbS_2$ for Use as a PV Absorber

Solar power is the last energy resource that isn't owned yet—nobody taxes the sun yet.

Bonnie Raitt, Musician, 1978

This chapter involves measurements of thin-film samples of $CuSbS_2$ (CAS), and the electronic characterisation of this material for use in PV devices. The majority of the results presented here are currently unpublished, but are in the final stages of manuscript preparation as:

Whittles, T. J.; Veal, T. D.; Savory, C. N.; Welch, A. W.; de Souza Lucas, F. W.; Gibbon, J. T.; Birkett, M.; Potter, R. J.; Scanlon, D. O.; Zakutayev, A.; Dhanak, V. R. Core-Levels, Band Alignments and Valence Band States in $CuSbS_2$. *Manuscript in Preparation* **2017**.

Related measurements of other thin-film samples of CAS, have been published as:

The original version of this chapter was revised: Table was reformatted and the references to the table were cited. The correction to this chapter is available at https://doi.org/10.1007/978-3-319-91665-1_8

© Springer International Publishing AG, part of Springer Nature 2018

T. J. Whittles, *Electronic Characterisation of Earth-Abundant Sulphides for Solar Photovoltaics*, Springer Theses, https://doi.org/10.1007/978-3-319-91665-1_3

Peccerillo, E.; Major, J.; Phillips, L.; Treharne, R.; Whittles, T. J.; Dhanak, V.; Halliday, D.; Durose, K. Characterization of Sulfurized $CuSbS_2$ Thin Films for PV Applications. In *2014 IEEE 40th Photovoltaic Specialist Conference (PVSC)*; IEEE, 2014; pp 0266–0269.

To date, the earth-abundant material CAS has seen relatively poor solar cell performance despite demonstrating good properties as a solar PV absorber material. Here, phase-pure CAS, confirmed by XRD and Raman spectroscopy, demonstrates a solar-matched bandgap of 1.55 eV, which agrees with theoretical calculations. Also, an analysis of the CL spectra shows how XPS can play a valuable role in characterising CAS films, and determining the effects of contaminants. Furthermore, a comparison between the VB XPS spectra and theoretical DoS calculations shows how states arising from the antimony cation lay beneath states from the copper cation. Together, these raise the level of the VBM, giving IP and EA values for this material[1] that are lower than other common absorbers, including CIGS. The subsequent conduction band misalignment with CdS is a reason for the poor performance of cells incorporating a CAS/CdS heterojunction, supporting the idea that using a cell design analogous to CIGS is unhelpful. Thus, with a greater knowledge of the underlying electronic structure, cell architectures can be redesigned.

3.1 $CuSbS_2$: The Material

The material CAS is currently being studied as a PV absorber material. Research on this material is however, new and its full potential may still yet to be realised.

3.1.1 History

As introduced in Sect. 1.4.2, CAS is a member of the sulphosalts whose elemental constituents are considered earth-abundant. Within the Cu–Sb–S family, several naturally occurring minerals are known and have been studied. Apart from $CuSbS_2$, these include: tetrahedrite [1] $Cu_{12}Sb_4S_{13}$, famatinite [2] Cu_3SbS_4, and skinnerite [3] Cu_3SbS_3, the structures of which have all been determined, but have not seen research outside of the mineralogical community, excepting skinnerite, which has been synthesised [4], and is now being explored as a potential PV material as well [5, 6].

$CuSbS_2$ on the other hand, also known in its mineral form chalcostibite [7], was studied synthetically for its semiconducting properties as early as 1980 by Wachtel and Noreika [8], who grew single crystals of CAS and found it to demonstrate strong p-type conductivity with low mobility, high carrier concentrations and a narrow bandgap of 0.28 eV; these results confirming previous predictions for this material [9]. Because these findings are not promising for PV use, CAS was largely forgotten about

[1] 4.98 eV and 3.43 eV, respectively.

until recently, when reassessments following advances in characterisation methods saw it demonstrate properties that are PV relevant. Interest has been maintained in CAS also because of its potential use in other areas of semiconductor applications, such as supercapacitors [10], dye-sensitised solar cells [11, 12], or electrodes in batteries [13].

3.1.2 $CuSbS_2$ as a Solar Absorber

The use of indium in CIGS cells poses potential economic and environmental problems, and as such, viable replacements are being searched for. The ionic radius of antimony is almost equivalent to that of indium [14–17], but the scarcity and world demand for antimony is lower than for indium [18], and as such, could prove a good substitute, both economically and in terms of PV properties. Furthermore, the addition of trivalent indium or gallium reduced the copper ion mobility in CIGS over Cu_2S [19–21], and a similar effect is expected for other trivalent elements, such as antimony in CAS.

CAS was not realised as a solar absorber until 2001, where it was found to demonstrate strong p-type conductivity and a suitable bandgap of 1.52 eV [21], and from there, the first cell was constructed in 2005 [22]. Since then it has continued to demonstrate attractive PV properties [23]: a solar-matched bandgap of ~1.5 eV [15, 21, 24], inherent p-type conductivity due to the dominant copper vacancy [20, 21, 25], and absorption stronger than both CIGS and CZTS [26].

As well as attractive properties, CAS has been grown in thin-film and nanoscale form by a variety of methods, both physical[2] and chemical[3]: beneficial for PV use. Also, with a relatively low melting point of ~540 °C [25, 27], CAS is amenable to crystallisation at lower temperatures.

Despite these benefits, solar cell performance thus far has been limited and shortcomings of the literature reports on device properties have been ackowledged [29]. There have been very few reports of fully built PV devices, reporting no [22, 32, 33], or very low[4] [16, 25, 27, 30] efficiencies, which are detailed in Table 3.1. The efficiency record stands at 3.22% [39], and even after over a decade of research, recent reports still acknowledge that further investigation into device fabrication and characterisation is required [16, 26].

$CuSbS_2$ was chosen for study here, over the Cu_3SbS_3 phase because although both of these materials benefit from the same earth-abundant composition properties, Cu_3SbS_3 was shown to have a bandgap around 1.83 eV [40], too high for effective PV devices. The selenium equivalent of CAS, that is $CuSbSe_2$ (CASe), has also begun to be studied for its use in PV and it is believed that a mixed sulphide-selenide would

[2]Thermal evaporation [16, 17, 27, 28] and sputtering [19, 29, 30].

[3]Spray pyrolysis [14, 20, 31], CBD [15, 21, 22, 32, 33], spin coating [25], electrodeposition [24, 28, 34], solution processing [35] and solvo-/hydro-thermal [36–38].

[4]$< 2\%$.

Table 3.1 Performance of CAS-based solar PV devices

Absorber Growth Method	Device Architecture	Efficiency	Year	Reference
CBD & Evaporation	SLG/FTO/CdS/CAS/C/Ag	None reported	2015	Ornelas-Acosta et al. [32]
CBD	SLG/FTO/CdS:In/Sb_2S_3/CAS/Ag	None reported	2005	Rodríguez-Lazcano et al. [22]
CBD & Evaporation	SLG/FTO/CdS/CAS/C/Ag	None reported	2014	Ornelas-Acosta et al. [33]
Spin-Coating	FTO/CAS/CdS/ZnO/AZO/Au	0.50%	2014	Yang et al. [25]
Sputtering	SLG/Mo/CAS/CdS/ZnO/AZO/Al	0.86%	2016	Welch et al. [30]
CBD & Evaporation	SLG/FTO/CdS/Sb_2S_3/CAS/C/Ag	1%	2015	Krishnan et al. [16]
Evaporation	Mo/CAS/CdS/ZnO/AZO/Ag	1.9%	2016	Wan et al. [27]
Electro-deposition	SLG/Mo/CAS/CdS/AZO	3.13%	2014	Septina et al. [24]
Spin-Coating	Mo/CAS/CdS/ZnO/AZO	3.22%	2016	Banu et al. [39]

allow tuning of the bandgap [41], similar to the case with CZTS.[5] However, due to the problems associated with selenium use,[6] this material is not considered here.

3.1.3 Structure

Although CAS was developed with the aim of analogously replacing CIGS, there are several apparent differences between the two materials. The mineral form of CAS is called chalcostibite, named after the Greek names for copper and antimony, and was first discovered at the Graf Jost-Christian mine in Wolfsberg, Harz, Sachsen-Anhalt, Germany in 1835 [42], which led to its original name, wolfsbergite [43]. Although antimony takes the trivalent oxidation state in CAS, like gallium and indium do in CIGS, CAS does not form in the tetrahedral chalcopyrite crystal structure like CIGS [44, 45]. Instead, CAS forms in the structure shown in Fig. 3.1, which is somewhat distorted by the stereochemically active antimony lone-pair electrons, which do not take part in bonding [46]. The copper atoms are 4-fold coordinated to sulphur in almost regular tetrahedra, whereas the antimony atoms are 5-fold coordinated to sulphur in a distorted square-based pyramid arrangement. The lone-pair electron density is then directed into the void between the SbS_5 pyramid units [47]. With two

[5] See Sect. 6.1.3.

[6] See Sect. 1.4.

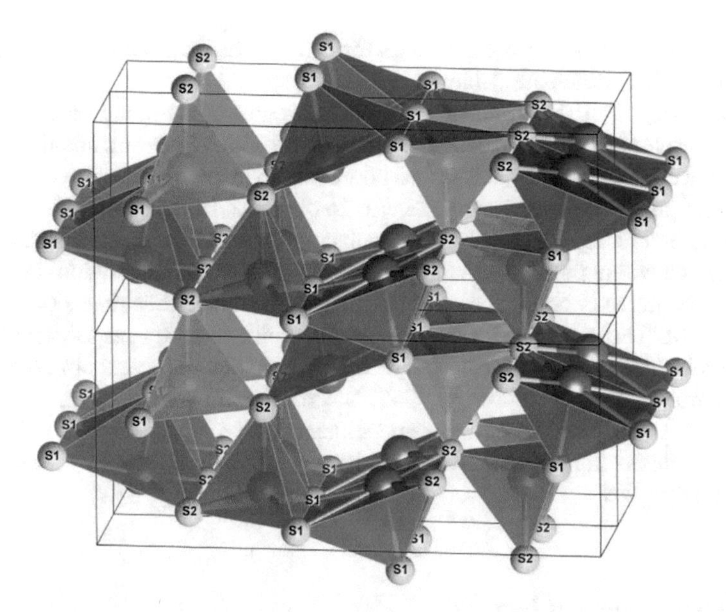

Fig. 3.1 Crystal structure of chalcostibite [48] $CuSbS_2$, showing the individual atoms of Cu (orange), Sb (purple) and S (yellow). The inequivalent sulphur coordination sites are marked

Sb–S bonds being much longer than the other three[7] [48], the crystal structure is layered through the plane intersecting these bonds [30]. Consequently, there are two inequivalent sulphur sites, one which is coordinated to two Cu and three Sb atoms, and the other which is coordinated to two Cu and two Sb atoms[8] [49].

3.1.4 Motivation and Scope of This Study

The low device efficiencies despite the attractive PV properties of CAS is the main motivation for this study, which demands a more fundamental understanding of it, in order to drive progress into researching novel ideas for its development. Indeed, a continued heavy reliance on analogies with CIGS could cause CAS development to stagnate, as was the case with CZTS [50], even though initial development was accelerated using these analogies. It is therefore important to acknowledge and address the differences between CAS and CIGS.

The crystal structure of CAS is very dissimilar from the tetrahedral structures of CIGS or CZTS[9] [29], and leads to some differing characteristics, which are thought to

[7]Shown in grey in Fig. 3.1.

[8]Labelled S1 and S2 in Fig. 3.1, respectively.

[9]See Fig. 6.1.

be a possible cause of the poor device performances. It is the details of the electronic structure of CAS which are explored in this chapter.

Specifically, a full XPS analysis of CAS is presented, including core-levels, the effect of surface cleaning and contamination, along with VB spectra and the position of the band edges. From this, it is shown how complexities in the spectra can make the surface contamination of this material easy to overlook, but also why it is important to identify and recognise the effects of such contamination. These findings are then corroborated within the VB with theoretical density of states calculations in order to further explain the bonding nature of this material and how this affects the position of the VBM. This then gives rise to a low IP, which is partly responsible for the poor efficiencies seen so far because of the poor conduction band alignment with the commonly used window layer, CdS. Ultimately, the poor performance of this material is ascribed not to an inherent fundamental problem, but to a reliance on comparing this material with CIGS, which then results in outcomes which are not beneficial for CAS.

3.2 Experimental Details

Complete details of the experimental systems used here can be found in Chap. 2 and the specific details pertinent to this study, including a full description of the cleaning procedure used for the thin-films, are detailed in Appendix C.

3.2.1 Growth of $CuSbS_2$ Films

The samples of CAS studied throughout this chapter were donated by the Zakutayev Group, at NREL, USA. Thin-films of CAS were grown by sputtering simultaneously from targets of Cu_2S and Sb_2S_3. Following this, the thin-films underwent a thermal treatment in an Sb_2S_3 atmosphere. Previous characterisations of these films can be found in the appropriate literature [19, 29, 51], along with more details of the growth methods.

Thin-films were provided before (no-TT) and after (TT) the thermal treatment was performed in order to determine the effect of this on the electronic structure of the material, if any. Also, for the different characterisations, thin-films were deposited on molybdenum-coated glass for XPS measurements, and on plain glass for optical measurements.

3.2.2 *Characterisation of $CuSbS_2$ Films*

Raman Spectroscopy

Raman spectra were acquired using a Horiba Scientific Jobin-Yvon LabRam HR system, using an exciting wavelength of 514.5 nm.

X-Ray Photoemission Spectroscopy

XPS was used to study the core-level electronic structure, SEC, VBM and occupied VBDoS of the CAS samples. They were affixed to UHV sample plates by spot welding tantalum straps across the sample edges. This also provided an electrical connection between the CAS films and the spectrometer. Ar^+ ion sputtering at 500 eV and radiative heating at 200 °C were utilised to clean contamination from the sample surfaces, the full details of which are given in Appendix C, Sect. C.5.1.

Prior to cleaning, the main C 1*s* signal, due to contaminant carbon, was found to have a BE of 284.72 eV and was deemed to have either experienced no, or very minimal charging due to this BE being observed for contaminant carbon on this material previously [27, 32, 33, 52, 53]. However, after subjecting the sample to surface cleaning, the very poor SNR for the C 1*s* peaks meant that this peak could no longer be used for charge correction. Instead, as the Cu 2*p* region doesn't suffer from any overlapping features, the spectra were calibrated using the Cu $2p_{3/2}$ BE of the sample prior to cleaning, which was measured to be 931.97 eV.

All of the fitted synthetic peaks in this study were Voigt profiles, fitted after subtraction of a Shirley background, unless otherwise stated. Cu 2*p* doublets were fitted with a separation of 19.80 eV [54] and an area ratio of 1:2. The FWHM of the Cu $2p_{1/2}$ peak is somewhat broader than that of the Cu $2p_{3/2}$ peak due to Coster–Kronig effects [55, 56]. Sb 3*d* doublets were fitted with a separation of 9.34 eV [57], an area ratio of 2:3 and equal FWHM. Fitted S 2*p* doublets were separated by 1.20 eV [58], with an area ratio of 1:2, and equal FWHM.

Optical Absorption Spectroscopy

Values of the direct and indirect bandgaps were determined from extrapolation of the absorption spectra derived from reflectance and transmittance measurements via FTIR spectroscopy at 300 and 4 K. The data reduction and fitting procedures are detailed in Sect. 2.5.2.

Density Functional Theory

Electronic-structure calculations were performed through the VASP code [59–62], starting from the atomic structure of CAS, obtained through XRD analysis. The band structure, pDoS in the valence and conduction bands, and bandgaps were determined.

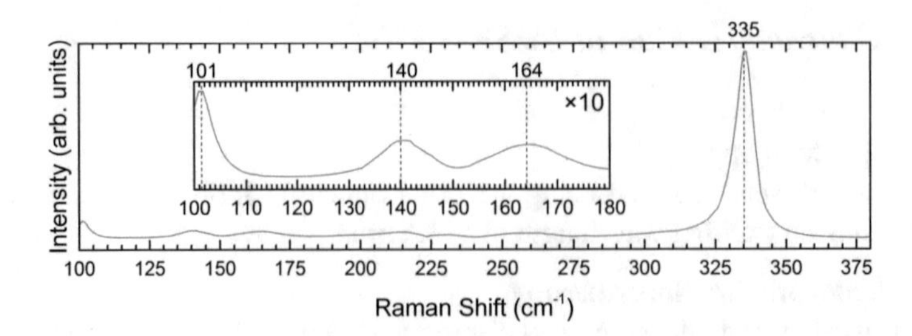

Fig. 3.2 Raman spectrum for the TT CAS sample. Strongest peak positions marked

3.3 Electronic Structure Studies of High-Quality $CuSbS_2$ Films

This section presents characterisations of the thermally treated sample of CAS for analysis of the potential of this material to be used as a solar absorber. The TT sample was chosen for these 'standard' characterisations as it has been shown to be of superior quality than the non-TT sample. Differences stemming from the thermal treatment shall be explored in Sect. 3.4.

3.3.1 Phase-Purity

The Raman spectrum for the TT CAS sample is shown in Fig. 3.2 and demonstrates a single intense peak at 335 cm^{-1}, which is indicative of the chalcostibite phase [63]. Also observed are the weaker Raman modes at 101, 140 and 164 cm^{-1} which have been reported previously for this material [27]. There are minor peaks around 215 and 260 cm^{-1} as well, which have been observed in previous spectra [27, 64]. The Raman modes corresponding to Sb_2S_3 [65], CuS [66, 67] and the other stoichiometries of copper antimony sulphide, Cu_3SbS_3 [68], Cu_3SbS_4 [69] and $Cu_{12}Sb_4S_{13}$ [63], are not present in the spectrum. This, and previous XRD measurements of samples prepared using an identical method [29], suggest that the $CuSbS_2$ is phase-pure.

3.3.2 XPS Core-Level Analysis

The Necessity for Rigorous Core-Level Analysis and the Potential Problems with XPS Spectra of CAS

The complexity and variety of compounds that can be formed increases when working with ternary, over binary systems, as is the case with CAS [70]. It is therefore imperative to be able to control and identify the phase of the resulting film in order that the correct compound is formed. With regards to CAS, a line compound that should be intolerant to stoichiometric nonconformity [19, 30], this is especially important. Beyond the standard techniques of XRD and RS[10] for phase identification, XPS is commonly used to confirm the expected oxidation states of the elements in solar absorber compounds [71–73]. Unfortunately, due to the subtle nuances that can be present in XPS spectra, coupled with inexpert analysis, it is common for important details to be either unnoticed or misinterpreted, as discussed in Sect. 2.3.4. One such detail which is overlooked is the inevitable formation of oxides at the surface of samples which have been exposed to atmosphere. If this happens during the transfer to the spectrometer, the effects of a surface oxide may be trivial, causing a slight shift in the measured properties; however, if this occurs during a processing step of solar cell production, then an extra layer will be included into the device stack, which may be a culprit for performance degradation. Whatever the circumstances, the formation of oxides on absorber materials must be acknowledged and addressed.

For CAS, it is possible for both the metals and the sulphur to oxidise, and as it is generally the case that metal-oxide peaks have higher BE than metal-sulphide peaks due to the higher electronegativity of oxygen over sulphur [74], this makes assignment of oxide peaks somewhat easier. For copper, the more likely to form cupric oxide,[11] [75], is aided in identification by the presence of characteristic shake-up lines in the Cu $2p$ spectra,[12] associated with the Cu^{2+} oxidation state [76], as well as a shift in BE from the sulphide [77]. However, an overreliance on the qualitative detection of such shake-up lines for confirmation or exclusion of the presence of copper oxide has resulted in incorrectly drawn conclusions. Indeed, in studies by Suriakarthick et al. [36], Jiasong et al. [78], and Su et al. [37], these satellite features are present but no fitting has been performed and they are not acknowledged in the discussion. In other studies, the satellites are explicitly excluded in the discussion, leading to the conclusion that no CuO is present; however, the data presented by Sarswat and Free [79] has such a poor SNR, that this conclusion cannot be accurately drawn, and the presence of these satellites is apparent in data presented by Wang et al. [80], even though this conclusion was explicitly stated as well.

The oxidation of sulphur is very clear in XPS spectra of materials where sulphur is usually present as an anion, and is characterised by intensity around 6 eV higher BE than the corresponding sulphide species [81], making detection of this species easier.

Oxidation, if present, is not guaranteed of all constituent elements, as has been assumed previously [82], and rigorous fitting of all elements should be performed in order to acquire a correct analysis. Difficulties in assessing the surface oxidation of antimony are compounded by the overlap of the O $1s$ and Sb $3d_{5/2}$ regions [83]. A fitting procedure is therefore required to determine the antimony and oxygen species. Here, a commonly followed procedure was utilised [83]: because any O $1s$ peaks would overlap only the Sb $3d_{5/2}$ peaks, the Sb $3d_{3/2}$ peaks were fitted independently

[10]See Sect. 3.3.1.

[11]CuO.

[12]See Sect. 2.3.1.

with the knowledge that only intensity contribution from antimony could be found there. Following that, the corresponding peaks for Sb $3d_{5/2}$ were constrained using the parameters given in Sect. 3.2.2. The remaining intensity in this region was then fitted, with confidence that this was due solely to O $1s$ photoelectrons.

In the literature, unacknowledged intensity arising from antimony oxides is commonplace, but the presented spectra can be critically assessed and even a qualitative analysis can provide a better understanding. Some studies, like those from Ornelas-Acosta et al. [32, 33], and Sarswat and Free [79], present good quality data for the overlap region, but because there is no discussion of the possibility of the overlapping peaks, their presence cannot be excluded. Whereas with others, like those from Ramasamy et al. [10], Suriakarthick et al. [36], Su et al. [37], Wang et al. [80], and Jiasong et al. [78], the obvious presence of shoulder peaks in the overlap region suggests that oxides are present and have been overlooked, especially where the main antimony peaks have been fitted, but there is still unaccounted-for intensity in this region, as is the case with studies by van Embden and Tachibana [82], Han et al. [84], and Ramos Aquino et al. [31]. Unfortunately, a retrospective analysis is sometimes not possible, for example in a study by An et al. [38], where only the Cu $2p$ region was shown, or another by Lou et al. [85], where the Sb $3d$ region is cropped in the middle of a peak.

It is noted that other effective methods of bypassing the analysis of overlapping peaks are possible, such as measuring peaks other than those of strongest intensity[13]; however, this is not recommended here as it does not address any oxygen present in the sample. It is believed that surface oxidation has previously, at least in part, been responsible for poor performing devices [52], the effects of which will be discussed in Sect. 3.4.1. It is therefore imperative that such contamination is identified in order that it can be removed, reduced, or at the least acknowledged. To this extent, XPS can be used for such a purpose, but the data must be of sufficiently high quality and thoroughly analysed.

Core-Level Analysis of the TT Sample: Before and After Surface Cleaning
The survey spectra for the sample, both before and after surface cleaning, are shown in Fig. 3.3, and demonstrate peaks for the expected elements of copper, antimony and sulphur from the material. Peaks from carbon and oxygen are also present; not unexpected because of hydrocarbon contamination from exposure to atmosphere, which then reduced after cleaning. Sodium was also found, which did not decrease after cleaning, suggesting that it has diffused through from the glass, as has been seen in similar materials [86]. Further discussion of the nature and effects of sodium with the corresponding spectra is given in Sect. 3.4.1.

The Cu $2p$ spectra were fitted with one doublet, which was attributed to Cu^+ in CAS. Because charge correction was achieved using the Cu $2p_{3/2}$ peak, the spectra are similar, and a comparison before and after cleaning is shown in Fig. 3.4. It can be seen that no discernible change occurred during the cleaning process, showing no oxidation of the copper in the material after growth.

[13] Such as Sb $4d$ instead of Sb $3d$ [25].

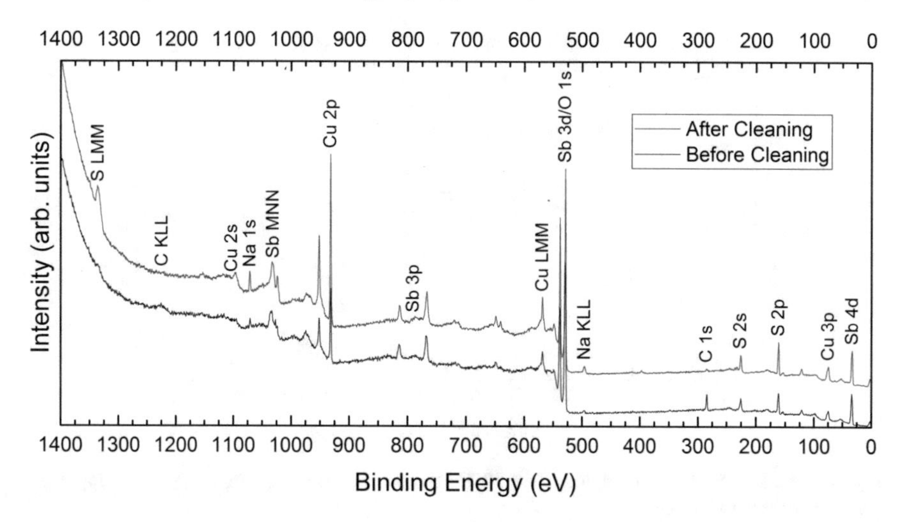

Fig. 3.3 XPS survey spectra for the TT CAS sample before and after surface cleaning

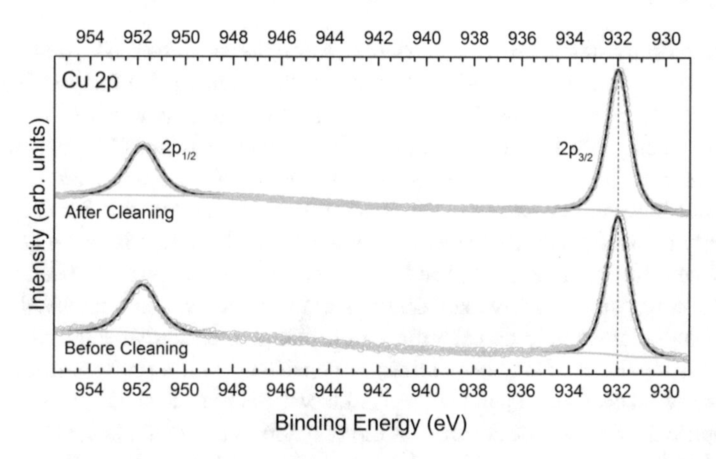

Fig. 3.4 XPS Cu 2*p* region of the TT CAS sample before and after surface cleaning. Fitted peaks shown in red and peak envelope shown in black

The overlapping regions of Sb 3*d* and O 1*s* were fitted according to the procedure detailed above, and the spectra before and after cleaning are shown in Fig. 3.5. It can be seen that prior to cleaning, two distinct peaks are visible in the Sb $3d_{3/2}$ region, with a shoulder to the lower BE side of the lower peak.[14] These features were fitted with three Sb 3*d* doublets, and the remaining intensity in the Sb $3d_{5/2}$ region, relating to O 1*s* emission, was fitted with two peaks. The strongest, sharpest antimony doublet (red dash) was attributed to Sb^{3+} in CAS, sharpest because of the

[14]~537 eV.

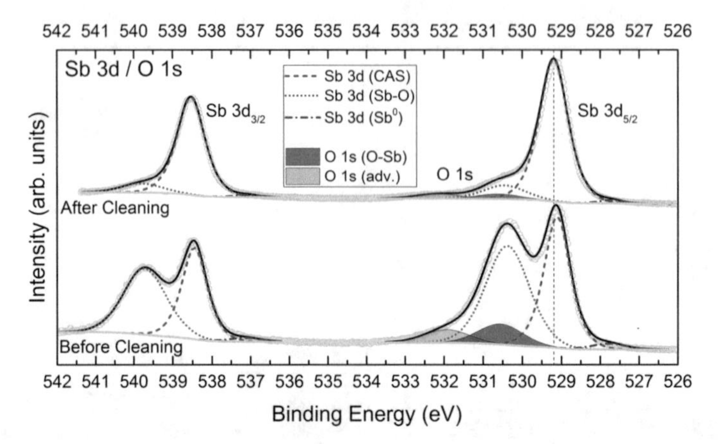

Fig. 3.5 XPS Sb 3*d* and O 1*s* region of the TT CAS sample before and after surface cleaning. Peak envelope shown in black

high crystallinity of the grown material. The doublet at highest BE (pink dot) was attributed to antimony in a Sb–O environment. It is thought that this oxide consists predominantly of Sb_2O_3, as the trivalency of antimony is maintained and the BE is in agreement with literature values [87]; however, it is possible that the oxide is a mixture of Sb_2O_3 and Sb_2O_5 because the literature is unclear on the spectral differences between the two oxides [83, 88–90]. Nevertheless, this Sb–O species is drastically reduced on surface cleaning, showing that the oxide is present mainly at the surface of the sample and formed during atmospheric exposure. After cleaning, a small amount of oxide remained on the surface, which was not removed in order to avoid inducing sample damage through prolonged, excessive sputtering. The O 1*s* peak at higher BE (orange shading) is consistent with an adventitious species associated with carbon [91] and the peak at lower BE (purple shading), almost fully overlapping the Sb $3d_{5/2}$ peak from antimony oxide, is attributed to the oxygen from within this antimony oxide [83], which reduced upon cleaning as well.

The third, small antimony doublet (blue dot dash), present at lower BE is attributed to metallic antimony [88]. It is believed that the metallic antimony is not created by the preferential sputtering of sulphur, nor by the dissociation of sulphur during the surface cleaning for several reasons. First, the amount of metallic antimony relative to the antimony from CAS decreases after surface cleaning, whereas if it were created by the surface cleaning, one would expect it to increase. Second, surface cleaning-induced damage to the sample was minimised by using low sputtering energies and anneal temperatures,[15] because in a study by Ornelas-Acosta et al. [32, 33], where high energy sputtering was used,[16] the amount of sputter-induced antimony was markedly higher.

[15] As detailed in Appendix C, Sect. C.5.1.

[16] 3 keV.

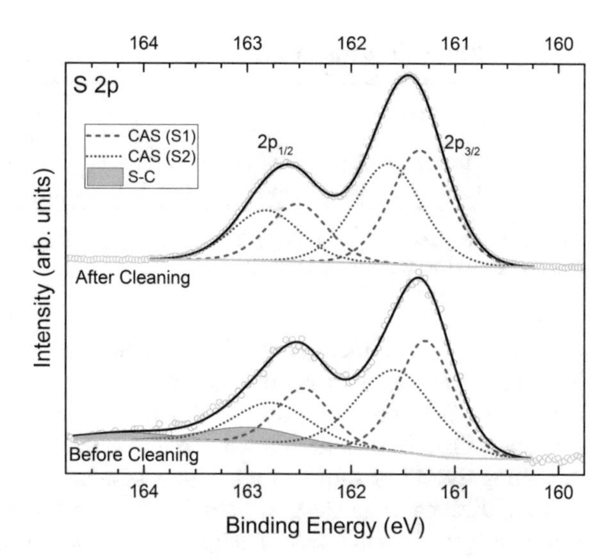

Fig. 3.6 XPS S 2*p* region of the TT CAS sample before and after surface cleaning. Peak envelope shown in black

It was also a concern during growth, that after the thermal treatment, excess Sb_2S_3 may remain on the surface of the sample. However, due to the expected BE of Sb $3d_{5/2}$ from Sb_2S_3,[17] and the lack of difference in spectral features of the Sb 3*d* region between the no-TT and TT samples,[18] it was concluded that no remnant Sb_2S_3 was present.

A comparison of the S 2*p* spectra before and after surface cleaning is shown in Fig. 3.6. Before cleaning, the spectra were fitted with three doublets, and after cleaning, were fitted with two doublets. The species which was eliminated after cleaning (green shading) is attributed to sulphur-containing surface contamination [93], as seen in other sulphide materials [72]. The two strongest doublets (red dash and blue dot), separated by 0.30 eV are assigned to S^{2-} in CAS: one for each of the coordination sites, as described in Sect. 3.1.3, and shown in Fig. 3.1. In the literature, little to no attempts at fitting the S 2*p* region have been made, with only two studies by Ornelas-Acosta et al. [32, 33] fitting this region with more than one doublet, but the SNR is poor and there the authors state that the extra doublets are due to unreacted precursor. However, it is believed that the extra sulphur peaks in that study were probably due to sputter damage, affirmed by the sputter profile shown.

The viability of the presence of two doublets in Fig. 3.6 is initially confirmed by the poor quality of fit when using one doublet, and is strengthened by: the area ratio between the doublets being 1:1, as expected from the crystal structure; that previously measured samples by the author fit this way equally well; and because of the large difference in coordination environments for the sulphur. These reasons are explored further in Appendix D, Sect. D.1.1. Despite this, the specific assignment of the S 2*p* doublets to their corresponding coordination environment is unclear, and

[17] 529.7 eV [88, 92].

[18] Which will be presented in Sect. 3.4.2, specifically Fig. 3.18.

Table 3.2 XPS BE for the main peaks of the TT CAS sample before and after surface cleaning

Sample	$CuSbS_2$			
	Cu $2p_{3/2}$	Sb $3d_{5/2}$	S $2p_{3/2}$ (S1)	S $2p_{3/2}$ (S2)
Before Cleaning	931.97 (1.13)	529.11 (0.80)	161.28 (0.62)	161.58 (0.88)
After Cleaning	931.97 (1.13)	529.19 (0.86)	161.33 (0.66)	161.64 (0.76)

All values are given in eV and the peak FWHM are given in parentheses. BE of all fitted peaks in this chapter can be found in Tables 3.3 and 3.4

the labelling of the doublets in Fig. 3.6 is arbitrary. Although the S1 environment is coordinated to an extra cation, two of the bonds with antimony are longer[19] than the others[20] [48], and as such are believed to be governed by van der Waals forces [6]. Given this, the distorted structure of CAS, and the complications involved with predicting the BE of sulphur for complex systems [94], then the amount of charge transfer, differences in electronegativities of the cations, the effect of bond lengths and the effect of second-nearest neighbours all contribute to the binding energy, and as such, no definitive assignment is given here. Prior to cleaning, the sample showed no presence of sulphate species, as has been seen in previous reports of this material [37, 52].

A summary of the fitted BE and FWHM of the core-levels described above is given in Table 3.2. The shifts in BE before and after surface cleaning are rather small and therefore it can be concluded that the surface cleaning employed here did not significantly affect the underlying material. Further discussion of the reported BE, including those from contamination, can be found in Sect. 3.4, along with all fitted BE in Tables 3.3 and 3.4. Also presented in Sect. 3.4.2 are the XPS-derived surface stoichiometry values, and a discussion and comparison of the effect of the thermal treatment stage of the growth with respect to the electronic structure and sample quality.

Comparison to the Literature

The BE for the Cu $2p_{3/2}$, Sb $3d_{5/2}$ and S $2p_{3/2}$ peaks found here for the TT sample are shown in Fig. 3.7 along with literature values of other XPS-measured samples of CAS. Generally, the binding energies reported in the literature for this material are in poor agreement with the values measured here. Exceptions to this are studies by Wan et al. [27], Peccerillo et al.[21] [52], and Ramos Aquino et al. [31], where the BE of all the elements of CAS are in decent agreement and all of these studies observed and acknowledged the presence of antimony oxide. Yang et al. [25] reported Cu $2p_{3/2}$ and S $2p_{3/2}$ BE that are in agreement with those measured here, but they chose to measure the Sb 4*d* region rather than the Sb 3*d* region, presumably to avoid the overlap problem with O 1*s*. Whilst this will provide the same information regarding

[19] 3.12 Å.

[20] 2.30–2.57 Å.

[21] The XPS measurements in this study were performed by the author.

Table 3.3 XPS BE for the CAS-relevant peaks of the CAS samples before and after surface cleaning

Sample		$CuSbS_2$			
		Cu $2p_{3/2}$	Sb $3d_{5/2}$	S $2p_{3/2}$ (S1)	S $2p_{3/2}$ (S2)
No-TT	Before Cleaning	931.97 (1.18)	529.11 (0.81)	161.31 (0.67)	161.61 (0.81)
	After Cleaning	931.97 (1.11)	529.11 (1.03)	161.17 (0.72)	161.47 (0.78)
TT	Before Cleaning	931.97 (1.13)	529.11 (0.80)	161.28 (0.62)	161.58 (0.88)
	After Cleaning	931.97 (1.13)	529.19 (0.86)	161.33 (0.66)	161.64 (0.76)

All values are given in eV and the peak FWHM are given in parentheses

the state of the antimony, it neglects the O 1*s* region, and cannot account for the presence of oxygen.

The BE for the cations reported by Ramasamy et al. [10] are all lower than those measured here, suggesting the misapplication of a charge correction; however, the agreement of the reported S $2p_{3/2}$ BE would undermine this theory. Nevertheless, studying the presented S 2*p* spectrum of that study suggests that the S $2p_{3/2}$ peak actually has a much lower BE than that quoted in the discussion, and therefore means that the BE of all the elements are indeed lower, and that relatively, they are in agreement with those given here. This misapplied charge correction is further evidenced by the confusion caused by the low BE of the Sb $3d_{5/2}$ peak, which caused the authors to conclude that Sb^{2+} is present: a very uncommon oxidation state [95].

Other XPS studies of CAS are limited by either the quality, or lack of data. Su et al. [37] presented XPS data for the regions of CAS, yet the analysis consisted of one sentence, stating that the material was pure and of correct composition. The

Table 3.4 XPS BE for the contamination-relevant peaks of the CAS samples before and after surface cleaning

Sample		Sb_2O_3		Contamination				
		Sb $3d_{5/2}$	O 1*s*	Sb $3d_{5/2}$ (Sb^0)	O 1*s* (adv.)	S $2p_{3/2}$ (S-C)	Na 1*s* (1)	Na 1*s* (2)
No-TT	Before Cleaning	530.61 (1.39)	530.61 (1.44)	—	531.82 (1.44)	162.72 (1.30)	1071.62 (1.53)	—
	After Cleaning	530.61 (1.35)	530.46 (0.98)	527.55 (0.85)	532.06 (0.98)	—	1071.30 (1.24)	1072.27 (1.24)
TT	Before Cleaning	530.38 (1.31)	530.55 (1.35)	527.78 (1.03)	531.93 (1.35)	162.97 (1.10)	1071.48 (1.58)	—
	After Cleaning	530.46 (1.24)	530.60 (1.33)	527.74 (0.68)	532.15 (1.33)	—	1071.54 (1.40)	1072.44 (1.40)

All values are given in eV and the peak FWHM are given in parentheses

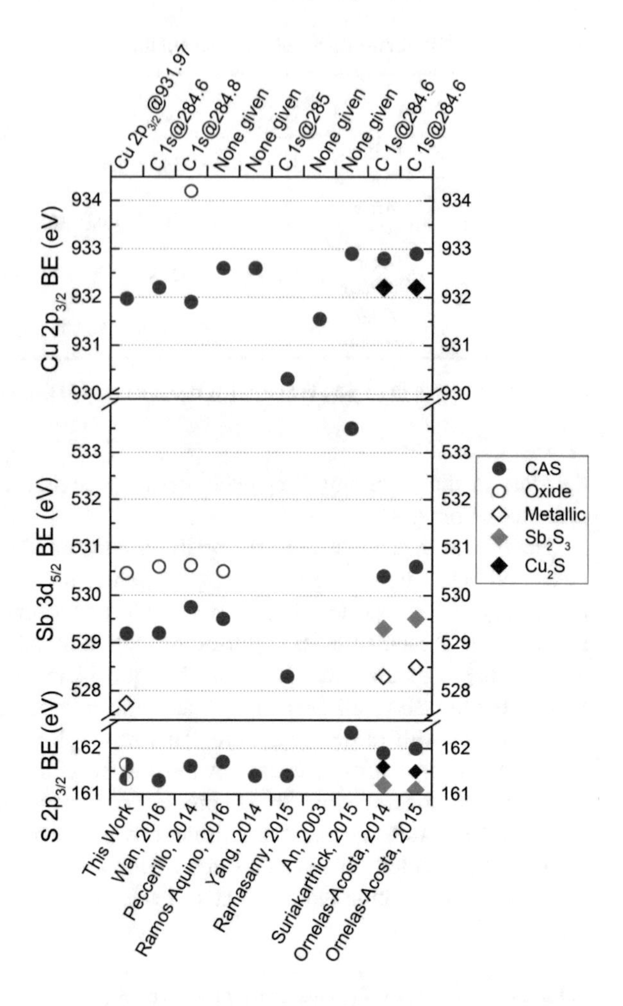

Fig. 3.7 Comparison between the XPS BE measured in this study for the clean TT CAS sample, and those reported previously in the literature for CAS. Literature references are given on the bottom x-axis [10, 25, 27, 31–33, 36, 38, 52], and the quoted charge referencing method employed in the corresponding study are given on the top x-axis. Species assignments are as given in the corresponding reference, see text for discussion of the agreements and more details

lack of explicitly reported BE weakens this conclusion, especially given that the SNR of the data is poor and there is evidence of several secondary peaks. An et al. [38] used XPS to characterise CAS, but only presented the Cu 2*p* region with very poor SNR, and they state that the BE is in agreement with literature. These studies are especially limiting as they are some of the first XPS measurements of CAS. Furthermore, the stated conclusions in these studies, that the material is pure and as expected, is misleading, given there are no previous studies of this material.

Those studies presented in Fig. 3.7, which are not in agreement with the BE values measured here are believed to be due to either a lack of fitting procedure, else, where fitting was performed, an incorrect assignment of the peaks. Suriakarthick et al. [36] studied samples of CAS grown at different temperatures, and the presented XPS spectra show drastic changes; however, with no applied fitting or discussion of these differences, it is difficult to determine their origin. Nevertheless, there is strong evidence suggesting the presence of oxide components, due to the wide features in the Sb 3*d* spectra and the presence of satellites in the Cu 2*p* spectra, which could be the cause of the higher BE reported there. Ornelas-Acosta et al. presented two studies of CAS which utilise XPS [32, 33], and they found similar results in both cases. They fitted the Cu 2*p* region with two doublets, assigning the one with higher BE to CAS, and the lower to Cu_2S, claiming that it formed during the growth. Three doublets were fitted to the Sb 3*d* region, the one with highest BE assigned to CAS, the intermediary one to Sb_2S_3; a precursor, and the lowest BE doublet to metallic antimony, formed due to the cleaning method employed. The S 2*p* region was fitted with three doublets, corresponding to the three sulphides mentioned above. Comparing these assignments with those reported by others[22] would suggest a mistake, with the likelihood being that the middle antimony doublet originated from CAS, and the higher BE antimony doublet originated from an oxide layer, which would be in very good agreement with the other studies. Another problem with these studies is the amount of metallic antimony present in the samples, resultant probably because of the high energy sputtering which was used in the cleaning. This endorses the caution paid in the current study: that more aggressive sputtering would cause damage to the sample, leading to an unrepresentative measurement of the material.

After accounting for all these differences, there is general agreement between the few reported XPS measurements of CAS.

3.3.3 *Natural Band Alignments*

When considering solar cell materials, another often overlooked point to consider is the position of the band edges and how this will affect the architecture of the cell, as was discussed in Sect. 1.2.2. An oft quoted property of absorber materials is the bandgap, and whilst a relevant bandgap value is necessary for decent efficiencies, this can be compromised if the band alignments hinder charge carrier extraction.

The bandgaps of the CAS sample were determined from the absorption spectra, shown in the Tauc plots in Fig. 3.8, using the method described in Sect. 2.5.2. Room temperature[23] measurements allowed the extraction of both the direct (1.60 eV) and indirect (1.55 eV) bandgaps. That the fundamental gap of this material is indirect is not necessarily a problem, as the direct gap still has a value suitable for PV use, and the measurements here are in agreement with previous studies, which found similar values [15, 17, 21, 22, 24, 25, 27, 29, 31–34, 41, 96].

The direct and indirect gaps at cryogenic temperature[24] were also determined for comparison with the DFT calculations and were found to be 1.69 and 1.59 eV, respectively. As the direct and indirect bandgaps from DFT[25] were found to be 1.82

[22]See Fig. 3.7.

[23]300 K.

[24]4 K.

[25]0 K.

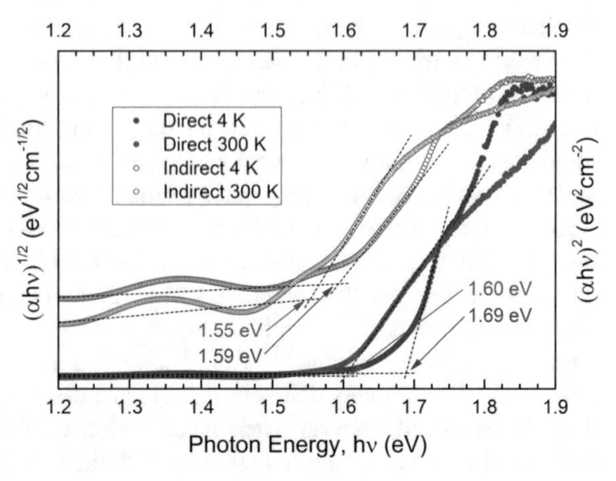

Fig. 3.8 Optical absorption data from the TT CAS sample. Tauc plot fittings for the direct and indirect bandgaps of the TT CAS sample at 4 and 300 K

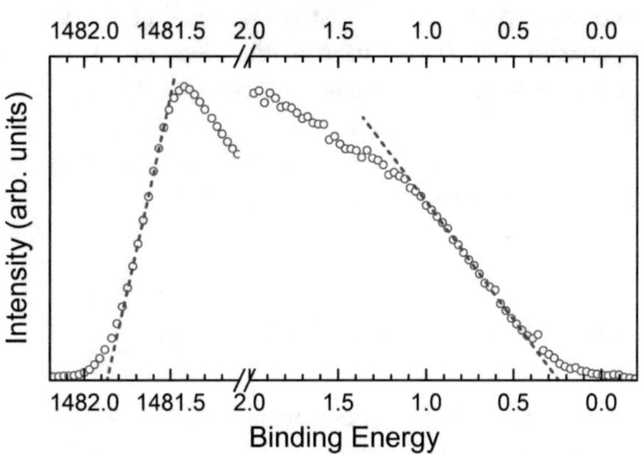

Fig. 3.9 VBM and SEC fittings for the clean TT CAS sample from XPS data

and 1.67 eV, respectively, this is in satisfactory agreement with the experimental results at 4 K, with the slight discrepancy occurring as a result of error in the fitting method or calculations. These calculated values are in agreement with previous calculations as well [5, 26, 47].

As was discussed in Sect. 1.6.2, it is important to know the IP and EA of PV absorber materials. The IP and WF values were extracted for the TT sample after cleaning, and were found to be 4.98 and 4.73 eV, respectively, from the fittings of the VBM and SEC shown in Fig. 3.9. Then, by using a vacuum alignment procedure [97] and taking the measured indirect 300 K bandgap value of 1.55 eV from Fig. 3.8, this is shown for comparison with CZTS, CIGS and CdS in Fig. 3.10. Consequently, the EA was calculated to be 3.43 eV by taking the difference between the measured IP and bandgap.

Although XPS can be used to great effect in determining band positions [100] and also performing band alignments at junction interfaces [101], it is not widely implemented, especially for new materials. Hence, these measurements are absent from the majority of the literature on CAS. However, Yang et al. [25] measured a

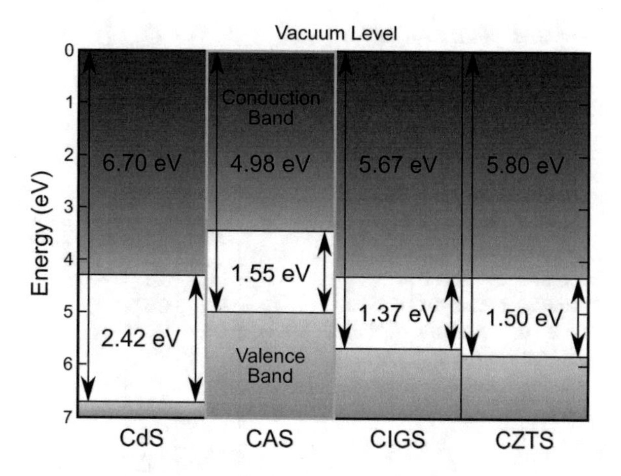

Fig. 3.10 Vacuum-aligned band diagram for CAS from IP measurements using XPS and bandgap from optical absorption measurements. Comparison with other common absorbers and the common n-type window layer material, CdS. Literature values of IP and bandgap are taken for CdS [98], CIGS [99] and CZTS [98]

WF of 4.86 eV using UPS, in good agreement with the value presented here, and an IP of 5.25 eV, different to the value measured here because the VBM fitting in that study is unclear. Measurements of a CAS sample, where there was heavy antimony oxide contamination present, resulted in a very high IP [52], thought to be due to the oxide. Further discussion of the effects of the oxide is given in Sect. 3.4.1.

It has been previously suggested that the poor V_{OC} observed in CAS cells arises partly from a poor CB alignment with the n-type layer [30], and this is confirmed here in Fig. 3.10, with the CBM of CAS lying 0.85 eV higher than that of CdS, which is in agreement with a theoretical study [30]. With such a high CBO, this would suggest that a recombination centre would be present at this interface [102]. In fact, in order to stop the formation of an electron barrier at the interface, the CBM of the n-type layer should be 0–0.4 eV higher than that of the absorber [103], as is the case with both CIGS/CdS and CZTS/CdS as seen in Fig. 3.10.

All CAS device structures so far use CdS as the window layer.[26] It is therefore obvious that utilising traditional CIGS cell architectures for CAS is unfavourable for device performance. The use of CdS as the n-type layer not only creates the cliff-like CBO [30], but also undermines the goal of developing environmentally friendly structures. It may also be the case that along with replacing the n-type layer, the back contact may also need replacing, as the WF of typically used molybdenum may not be favourable for charge extraction. It is therefore apparent that further research is needed to identify more ideal partner materials for CAS in a cell [104], especially with regards to measuring band alignments in real systems.

[26]See Table 3.1.

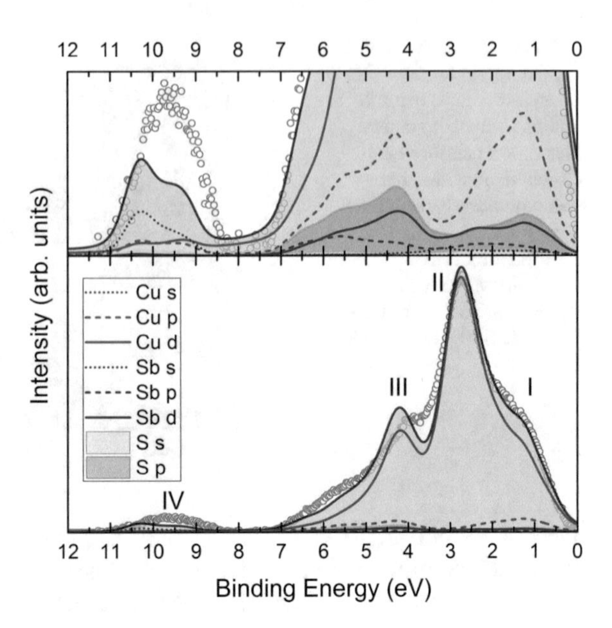

Fig. 3.11 Simulated and measured VB spectra for CAS with respect to the Fermi-level at 0 eV. Background subtracted XPS data is compared with broadened and corrected partial DoS curves. Top spectra shows intensity zoomed region to better show the underlying curves. Green data are from XPS and the black curve with grey shading is the total summed DoS

3.3.4 Density of States Analysis

The low IP observed here could be explained through study of the bonding mechanism of this material, and as such, DFT calculations resulted in pDoS curves for the VB and CB. The pDoS in the VB, and VB spectra from XPS,[27] are shown in Fig. 3.11. The corrected[28] pDoS curves were summed, and the total curve aligned to the XPS data for direct comparison. There is good agreement between the calculated DoS and the XPS spectra, with all features accounted for and the relative intensities of the correct order. Especially, the Cu 3*d* states of the main feature (II) and the leading edge at the top of the valence band (I) are reproduced well. The only differences arise at features III and IV, and are probably caused by final-state relaxation effects in XPS, which are known to shift features that are deeper in the VB to lower BE [105].

The large photoionisation cross section of the Cu 3*d* orbital means that this dominates the corrected DoS curve and it is therefore prudent, given the excellent experimental corroboration shown in Fig. 3.11, to discuss the pDoS contributions with respect to the uncorrected curves. These are shown in Fig. 3.12a for both the VB and CB. Without the corrections, the VBDoS is still dominated by Cu 3*d* states, and the valence band shows three distinct features (V, VI, and V*). Similar to both CIGS and CZTS, the copper in CAS is tetrahedrally coordinated to sulphur in the crystal structure and therefore the crystal field splits the Cu 3*d* orbitals into a non-bonding e doublet, and a t_2 triplet, which is able to bond [106]. These features are present

[27] For the TT sample after surface cleaning, and after having subtracted a Shirley background.

[28] Using the parameters described in Appendix C, Sect. C.3.

in Fig. 3.12 as the non-bonding e states (VI), and the bonding (V) and antibonding (V*) hybridisation of the t_2 states with S 3*p* states. The bottom of the highest VB is comprised of bonding orbitals of S 3*p* states with a slight contribution from Sb 5*p* states (VII), and the antibonding states of this interaction form the main contribution to the CB (VII*).

Further insights into the bonding nature are gained when studying the states normally overwhelmed by intensity from Cu *d*/S *p* hybridisation and to this end, the zoomed region of the valence and conduction band edges in Fig. 3.12b is considered. At first glance, it would seem that the Sb 5*s* states are localised in feature VIII, which agrees with the classical model of lone-pair electrons: that they are too tightly bound and therefore non-bonding. However, there are also Sb *s* states at the top of the valence band (X) and the bottom of the CB (X*), which suggests that the contribution here is due to interactions via the revised lone-pair model [46]. Here, the lone-pair electrons first interact with S 3*p* electrons and the full antibonding orbital of this interaction is then sufficiently high, that it can interact with empty Sb 5*p* states, resulting in bonding states in the valence band (X) and antibonding states at the bottom of the CB (X*). These findings, in agreement with previous analyses of the VBDoS of CAS [47, 49], are in direct opposition to a more recent study which claimed that the Sb 5*s* lone-pair is inert, localised and contributes nothing to bonding in the VB [45]. It is believed that this interpretation is incorrect for two main reasons: first, because of the excellent agreement between the XPS and DoS, the underlying bonding nature is supported; and second, due to the distorted nature of the SbS_5 pyramids in the crystal structure, the revised lone-pair model suggests that these interactions should indeed take place, whereas in undistorted structures, this interaction is symmetry forbidden [47].

The bonding mechanisms described above and the other main contributions are shown in Fig. 3.12c to more clearly demonstrate the hybridisations from the configuration energies (CE), most importantly, the lone-pair effects, and the nature of the Cu 3*d* states in the VB.

The described features are in agreement with previous analyses of CAS DoS calculations [26, 45, 47, 49]; however, no solid link has been previously made to the effect of these features regarding the electronic properties. As discussed in Sect. 3.1.4, much initial progress was made likening CAS to CIGS and also to CZTS, and the comparisons are still drawn in analyses of the DoS [47], especially regarding the Cu 3*d*/S 3*p* states in the VB [26], which are indeed very similar to the tetrahedral environment also present in CIGS. However, the full merit of this practice is questioned when it is clear that the crystal structure of CAS is so different from CIGS, then so should be the DoS, as the bonding is what leads to the adopted crystal structure. Clearly, from the CBDoS, Sb 5*p* states are the main cation contributor in CAS, whereas in CIGS, the bottom of the CB is mainly In/Ga *s* states [106–111]. This has been recognised previously and found to be the reason why CAS shows superior absorption to CIGS [5], but no further discussions have been made [45]. Here, it is posited that this difference is also the cause of the low IP found previously and discussed above. In CIGS, the In/Ga are in the 3+ oxidation state, as is Sb in CAS; however, being group III elements, In/Ga have empty valence *s* orbitals

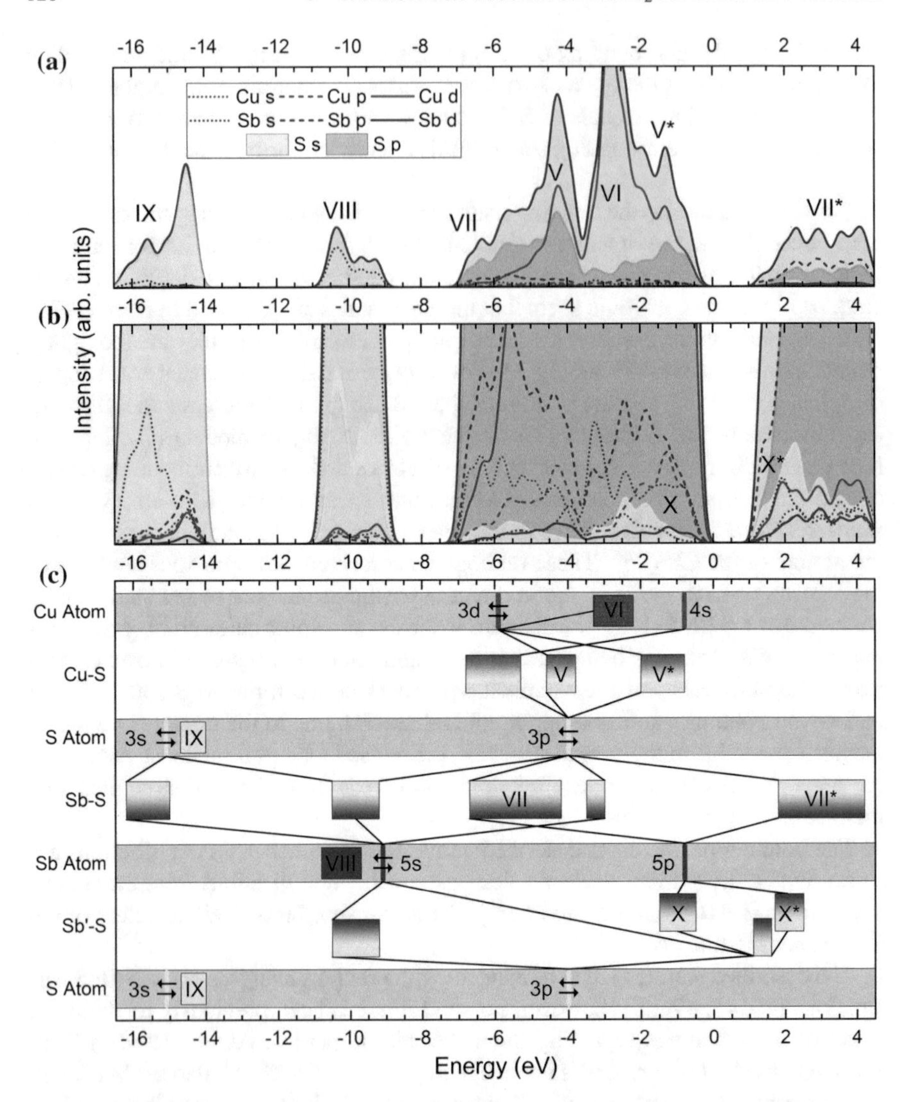

Fig. 3.12 **a** Total and partial electronic DoS curves for CAS, with **b** intensity zoomed region to more clearly show the underlying orbitals. Curves have been convolved with a Gaussian function (0.3 eV FWHM) in order to better distinguish features. DoS curves are aligned to the VBM. Black curve with grey shading is the total summed DoS. **c** CE for the valence orbitals [107, 108] displayed with a schematic of the bonding hybridisations as discussed in the text. Antimony-sulphur bonding is separated into regular bonding (Sb–S) and lone-pair bonding (Sb′–S). It is noted that the CE values do not take ionisation, multi-electron occupancy or hybridisation into account and are shown only as a schematic guide. Part labels are discussed and referred to throughout the text

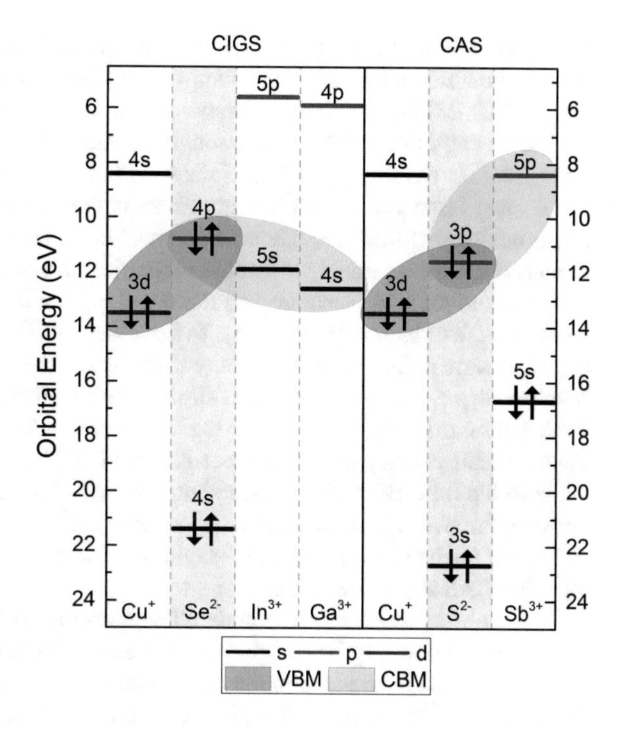

Fig. 3.13 CE [107, 108] for the valence orbitals of CIGS and CAS. The formal ionic occupancies of the orbitals within the materials are marked and the main orbital contributions to the VBM and CBM are highlighted

for the 3+oxidation state,[29] whereas in group V Sb, it is full.[30] This means that in CIGS, empty cation *s* orbitals bond with full anion *p* orbitals to form regular, full bonding states at the bottom of the VB and empty, antibonding states at the bottom of the CB, whereas in CAS the full Sb 5*s* orbital is available for bonding, and due to the mechanism described above, these states are found throughout the VB and at the bottom of the CB. It is unsurprising that the first CB of CAS is dominated by antibonding Sb 5*p*/S 3*p* states as Sb 5*p* are the first empty orbitals that are within proximity of the full S 3*p* orbitals. This is shown, in contrast to the empty states of CIGS, in the schematic CE diagram in Fig. 3.13, with the main contributions to the VBM and CBM marked. From this, it can also be seen that the empty cation *s* states in CIGS are closer to the anion *p* orbital than the also empty cation *p* orbitals, and therefore the cation *s* nature of the CBM in CIGS is explained as well.

Although it is believed that the Sb lone-pairs play some effect in raising the energy levels of the bands in CAS, the Cu *d* states must also have an effect due to their dominating presence in the VB. In other *d*–orbital dominant semiconductors, the chemical trend is that with increasing separation between the cation *d* and anion *p* levels, the VBM rises, given a common-cation [112, 113]. If one accepts this, then by studying Fig. 3.13, the CAS analogue replacing S with Se, that is CASe, should yield a lower IP even than CAS, and such has been reported previously [49, 114].

[29] $[Kr]4d^{10}5s^{0}5p^{0}$ for In^{3+}. $[Ar]3d^{10}4s^{0}4p^{0}$ for Ga^{3+}.

[30] $[Kr]4d^{10}5s^{2}5p^{0}$ for Sb^{3+}.

This then would suggest that CIGS, with common Cu-cation, and Se as an anion, should show a similar trend when compared to CAS. As this is not the case, it is then evident that it is the presence of antimony over indium/gallium which is responsible for the raising of the VBM, both through the addition of Sb 5*s* states to the valence band, but also the shifting of Cu 3*d* states within the VBM.

Another issue pertinent to PV absorbers is the formation of defects, which can affect device performance. 'Defect tolerance' is a term used when the formation of defects does not negatively affect the properties of a material and occurs when the VBM is antibonding in nature and the CBM is bonding in nature, so that defects would produce shallow levels [115]. To this extent, CAS should benefit from at least some measure of this property because even though the CBM is predominantly antibonding Sb *p*/S *p* states, the VBM is strongly antibonding Cu *d*/S *p* with also a slight contribution from Sb *s* states due to the lone-pair bonding mechanism. Furthermore, the dominant vacancy in CAS has been shown to be V_{Cu} [25] which, as an acceptor, leads to the inherent p-type conductivity and also the crystallographic differences between the two cation-anion structural motifs makes CAS less prone to cation disorder, and the band tailing associated with it [29], both of which further support this idea that CAS should be defect tolerant.

These observations and reasoning all support the postulate that advancement in CAS research now requires a definitive move away from CIGS analogies. The lone-pair of electrons from the antimony cation provides extra states to both the valence and conduction bands of CAS, which are not present in CIGS. This leads to a higher VBM level, resulting in a low IP for CAS, which results in a performance-damaging CBO with CdS; commonly used in CIGS cells. Therefore, CAS should be developed within its own class of earth-abundant absorbers.

3.4 Further Effects of Contamination and the Thermal Treatment of $CuSbS_2$

Throughout Sect. 3.3, the properties of CAS were studied using the thermally treated sample as a standard, representative of the material, after it had undergone surface cleaning. Such measurements are beneficial in determining the fundamental properties of materials, but it is also useful, given the breadth of applications for which XPS can be used, to study the effects that the contamination of the sample could have on the properties, and how this could affect device performance. These results are presented in this section, along with comparative measurements of the no-TT sample, in order to determine the effects of this thermal treatment upon the electronic structure, and also the surface stoichiometry of the sample. All fitted BE from both samples can be found in Table 3.3 for the CAS-relevant peaks, and Table 3.4 for the contamination-relevant peaks.

3.4.1 *The Effects of Contamination*

The presence of contamination on the TT sample before surface cleaning was acknowledged in Sect. 3.3.2, especially the presence of sodium in Fig. 3.3, and the antimony oxide evident in Fig. 3.5. The effects of these contaminants were touched upon there, but now a discussion of the ramifications of these will be given.

Oxidation
In Sect. 3.3.2, only the antimony appeared to have oxidised, and from the comparison with the literature in Fig. 3.7, it would seem that this observation is commonplace with regards to CAS. Although this oxidation reduced with surface cleaning, suggesting it is present at the surface as a result of atmospheric handling, and not within the bulk of the material, it is still important to acknowledge its presence and effects, especially if there is the possibility of its formation midway through device fabrication, as this will lead to the incorporation of this oxide layer into the device stack, possibly contributing to the poor performance of CAS so far.

This oxide is likely to be Sb_2O_3, which has n-type conductivity [116] and a large bandgap,[31] and therefore would act as an insulator within the cell, inhibiting charge transport and causing detriment to the efficiency. Also, it is posited that only with the use of surface sensitive techniques such as XPS, can the true ramifications of this oxide formation be determined, because as the oxide is very thin[32] and the crystallinity probably poor, optical measurements of the bandgap and XRD analysis may not be affected by it. Indeed, in a previous measurement of a CAS sample [52], no crystalline oxide phase was detected by XRD, and optical measurements resulted in a bandgap value of 1.5 eV, but XPS analysis revealed a large level of antimony oxide at the surface and combined UPS and IPES measurements resulted in a bandgap value at the surface of 2.75 eV, thought to be from the oxide. In this respect, surface sensitive XPS can provide indispensable information: identifying unwanted contaminants and also measuring the effect they have on the electronic structure.

A discussion of the differences in the level of oxide contamination for the TT and no-TT samples will be given in Sect. 3.4.2.

Sodium
As well as being detected on the TT sample in Fig. 3.3, sodium was also found on the no-TT sample[33] and in both cases did not dramatically decrease after surface cleaning, suggesting that it is present not only at the surface, like the contaminant oxygen and carbon, but that it is within the bulk of the samples. The Na 1*s* spectra for the samples before and after cleaning are shown in Fig. 3.14. Before cleaning, the spectra for both samples were fitted with one peak, and with two peaks after cleaning. Sodium diffusion from the glass is common in solar absorber materials, where it has been shown to play a beneficial role for CZTS [118] and CIGS [119],

---

[31] ~3.8 eV for the amorphous phase [117].

[32] Evidenced from the analysis depth and the fact that the underlying material is observed in Fig. 3.5.

[33] See later survey spectra presented in Fig. 3.15.

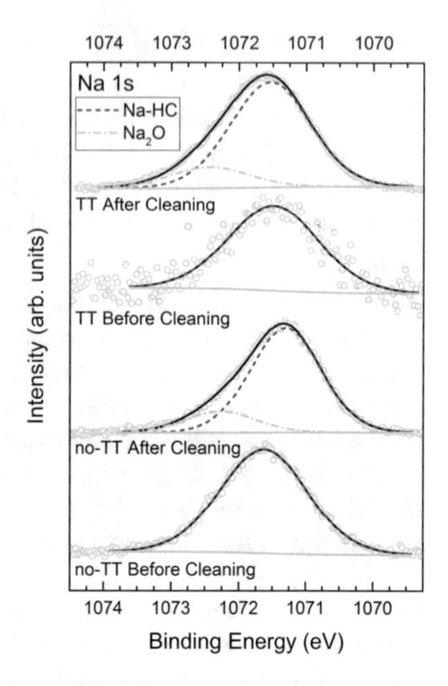

Fig. 3.14 XPS Na 1*s* region of the CAS samples before and after surface cleaning. Peak envelope shown in black

yet be a detriment to CdTe [120]. An in-depth study of the effect of sodium on this material is therefore warranted, especially as the nature of the species observed here is difficult to interpret.

Nevertheless, it is believed that the sodium diffuses from the glass substrate during growth to the surface of the CAS thin-film as Na_2O,[34] the literature value of which [121] is in agreement with the higher BE peak from the clean samples (blue dash dot) in Fig. 3.14. Then, when the sample is exposed to atmosphere, the sodium reacts with atmospheric hydrocarbons, producing the peak at lower BE (red dash), which probably consists of a combination of Na_2CO_3, $NaHCO_3$, and NaOH, and is in agreement with literature values [122, 123].

It is believed that the two species are present even before surface cleaning; however, because the shift between the peaks is small and the signal from the underlying Na_2O is obscured by the contaminants, the peak from the contaminants takes precedent in the samples before cleaning, and only after cleaning is some of the oxide revealed. The O 1*s* peaks associated with these compounds fall within the overlapping region and therefore make it difficult to draw conclusions as to the exact nature of the sodium compounds present.

This sodium presence at the surface of the CAS film may be part of the reason for the surface oxidation of the antimony observed in the XPS spectra. The catalytic effect of alkali-metals on the surface oxidation of absorber materials is well known [124–127], and with CAS, the probable reaction site is the antimony, because it is observed that surface oxidation takes place here,[35] and also it is thermodynamically

[34]Which is the form of sodium within SLG.

[35]See Fig. 3.5.

more favourable for the antimony to oxidise when compared with the copper [128, 129].

3.4.2 *The Effects of the Thermal Treatment*

The thermal treatment in an Sb_2S_3 atmosphere was found to produce CAS films of a superior quality than without it [29], in terms of structural and optoelectronic properties, resulting in better reproducibility. Here, the electronic structures of CAS samples with and without the thermal treatment are compared, measured by XPS, in order to determine the effect of it.

XPS Core-Level Comparison

The survey spectra for the samples are shown in Fig. 3.15, and between the samples appear the same; however, for the no-TT sample, both before and after cleaning, molybdenum was detected in the spectra, shown in Fig. 3.15. As the substrate consists of molybdenum, it is believed that the molybdenum signal arises from the underlying substrate itself, because, when the samples were observed by optical microscopy,[36] as shown in Fig. 3.16, the no-TT sample (a and c) appeared to contain voids in places, which are believed to be holes in the CAS film, allowing unimpeded ejection of electrons from the substrate. That the TT sample did not have these features (b and d), and that molybdenum is not visible in the spectra of this sample, suggests that the thermal treatment has produced better quality films with greater uniformity.

The Cu 2*p* and S 2*p* spectra for the samples are shown in Fig. 3.17. Because the spectra were calibrated using the Cu $2p_{3/2}$ peak, these spectra appear the same, displaying a single Cu 2*p* doublet common to the TT sample. The same is true for the S 2*p* region with three doublets before cleaning and two after surface cleaning, with the assignments the same as in Sect. 3.3.2.

The overlapping Sb 3*d* and O 1*s* spectral regions are shown in Fig. 3.18. The no-TT sample shows almost the same spectral features as the TT sample,[37] except before cleaning, the no-TT sample showed no evidence of metallic antimony. It is possible that this is because the sputter cleaning is preferentially removing sulphur [32, 130, 131], leaving behind metallic antimony, else the anneal is causing the sulphur to dissociate from the antimony. This latter point is more likely, because metallic antimony was also observed on the TT sample before surface cleaning. As this sample had already undergone a thermal treatment, it is likely that this is where this species originated from. It is noted however, from the surface stoichiometry values that will be presented in Table 3.6, that the level of metallic antimony observed here is very slight, at around 2% the amount of antimony from CAS in the clean samples. It is further noted that metallic peaks should have asymmetric line shapes in XPS spectra [132], but because of the low intensity of the peaks from metallic antimony, they are fit satisfactorily with symmetric peaks.

---

[36]Using a Nikon Eclipse LV100 microscope.

[37]Assigned in Sect. 3.3.2.

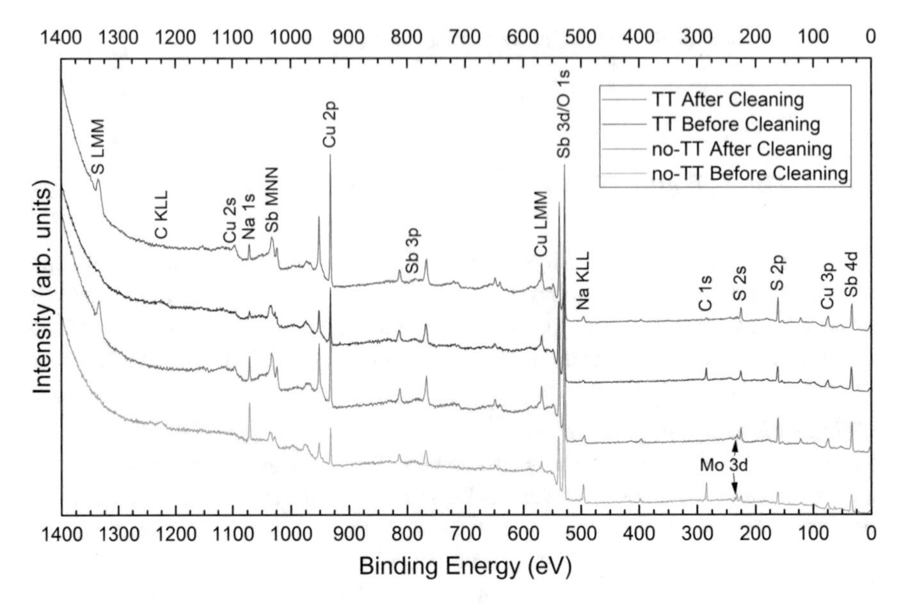

Fig. 3.15 XPS survey spectra for the CAS samples before and after surface cleaning

Both with and without the thermal treatment, the samples before cleaning showed significant presence of antimony oxide (pink dot). However, both from the spectra and also the stoichiometry values that will be presented in Table 3.6, it is clear that the amount of oxide present on the no-TT sample is greater than on the TT sample. Before cleaning, the amount of antimony from antimony oxide compared with the amount from CAS was 172% for the no-TT sample, compared with 118% for the TT sample. This suggests that the thermal treatment has somewhat suppressed the oxide formation if it was to occur after the thermal treatment, or that it has removed some of the oxide if the oxidation occurred before the thermal treatment.

One possible explanation for the differences in oxidation observed here relates to the presence of sodium as was discussed in Sect. 3.4.1. If it can be taken that the sodium acts catalytically at the surface and is part of the reason as to why the antimony has oxidised [124], and given that there is a greater amount of sodium present at the surface of the no-TT sample compared with the TT sample prior to cleaning, then this is why the TT sample has a lower level of oxidation: because the thermal treatment appears to have suppressed the sodium diffusion. The mechanism for this is unclear; however, it is possible that the thermal treatment has produced some form of 'passivating layer', which suppresses sodium diffusion to the surface and may even in and of itself, inhibit the oxide formation.

It is possible that this 'passivating layer' would then either be damaged or removed after the surface cleaning that took place in vacuo. If this is the case, then by exposing the sample after surface cleaning once again to atmospheric conditions, one should observe different results than before the sample received the surface cleaning. A comparison of the spectra for the TT sample before the initial cleaning and after

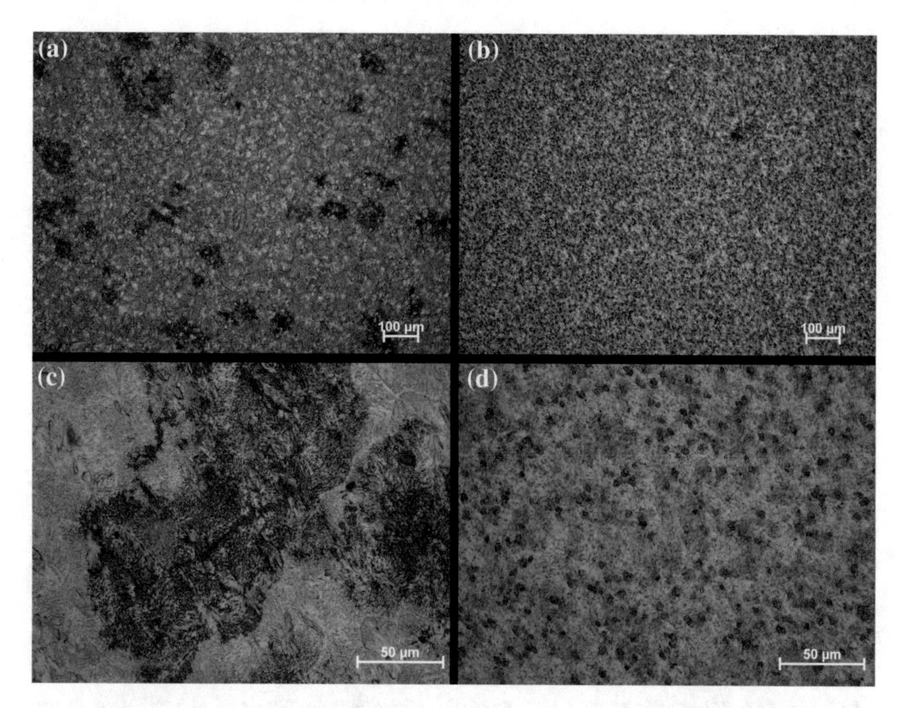

Fig. 3.16 Typical optical micrographs for the CAS samples. **a** and **c** no-TT sample at 10× and 50× zoom respectively. **b** and **d** TT sample at 10× and 50× zoom respectively

being re-exposed to atmosphere, is presented in Fig. 3.19. In the survey of the sample after re-exposure,[38] a very large Na 1*s* peak is clear above the other peaks, which suggests that after the in vacuo cleaning and re-exposure, much more sodium was able to diffuse to the surface, supporting the idea of the formation of a 'passivating layer'. The size of the Sb $3d_{5/2}$ peak originating from antimony oxide is also much larger on the re-exposed spectrum,[39] further supporting this idea. It is also clear that after re-exposing the sample to the atmosphere, the other constituent elements have oxidised as well, evidenced by the extra doublet in the Cu 2*p* spectrum at higher BE,[40] the shake-up lines between 945–940 eV, which are characteristic of CuO [76], and the extra doublet in the S 2*p* spectrum at higher BE,[41] characteristic of sulphate species [81], which could have formed because of the increased amount of sodium, resulting a compound such as Na_2SO_4.

These observations detailed above, and the interpretations thereof, are in agreement with a previous study of the effect of the thermal treatment of these samples [29], in which it was found that the thermally treated samples showed improved

[38]Figure 3.19a.

[39]Figure 3.19c.

[40]Pink dot in Fig. 3.19b.

[41]Purple shading in Fig. 3.19d.

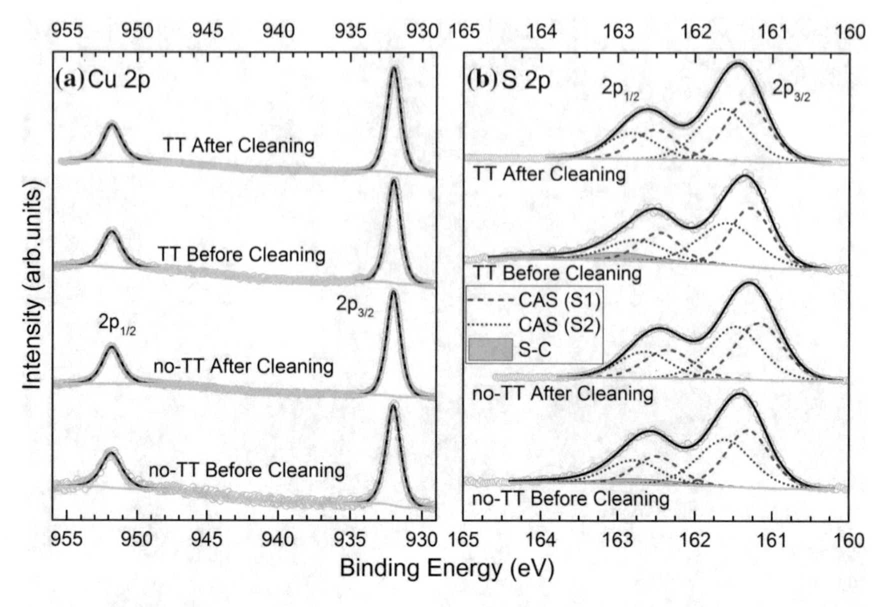

Fig. 3.17 XPS **a** Cu 2*p* and **b** S 2*p* regions of the CAS samples before and after surface cleaning. Peak envelope shown in black

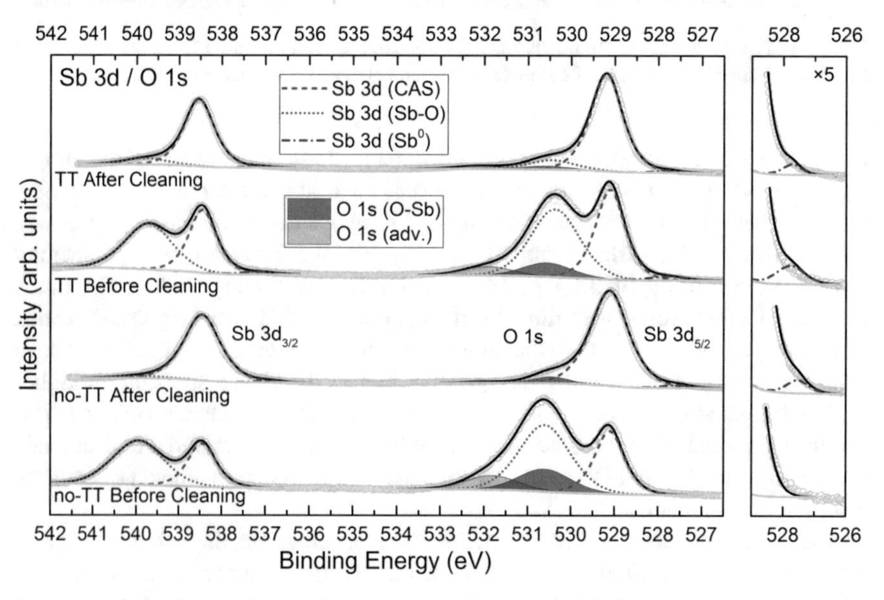

Fig. 3.18 XPS Sb 3*d* and O 1*s* region of the CAS samples before and after surface cleaning. Peak envelope shown in black. Inset to the right is zoomed to better show the metallic antimony species

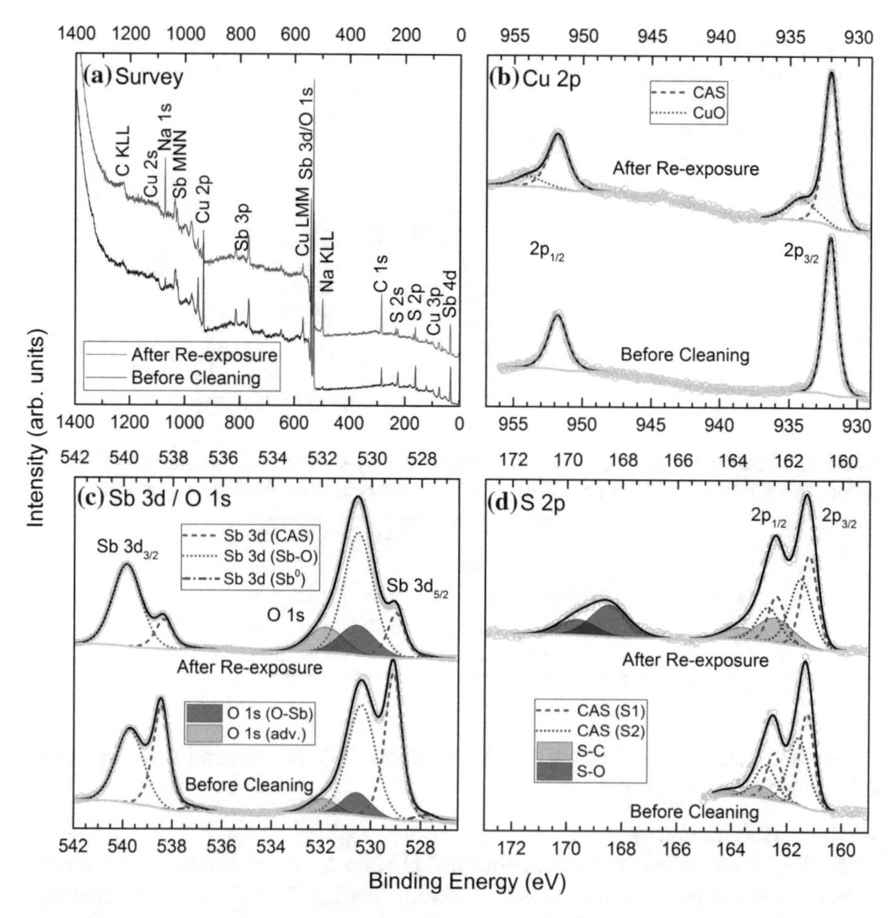

Fig. 3.19 XPS spectra for the TT CAS sample before surface cleaning and after re-exposure to atmospheric conditions. **a** Survey spectra, **b** Cu 2*p* region, **c** overlapping Sb 3*d* and O 1*s* region, **d** S 2*p* region. Fitted peak envelopes shown in black

structural quality, cell-quality reproducibility, and optoelectronic properties, with no real change electronically. From XPS, it is shown that the thermal treatment had little effect on the surface chemistry of the sample itself, given that the BE shifts in Tables 3.3 and 3.4 are small. The thermal treatment appears to have suppressed the formation of oxidation on the surface of the samples which agrees with the improved cell-quality reproducibility seen previously, because it has been shown how the oxide formation can be detrimental to cell performance.

Comparison of Band Levels

The fittings of the VBM and SEC are shown in Fig. 3.20, with the extracted values shown in Table 3.5, and it was determined that the IP of the samples with and without the thermal treatment are the same. This is in agreement with the observation that

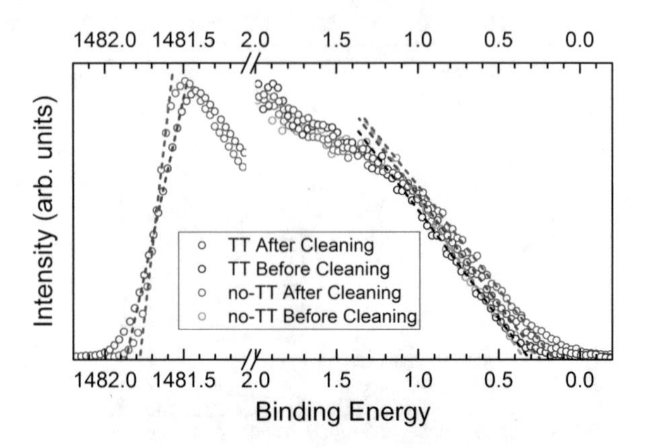

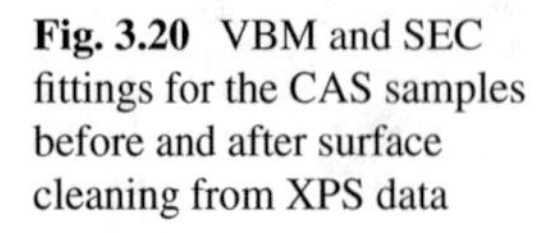
Fig. 3.20 VBM and SEC fittings for the CAS samples before and after surface cleaning from XPS data

Table 3.5 VBM to Fermi-level separations and extracted values of work function (WF) and ionisation potential (IP) from XPS data for the CAS samples before and after surface cleaning

	Sample	VBM–E_F	WF	IP
No-TT	Before Cleaning	0.34	—	—
	After Cleaning	0.16	4.82	4.98
TT	Before Cleaning	0.31	—	—
	After Cleaning	0.24	4.73	4.98

the thermal treatment did not affect the bandgap [29], and hence the energy levels remain in the same position.

Surface Stoichiometry

The XPS spectra presented throughout this chapter have been normalised in order to better demonstrate the presence of different species. Therefore, for comparisons between the amount of different species, reference is made to the XPS-derived surface stoichiometry values, detailed in Table 3.6. As no transmission function correction was available for the instrumentation used here, absolute values are not exact and should be interpreted with caution.[42] Nevertheless, a correction for the Cu/S ratios was applied, which derived from independent stoichiometry measurements of standard CuS and Cu_2S powders.[43] This means that the Cu/S ratios presented here are more representative than the Sb/S ratios, because the applied correction should, to some extent, account for systematic instrumental offsets.

In CAS, the ratio of Cu:Sb:S should be 1:1:2. By normalising to a copper content of 1.0, after cleaning the ratio was found to be 1.0:1.9:2.4 for the no-TT sample and 1.0:1.4:2.0 for the TT sample. As the Cu/S ratio is more representative than the Sb/S ratio as described above, but the trend of the Sb/S ratio should also be of value, it is concluded that the TT sample stoichiometry is in agreement with the phase determined from Raman and XRD measurements [29], but the no-TT sample

[42]For example, the apparently very high level of sodium.

[43]Further details of this derivation are given in Sect. **6.5.1**.

Table 3.6 Atomic percentages derived from XPS measurements for the CAS samples before and after surface cleaning

	Sample	CuSbS$_2$			Sb$_2$O$_3$		Contamination		
		Cu	Sb	S	Sb	O	Sb^0	S (S-C)	Na
No-TT	Before Cleaning	4.7	6.9	15.4	11.9	30.4	—	1.2	29.4
	After Cleaning	12.6	24.4	30.8	1.1	11.2	0.6	—	19.3
TT	Before Cleaning	9.2	14.5	23.6	17.1	25.9	0.8	3.5	5.5
	After Cleaning	16.6	23.4	32.8	3.1	8.5	0.4	—	15.3

is antimony and sulphur rich, giving further evidence that the thermal treatment has improved the sample quality.

3.5 Summary

The VBDoS and the size and nature of the bandgaps of the promising absorber material, CAS have been compared experimentally and theoretically for the first time in order to develop a better understanding of the electronic structure of this unusually structured material, and also how this relates to the measured electronic properties, with the hope of explaining why CAS cells have, so far, demonstrated poor efficiencies.

It has been shown that the strong Cu *d* nature of the VB obscures states arising from antimony. However, these states also play a role in determining the properties of the material. In contrast with CIGS, the CBM of CAS consists of second-cation *p*/anion *p* states rather than second-cation *s*/anion *p* states: a result of Sb^{3+} being a group V element rather than the group III $(In/Ga)^{3+}$ in CIGS. The full Sb 5*s* states therefore also feature at the top of the VB through the revised lone-pair model, along with Sb 5*p* states. These extra contributions to the VB play some role in raising the VBM and causing the comparatively low IP of 4.98 eV observed here.

The measured band levels support the reasoning that CdS is an inappropriate n-type window layer for CAS because of the large CBO, resulting from the low EA,[44] which was determined using a combination of IP measurements from XPS and a bandgap measurement of 1.55 eV from the absorption spectrum. The premise of utilising CdS within cell design, which arises from CIGS development, is part of the reason that CAS has seen poor performance. It is therefore clear that the use of CdS as the n-type layer in CAS cells is less favourable than for CIGS cells. However, with an alternative n-type material with better band alignments, the potential of CAS as a solar absorber remains.

A thorough core-level XPS analysis of CAS has also been shown, which reveals that the two sulphur coordination environments can be determined by XPS, and also

[44] 3.43 eV.

the effect that contamination can have, which is crucial given the complexities of the spectra of this material. Following a thorough literature survey of CAS, it is believed that these complexities have led to misinterpretations of the spectra in the past, and therefore this could have contributed to the poor performance of CAS so far as well.

The thermal treatment applied during growth, which produces higher quality thin-films, was shown to have little effect in terms of electronic structure differences; however, it is believed that this thermal treatment caused to improve the quality of the films by discouraging the oxidation and diffusion of sodium at the surface of the material, hence providing a more reproducible surface onto which other solar cell layers can be grown. Nevertheless, it is clear that further growth initiatives are also required in order to further the advancement of CAS.

References

1. Wuensch BJ. The crystal structure of tetrahedrite, $Cu_{12}Sb_4S_{13}$. Zeitschrift für Krist. 1964;119(5–6):437–53.
2. Pfitzner A, Bernert T. The system Cu_3AsS_4–Cu_3SbS_4 and investigations on normal tetrahedral structures. Zeitschrift für Krist. - Cryst. Mater. 2004;219(1):20–6.
3. Karup-Møller S, Makovicky E. Skinnerite, Cu_3SbS_3, A New Sulfosalt from the Llimaussaq Alkaline Intrusion, South Greenland. Am. Mineral. 1974;59:889–95.
4. Skinner BJ, Luce FD, Makovicky E. Studies of the sulfosalts of copper III; phases and phase relations in the system Cu-Sb-S. Econ Geol. 1972;67(7):924–38.
5. Yu L, Kokenyesi RS, Keszler DA, Zunger A. Inverse design of high absorption thin-film photovoltaic materials. Adv Energy Mater. 2013;3(1):43–8.
6. Ramasamy K, Sims H, Butler WH, Gupta A. Selective nanocrystal synthesis and calculated electronic structure of all four phases of copper–antimony–sulfide. Chem Mater. 2014;26(9):2891–9.
7. Hofmann W. Strukturelle und morphologische Zusammenhänge bei erzen vom formeltyp ABC_2. Zeitschrift für Krist. Cryst Mater. 1933;84(1–6):177–203.
8. Wachtel A, Noreika A. Growth and characterization of $CuSbS_2$ crystals. J Electron Mater. 1980;9(2):281–97.
9. Wernick JH, Benson KE. New semiconducting ternary compounds. J Phys Chem Solids. 1957;3(1–2):157–9.
10. Ramasamy K, Gupta RK, Sims H, Palchoudhury S, Ivanov S, Gupta A. Layered ternary sulfide $CuSbS_2$ nanoplates for flexible solid-state supercapacitors. J Mater Chem A. 2015;3(25):13263–74.
11. Ramasamy K, Tien B, Archana PS, Gupta A. Copper antimony sulfide ($CuSbS_2$) mesocrystals: a potential counter electrode material for dye-sensitized solar cells. Mater Lett. 2014;124:227–30.
12. Choi YC, Yeom EJ, Ahn TK, Seok SL. $CuSbS_2$-sensitized inorganic-organic heterojunction solar cells fabricated using a metal-thiourea complex solution. Angew Chemie Int Ed. 2015;54(13):4005–9.
13. Marino C, Block T, Pöttgen R, Villevieille C. $CuSbS_2$ as a negative electrode material for sodium ion batteries. J Power Sources. 2017;342:616–22.
14. Manolache S, Duta A, Isac L, Nanu M, Goossens A, Schoonman J. The influence of the precursor concentration on $CuSbS_2$ thin films deposited from aqueous solutions. Thin Solid Films. 2007;515(15):5957–60.
15. Garza C, Shaji S, Arato A, Perez Tijerina E, Alan Castillo G, Das Roy TK, Krishnan B. P-Type $CuSbS_2$ thin films by thermal diffusion of copper into Sb_2S_3. Sol Energy Mater Sol Cells. 2011;95(8):2001–5.

16. Krishnan B, Shaji S, Ernesto Ornelas R. Progress in development of copper antimony sulfide thin films as an alternative material for solar energy harvesting. J Mater Sci: Mater Electron. 2015;26(7):4770–81.
17. Rabhi A, Kanzari M, Rezig B. Optical and structural properties of $CuSbS_2$ thin films grown by thermal evaporation method. Thin Solid Films. 2009;517(7):2477–80.
18. Kehoe AB, Temple DJ, Watson GW, Scanlon DO. Cu_3MCh_3 (M = Sb, Bi; Ch = S, Se) as candidate solar cell absorbers: insights from theory. Phys Chem Chem Phys. 2013;15(37):15477.
19. Welch AW, Zawadzki PP, Lany S, Wolden CA, Zakutayev A. Self-regulated growth and tunable properties of $CuSbS_2$ solar absorbers. Sol Energy Mater Sol Cells. 2015;132:499–506.
20. Popovici I, Duta A. Tailoring the composition and properties of sprayed $CuSbS_2$ thin films by using polymeric additives. Int J Photoenergy. 2012;2012:1–6.
21. Rodríguez-Lazcano Y, Nair MTS, Nair PK. $CuSbS_2$ thin film formed through annealing chemically deposited Sb_2S_3–CuS thin films. J Cryst Growth. 2001;223(3):399–406.
22. Rodríguez-Lazcano Y, Nair MTS, Nair PK. Photovoltaic P-I-N structure of Sb_2S_3 and $CuSbS_2$ absorber films obtained via chemical bath deposition. J Electrochem Soc. 2005;152(8):G635–8.
23. Ganose AM, Savory CN, Scanlon DO. Beyond methylammonium lead iodide: prospects for the emergent field of ns^2 containing solar absorbers. Chem Commun. 2017;53(1):20–44.
24. Septina W, Ikeda S, Iga Y, Harada T, Matsumura M. Thin film solar cell based on $CuSbS_2$ absorber fabricated from an electrochemically deposited metal stack. Thin Solid Films. 2014;550:700–4.
25. Yang B, Wang L, Han J, Zhou Y, Song H, Chen S, Zhong J, Lv L, Niu D, Tang J. $CuSbS_2$ as a promising earth-abundant photovoltaic absorber material: a combined theoretical and experimental study. Chem Mater. 2014;26(10):3135–43.
26. Kumar M, Persson C. $CuSbS_2$ and $CuBiS_2$ as potential absorber materials for thin-film solar cells. J Renew Sustain Energy. 2013;5(3):31616.
27. Wan L, Ma C, Hu K, Zhou R, Mao X, Pan S, Wong LH, Xu J. Two-stage co-evaporated $CuSbS_2$ thin films for solar cells. J Alloys Compd. 2016;680:182–90.
28. Colombara D, Peter LM, Rogers KD, Painter JD, Roncallo S. Formation of $CuSbS_2$ and $CuSbSe_2$ thin films via chalcogenisation of Sb–Cu metal precursors. Thin Solid Films. 2011;519(21):7438–43.
29. de Souza Lucas FW, Welch AW, Baranowski LL, Dippo PC, Hempel H, Unold T, Eichberger R, Blank B, Rau U, Mascaro LH, Zakutayev A. Effects of thermochemical treatment on $CuSbS_2$ photovoltaic absorber quality and solar cell reproducibility. J Phys Chem C. 2016;120(33):18377–85.
30. Welch AW, Baranowski LL, Zawadzki P, DeHart C, Johnston S, Lany S, Wolden CA, Zakutayev A. Accelerated development of $CuSbS_2$ thin film photovoltaic device prototypes. Prog Photovoltaics Res Appl. 2016;24(7):929–39.
31. Ramos Aquino JA, Rodriguez Vela DL, Shaji S, Avellaneda DA, Krishnan B. Spray pyrolysed thin films of copper antimony sulfide as photovoltaic absorber. Phys Status Solidi. 2016;13(1):24–9.
32. Ornelas-Acosta RE, Shaji S, Avellaneda D, Castillo GA, Das Roy TK, Krishnan B. Thin films of copper antimony sulfide: a photovoltaic absorber material. Mater Res Bull. 2015;61:215–25.
33. Ornelas-Acosta RE, Avellaneda D, Shaji S, Castillo GA, Das Roy TK, Krishnan B. $CuSbS_2$ thin films by heating Sb_2S_3/Cu layers for PV applications. J Mater Sci Mater Electron. 2014;25(10):4356–62.
34. Rastogi AC, Janardhana NR. Properties of $CuSbS_2$ thin films electrodeposited from ionic liquids as P-Type absorber for photovoltaic solar cells. Thin Solid Films. 2014;565:285–92.
35. Suehiro S, Horita K, Yuasa M, Tanaka T, Fujita K, Ishiwata Y, Shimanoe K, Kida T. Synthesis of copper–antimony-sulfide nanocrystals for solution-processed solar cells. Inorg Chem. 2015;54(16):7840–5.
36. Suriakarthick R, Nirmal Kumar V, Shyju TS, Gopalakrishnan R. Effect of substrate temperature on copper antimony sulphide thin films from thermal evaporation. J Alloys Compd. 2015;651:423–33.

37. Su H, Xie Y, Wan S, Li B, Qian Y. A novel one-step solvothermal route to nanocrystalline $CuSbS_2$ and Ag_3SbS_3. Solid State Ionics. 1999;123(1–4):319–24.
38. An C, Liu Q, Tang K, Yang Q, Chen X, Liu J, Qian Y. The influences of surfactant concentration on the quality of chalcostibite nanorods. J Cryst Growth. 2003;256(1–2):128–33.
39. Banu S, Ahn SJ, Ahn SK, Yoon K, Cho A. Fabrication and characterization of cost-efficient $CuSbS_2$ thin film solar cells using hybrid inks. Sol Energy Mater Sol Cells. 2016;151:14–23.
40. Maiello P, Zoppi G, Miles RW, Pearsall N, Forbes I. Chalcogenisation of Cu–Sb metallic precursors into $Cu_3Sb(Se_xS_{1-x})_3$. Sol Energy Mater Sol Cells. 2013;113:186–94.
41. Zhou J, Bian G-Q, Zhu Q-Y, Zhang Y, Li C-Y, Dai J. Solvothermal crystal growth of $CuSbQ_2$ (Q = S, Se) and the correlation between macroscopic morphology and microscopic structure. J Solid State Chem. 2009;182(2):259–64.
42. Irvine SJC. Materials challenges: inorganic photovoltaic solar energy. In: Irvine SJC, editor. RSC energy and environment series. Cambridge: Royal society of chemistry; 2014.
43. Hofmann, W. Ergebnisse Der Strukturbestimmung Komplexer Sulfide. Zeitschrift für Krist.—Cryst. Mater. **1935**, *92* (1–6).
44. Makovicky E. Crystal structures of sulfides and other chalcogenides. Rev Mineral Geochemistry. 2006;61(1):7–125.
45. Maeda T, Wada T. First-principles study of electronic structure of $CuSbS_2$ and $CuSbSe_2$ photovoltaic semiconductors. Thin Solid Films. 2015;582:401–7.
46. Walsh A, Payne DJ, Egdell RG, Watson GW. Stereochemistry of post-transition metal oxides: revision of the classical lone pair model. Chem Soc Rev. 2011;40(9):4455–63.
47. Dufton JTR, Walsh A, Panchmatia PM, Peter LM, Colombara D, Islam MS. Structural and electronic properties of $CuSbS_2$ and $CuBiS_2$: potential absorber materials for thin-film solar cells. Phys Chem Chem Phys. 2012;14(20):7229.
48. Kyono A. Crystal structures of chalcostibite ($CuSbS_2$) and emplectite ($CuBiS_2$): structural relationship of stereochemical activity between chalcostibite and emplectite. Am Mineral. 2005;90(1):162–5.
49. Temple DJ, Kehoe AB, Allen JP, Watson GW, Scanlon DO. Geometry, electronic structure, and bonding in $CuMCh_2$ (M = Sb, Bi; Ch = S, Se): alternative solar cell absorber materials? J Phys Chem C. 2012;116(13):7334–40.
50. Siebentritt S. Why are kesterite solar cells not 20% efficient? Thin Solid Films. 2013;535(1):1–4.
51. Willian de Souza Lucas F, Welch AW, Baranowski LL, Dippo PC, Mascaro LH, Zakutayev A. Thermal treatment improvement of $CuSbS_2$ absorbers. In: 2015 IEEE 42nd photovoltaic specialist conference (PVSC); IEEE; 2015. pp 1–5.
52. Peccerillo E, Major J, Phillips L, Treharne R, Whittles TJ, Dhanak V, Halliday D, Durose K. Characterization of sulfurized $CuSbS_2$ thin films for PV applications. In: 2014 IEEE 40th photovoltaic specialist conference (PVSC); IEEE, 2014; pp 0266–9.
53. Zhong J, Xiang W, Cai Q, Liang X. Synthesis, characterization and optical properties of flower-like Cu_3BiS_3 nanorods. Mater Lett. 2012;70:63–6.
54. Lebugle A, Axelsson U, Nyholm R, Mårtensson N. Experimental L and M core level binding energies for the metals ^{22}Ti to ^{30}Zn. Phys Scr. 1981;23(5A):825–7.
55. Coster DL, Kronig RD. New type of auger effect and its influence on the X-ray spectrum. Physica. 1935;2(1–12):13–24.
56. Nyholm R, Martensson N, Lebugle A, Axelsson U. Auger and Coster-Kronig broadening effects in the 2p and 3p photoelectron spectra from the metals ^{22}Ti-^{30}Zn. J Phys F: Met Phys. 1981;11(8):1727–33.
57. Nyholm R, Martensson N. Core level binding energies for the elements Zr-Te (Z = 40–52). J Phys C: Solid State Phys. 1980;13(11):L279–84.
58. Barrie A, Drummond IW, Herd QC. Correlation of calculated and measured 2p spin-orbit splitting by electron spectroscopy using monochromatic X-radiation. J Electron Spectros Relat Phenomena. 1974;5(1):217–25.
59. Kresse G, Hafner J. Ab Initio molecular dynamics for liquid metals. Phys Rev B. 1993;47(1):558–61.

60. Kresse G, Hafner J. Ab Initio molecular-dynamics simulation of the liquid-metal–amorphous-semiconductor transition in germanium. Phys Rev B. 1994;49(20):14251–69.
61. Kresse G, Furthmüller J. Efficiency of ab-initio total energy calculations for metals and semiconductors using a plane-wave basis set. Comput Mater Sci. 1996;6(1):15–50.
62. Kresse G, Furthmüller J. Efficient iterative schemes for ab initio total-energy calculations using a plane-wave basis set. Phys Rev B. 1996;54(16):11169–86.
63. Rath T, MacLachlan AJ, Brown MD, Haque SA. Structural, optical and charge generation properties of chalcostibite and tetrahedrite copper antimony sulfide thin films prepared from metal xanthates. J Mater Chem A. 2015;3(47):24155–62.
64. Baker J, Kumar RS, Sneed D, Connolly A, Zhang Y, Velisavljevic N, Paladugu J, Pravica M, Chen C, Cornelius A, Zhao Y. Pressure induced structural transitions in $CuSbS_2$ and $CuSbSe_2$ thermoelectric compounds. J Alloys Compd. 2015;643:186–94.
65. Efthimiopoulos I, Buchan C, Wang Y. Structural properties of Sb_2S_3 under pressure: evidence of an electronic topological transition. Sci Rep. 2016;6(April):24246.
66. Pronti L, Felici AC, Alesiani M, Tarquini O, Bracciale MP, Santarelli ML, Pardini G, Piacentini M. Characterisation of corrosion layers formed under burial environment of copper-based greek and roman coins from pompeii. Appl Phys A. 2015;121(1):59–68.
67. Mernagh TP, Trudu AG. A laser raman microprobe study of some geologically important sulphide minerals. Chem Geol. 1993;103(1–4):113–27.
68. Qiu XD, Ji SL, Chen C, Liu GQ, Ye CH. Synthesis, characterization, and surface-enhanced raman scattering of near infrared absorbing Cu_3SbS_3 nanocrystals. CrystEngComm. 2013;15(48):10431.
69. Aup-Ngoen K, Thongtem T, Thongtem S. Characterization of Cu_3SbS_4 microflowers produced by a cyclic microwave radiation. Mater Lett. 2012;66(1):182–6.
70. Zakutayev A, Baranowski LL,. Welch AW, Wolden, CA, Toberer ES. Comparison of Cu_2SnS_3 and $CuSbS_2$ as potential solar cell absorbers. In: 2014 IEEE 40th photovoltaic specialist conference (PVSC); IEEE, 2014; vol. 3, pp 2436–8.
71. Scragg JJ, Dale PJ, Colombara D, Peter LM. Thermodynamic aspects of the synthesis of thin-film materials for solar cells. ChemPhysChem. 2012;13(12):3035–46.
72. Whittles TJ, Burton LA, Skelton JM, Walsh A, Veal TD, Dhanak VR. Band alignments, valence bands, and core levels in the tin sulfides SnS, SnS_2, and Sn_2S_3: experiment and theory. Chem Mater. 2016;28(11):3718–26.
73. Bär M, Schubert BA, Marsen B, Wilks RG, Blum M, Krause S, Pookpanratana S, Zhang Y, Unold T, Yang W, Weinhardt L, Heske C, Schock HW. Cu_2ZnSnS_4 Thin-film solar cell absorbers illuminated by soft X-rays. J Mater Res. 2012;27(8):1097–104.
74. Morgan WE, Van Wazer JR. Binding energy shifts in the X-ray photoelectron spectra of a series of related group IVa compounds. J Phys Chem. 1973; 77(7):964–9.
75. Lenglet M, Kartouni K, Machefert J, Claude JM, Steinmetz P, Beauprez E, Heinrich J, Celati N. Low temperature oxidation of copper: the formation of CuO. Mater Res Bull. 1995;30(4):393–403.
76. Biesinger MC, Lau LWM, Gerson AR, Smart RSC. Resolving surface chemical states in XPS analysis of first row transition metals, Oxides and Hydroxides: Sc, Ti, V, Cu and Zn. Appl Surf Sci. 2010;257(3):887–98.
77. Goh SW, Buckley AN, Lamb RN, Rosenberg RA, Moran D. The oxidation states of copper and iron in mineral sulfides, and the Oxides formed on initial exposure of chalcopyrite and bornite to air. Geochim Cosmochim Acta. 2006;70(9):2210–28.
78. Jiasong Z, Weidong X, Huaidong J, Wen C, Lijun L, Xinyu Y, Xiaojuan L, Haitao L. A simple L-cystine-assisted solvothermal approach to Cu_3SbS_3 nanorods. Mater Lett. 2010;64(13):1499–502.
79. Sarswat PK, Free ML. Enhanced photoelectrochemical response from copper antimony zinc sulfide thin films on transparent conducting electrode. Int J Photoenergy. 2013;2013:1–7.
80. Wang MX, Yue GH, Fan XY, Yan PX. Properties and characterization of Cu_3SbS_3 nanowires synthesized by solvothermal route. J Cryst Growth. 2008;310(12):3062–6.

81. Peisert H, Chassé T, Streubel P, Meisel A, Szargan R. Relaxation energies in XPS and XAES of solid sulfur compounds. J. Electron Spectros. Relat. Phenomena. 1994;68(C):321–8.
82. van Embden J, Tachibana Y. Synthesis and characterisation of famatinite copper antimony sulfide nanocrystals. J Mater Chem. 2012;22(23):11466.
83. Garbassi F. XPS and AES Study of Antimony Oxides. Surf Interface Anal. 1980;2(5):165–9.
84. Han Q, Chen L, Zhu W, Wang M, Wang X, Yang X, Lu L. Synthesis of Sb_2S_3 peanut-shaped superstructures. Mater Lett. 2009;63(12):1030–2.
85. Lou W, Chen M, Wang X, Liu W. Novel single-source precursors approach to prepare highly uniform Bi_2S_3 and Sb_2S_3 nanorods via a solvothermal treatment. Chem Mater. 2007;19(4):872–8.
86. Mesa F, Chamorro W, Vallejo W, Baier R, Dittrich T, Grimm A, Lux-Steiner MC, Sadewasser S. Junction formation of Cu_3BiS_3 investigated by kelvin probe force microscopy and surface photovoltage measurements. Beilstein J Nanotechnol. 2012;3(1):277–84.
87. Delobel R, Baussart H, Leroy J-M, Grimblot J, Gengembre L. X-Ray photoelectron spectroscopy study of uranium and antimony mixed metal-oxide catalysts. J Chem Soc Faraday Trans 1 Phys Chem Condens Phases. 1983;79(4):879.
88. Morgan WE, Stec WJ, van Wazer JR. Inner-orbital binding energy shifts of antimony and bismuth compounds. Inorg Chem. 1972;12(4):953–5.
89. Tang X, Welzenis RG, van Setten FM, van Bosch AJ. Oxidation of the InSb surface at room temperature. Semicond Sci Technol. 1986;1(6):355–65.
90. Petit EJ, Riga J, Caudano R. Surface and interface XPS characterization of the oxide layer grown on antimony under UV laser irradiation. Surf Sci. 1991;251–252:529–34.
91. Payne BP, Biesinger MC, McIntyre NS. X-ray photoelectron spectroscopy studies of reactions on chromium metal and chromium oxide surfaces. J Electron Spectros Relat Phenomena. 2011;184(1–2):29–37.
92. Vasquez RP, Grunthaner FJ. Chemical composition of the SiO_2/InSb interface as determined by X-ray photoelectron spectroscopy. J Appl Phys. 1981;52(5):3509–14.
93. Lindberg BJ, Hamrin K, Johansson G, Gelius U, Fahlman A, Nordling C, Siegbahn K. Molecular spectroscopy by means of ESCA II. sulfur compounds. correlation of electron binding energy with structure. Phys Scr. 1970;1(5–6):286–98.
94. Gerson AR, Bredow T. Interpretation of sulphur 2p XPS spectra in sulfide minerals by means of ab initio calculations. Surf Interface Anal. 2000;29(2):145–50.
95. Breunig HJ, Rösler R. Organoantimony compounds with element—element bonds. Coord Chem Rev. 1997;163:33–53.
96. Rabhi A, Kanzari M, Rezig B. Growth and vacuum post-annealing effect on the properties of the new absorber $CuSbS_2$ thin films. Mater Lett. 2008;62(20):3576–8.
97. Klein A. Energy band alignment in chalcogenide thin film solar cells from photoelectron spectroscopy. J Phys: Condens Matter. 2015;27(13):134201.
98. Burton LA, Walsh A. Band alignment in SnS thin-film solar cells: possible origin of the low conversion efficiency. Appl Phys Lett. 2013;102(13):132111.
99. Hinuma Y, Oba F, Kumagai Y, Tanaka I. Ionization potentials of (112) and $(11\bar{2})$ facet surfaces of $CuInSe_2$ and $CuGaSe_2$. Phys Rev B. 2012;86(24):245433.
100. Burton LA, Whittles TJ, Hesp D, Linhart WM, Skelton JM, Hou B, Webster RF, O'Dowd G, Reece C, Cherns D, Fermin DJ, Veal TD, Dhanak VR, Walsh A. Electronic and optical properties of single crystal SnS_2: an earth-abundant disulfide photocatalyst. J Mater Chem A. 2016;4(4):1312–8.
101. Durose K, Asher SE, Jaegermann W, Levi D, McCandless BE, Metzger W, Moutinho H, Paulson PD, Perkins CL, Sites JR, Teeter G, Terheggen M. Physical characterization of thin-film solar cells. Prog Photovoltaics. 2004;12(2–3):177–217.
102. Sinsermsuksakul P, Hartman K, Bok Kim S, Heo J, Sun L, Hejin Park H, Chakraborty R, Buonassisi T, Gordon RG. Enhancing the efficiency of SnS solar cells via band-offset engineering with a zinc oxysulfide buffer layer. Appl Phys Lett. 2013;102(5):53901.
103. Minemoto T, Matsui T, Takakura H, Hamakawa Y, Negami T, Hashimoto Y, Uenoyama T, Kitagawa M. Theoretical analysis of the effect of conduction band offset of window/CIS

layers on performance of CIS Solar cells using device simulation. Sol Energy Mater Sol Cells. 2001;67(1–4):83–8.
104. Baranowski LL, Christensen S, Welch AW, Lany S, Young M, Toberer ES, Zakutayev A. Conduction band position tuning and Ga-doping in (Cd, Zn)S alloy thin films. Mater Chem Front. 2017.
105. Ley L, Pollak RA, McFeely FR, Kowalczyk SP, Shirley DA. Total valence-band densities of states of III-V and II-VI compounds from X-ray photoemission spectroscopy. Phys Rev B. 1974;9(2):600–21.
106. Zhang Y, Yuan X, Sun X, Shih B-C, Zhang P, Zhang W. Comparative study of structural and electronic properties of Cu-based multinary semiconductors. Phys Rev B. 2011;84(7):75127.
107. Mann JB, Meek TL, Knight ET, Capitani JF, Allen LC. Configuration energies of the D-block elements. J Am Chem Soc. 2000;122(21):5132–7.
108. Mann JB, Meek TL, Allen LC. Configuration energies of the main group elements. J Am Chem Soc. 2000;122(12):2780–3.
109. Maeda T, Wada T. Characteristics of chemical bond and vacancy formation in chalcopyrite-type $CuInSe_2$ and related compounds. Phys Status Solidi. 2009;6(5):1312–6.
110. Chen X-D, Chen L, Sun Q-Q, Zhou P, Zhang DW. Hybrid density functional theory study of $Cu(In_{1-x}\ Ga_x)Se_2$ band structure for solar cell application. AIP Adv. 2014;4(8):87118.
111. Bekaert J, Saniz R, Partoens B, Lamoen D. Native point defects in $CuIn_{1-x}Ga_xSe_2$: hybrid density functional calculations predict the origin of P- and N-type conductivity. Phys Chem Chem Phys. 2014;16(40):22299–308.
112. Wei SH, Zunger A. Calculated natural band offsets of all II-VI and III-V semiconductors: chemical trends and the role of cation D orbitals. Appl Phys Lett. 1998;72(16):2011–3.
113. Wei S-H, Zunger A. Role of metal D states in II–VI semiconductors. Phys Rev B. 1988;37(15):8958–81.
114. Xue D-J, Yang B, Yuan Z-K, Wang G, Liu X, Zhou Y, Hu L, Pan D, Chen S, Tang J. $CuSbSe_2$ as a potential photovoltaic absorber material: studies from theory to experiment. Adv Energy Mater. 2015;5(23):1501203.
115. Zakutayev A, Caskey CM, Fioretti AN, Ginley DS, Vidal J, Stevanovic V, Tea E, Lany S. Defect tolerant semiconductors for solar energy conversion. J Phys Chem Lett. 2014;5(7):1117–25.
116. Allen JP, Carey JJ, Walsh A, Scanlon DO, Watson GW. Electronic structures of antimony oxides. J Phys Chem C. 2013;117(28):14759–69.
117. Wood C, van Pelt B, Dwight A. The optical properties of amorphous and crystalline Sb_2O_3. Phys Status Solidi. 1972;54(2):701–6.
118. Prabhakar T, Jampana N. Effect of sodium diffusion on the structural and electrical properties of Cu_2ZnSnS_4 thin films. Sol Energy Mater Sol Cells. 2011;95(3):1001–4.
119. Rudmann D, Bilger G, Kaelin M, Haug FJ, Zogg H, Tiwari AN. Effects of NaF coevaporation on structural properties of $Cu(In, Ga)Se_2$ thin films. Thin Solid Films. 2003;431–432(3):37–40.
120. Kranz L, Perrenoud J, Pianezzi F, Gretener C, Rossbach P, Buecheler S, Tiwari AN. Effect of sodium on recrystallization and photovoltaic properties of CdTe solar cells. Sol Energy Mater Sol Cells. 2012;105:213–9.
121. Barrie A, Street FJ. An auger and X-ray photoelectron spectroscopic study of sodium metal and sodium oxide. J Electron Spectros Relat Phenomena. 1975;7(1):1–31.
122. Würz R, Rusu M, Schedel-Niedrig T, Lux-Steiner MC, Bluhm H, Hävecker M, Kleimenov E, Knop-Gericke A, Schlögl R. In Situ X-ray photoelectron spectroscopy study of the oxidation of $CuGaSe_2$. Surf Sci. 2005;580(1–3):80–94.
123. Zhu C, Osherov A, Panzer MJ. Surface chemistry of electrodeposited Cu_2O films studied by XPS. Electrochim Acta. 2013;111:771–8.
124. Kronik L, Cahen D, Schock HW. Effects of sodium on polycrystalline $Cu(In, Ga)Se_2$ and its solar cell performance. Adv Mater. 1998;10(1):31–6.
125. Soukiassian P, Gentle TM, Bakshi MH, Hurych Z. SiO_2-Si interface formation by catalytic oxidation using alkali metals and removal of the catalyst species. J Appl Phys. 1986;60(12):4339–41.

126. Starnberg HI, Soukiassian P, Hurych Z. Alkali-metal-promoted oxidation of the Si(100)2 × 1 surface: coverage dependence and nonlocality. Phys Rev B. 1989;39(17):12775–82.
127. Ding X, Dong G, Hou X, Wang X. The adsorption of oxygen on alkali metal covered GaAs (1 1 1) surfaces. Solid State Commun. 1987;61(6):391–3.
128. Lide DR. CRC Handbook of Chemistry and Physics 84th Edition; 2003.
129. Kemori N, Denholm W, Saunders S. Measurements of standard gibbs energies of formation of SbO, MgSbO and CaSbO by an Emf method. Can Metall Q. 1996;35(3):269–74.
130. Sundberg J, Lindblad R, Gorgoi M, Rensmo H, Jansson U, Lindblad A. Understanding the effects of sputter damage in W-S thin films by HAXPES. Appl Surf Sci. 2014;305:203–13.
131. Loeffler MJ, Dukes CA, Chang WY, McFadden LA, Baragiola RA. Laboratory simulations of sulfur depletion at eros. Icarus. 2008;195(2):622–9.
132. Doniach S, Sunjic M. Many-electron singularity in X-ray photoemission and X-ray line spectra from metals. J Phys C: Solid State Phys. 1970;3(2):285–91.

Chapter 4
The Electronic Structure of Cu_3BiS_3 for Use as a PV Absorber

The use of solar energy has not been opened up because the oil industry does not own the sun.

Ralph Nader, Activist, 1980

This chapter involves measurements of thin-film Cu_3BiS_3 (CBS) and the electronic characterisation of this material for use in PV devices. The majority of the results presented here are unpublished, but are in the final stages of manuscript preparation as:

Whittles, T. J.; Veal, T. D.; Yates, P.; Savory, C. N.; Scanlon, D. O.; Major, J. D.; Durose, K.; Dhanak, V. R. Core-levels, Band Alignments and Valence Band States in Cu_3BiS_3 for its use as a PV Absorber. *Manuscript in Preparation* **2017**.

The earth-abundant material CBS has shown many exciting photovoltaic-relevant properties and research interest is increasing. However, to-date, no devices have been tested. A thorough fitting model is developed and implemented in a full analysis of the XPS spectra of CBS. This overcomes previous problems in the literature, demonstrating the comprehensive potential of this analysis technique when properly applied and analysed. Not only does it allow the oxidation states of the elements to be determined, aiding with phase identification, but it can also be used to identify and analyse the effects of contamination present on the material. This is beneficial, as such contamination can be detrimental to device performance. Agreement between

© Springer International Publishing AG, part of Springer Nature 2018

T. J. Whittles, *Electronic Characterisation of Earth-Abundant Sulphides for Solar Photovoltaics*, Springer Theses, https://doi.org/10.1007/978-3-319-91665-1_4

the experimentally measured XPS valence band spectra and electronic-structure calculations from DFT reveals the bonding nature within CBS to be superficially similar to CIGS and CZTS, from which, analogies are often drawn. However, the bonding nature is fundamentally different because of the differing structure caused by the Bi^{3+} cation and the occupied Bi 6*s* orbital. This structure gives rise to the low IP measured here,[1] resulting in poor conduction band alignment with CdS, which suggests that in order for CBS devices to progress, established device architectures for CIGS and CZTS should be rejected in favour of better band-aligned and environmentally friendly materials.

4.1 Cu_3BiS_3: The Material

The material CBS has recently been acknowledged for its potential as an earth-abundant solar absorber, and therefore is not fully established in the research community with regards to properties and characteristics.

4.1.1 History and Uses

The sulphosalts, as introduced in Sect. 1.4.2, show potential for use as solar absorbers, if their constituents are earth-abundant. The Cu–Bi–S family fulfils this criterion, akin to the Cu–Sb–S family in Chap. 3, where the problematic In/Ga in CIGS is replaced with Bi. Compared with indium, bismuth sees much greater ore extraction rates, has greater world reserves and is therefore much cheaper [1]: all beneficial for use in an earth-abundant material.

In this chapter, the compound Cu_3BiS_3 is of interest. Occurring naturally as the mineral wittichenite[2]; known since 1805 [2, 3], and first synthesised in 1947 [4], CBS was noted for its potential as a solar PV absorber in 1997 [5]. Interest in CBS is also maintained because of its potential use in other areas, such as: photoelectrochemistry [6], anode materials in batteries [7], near-infrared detectors [8], and cancer diagnosis [9].

4.1.2 Cu_3BiS_3 as a Solar Absorber

Since the recognition of the potential of CBS as a PV absorber material, the measured properties have been promising: inherent p-type conductivity [10]; an appropriate

[1] 5.18 eV.

[2] Named after Wittichen, Baden, Germany; the locality of the mine in which it was discovered.

bandgap measured at ~1.4 eV [8, 11–14]; and strong absorption [12, 14, 15], calculated to be higher than both CIGS and CZTS [16].

CBS has also been synthesised by many routes typical of other thin-film absorbers[3]. This has enabled different aspects of the material to be studied, such as the effect of post-deposition annealing on crystallite size and resistivity [12], the passivation of defects [28], the effect of substrate temperature on growth quality [27], the temperature dependence of the bandgap [17], the effect of temperature on phase segregation [11], and the effect of grain size on carrier transport [8].

Despite these promising initial results, research into CBS as a solar absorber has nevertheless been somewhat scarce, and as such, no devices have yet been tested [18, 29–31].

4.1.3 Structure

In Chap. 3, it was discussed how the analogies drawn between $CuSbS_2$ and CIGS had hindered the development of that material. This matter, with regards to CBS, will be introduced in Sect. 4.1.4, but for now, similar to in Sect. 3.1.3, although CBS seems similar to CIGS electronically, the adopted crystal structures are different.

The crystal structure of CBS is shown in Fig. 4.1a and demonstrates the complexity and distortion of the structure compared to the almost regular tetrahedra seen in the crystal structure of CIGS [32]. In CBS, the Bi atoms are 3-fold coordinated to sulphur in distorted tetrahedral BiS_3 units with the Bi atom at one vertex, which is similar to one coordination environment seen in the structure of bismuthinite Bi_2S_3, where the ribbon ends consist of this type of polyhedra [33, 34]. These are derived from regular octahedra by the Bi lone-pair first replacing one sulphur atom at an octahedral vertex, creating distorted square-based pyramids[4] [11], and then the displacement of the Bi atom due to the copper presence, which elongates two Bi–S bonds to an extent where they are not present in the wittichenite structure [3, 10]. The Cu atoms are also three-fold coordinated to S, but in distorted trigonal near-planar CuS_3 units, which can be considered distorted tetrahedra; such polyhedra are seen in the structure of low chalcocite Cu_2S[5] [35]. The individual crystal structure units are shown in Fig. 4.1b.

This crystal structure, akin to a mixture of Cu_2S and Bi_2S_3, is expected as CBS was first synthesised from the dry fusion of these two compounds [4]. There are three different copper sites and three different sulphur sites, which are labelled in Fig. 4.1a. However, the coordination remains the same between them with bond lengths and angles differing only slightly. This unusual structure, further details of which can be found in the appropriate literature [3, 16], demands a greater understanding of

[3]CBD [5, 15], sputtering [12, 17–19], evaporation [8, 13, 14], hydro-/solvo-thermal [6, 7, 20–25], electrodeposition [11], screen printing [26], spray pyrolysis [27].

[4]The other coordination environment in Bi_2S_3 [33].

[5]See Fig. 6.2.

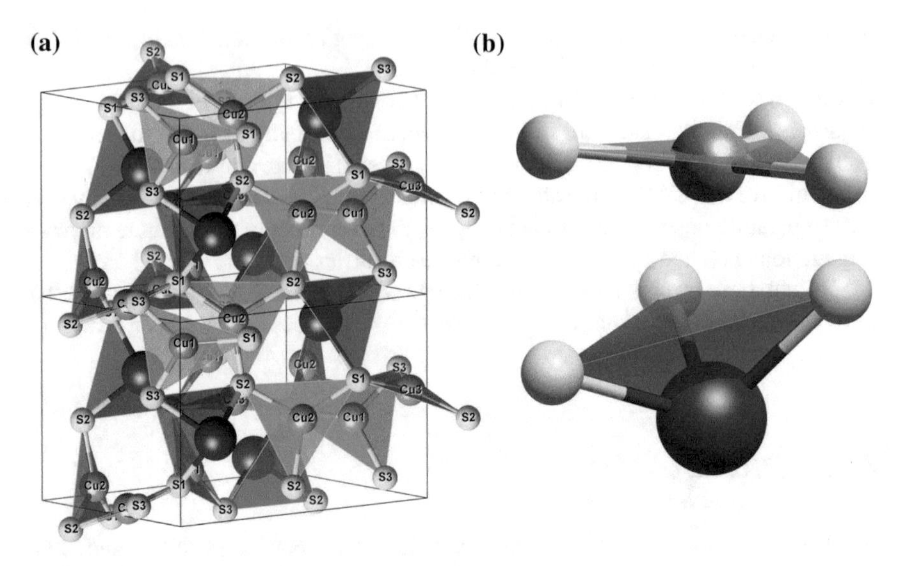

Fig. 4.1 **a** Crystal structure of wittichenite [3] Cu_3BiS_3, showing the individual atoms of Cu (orange), Bi (purple) and S (yellow). The inequivalent copper and sulphur sites are marked. **b** Individual BiS_3 and typical CuS_3 structural units

its effects on the properties of the material in order that development can progress further.

4.1.4 Motivation and Scope of This Study

The lack of reported device efficiencies regardless of the promising properties is the main motivation for this study, in order to better understand the material and assess the suitability of using CBS as a PV absorber.

As with all fledgling materials for solar PV use, there are still many issues to overcome and questions to be raised before the material is device ready, let alone record efficiency setting. Two of the main issues are developing the growth of the material so that it is single phase and of reproducible quality, and determining the true nature of the electronic structure. Rigorous characterisation of high-quality material, coupled with knowledge of the electronic structure will help enable this material to be incorporated correctly into device structures.

The related family of materials that were presented in Chap. 3, with antimony instead of bismuth, have also been shown to be relevant for PV [36], and were developed directly from the motivation to replace In/Ga in CIGS [37]. Conversely, it would appear that CBS advanced independently of this, even though replacing trivalent In/Ga with trivalent Bi is analogous [10], instead developing from the motivation to synthesise the natural mineral wittichenite [5, 26, 29] and discovering PV relevant properties [15]. The research which followed saw the likening of CBS to CIGS [11, 22, 27] in terms of growth [12, 18, 19] and cell architecture [30, 31, 38] approaches, as was the case with CAS [39] from Chap. 3. It is not yet clear in the stages of development for CBS whether this analogy with CIGS is beneficial or not, similar to the cases of CZTS [40] and CAS.[6] Although the bismuth in CBS is trivalent as is indium and gallium in CIGS, CBS does not adopt the usual tetrahedral crystal structure of solar absorbers derived from the Si structure[7] [16, 32]. This is due to the lattice distortion, but also to the differing stoichiometry.

In this chapter, the electronic characterisation of CBS is presented, shown to be phase-pure by XRD and RS. A full XPS analysis, including core-levels and the effects of cleaning, the position of the band edges, and VB spectra, demonstrates the full potential of this technique applied to this material. It shows not only how it can be used to identify the oxidation states of the elements and study the electronic structure in the VB, but also how it can detect the presence and investigate the effects of surface contamination. Indeed, added complexity to the XPS spectra specific to this material requires particular care when assessing such analyses and advice is given here on how to apply procedures and methods that can aid with the analysis. These findings are then corroborated with theoretical DoS calculations, which relate the bonding nature within CBS to the measured electronic properties. This, therefore, gives insights insofar as how to design cell architectures incorporating this material. A comparison between CBS and other common absorber materials is given in terms of band edge positions, and shows how established thin-film architectures are unsuitable. These properties are then discussed with relation to the underlying crystal and band structures, showing how the material properties are dependent upon these features, with the major differences arising as a result of the incorporation of the Bi^{3+} cation. Therefore, analogies drawn between CBS and tetrahedral solar absorbers are shown to be irrelevant to the point where they could cause a hindrance to development.

4.2 Experimental Details

Complete details of the experimental systems used here can be found in Chap. 2 and the specific details pertinent to this study, including a full description of the cleaning procedure used, are detailed in Appendix C.

[6]See Chap. 3.

[7]See Sect. 4.1.3.

4.2.1 Growth of Cu_3BiS_3 Films

The sample of CBS studied throughout this chapter was donated by the Durose Group, in the Stephenson Institute for Renewable Energy, at the University of Liverpool, UK. The thin-film of CBS was grown by evaporating a layer of copper, followed by a layer of bismuth, onto a glass substrate, which was then annealed in a sulphur atmosphere in order to crystallise the precursors. The relative thicknesses of the precursor layers determined the stoichiometry of the final film after annealing.

4.2.2 Characterisation of Cu_3BiS_3

X-Ray Diffraction
Thin-film XRD measurements were performed on the CBS sample using a Philips X'Pert PRO diffractometer equipped with a Cu Kα x-ray source running at 40 kV and 40 mA.

Raman Spectroscopy
Raman spectra were acquired using a Horiba Scientific Jobin-Yvon LabRam HR system, using an exciting wavelength of 514.5 nm.

X-Ray Photoemission Spectroscopy
XPS was used to study the core-level electronic structure, SEC, VBM and occupied VBDoS of the CBS sample. It was affixed to a UHV sample plate by spot welding tantalum straps across the sample edges. This also provided an electrical connection between the CBS film and the spectrometer. Ar^+ ion sputtering at 500 eV and radiative heating up to 250 °C were utilised to clean contamination from the sample surface, the full details of which are given in Appendix C, Sect. C.5.2.

In order to account for charging of the semiconducting sample, the spectra were aligned so that the C 1*s* signal, which was collected over many hours in order to achieve a good SNR, had a BE of 285.0 eV. As it has been shown that the use of the C 1*s* signal as a charge correction method is material dependent [41], and as this value has been used previously for this and similar materials [13, 42], its use here is therefore validated.

All of the fitted synthetic peaks in this study were Voigt profiles, fitted after subtraction of a Shirley background, unless otherwise stated. Cu 2*p* doublets were fitted with a separation of 19.80 eV [43] and an area ratio of 1:2. The FWHM of the Cu $2p_{1/2}$ peak is somewhat broader than that of the Cu $2p_{3/2}$ peak due to Coster–Kronig effects [44, 45]. Bi 4*f* doublets were fitted with a separation of 5.31 eV [46], an area ratio of 3:4 and equal FWHM. Bi 4*d* doublets were fitted with a separation of 23.70 eV [46, 47], an area ratio of 2:3 and equal FWHM. Fitted S 2*p* doublets were separated by 1.18 eV [48], with an area ratio of 1:2, and equal FWHM.

Density Functional Theory
Electronic-structure calculations were performed through the VASP code [49–52], starting from the atomic structure of CBS, obtained through XRD analysis. The band structure, and pDoS in the valence and conduction bands were determined.

4.3 Electronic Structure Studies of Cu_3BiS_3

This section presents characterisations of the CBS sample, in order to determine the suitability of this material to be used as a solar absorber. Complete analysis of the other aspects of this material found using XPS, including the effects of contamination, will be presented in Sect. 4.4.

4.3.1 *Phase-Purity*

XRD and RS are commonly employed techniques for solar absorbers because together they can help determine the phase of the grown material and can also be used to exclude and detect the formation of other phases.

The measured XRD pattern for the CBS sample is shown in Fig. 4.2 along with the theoretical diffraction pattern, calculated from the wittichenite crystal structure [3] using the method in Sect. 2.6.2.[8] The data matches the theoretical pattern well, with no peaks unaccounted for. The Raman spectrum for the CBS sample is shown in Fig. 4.3, with the identified peaks, representing the modes of CBS, marked.

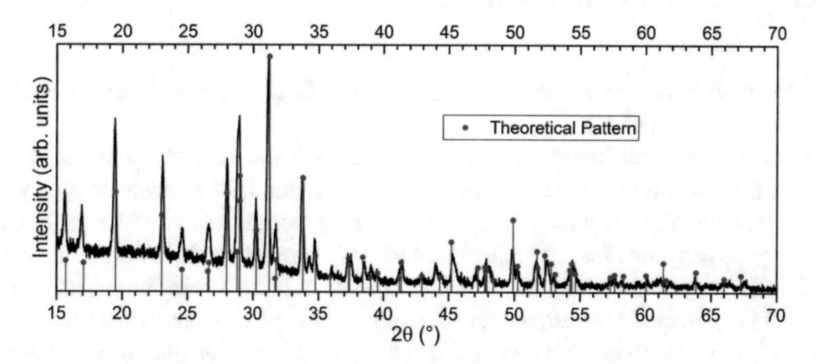

Fig. 4.2 XRD pattern for the CBS sample, compared with the ideal pattern, calculated theoretically from the crystal structure [3]

[8]With the calculated pattern data shown in Appendix D, Sect. D.2.3.

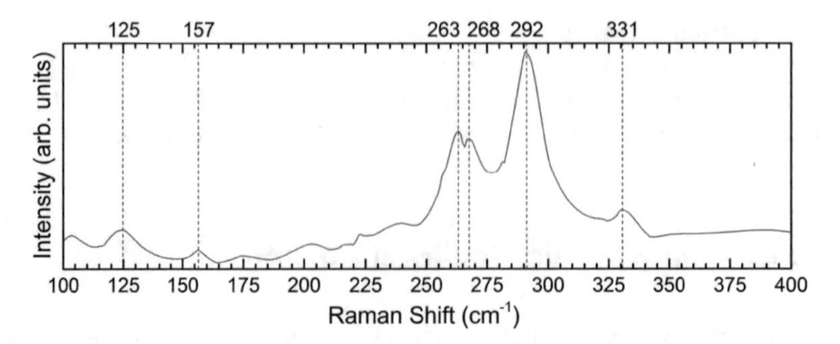

Fig. 4.3 Raman spectrum for the CBS sample. Strongest peak positions marked

Although the literature on CBS is not extensive, most growth studies do report measured XRD patterns, which are in agreement with that measured here [11–14, 17, 19, 21–24, 26, 53, 54]. The many peaks in the pattern, arising from the lack of symmetry in the crystal structure, make it easy to determine the correct phase, and the good match with the theoretical pattern supports the suggestion that the CBS sample here is phase-pure. RS, on the other hand, has not been widely employed in growth studies of CBS, but the spectrum measured here is thought to support the phase-purity of the material because it is in agreement with the few other reported RS spectra of CBS [17, 53, 55, 56]. Indeed, the superior resolution of the spectrum in Fig. 4.3, reveals individual peaks, where previously it has displayed a wide, broad feature.

4.3.2 XPS Core-Level Analysis

The Necessity for Rigorous Core-Level Analysis and the Potential Problems with XPS Spectra of CBS

Many of the materials being studied for their use as solar absorbers consist of three or more constituent elements. As such, there is opportunity for the associated phase diagrams, including secondary phases, to be expansive, and such is the case for CBS. There are possibilities for the binary sulphides [57] and other stoichiometries within the Cu–Bi–S system [58] to form during growth. It is therefore important that the material is characterised sufficiently to determine the phases present. Beyond XRD and RS[9]; in order to confirm the oxidation states of the elements in the material, core-level XPS analyses are frequently applied to solar absorbers [59]. Furthermore, XPS can be used with great effect to study the vacuum-aligned band positions of semiconductors [60], the band alignments of heterojunctions [61, 62], or to determine

[9] As shown in Sect. 4.3.1.

Table 4.1 XPS BE for the main peaks of CBS before and after surface cleaning with values for work function (WF) and ionisation potential (IP) also determined from XPS, shown in Sects. 4.3.3 and 4.4.2

Sample	Cu_3BiS_3				
	Cu $2p_{3/2}$	Bi $4f_{7/2}$	S $2p_{3/2}$	WF	IP
Before Cleaning	932.01 (1.16)	157.89 (0.65)	161.28 (0.81)	4.56	4.62
After Cleaning	932.23 (1.05)	158.19 (0.77)	161.47 (0.78)	4.79	5.18

All values are given in eV and the peak FWHM are given in parentheses. BE of all fitted peaks in this chapter can be found in Tables 4.2, 4.3, and 4.4

the presence of contaminants at the surfaces of solar absorber materials[10] [63], and also as a means of experimentally studying the VBDoS [64]. However, it is rare to find reports where this complete multitude of applications is exploited. This is especially the case for CBS, not only because it is a relatively understudied PV absorber, but also because of problems associated with the XPS spectra of materials containing both Bi and S, which will be further explained later. What follows therefore in this section, is a full dissemination of the uses of XPS, applied to CBS.

When considering the core-levels of CBS, one major issue is the direct overlap of the strongest peaks for Bi and S, that is the Bi $4f$ and S $2p$ regions [65]. The difference in spin-orbit splitting of these two doublets should help with confidently fitting the spectra; however, this is complicated by the fact that even though, stoichiometrically speaking, the sulphur signal should be three times greater than the bismuth signal, it will in fact be much[11] weaker than the bismuth signal when taking the differences in photoionisation cross sections into account [66]. This problem is further compounded by the likelihood of oxidation present on the surface of the material, which would cause further overlap within the region. The possibility of oxidation also prevents the centroid value of the peak from being representative of the material, as it could cause it to skew. Therefore, in order to achieve a representative fit for the peaks, a robust fitting model is required. Survey spectra and the main regions of interest are presented below, with a summary of fitted peak energies in Table 4.1. A full description of the fitting procedure applied here can be found in Appendix D, Sect. D.2.1 with the corroborating secondary regions of interest. Tables 4.2, 4.3, and 4.4, show the BE for all of the fitted peaks in this study, including those from contamination. It is noted that the spectra presented here have been normalised for better clarity of the different species and so, for intensity comparisons, the stoichiometry values of Table 4.5 are used, which will be discussed in Sect. 4.4.3.

[10] Which could be overlooked by XRD because of either the small quantity or amorphousness of the contaminant.

[11] 5.5 times.

Core-Level Analysis: Before and After Surface Cleaning

The survey spectra for CBS before and after surface cleaning are shown in Fig. 4.4 and show the expected peaks for Cu, Bi and S, as well as C and O present from contamination due to atmospheric exposure, which decrease after cleaning. Sodium is also present in the sample, and decreases after cleaning as well. This is possibly due to sodium diffusion from the glass because it is still present after cleaning and has been reported previously [38] for CBS. The presence of sodium here is noteworthy; however, further study of its effects on CBS is warranted, as sodium is known to be beneficial to CZTS [67] and CIGS [68], but detrimental to CdTe [69]. The associated spectra for sodium and the other contaminants will be presented in Sect. 4.4.1.

The Cu 2*p* spectra, shown in Fig. 4.5, were fitted with a single doublet (red dash) both before and after cleaning, which is assigned to the Cu^+ in CBS. As shown in Fig. 4.1a, there are three different copper sites in the CBS crystal structure; however, the coordination is the same and the bond lengths and angles are only slightly different, so it is expected that the copper in CBS will be well represented by a single doublet.

It is possible that the copper in CBS could be susceptible to surface oxidation, taking the form of cupric oxide[12] [70], which is more readily identified by characteristic shake-up lines present in the Cu 2*p* spectra[13] [71]. Because these are lacking and there is little difference in the Cu 2*p* spectra before and after cleaning, save for a shift in BE which will be discussed later, it was concluded that no oxidation of the copper had taken place. It is also noted that the Bi 4*s* peak is estimated to be located around 939 eV [46], but there is no evidence of this peak because it is believed that the signal is too weak and wide compared to the signal from Cu 2*p*.

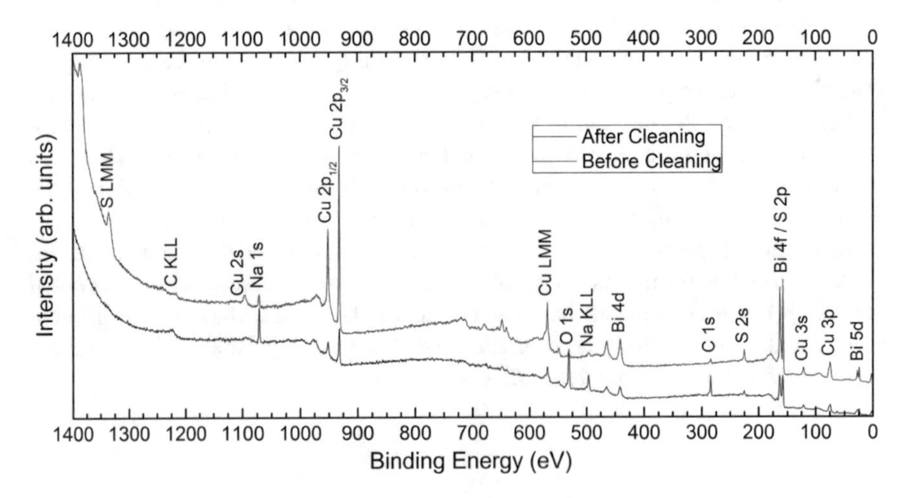

Fig. 4.4 XPS survey spectra for the CBS sample before and after surface cleaning

[12]CuO.

[13]See Sect. 2.3.1.

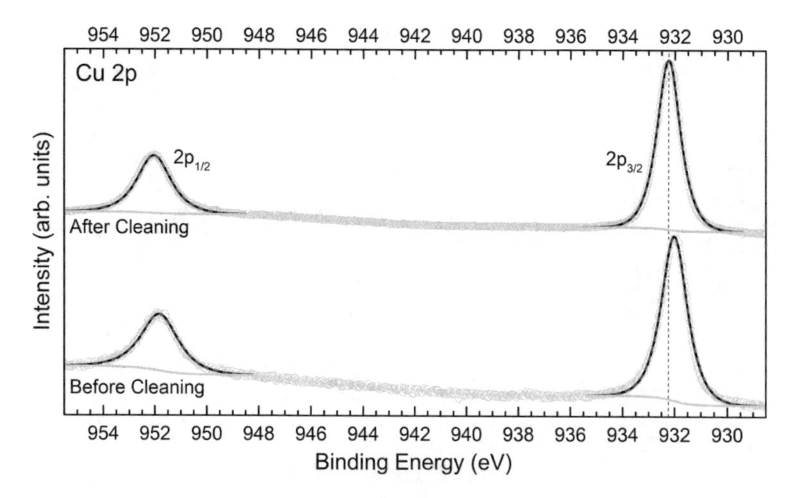

Fig. 4.5 XPS Cu 2*p* region of the CBS sample before and after surface cleaning. Fitted peaks shown in red and peak envelope in black

The fitted spectral region of Bi 4*f* and S 2*p* before and after cleaning the CBS sample is shown in Fig. 4.6 after applying the fitting procedure described in Appendix D, Sect. D.2.1. Three bismuth species are present in the sample before and after cleaning. The strongest, sharpest doublet (red dash) is attributed to Bi^{3+} in the crystalline CBS. After cleaning, this peak is determined to have a broader FWHM, which is believed to arise from damage to the Bi coordination environment in the crystalline material, induced by the sputtering stage of the cleaning [72, 73]. The wider Bi doublet located at higher BE (pink dot) is attributed to Bi_2O_3, which depletes significantly after cleaning, suggesting that it is present in the form of surface oxidised bismuth, rather than bulk oxidised bismuth, formed during growth. The shift in BE between the CBS and Bi_2O_3 arises from the greater electronegativity of oxygen over sulphur [74, 75], rather than a change in oxidation state of the bismuth. The assignment of Bi_2O_3 is further corroborated by the presence of a corresponding oxygen peak in the O 1*s* spectra.[14] The oxide peak in the Bi 4*f* region could contain a different oxide, or a non-stoichiometric mixture of oxides, but the measured BE of 159.06 eV is in agreement with other studies of Bi_2O_3 [75–77] and although pentavalent bismuth is a possibility [78], Bi_2O_5 has been shown not to form during the air-oxidation of bismuth [79, 80].

The small Bi doublet at lower BE (blue dash dot) to that of CBS, which depletes after cleaning is assigned to metallic Bi.[15] The BE of the Bi $4f_{7/2}$ peak of 156.94 eV is in agreement with literature values for Bi metal [42, 46, 47, 76, 81] and also an independently measured Bi foil.[16] The fact that the peak for this metallic bismuth

[14]Presented in Sect. 4.4.1, specifically Fig. 4.13b (pink dot).

[15]The fitting of which is justified in Appendix D, Sect. D.2.1 when describing the fitting procedure.

[16]Presented in Appendix D, Sect. D.2.1, specifically Figure D.4 (blue data).

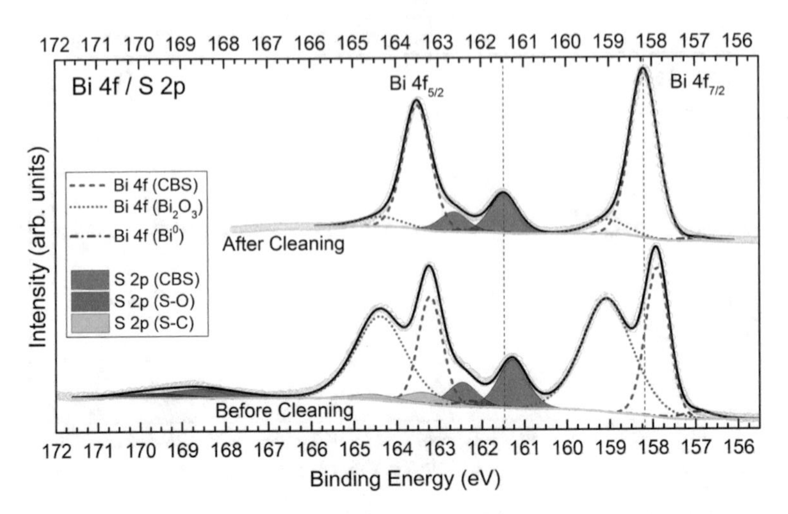

Fig. 4.6 XPS Bi 4*f* & S 2*p* region of the CBS sample before and after surface cleaning. Peak envelope shown in black

reduces after cleaning suggests that it is unreacted precursor, remnant on the surface: subsequently sputtered away or incorporated during the annealing. This is instead of sputter-reduced or sulphur-dissociated bismuth, which if present, one would expect to increase on sputtering, but this was not the case, as can be seen between the ratios of Bi from CBS and metallic Bi, which are given in Table 4.5.

Three sulphur species were fitted to the sample spectrum before cleaning, which reduced to two species after cleaning. The doublet at high BE which is not present after cleaning (purple shading) is attributed to sulphur bonded to oxygen, as a cation in the 6+ oxidation state, with a large shift[17] to higher BE than for sulphide species [82, 83]. This is present because of the surface oxidation of the sulphur and is removed after cleaning. Such a sulphur species has been observed in previous reports of CBS [5, 8] and other sulphide absorber materials [63, 84], but was not observed in CAS in Sect. 3.3.2, where it was determined to be lacking as a result of superior growth. Hence, the formation of sulphate species is believed to be a product of imperfect growth conditions, most probably when there is an excess of unreacted sulphur, which is available to oxidise.

The doublet with BE around 164 eV (green shading), which reduces significantly on cleaning, is attributed to sulphur-containing surface contamination (S–C) [85], as has been seen in other sulphides [64]. The strongest S doublet (red shading) is attributed to S^{2-} in CBS. From Fig. 4.1a, it can be seen that there are three inequivalent sulphur sites in the crystal structure of CBS, suggesting that there should be three S doublets attributable to CBS: one for each of the sites. However, at the resolution of this study, one doublet is reasoned to represent the sulphur environments, because the sulphur sites differ only slightly in terms of bond angles and lengths, without

[17] ~6 eV.

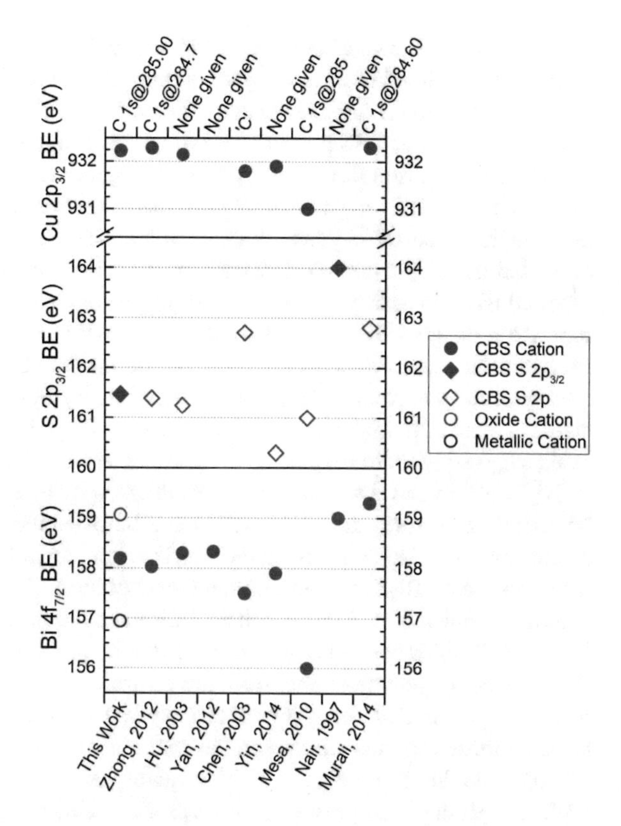

Fig. 4.7 Comparison between the XPS BE measured in this study for the clean CBS sample, and those reported previously in the literature for CBS. Literature references are given on the bottom x-axis [5, 6, 8, 13, 21, 22, 24, 25], and the quoted charge referencing method employed in the corresponding study are given on the top x-axis. Species assignments are as given in the corresponding reference, see text for discussion of the agreements and more details

changes in the coordination [3, 10]. It is noted that the above discussion in terms of peak trends is corroborated by the secondary regions that were recorded for Bi and S, which are further explored in Appendix D, Sect. D.2.1.

Comparison to the Literature

Table 4.1 shows the BE for the peaks associated with CBS that were presented above. As a relatively understudied material in terms of applications, and due to the complex nature of the XPS spectra, few studies have presented XPS results from CBS. Nevertheless, the values measured above are shown in Fig. 4.7 along with the reported literature values for other XPS-measured samples of CBS, which allows a critical analysis of these reports.

The measured BE of the Cu $2p_{3/2}$ peak, which is easy to analyse when compared to the Bi $4f$ and S $2p$ overlapping region, is in agreement with some other studies of CBS. Unfortunately however, these quoted BE are limited. Chen et al. [24], Hu et al. [22], and Yin and Jia [6] presented Cu $2p$ regions with poor SNR, whereas Murali et al. [8] and Zhong et al. [21] presented Cu $2p$ regions that have clear presence of an extra copper species,[18] which is overlooked because no fitting was performed.

[18]Most likely copper oxide.

As has been shown above, the Bi 4*f* and S 2*p* region can contain peaks which cannot be accounted for without peak fitting. Therefore, the BE previously reported for the Bi $4f_{7/2}$ and S $2p_{3/2}$ peaks fall within a substantial range, as shown in Fig. 4.7. Only two studies of CBS presented XPS spectra of the overlap region where a fitting model was used. Murali et al. [8] presented high quality data in the overlap region, but the Bi $4f_{7/2}$ BE does not agree with the value found in this study. This may be because the fitted Bi 4*f* peaks in that study appear very wide, suggesting that they consist of more than one species, with the possible evidence of an oxide because the reported BE is in agreement with the BE for the oxide measured here. Yan et al. [25] reported a Bi $4f_{7/2}$ BE which is in agreement with the value here, but the analysis found in that report appears confused regarding the nature of the spectra, and the quality of fit is unclear because of the poor quality figure presented. Both of these studies are also limited because they model the S 2*p* doublet as a very wide, single peak, suggesting there may be other species present, which are unaccounted for.

Bi_2S_3 is a material which is better established in the research community, having been studied longer than CBS. As it contains both bismuth and sulphur, XPS analyses of this material also suffer from the overlap problem, and it is common throughout the literature on Bi_2S_3 to find measurements of the S 2*s* region [65, 75, 86–95]. Yin and Jia [6] took this same approach with CBS and measured the S 2*s* region; however, to ignore the S 2*p* contribution detracts from the problem that underlying peaks can skew the Bi 4*f* peak positions from the centroid values. This, and the poor quality data presented in that study, is believed to be part of the reason for the disagreement between those reported values and the ones shown in Fig. 4.7.

Some studies of CBS present high quality XPS data, but the analysis is often limited. Nair et al. [5] presented depth profiles of different growths of CBS and the spectra show multiple peaks in the different regions. But, with no fitting applied, it cannot be determined what these originate from, and also, the BE for Cu $2p_{3/2}$ was not reported. A similar situation exists where the presence of other species is explicitly excluded in the discussion without corroborative fitting evidence. Yin and Jia [6] excluded the presence of Cu^{2+} because of the lack of satellite features; however, the lack of fitting and poor quality of data presented there means that this cannot be fully assessed. Mesa et al. [38] acknowledged the presence of oxidised bismuth by the shape of the spectra and excluded the formation of metallic bismuth, but with no fitting of this complex spectral region, these conclusions are weakened. Reports by Mesa et al. [13, 38] are also limited by the incomplete reporting of spectra, choosing only to show the survey spectrum, or to not show the Cu 2*p* region.

Finally, some studies present spectra which question the legitimacy of the data. Chen et al. [24], Hu et al. [22] and Zhong et al. [21] all present the Bi 4*f* and S 2*p* regions separately. If the overlapping region is fitted well, then it is possible to show these features individually: by subtracting the spectra. However, because there is discussion neither of the overlapping features, nor of any applied fitting procedures, one cannot draw the conclusion that this has been applied, and the reported BE are brought into question.

Given these severe weaknesses in the literature for CBS, it is believed that the spectra presented here, having taken all features into account, can provide a foundation onto which further studies of CBS can be supported. This can then allow further development of this material, especially as the BE reported here are in agreement to a certain degree with the literature. Also, because of the similar Bi–S crystal structure units for CBS and Bi_2S_3 [33, 34], it is expected that Bi $4f_{7/2}$ and S $2p_{3/2}$ BE should be comparable for both materials and the values measured here are in general agreement with representative studies of Bi_2S_3 in the literature [65, 75, 89, 95–97]. Furthermore, the fitting procedure applied here can be used for any other materials which contain both bismuth and sulphur, especially Bi_2S_3, which finds technological use in photocatalysis, optoelectronics, thermoelectrics, and sensors [87, 91, 93, 95, 96, 98].

4.3.3 *Natural Band Alignments*

The bandgap of CBS is somewhat contested in the literature as to its nature and size. Estrella et al. [15], Gerein and Haber [12] and Zeng et al. [7] all measured the bandgap of CBS, finding a linear representation of the Tauc plot when using an exponent value of $n = 3/2$ in Eq. (2.37),[19] which corresponded to a direct forbidden bandgap [99] of 1.28, 1.2, and 1.4 eV for the different studies, respectively. Other experimental measurements of the bandgap of CBS from absorption spectra by Mesa et al. [13], Viezbicke and Birnie [23], Mesa and Gordillo [14], Liu et al. [27], Murali et al. [8] and Yan et al. [100] fitted a linear region to a Tauc plot relating to a direct allowed bandgap, resulting in values from 1.39–1.72 eV. However, these have poor fittings, and there is still significant absorption below the bandgap value, suggesting there is another onset present in the spectrum.

The true nature of the bandgap of CBS is thought to be different to both of these interpretations. Yakushev et al. [17] presented photoreflectance measurements of CBS which demonstrated the presence of two resonances, attributed to two different bandgaps[20] due to a splitting of the VB. From absorption measurements corroborated by calculated absorption spectra,[21] it is believed that the fundamental bandgap of CBS is around 1.2 eV and indirect, but because of the low density of states here,[22] the absorption is dominated by higher energy direct transitions to the split VB around 1.4 eV, corresponding with the values found in the literature. Both of these transitions are still suitable for PV use, but because of the presence of the lower energy transition, this could limit the V_{OC} of CBS cells below the accepted bandgap of 1.4 eV.

Although there are no reports on the band offsets of a junction formed with CBS, and this cannot be performed without the growth of the partner material, there is still

[19] See Sect. 2.5.2.

[20] Of the order 1.2 and 1.5 eV.

[21] Data not shown, but manuscript in preparation.

[22] See Fig. 4.11b.

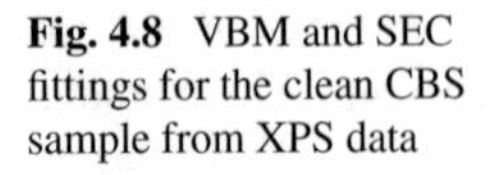
Fig. 4.8 VBM and SEC fittings for the clean CBS sample from XPS data

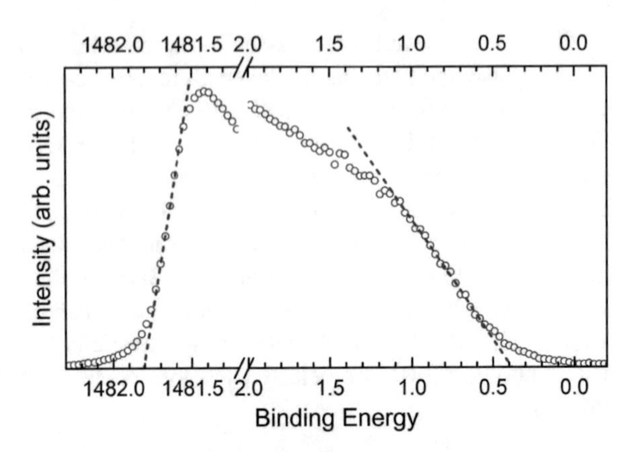

merit in measuring the natural band positions of a material[23] [101, 102], especially at this early stage in development, in order to help choose suitable partner materials.

XPS was used to measure the IP and WF of the clean CBS sample, the fittings of the VBM and SEC are shown in Fig. 4.8, and the IP and WF were determined to be 5.18 and 4.79 eV, respectively, which are also shown in Table 4.1. There are few reports on the band levels of this material in the literature. Only Mesa et al. [31, 38, 103] have published results of the WF of CBS and these values are lower than the values measured here. These reports utilised KPFM to measure the WF of CBS, but no in vacuo cleaning procedure was described. One report utilised an ex vacuo chemical etch, but substantial contamination still remained after the etch and was revealed by XPS [38]. It is therefore concluded that the lower WF reported in these articles were caused by the contamination, especially as KPFM takes an average value of WF measured over a sampling area, whereas XPS measures the lowest WF within a sampling area, no matter how small this region [104]. As of yet, no studies have reported IP values for CBS. It is believed that the slight remnant bismuth oxide on the surface after cleaning will have little to no effect on the values measured here. This is because Bi_2O_3 has been shown to have a very high IP[24] compared to the materials under consideration here and therefore will not affect the measurements from XPS: it is not detected, even before cleaning when the oxide layer was large. Therefore, after cleaning, it is believed that the IP measured here is representative of CBS.

By using the IP value for the sample after cleaning, taking the bandgap of CBS to be 1.40 eV from the discussion above, and by following a vacuum alignment procedure [107], the natural band offsets between CBS after cleaning, CAS, CdTe, CIGS, CZTS, and the common n-type window layer material CdS, are shown in Fig. 4.9.

---

[23] See Sect. 1.6.2.

[24] >7.5 eV [105, 106].

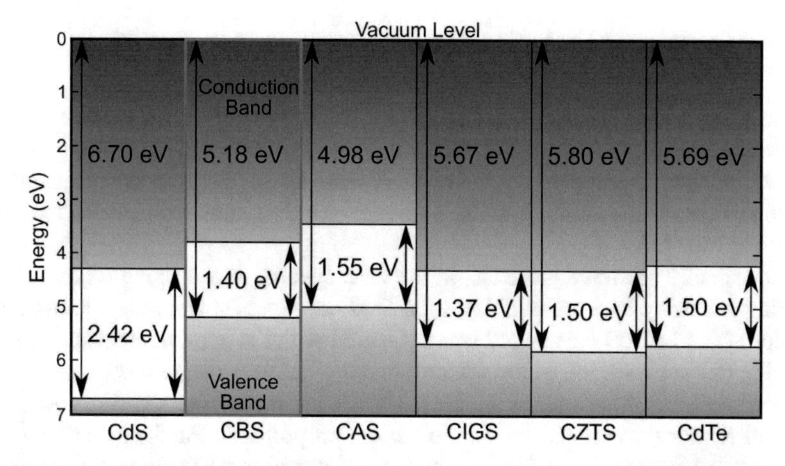

Fig. 4.9 Vacuum-aligned band diagram for CBS from IP measurements using XPS and bandgap from the literature. Comparison with other common absorbers and the common n-type window layer material, CdS. Literature values of IP and bandgap are taken for CdS [108], CIGS [109], CZTS [108] and CdTe [110], with the values for CAS taken from Chap. 3

New absorber materials for solar cells can suffer from inappropriate cell-architecture design in terms of band misalignments [108]. Commonly appropriated into cell design, no doubt because of the success with CIGS [111] and CdTe [112], CdS as the n-type window layer would be an inappropriate choice for CBS cells because, from Fig. 4.9, it can be seen that there is a negative CBO for CdS/CBS, present because of the low IP compared to CIGS and CZTS, which is in a similar vein to the case with CAS.[25] Furthermore, any concerns of the environmental impact associated with the use of CdS are eradicated if this material is not used at all. It has been previously shown that optimal efficiencies are produced in type I hetero-junctions which have the CBM of the n-type layer 0–0.4 eV higher than that of the absorber layer [114, 113], and that with a large CBO of the type shown in Fig. 4.9 between CdS and CBS, a recombination centre could be present [115]. Therefore, it is likely that a material with a higher CBM and more environmentally friendly elements than CdS is required, and on the other side of the junction, a lower WF contact metal may also be required. Such a strategy was also necessitated for SnS solar cells[26] [64, 108], where the use of alternative materials yielded promising results [113, 116].

[25] See Chap. 3.

[26] See Chap. 5.

4.3.4 Density of States Analysis

It is important for solar absorbers that the electronic structure in the valence and conduction bands are studied, so that the bonding mechanisms and electronic properties can be understood and predictions made. A comparison between the measured VB spectra of the clean sample from XPS, along with the simulated pDoS in the VB is shown in Fig. 4.10.

Inelastically scattered electrons in the XPS spectra were accounted for by subtracting a Shirley background. The total DoS curve arises from the summation of the corrected[27] pDoS curves and was compared with the experimental spectrum by aligning to the maximum of the peak around 3 eV. The agreement between the calculated DoS and XPS spectra in the region at the top of the VB (I) is excellent, with all features accounted for, in the correct proportions. Final state effects were unaccounted for, but it is noted that final-state relaxation will tend to cause features at the bottom of the VB to be shifted nearer to the top in XPS spectra [117]. This explains why the features lower in the VB (II & III) are present with approximately the correct intensities, but are shifted to lower BE in the XPS spectra.

The agreement shown between the experiment and theory in Fig. 4.10 is supportive of the results. However, it is clear that the XPS spectra, at least at the top of the VB, is

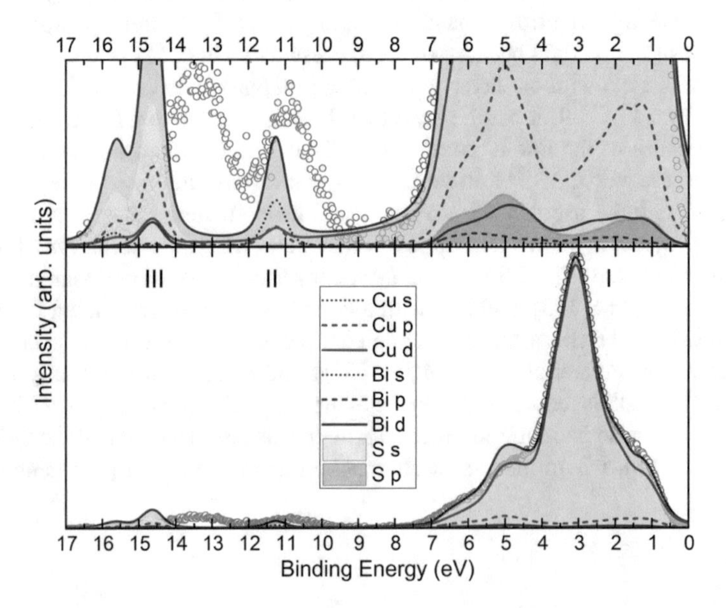

Fig. 4.10 Simulated and measured VB spectra for CBS with respect to the Fermi-level at 0 eV. Background subtracted XPS data is compared with broadened and corrected partial DoS curves. Top spectra shows intensity zoomed region to better show the underlying curves. Green data are from XPS and the black curve with grey shading is the total summed DoS

[27] Using the parameters described in Appendix C, Sect. C.3.

dominated by Cu 3*d* states because of the larger cross-section of *d* states compared with other orbitals [66]. Therefore, to better demonstrate the interactions between the orbitals present, the pDoS curves before photoemission corrections were applied are presented in Fig. 4.11a, both for the valence and conduction bands. Without these corrections, Cu 3*d* states are still the majority contribution to the VB, but other features are now also discernible. The Cu 3*d* states in the VB are split into three distinct regions (IV, V, IV*), which result from the tetrahedra-based Cu–S bonding units. In materials where the crystal structure consists of regular tetrahedral Cu-anion units, the crystal field splits the Cu *d* states into a non-bonding e_g doublet, and a t_{2g} triplet, which is able to bond with the anion *p* states. As the distorted Cu–S units in CBS are derived from tetrahedra, the form of the DoS observed here is similar to that of regular tetrahedral materials [118, 119]. That is, there is a large central feature (V) consisting of the non-bonding e states straddled in energy by two weaker features (IV, IV*), representing the t_2 orbitals hybridised with S *p* orbitals.[28] However, because the Cu–S units in CBS are distorted, the ratios between the bonding and non-bonding states are different to those from regular tetrahedra and there are other features[29] present in the DoS which are attributed to the crystal distortion, allowing some hybridisation which is symmetry forbidden in regular structures, and vice versa. Such an observation is further backed by the top of the VB of CBS more closely resembling that of Cu_2S[30] [120], rather than that of CAS[31] [36].

Below the VB are two distinct shallow CL regions: one (VI) consisting mainly of Bi *s* states, and another (VII) consisting of mostly S *s* states. The configuration energies (CE) of the orbitals are shown in Fig. 4.11c and from this, it is concluded that the Bi 6*s* and S 3*s* orbitals are semi-localised because of the strength of features VI and VII and also the lack of contribution of these orbitals to the top of the VB. At the CBM and encompassing most of the first CB are antibonding states of Bi *p* mixed with S *p* (IX*). The corresponding bonding states are dispersed throughout the VB, supplementary to the Cu *d* states (IX).

Further comprehension of the bonding nature is gained from studying the states from the Bi cation at the VBM and CBM, which is shown more clearly in the zoomed DoS, shown in Fig. 4.11b. Due to the stoichiometry of CBS and the prominence of Cu states, those states arising from Bi hybridisation are somewhat suppressed; however, it is important to note that their presence may have some effect on the properties. It has been suggested that in the family of materials to which CBS belongs, the presence of cation *s* states at the top of the VB becomes less significant as the second cation moves down group V [10]. This can be explained by the energy separation of the cation *s* and anion *p* orbitals becoming larger as one moves down the group and the interaction between these orbitals becomes less significant. At first it appears that Bi *s* states are present in the VB only as the anti-bonding hybridisation with S *p* states (X*). However, upon closer inspection, it is found that there is a contribution from

[28] Bonding and antibonding states, respectively.

[29] Shoulders and peaks.

[30] Which also has highly distorted tetrahedral structure units. See Fig. 6.2.

[31] Which has near-regular Cu–S tetrahedral structure units. See Fig. 3.1.

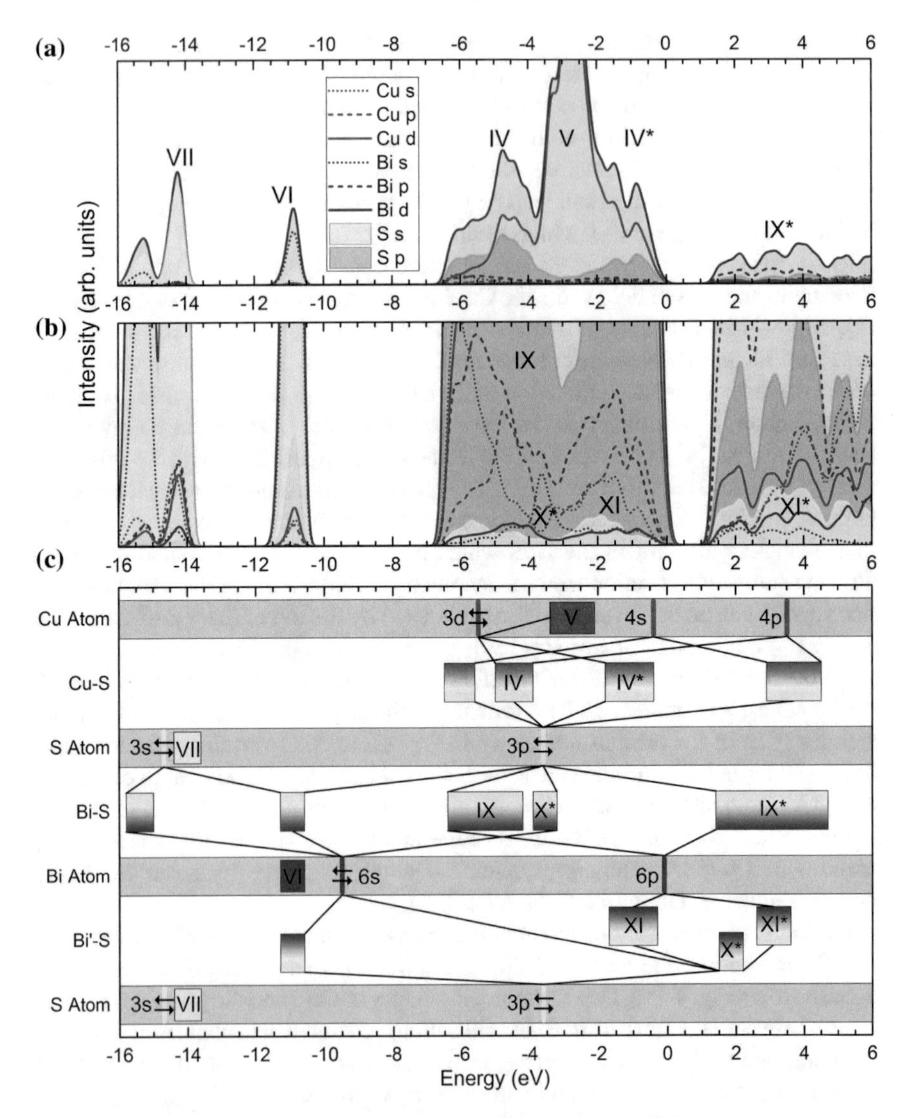

Fig. 4.11 **a** Total and partial electronic DoS curves for CBS, with **b** intensity zoomed region to more clearly show the underlying orbitals. Curves have been convolved with a Gaussian function (0.3 eV FWHM) in order to better distinguish features. DoS curves are aligned to the VBM. Black curve with grey shading is the total summed DoS. **c** CE for the valence orbitals [121, 122] displayed with a schematic of the bonding hybridisations as discussed in the text. Bismuth-sulphur bonding is separated into regular bonding (Bi–S) and lone-pair bonding (Bi′–S). It is noted that the CE values do not take ionisation, multi-electron occupancy or hybridisation into account and are shown only as a schematic guide. Part labels are discussed and referred to throughout the text

Bi *s* states in the CB (XI*), of an intensity which is comparable to that near the VBM (XI). It is therefore concluded that this intensity shows that the Bi *s* states hybridise somewhat via the revised lone-pair model [123]: the antibonding Bi *s*/S *p* states (X*) are in close energy proximity to the Bi *p* orbitals and therefore further hybridise with these, leading to full bonding states close to the VBM (XI). This bonding regime is illustrated on the orbital diagram in Fig. 4.11c, but this interaction is weaker than observed in other materials with lone-pair activity, which is expected, and thought to pose little significance to the properties of this material. This is reflected in the crystal structure: when compared to materials with significant lone-pair activity [123], the void space into which the lone-pair electron density is projected is not as substantial in CBS. The remainder of the majority states present in the valence and conduction bands are further summarised in the bonding schematic in Fig. 4.11c.

Previous theoretical calculations of the valence and conduction bands of CBS focussed on the relationship between the band structure and the optical absorption spectra [16, 17, 124], with only one study by Kehoe et al. [10] discussing the bonding nature, but not in relation to the band levels. Nevertheless, the general forms of the pDoS curves are in agreement with those calculated here. It was noted by Yu et al. [124], how the *s*-like nature of the VBM for this family of materials with a group V^{3+} cation gives stronger absorption than a group V^{5+} cation, and consequently, higher absorption than CIGS. This is one of the reasons why this family of materials is receiving increased research interest.

Although the bandgap and absorption characteristics have been studied both experimentally and theoretically, and have been shown to be PV suitable, the band positions, which facilitate cell-architecture design, have been somewhat disregarded. It is believed that the measurements of IP found here, coupled with the nature of the bonding in the valence and conduction bands, are in agreement and can be used to aid the design of future cells. The IP measured here is somewhat lower than that of other common absorbers[32], and will therefore require more suitable partner materials within a solar cell structure. From an orbital bonding point of view, it is believed that this low IP is inherent to this family of materials and arises from the unusual bonding situation when compared to other absorbers, created mainly by the inclusion of the group V second cation. Reference is henceforth made to Fig. 4.12 for the relative positions of the valence orbitals of the constituent elements of common solar absorbers and their occupancy in the materials, given the formal ionic oxidation states.

CdTe is an 'archetypal II–VI' semiconductor which has a VBM consisting of bonding cation *p*/anion *p* states, with the corresponding antibonding states residing high in the CB, and a CBM consisting of antibonding cation *s*/anion *p* states, with the corresponding bonding states residing low in the VB [125, 126]. The cation *d* states in CdTe are too strongly bound to interact with valence states and so do not contribute to the top of the VB.

[32]Discussed in Sect. 4.3.3.

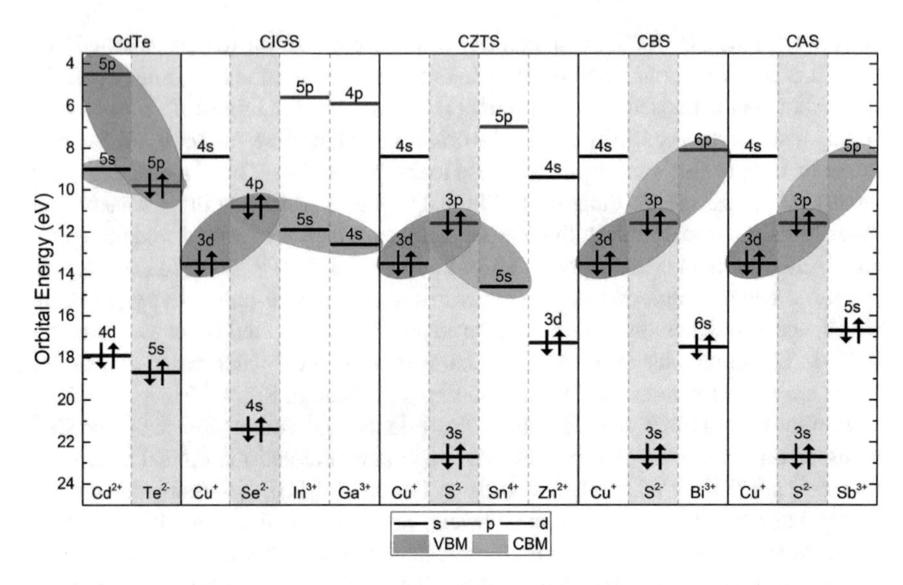

Fig. 4.12 CE [121, 122] for the valence orbitals of CdTe, CIGS, CZTS, CBS and CAS. The formal ionic occupancies of the orbitals within the materials are marked and the main orbital contributions to the VBM and CBM are highlighted. Cd 5*p* is placed at an estimated value following the trend from Sn & In

In the copper-containing compounds[33] however, the full Cu *d* orbitals are in energetic proximity to the full anion *p* orbitals and therefore the VB of these materials is dominated by Cu *d*/anion *p* states in the manner as described above. This is beneficial for PV absorbers and solar cell production for several reasons: the strong copper vacancy gives a shallow defect, which is responsible for the intrinsic p-type conductivity of these materials [10, 127–129]; the high-lying Cu *d* orbital produces a VBM level which is favourable for junction formation and ohmic contacting [60]; and the antibonding states at the top of the VB have been shown to produce materials which are defect-tolerant [130]. The binary Cu-chalcogenide Cu_2S suffered from instability, which was then inhibited by the addition of another cation with an unpopulated valence *d*-band, generating the CB and hence preventing copper self-reduction upon illumination [131]. These features should extend to, and be present in CBS as well.

Although there are differences between the VB of CdTe and the copper-containing compounds, the bottom of the CB of CdTe, CIGS and CZTS are similar. They are comprised of antibonding non-Cu cation *s*/anion *p* states, which are in close proximity.[34]

[33]CIGS, CZTS, CAS, and CBS.

[34]See Fig. 4.12.

CBS also has the same main Cu *d*/anion *p* contribution to the VB as CZTS and CIGS, and superficially the DoS reflect each other. However, the bonding within the CB is somewhat different, and there are underlying differences in the VB as well[35], which is the cause for the low IP compared with the other absorbers. The main reason for this is the electronic structure of the non-Cu cation. In CIGS, this is $(In/Ga)^{3+}$ and in CZTS is Sn^{4+}, which have unoccupied valence *s* orbitals close in energy to the occupied anion *p* orbitals. Antibonding states, resultant from the hybridisation of these levels make up the CBM, whereas the bonding states lie deeper in the VB, separate from the Cu *d* features at the VBM. However, in CBS, the non-Cu cation is Bi^{3+}, which has more tightly bound, occupied valence *s* orbitals, and it is the more weakly bound valence *p* orbitals that are closer in energy to the anion *p* orbitals. Thus, because of the differences in energy seen in Fig. 4.12, the Bi *p*/S *p* antibonding states form the CBM of CBS, and the bonding states lie shallower in the VB, overlapping somewhat with the Cu *d* states. This is further coupled with a slight contribution from Bi *s* states which lie at the VBM because of the model described above. These extra states influence the band levels, causing the VBM to rise, which results in a lower IP, and creates the band misalignment with CdS[36], which is thought to create a recombination centre in PV devices. In CIGS and CZTS, the bonding states from non-Cu cation *s*/anion *p* hybridisation lie much deeper in the VB and do not overlap with the Cu *d* states or contribute to the VBM, resulting in a higher IP [119, 132]. Due to the similar situation seen with CAS[37], this bonding regime is thought to be an inherent feature for this family of materials.

Previous studies found that this family of materials demonstrates much better absorption than CIGS [124]. This, and the insights into the band levels gained here, can be understood as a result of the bonding nature, mainly from the presence of second-cation *s* electrons, which are not present in CIGS. It is then concluded that the success of CIGS over CBS is not due to inherent problems with CBS as an absorber, but rather is due the timescale of development that CIGS has enjoyed, and that, with appropriate research into material growth and cell design, CBS maintains definite potential as a solar PV absorber.

4.4 The Effects of Contamination on Cu_3BiS_3

In Sect. 4.3, the focus was to present standard XPS spectra of CBS, relating this to the band structure and bonding nature. However, given that XPS can also be used to assess the level of surface contamination present on materials, and the effect that these have, it is now prudent to discuss these effects regarding CBS.

---

[35]Shown in Figs. 4.11 and 4.12.

[36]When compared to other absorbers, seen in Fig. 4.9.

[37]In Chap. 3.

Table 4.2 XPS BE for the CBS-relevant peaks of the CBS sample before and after surface cleaning

Sample	Cu_3BiS_3				
	Cu 2$p_{3/2}$	**Bi 4$f_{7/2}$**	**S 2$p_{3/2}$**	**Bi 4$d_{5/2}$**	**S 2s**
Before Cleaning	932.01 (1.16)	157.89 (0.65)	161.28 (0.81)	441.63 (3.54)	225.71 (1.87)
After Cleaning	932.23 (1.05)	158.19 (0.77)	161.47 (0.78)	441.47 (4.22)	225.84 (1.75)

All values are given in eV and the peak FWHM are given in parentheses

Table 4.3 XPS BE for the Bi_2O_3-relevant peaks of the CBS sample before and after surface cleaning

Sample	Bi_2O_3		
	Bi 4$f_{7/2}$	**Bi 4$d_{5/2}$**	**O 1s**
Before Cleaning	159.06 (1.37)	442.52 (7.47)	529.63 (0.99)
After Cleaning	159.06 (1.02)	442.51 (5.61)	529.63 (1.00)

All values are given in eV and the peak FWHM are given in parentheses

Table 4.4 XPS BE for the contamination-relevant peaks of the CBS sample before and after surface cleaning

Sample	Contamination							
	Bi 4$f_{7/2}$ (Bi^0)	**Bi 4$d_{5/2}$ (Bi^0)**	**O 1s (adv.)**	**S 2$p_{3/2}$ (S-C)**	**S 2s (S-C)**	**S 2$p_{3/2}$ (S-O)**	**S 2s (S-O)**	**Na 1s**
Before Cleaning	156.94 (0.82)	440.00 (4.49)	531.74 (1.84)	163.33 (1.23)	228.04 (2.84)	168.50 (2.04)	232.55 (2.97)	1071.68 (1.54)
After Cleaning	156.94 (0.82)	440.00 (4.51)	531.94 (2.19)	163.33 (1.26)	228.04 (2.84)	—	—	1071.93 (1.60)

All values are given in eV and the peak FWHM are given in parentheses

Table 4.5 Atomic percentages derived from XPS measurements for CBS before and after surface cleaning

Sample	Cu_3BiS_3			Bi_2O_3		Contamination					
	Cu	**Bi**	**S**	**Bi**	**O**	**Bi^0**	**O (adv.)**	**S (S-C)**	**S (S-O)**	**Na**	**C (adv.)**
Before Cleaning	3.9	1.5	7.3	2.4	1.9	0.1	27.7	2.8	3.7	20.9	27.8
After Cleaning	23.4	10.1	29.8	1.2	0.4	0.3	11.0	1.3	0.0	9.7	12.7

BE for all of the fitted peaks in this study are given in Table 4.2 for the peaks concerning CBS, Table 4.3 for the peaks concerning Bi_2O_3, and Table 4.4 for the peaks concerning the remainder of the contaminant species.

4.4.1 The Presence of Contaminants in the Core-Level XPS Spectra

After the presence of carbon, oxygen and sodium was observed in the CBS sample in Fig. 4.4, the C 1*s*, O 1*s* and Na 1*s* regions for the CBS sample before and after cleaning were also recorded, and are presented in Fig. 4.13a–c, respectively.

The C 1*s* spectra for the sample before cleaning shows three peaks (orange dash) which are all attributed to the different bonding regimes associated with adventitious hydrocarbon contamination [41]. These reduce to two peaks after cleaning. The prominent peak was used for the charge correction as described in Sect. 4.2.2 and is associated with C–H or C–C bonds [41].

The O 1*s* spectra both before and after surface cleaning show three features. The wide, low intensity feature at high BE (cyan dash dot) is due to Auger features from sodium. This complicates the fitting of this region because the spectral shape of this feature is unknown and therefore, there is less confidence in the further fitting of the O 1*s* peaks. Nevertheless, there are two definite O 1*s* peaks present both before and after cleaning. The main peak with highest intensity (orange dash) is attributed to adventitious oxygen [133] and the small peak at lower BE (pink dot) is the oxygen from oxidised bismuth[38], which is in agreement with the literature [76, 134], and the effects of which will be discussed in Sect. 4.4.2. Both O 1*s* peaks reduce on cleaning, suggesting that they are both present only at the surface of the sample. Although the adventitious oxygen has been fitted with a single peak, it is believed that before surface cleaning, the peak from the sulphur bonded to oxygen[39] also resides around

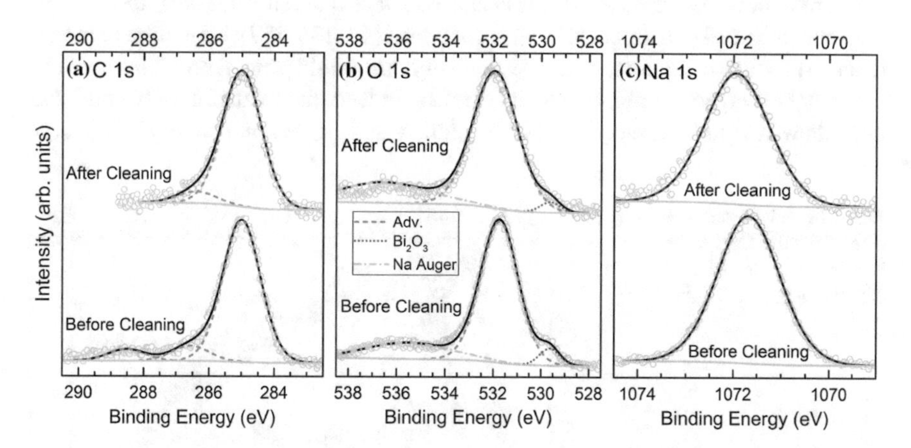

Fig. 4.13 XPS contamination regions of the CBS sample before and after surface cleaning. **a** C 1*s* region, **b** O 1*s* and Na Auger region, **c** Na 1*s* region. Peak envelope shown in black

[38] As discussed in Sect. 4.3.2.

[39] Evidenced from the S 2*p* spectrum in Fig. 4.6.

this energy [85]. This was not fitted further due to the complications associated with the sodium Auger features.

The Na 1*s* region was fitted with one peak both before and after cleaning (red dash); however, it is noted that any shifts in the Na 1*s* peak are very small so it is possible that more than one species is present. The shift seen in this peak before and after the surface cleaning is consistent with the shift of the VBM[40], suggesting that the sodium is incorporated with the material, because peaks associated with CBS also shift by this amount before and after cleaning. The impact of the presence of sodium was discussed in Sect. 4.3.2.

4.4.2 Band Level Effects

In Sect. 4.3.3, the IP and WF of the clean CBS were discussed in terms of band alignments, and the fitting of the SEC and VBM for the clean sample were shown in Fig. 4.8. Here, these are compared with the same values for the sample before surface cleaning, the fittings of the SEC and VBM for which are shown in Fig. 4.14. As can be seen in Table 4.1, after cleaning, the IP and WF are found to increase, this is because of the removal of carbon contamination, which is known to lower the IP and WF of materials [135].

The demonstrated readiness of the CBS film to oxidise, along with other reports in the literature of the oxidation of this material, shown and discussed in Sect. 4.3.2, are potential issues which must be acknowledged and overcome if devices incorporating CBS are to become practical. This is because surface oxidation has been shown to be detrimental to cell performance in other materials [63, 136, 137]. Oxidation of metals tends to produce insulators, which, even if very thin, could present problems between the solar cell layers, inhibiting charge carriers. Indeed, the oxidation of bismuth has been shown to produce many phases, with both n- and p-type conductivity [138], large

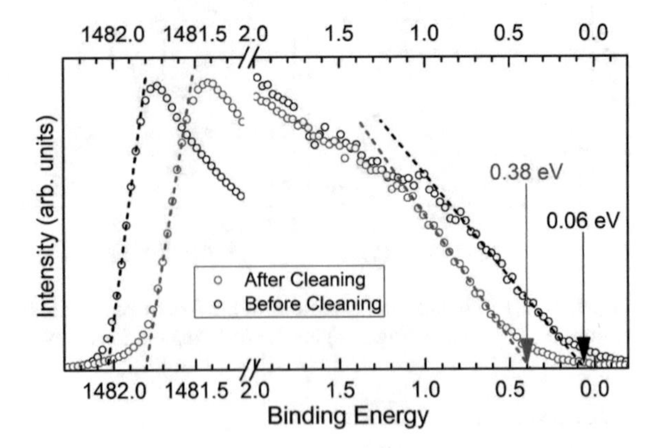

Fig. 4.14 VBM and SEC fittings for the CBS sample before and after surface cleaning from XPS data

[40] See Fig. 4.14.

bandgaps [79, 139] and high IP [140], all of which could pose problems within CBS when considering the band positions shown in Fig. 4.9. Such surface oxide formation can be studied using XPS because it is surface-sensitive, and does not discriminate with regards to crystallinity compared to XRD. Thus, it is posited that XPS can be used to great effect characterising CBS in conjunction with other techniques such as XRD, SEM, and RS.

4.4.3 Surface Stoichiometry and Binding Energy Shifts

The XPS spectra shown throughout this chapter have been normalised in order to more clearly show the different species present. For comparison between the amount of different species, reference is made to the XPS-derived surface stoichiometry values, detailed in Table 4.5. As no transmission function correction was available for the instrumentation used here, absolute values are not exact and should be interpreted with caution; however, comparisons between values and also trends before and after cleaning are more accurate. In order to compare the stoichiometry values of the CBS material, corrections were applied to the Cu/S ratios, which derived from independent stoichiometry measurements of standard CuS and Cu_2S powders[41]. Literature value corrections [141] were applied to the Bi/S ratios as the transmission function should not have much effect between the Bi 4*f* and S 2*p* regions as they are energetically similar.

After cleaning, it can be seen in Table 4.1 that there is a shift to higher BE for the peaks in the range 0.19–0.30 eV. All of the peaks shifting to higher BE suggest that this is due to Fermi-level shifting after cleaning, which is also evidenced by the shift in the position of the VBM, shown in Fig. 4.14. Also, because there are small differences between the shifts, this suggests that some slight stoichiometric changes have taken place before and after cleaning. Indeed, the surface stoichiometry values in Table 4.5 show that such changes have taken place. In CBS, the ratio of Cu:Bi:S should be 3:1:3. By normalising to a Bi content of 1.0, the stoichiometry of the CBS sample before cleaning is 2.6:1.0:4.9 and after cleaning is 2.3:1.0:3.0. Before cleaning, the surface of the sample appears to have an excess of sulphur. After cleaning, the Bi:S ratio is near stoichiometric but the surface is copper-poor. It is believed that because chemical shifts between sulphide species are small [142], this sulphur-rich surface before cleaning is due to the presence of partially reacted sulphide[42] formed during the growth, which is then either incorporated or lost during the anneal in the cleaning process.

The 'copper-poor' surface composition of the clean sample is not unexpected and has been observed in other studies of Cu-based solar absorbers, not least CIGS [143, 144] and CZTS [63, 136, 145]. Here, a similar situation could exist, where surface effects are responsible for the formation of small amounts of a Cu-free surface

[41]Further details of this derivation are given in Sect. 6.5.1.

[42]Consisting of a peak which overlaps that from CBS.

species. Such an explanation could be relevant when designing cell architectures and warrants further study because this surface species could form an unexpected junction, as was observed with CIGS [146, 147].

CIGS solar cells were developed by adding another element in order to overcome the instability problems occurring in the Cu_2S solar cells [37], which tended to degrade because of the ionic migration of copper at room temperature [148, 149]. A similar stability has been observed with the addition of antimony and bismuth to the base of Cu_2S [150], and thus, this family of ternary copper chalcogenides should not suffer from the same degradation problems that marred Cu_2S cells. However, it has been shown in crystallographic studies that CBS undergoes a series of phase transitions at low temperatures[43] [151], and after such, CBS is an ionic conductor with mobile copper ions [152, 153]. Therefore, it is possible that during solar cell fabrication steps that require the use of higher temperatures,[44] this mobility of copper ions could be detrimental and lead to cell degradation as in Cu_2S. Although the surface off-stoichiometry seen here is believed to arise as a product of the growth, it is possible that the annealing step of the cleaning process could have caused copper migration away from the surface, leading to the surface copper-deficiency. The wider ramifications of these observations are as yet unknown, as there exists no study of the defects in CBS [10]. Therefore, it is clear that in order to further progress CBS as an absorber, a full defect analysis is required with the associated defect levels and their dependency on growth conditions.

4.5 Summary

For the first time, the experimentally measured VB spectrum from XPS and theoretically calculated DoS for the potential solar absorber, CBS have been compared and the agreement substantiates the subsequent analysis of the bonding nature of this material.

The first experimentally measured values of IP for this material have also been presented, which show that the band levels are too high to be suited to CdS as a window layer and that alternative n-type materials are required. The low IP is part-explained by how the VB has a major contribution of Cu *d* states, but also a minority contribution of Bi *p* states, which is dissimilar to the cases of CIGS and CZTS. The effects of the lone-pair Bi 6*s* electrons were also discussed and the Bi *p* nature of the CB is shown to be due to the use of group V Bi^{3+} which has full valence *s* states and empty valence *p* states: again dissimilar to the non-Cu cations in CIGS or CZTS, which have empty valence *s* states.

Additionally, a thorough analysis of the core-level XPS spectra of CBS has been presented, and a fitting model developed that helps disambiguate the complexities associated with the spectra, arising from the strong overlap of the Bi 4*f* and S 2*p*

[43] <200 °C.

[44] Such as the window layer deposition [30].

regions. Thereby, the crucial role that XPS can play when analysing this material for use in PV has been demonstrated, both in terms of the characterisation of the material grown, but also with respect to identifying and analysing the presence and effects of surface contamination. This is especially the case for its susceptibility to bismuth oxidation, which could easily be overlooked.

So, with these understandings, the promising potential shown by CBS is not dampened, although direct analogies with long-established PV absorbers should be avoided and new development strategies be implemented.

References

1. U.S. Geological Survey. Mineral commodity summaries 2016; 2016.
2. Hintze CAF. Handbuch Der Mineralogie; Cambridge; 1904.
3. Kocman V, Nuffield EW. The crystal structure of Wittichenite, Cu_3BiS_3. Acta Crystallogr Sect B Struct Crystallogr Cryst Chem. 1973;29 (11):2528–35.
4. Nuffield EW. Studies of mineral sulpho-salts: Xi-Wittichenite (Klaprothite). Econ Geol. 1947;42(12):147–60.
5. Nair PK, Huang L, Nair MTS, Hu H, Meyers EA, Zingaro RA. Formation of P-type Cu_3BiS_3 absorber thin films by annealing chemically deposited Bi_2S_3–CuS thin films. J Mater Res. 1997;12(3):651–6.
6. Yin J, Jia J. Synthesis of Cu_3BiS_3 nanosheet films on TiO_2 nanorod arrays by a solvothermal route and their photoelectrochemical characteristics. Cryst Eng Comm. 2014;16(13):2795.
7. Zeng Y, Li H, Qu B, Xiang B, Wang L, Zhang Q, Li Q, Wang T, Wang Y. Facile synthesis of flower-like Cu_3BiS_3 hierarchical nanostructures and their electrochemical properties for lithium-ion batteries. Cryst Eng Comm. 2012;14(2):550–4.
8. Murali B, Madhuri M, Krupanidhi SB. Near-infrared photoactive Cu_3BiS_3 thin films by Co-evaporation. J Appl Phys. 2014;115(17):173109.
9. Liu J, Wang P, Zhang X, Wang L, Wang D, Gu Z, Tang J, Guo M, Cao M, Zhou H, Liu Y, Chen C. Rapid degradation and high renal clearance of Cu_3BiS_3 nanodots for efficient cancer diagnosis and photothermal therapy in vivo. ACS Nano. 2016;10(4):4587–98.
10. Kehoe AB, Temple DJ, Watson GW, Scanlon DO. Cu_3MCh_3 (M = Sb, Bi; Ch = S, Se) as candidate solar cell absorbers: insights from theory. Phys Chem Chem Phys. 2013;15(37):15477.
11. Colombara D, Peter LM, Hutchings K, Rogers KD, Schäfer S, Dufton JTR, Islam MS. Formation of Cu_3BiS_3 thin films via sulfurization of Bi–Cu metal precursors. Thin Solid Films. 2012;520(16):5165–71.
12. Gerein NJ, Haber JA. One-step synthesis and optical and electrical properties of thin film Cu_3BiS_3 for use as a solar absorber in photovoltaic devices. Chem Mater. 2006;18(26):6297–302.
13. Mesa F, Dussan A, Gordillo G. Study of the growth process and optoelectrical properties of nanocrystalline Cu_3BiS_3 thin films. Phys status solidi. 2010;7(3–4):NA-NA.
14. Mesa F, Gordillo G. Effect of preparation conditions on the properties of Cu_3BiS_3 thin films grown by a two-step process. J Phys: Conf Ser. 2009;167:12019.
15. Estrella V, Nair MTS, Nair PK. Semiconducting Cu_3BiS_3 thin films formed by the solid-state reaction of CuS and bismuth thin films. Semicond Sci Technol. 2003;18(2):190–4.
16. Kumar M, Persson C. Cu_3BiS_3 as a potential photovoltaic absorber with high optical efficiency. Appl Phys Lett. 2013;102(6):3–7.
17. Yakushev MV, Maiello P, Raadik T, Shaw MJ, Edwards PR, Krustok J, Mudryi AV, Forbes I, Martin RW. Electronic and structural characterisation of Cu_3BiS_3 thin films for the absorber layer of sustainable photovoltaics. Thin Solid Films. 2014;562:195–9.
18. Gerein N, Haber J. Synthesis and optical and electrical properties of thin films Cu_3BiS_3-a potential solar absorber for photovoltaic devices. In: 2006 IEEE 4th world conference on photovoltaic energy conference; IEEE, 2006; vol. 1. p. 564–6.

19. Gerein NJ, Haber JA. Synthesis of Cu_3BiS_3 thin films by heating metal and metal sulfide precursor films under hydrogen sulfide. Chem Mater. 2006;18(26):6289–96.
20. Zhou J, Bian G-Q, Zhu Q-Y, Zhang Y, Li C-Y, Dai J. Solvothermal crystal growth of $CuSbQ_2$ ($Q = S$, Se) and the correlation between macroscopic morphology and microscopic structure. J Solid State Chem. 2009;182(2):259–264.
21. Zhong J, Xiang W, Cai Q, Liang X. Synthesis, characterization and optical properties of flower-like Cu_3BiS_3 nanorods. Mater Lett. 2012;70:63–6.
22. Hu J, Deng B, Wang C, Tang K, Qian Y. Convenient hydrothermal decomposition process for preparation of nanocrystalline mineral Cu_3BiS_3 and $Pb_{1-x}Bi_{2x/3}S$. Mater Chem Phys. 2003;78(3):650–4.
23. Viezbicke BD, Birnie DP. Solvothermal synthesis of Cu_3BiS_3 enabled by precursor complexing. ACS Sustain Chem Eng. 2013;1(3):306–8.
24. Chen D, Shen G, Tang K, Liu X, Qian Y, Zhou G. The synthesis of Cu_3BiS_3 nanorods via a simple ethanol-thermal route. J Cryst Growth. 2003;253(1–4):512–6.
25. Yan J, Yu J, Zhang W, Li Y, Yang X, Li A, Yang X, Wang W, Wang J. Synthesis of Cu_3BiS_3 and $AgBiS_2$ crystallites with controlled morphology using hypocrellin template and their catalytic role in the polymerization of alkylsilane. J Mater Sci. 2012;47(9):4159–66.
26. Hu H, Gomez-Daza O, Nair PK. Screen-printed Cu_3BiS_3-polyacrylic acid composite coatings. J Mater Res. 1998;13(9):2453–6.
27. Liu S, Wang X, Nie L, Chen L, Yuan R. Spray pyrolysis deposition of Cu_3BiS_3 thin films. Thin Solid Films. 2015;585:72–5.
28. Mesa F, Dussan A, Paez-Sierra BA, Rodriguez-Hernandez H. Hall effect and transient surface photovoltage (SPV) study of Cu_3BiS_3 thin films abstract. Univ Sci. 2014;19(2):99–105.
29. Gerein NJ, Haber JA. Cu_3BiS_3, Cu_3BiS_4, Ga_3BiS_3 and $Cu_5Ga_2BiS_8$ as potential solar absorbers for thin film photovoltaics. In: Conference record of the thirty-first IEEE photovoltaic specialists conference, 2005; IEEE, 2005. p. 159–62.
30. Mesa F, Dussan A, Sandino J, Lichte H. Characterization of Al/Cu_3BiS_3/buffer/ZnO solar cells structure by TEM. J Nanopart Res. 2012;14(9).
31. Mesa F, Fajardo D. Study of heterostructures of Cu_3BiS_3—buffer layer measured by kelvin probe force microscopy measurements (KPFM) 1. Can J Phys. 2014;92(7/8):892–5.
32. Schorr S. Structural aspects of adamantine like multinary chalcogenides. Thin Solid Films. 2007;515(15):5985–91.
33. Lundegaard LF, Makovicky E, Boffa-Ballaran T, Balic-Zunic T. Crystal structure and cation lone electron pair activity of Bi_2S_3 between 0 and 10 GPa. Phys Chem Miner. 2005;32(8–9):578–84.
34. Caracas R, Gonze X. First-principles study of the electronic properties of A_2B_3 minerals, with $A = Bi$, Sb and $B = S$. Se Phys Chem Miner. 2005;32(4):295–300.
35. Evans HT. Crystal structure of low chalcocite. Nat Phys Sci. 1971;232(29):69–70.
36. Temple DJ, Kehoe AB, Allen JP, Watson GW, Scanlon DO. Geometry, electronic structure, and bonding in $CuMCh_2$ ($M = Sb$, Bi; $Ch = S$, Se): alternative solar cell absorber materials? J Phys Chem C. 2012;116(13):7334–40.
37. Rodríguez-Lazcano Y, Nair MTS, Nair PK. $CuSbS_2$ thin film formed through annealing chemically deposited Sb_2S_3–CuS thin films. J Cryst Growth. 2001;223(3):399–406.
38. Mesa F, Chamorro W, Vallejo W, Baier R, Dittrich T, Grimm A, Lux-Steiner MC, Sadewasser S. Junction formation of Cu_3BiS_3 Investigated by kelvin probe force microscopy and surface photovoltage measurements. Beilstein J Nanotechnol. 2012;3(1):277–84.
39. Septina W, Ikeda S, Iga Y, Harada T, Matsumura M. Thin film solar cell based on $CuSbS_2$ absorber fabricated from an electrochemically deposited metal stack. Thin Solid Films. 2014;550:700–4.
40. Siebentritt S. Why are kesterite solar cells not 20% efficient? Thin Solid Films. 2013;535(1):1–4.
41. Barr TL, Seal S. Nature of the use of adventitious carbon as a binding energy standard. J Vac Sci Technol A Vac Surf Film. 1995;13(3):1239.

42. Morgan WE, Stec WJ, van Wazer JR. Inner-orbital binding energy shifts of antimony and bismuth compounds. Inorg Chem. 1972;12(4):953–5.
43. Lebugle A, Axelsson U, Nyholm R, Mårtensson N. Experimental L and M core level binding energies for the metals ^{22}Ti to ^{30}Zn. Phys Scr. 1981;23(5A):825–7.
44. Coster D, De Kronig RL. New type of auger effect and its influence on the X-ray spectrum. Physica. 1935;2(1–12):13–24.
45. Nyholm R, Martensson N, Lebugle A, Axelsson U. Auger and Coster-Kronig broadening effects in the 2p and 3p photoelectron spectra from the metals ^{22}Ti-^{30}Zn. J Phys F: Met Phys. 1981;11(8):1727–33.
46. Nyholm R, Berndtsson A, Martensson N. Core level binding energies for the elements Hf to Bi (Z = 72–83). J Phys C: Solid State Phys. 1980;13(36):L1091–6.
47. Shalvoy RB, Fisher GB, Stiles PJ. Bond Ionicity and structural stability of some average-valence-five materials studied by X-ray photoemission. Phys Rev B. 1977;15(4):1680–97.
48. Barrie A, Drummond IW, Herd QC. Correlation of calculated and measured 2p spin-orbit splitting by electron spectroscopy using monochromatic X-radiation. J Electron Spectros Relat Phenom. 1974;5(1):217–25.
49. Kresse G, Hafner J. Ab Initio molecular dynamics for liquid metals. Phys Rev B. 1993;47(1):558–61.
50. Kresse G, Hafner J. Ab Initio molecular-dynamics simulation of the liquid-metal–amorphous-semiconductor transition in Germanium. Phys Rev B. 1994;49(20):14251–69.
51. Kresse G, Furthmüller J. Efficiency of Ab-Initio Total energy calculations for metals and semiconductors using a plane-wave basis set. Comput Mater Sci. 1996;6(1):15–50.
52. Kresse G, Furthmüller J. Efficient iterative schemes for Ab Initio total-energy calculations using a plane-wave basis set. Phys Rev B. 1996;54(16):11169–86.
53. Deshmukh SG, Patel SJ, Patel KK, Panchal AK, Kheraj V. Effect of annealing temperature on flowerlike Cu_3BiS_3 thin films grown by chemical bath deposition. J Electron Mater. 2017;2–4.
54. Nair PK, Nair SM, Hu H, Huang L, Zingaro RA, Meyers EA. New P-type absorber films formed by interfacial diffusion in chemically deposited metal chalcogenide multilayer films. In: Lampert CM, Deb SK, Granqvist C-G, editors. Proceedings of the SPIE; 1995, vol. 2531. p. 208–19.
55. Yakushev MV, Maiello P, Raadik T, Shaw MJ, Edwards PR, Krustok J, Mudryi AV, Forbes I, Martin RW. Investigation of the structural, optical and electrical properties of Cu_3BiS_3 semiconducting thin films. Energy Procedia. 2014;60(C):166–72.
56. Deshmukh SG, Panchal AK, Kheraj V. Development of Cu_3BiS_3 thin films by chemical bath deposition route. J Mater Sci Mater Electron. 2017;28.
57. Colombara D, Peter LM, Rogers KD, Hutchings K. Thermochemical and kinetic aspects of the sulfurization of Cu–Sb and Cu–Bi thin films. J Solid State Chem. 2012;186:36–46.
58. Wang N. The Cu–Bi–S System: results from low-temperature experiments. Min Mag. 1994;58(391):201–4.
59. Scragg JJ, Dale PJ, Colombara D, Peter LM. Thermodynamic aspects of the synthesis of thin-film materials for solar cells. Chem Phys Chem. 2012;13(12):3035–46.
60. Klein A. Energy band alignment at interfaces of semiconducting oxides: a review of experimental determination using photoelectron spectroscopy and comparison with theoretical predictions by the electron affinity rule, charge neutrality levels, and the common anion. Thin Solid Films. 2012;520(10):3721–8.
61. Bär M, Weinhardt L, Heske C. Soft X-ray and electron spectroscopy: a unique "Tool Chest" to characterize the chemical and electronic properties of surfaces and interfaces. In: Advanced characterization techniques for thin film solar cells. Wiley-VCH Verlag GmbH & Co. KGaA: Weinheim, Germany; 2011. p. 387–409.
62. Fritsche J, Klein A, Jaegermann W. Thin film solar cells: materials science at interfaces. Adv Eng Mater. 2005;7(10):914–20.
63. Bär M, Schubert B-A, Marsen B, Krause S, Pookpanratana S, Unold T, Weinhardt L, Heske C, Schock H-W. Native oxidation and Cu-poor surface structure of thin film Cu_2ZnSnS_4 solar cell absorbers. Appl Phys Lett. 2011;99(11):112103.

64. Whittles TJ, Burton LA, Skelton JM, Walsh A, Veal TD, Dhanak VR. Band alignments, valence bands, and core levels in the tin sulfides SnS, SnS_2, and Sn_2S_3: experiment and theory. Chem Mater. 2016;28(11):3718–26.
65. Grigas J, Talik E, Lazauskas V. X-ray photoelectron spectra and electronic structure of Bi_2S_3 crystals. Phys. status solidi. 2002;232(2):220–30.
66. Yeh JJ, Lindau I. Atomic subshell photoionization cross sections and asymmetry parameters: $1 \leqslant Z \leqslant 103$. At Data Nucl Data Tables. 1985;32(1):1–155.
67. Prabhakar T, Jampana N. Effect of sodium diffusion on the structural and electrical properties of Cu_2ZnSnS_4 thin films. Sol Energy Mater Sol Cells. 2011;95(3):1001–4.
68. Rudmann D, Bilger G, Kaelin M, Haug FJ, Zogg H, Tiwari AN. Effects of NaF coevaporation on structural properties of Cu(In, Ga)Se_2 thin films. Thin Solid Films. 2003;431–432(3):37–40.
69. Kranz L, Perrenoud J, Pianezzi F, Gretener C, Rossbach P, Buecheler S, Tiwari AN. Effect of sodium on recrystallization and photovoltaic properties of CdTe solar cells. Sol Energy Mater Sol Cells. 2012;105:213–9.
70. Lenglet M, Kartouni K, Machefert J, Claude JM, Steinmetz P, Beauprez E, Heinrich J, Celati N. Low temperature oxidation of copper: the formation of CuO. Mater Res Bull. 1995;30(4):393–403.
71. Biesinger MC, Lau LWM, Gerson AR, Smart RSC. Resolving surface chemical states in XPS analysis of first row transition metals, oxides and hydroxides: Sc, Ti, V, Cu and Zn. Appl Surf Sci. 2010;257(3):887–98.
72. Czanderna AW, Madey TE, Powell CJ. Beam effects, surface topography, and depth profiling in surface analysis. In: Czanderna AW, Madey TE, Powell CJ, editors. Methods of surface characterization. Kluwer Academic Publishers: Boston; 2002, vol. 5.
73. Hochella MF, Brown GE. Aspects of silicate surface and bulk structure analysis using X-ray photoelectron spectroscopy (XPS). Geochim Cosmochim Acta. 1988;52(6):1641–8.
74. Morgan WE, Van Wazer JR. Binding energy shifts in the X-ray photoelectron spectra of a series of related group IVa compounds. J Phys Chem. 1973;77(7):964–9.
75. Debies TP, Rabalais JW. X-ray photoelectron spectra and electronic structure of Bi_2X_3 ($X = O$, S, Se, Te). Chem Phys. 1977;20(2):277–83.
76. Dharmadhikari VS, Sainkar SR, Badrinarayan S, Goswami A. Characterisation of thin films of bismuth oxide by X-ray photoelectron spectroscopy. J Electron Spectros Relat Phenomena. 1982;25(2):181–9.
77. Schuhl Y, Baussart H, Delobel R, Le Bras M, Leroy J-M. Study of mixed-oxide catalysts containing bismuth, vanadium and antimony. J Chem Soc, Faraday Trans. 1983;1(79):2055–69.
78. Hutchins EB, Lenher V. Pentavalent Bismuth. J Am Chem Soc. 1907;29(1):31–3.
79. Leontie L, Caraman M, Alexe M, Harnagea C. Structural and optical characteristics of bismuth oxide thin films. Surf Sci. 2002;507–510:480–5.
80. George J, Pradeep B, Joseph KS. Oxidation of bismuth films in air and superheated steam. Thin Solid Films. 1986;144(2):255–64.
81. Powell CJ. Recommended Auger parameters for 42 elemental solids. J Electron Spectros Relat Phenomena. 2012;185(1–2):1–3.
82. Peisert H, Chassé T, Streubel P, Meisel A, Szargan R. Relaxation energies in XPS and XAES of solid sulfur compounds. J Electron Spectros Relat Phenomena. 1994;68(C):321–8.
83. Smart RSC, Amarantidis J, Skinner WM, Prestidge CA, La Vanier L, Grano SR. Surface analytical studies of oxidation and collector adsorption in sulfide mineral flotation. In: Scanning microscopy; 2003; vol. 12. p. 3–62.
84. Partain LD, Schneider RA, Donaghey LF, McLeod PS. Surface chemistry of Cu_xS and Cu_xS/CdS determined from X-ray photoelectron spectroscopy. J Appl Phys. 1985;57(11):5056.
85. Lindberg BJ, Hamrin K, Johansson G, Gelius U, Fahlman A, Nordling C, Siegbahn K. Molecular spectroscopy by means of ESCA II. sulfur compounds. correlation of electron binding energy with structure. Phys Scr. 1970;1(5–6):286–98.

86. Chen R, So MH, Che C-M, Sun H. Controlled synthesis of high crystalline bismuth sulfide nanorods: using bismuth citrate as a precursor. J Mater Chem. 2005;15(42):4540.
87. Fang Z, Liu Y, Fan Y, Ni Y, Wei X, Tang K, Shen J, Chen Y. Epitaxial growth of CdS nanoparticle on Bi_2S_3 nanowire and photocatalytic application of the heterostructure. J Phys Chem C. 2011;115(29):13968–76.
88. Liao X-H, Wang H, Zhu J-J, Chen H-Y. Preparation of Bi_2S_3 nanorods by microwave irradiation. Mater Res Bull. 2001;36(13–14):2339–46.
89. Liufu S-C, Chen L-D, Yao Q, Wang C-F. Bismuth sulfide thin films with low resistivity on self-assembled monolayers. J Phys Chem B. 2006;110(47):24054–61.
90. Panigrahi PK, Pathak A. The growth of bismuth sulfide nanorods from spherical-shaped amorphous precursor particles under hydrothermal condition. J Nanoparticles. 2013;2013:1–11.
91. Purkayastha A, Yan Q, Raghuveer MS, Gandhi DD, Li H, Liu ZW, Ramanujan RV, Borca-Tasciuc T, Ramanath G. Surfactant-directed synthesis of branched bismuth telluride/sulfide core/shell nanorods. Adv Mater. 2008;20(14):2679–83.
92. Tamašauskaitė Tamašiūnaitė L, Šimkūnaitė-Stanynienė B, Naruškevičius L, Valiulienė G, Žielienė A, Sudavičius A. EQCM study of electrochemical modification of Bi_2S_3 films in the Zn^{2+} -containing electrolyte. J Electroanal Chem. 2009;633(2):347–53.
93. Tamašauskaitė-Tamašiūnaitė L, Valiulienė G, Žielienė A, Šimkūnaitė-Stanynienė B, Naruškevičius L, Sudavičius A. EQCM study on the oxidation/reduction of bismuth sulfide thin films. J Electroanal Chem. 2010;642(1):22–9.
94. Wang H, Zhu J-J, Zhu J-M, Chen H-Y. sonochemical method for the preparation of bismuth sulfide nanorods. J Phys Chem B. 2002;106(15):3848–54.
95. Zhong J, Xiang W, Liu L, Yang X, Cai W, Zhang J, Liang X. Biomolecule-assisted solvothermal synthesis of bismuth sulfide nanorods. J Mater Sci Technol. 2010;26(5):417–22.
96. Lou W, Chen M, Wang X, Liu W. Novel single-source precursors approach to prepare highly uniform Bi_2S_3 and Sb_2S_3 nanorods via a solvothermal treatment. Chem Mater. 2007;19(4):872–8.
97. Yang X, Wang X, Zhang Z. Facile solvothermal synthesis of single-crystalline Bi_2S_3 nanorods on a large scale. Mater Chem Phys. 2006;95(1):154–7.
98. Li W. Synthesis and characterization of bismuth sulfide nanowires through microwave solvothermal technique. Mater Lett. 2008;62(2):243–5.
99. Viezbicke BD, Patel S, Davis BE, Birnie DP. Evaluation of the Tauc method for optical absorption edge determination: ZnO thin films as a model system. Phys status solidi. 2015;252(8):1700–10.
100. Yan C, Gu E, Liu F, Lai Y, Li J, Liu Y. Colloidal synthesis and characterizations of Wittichenite copper bismuth sulphide nanocrystals. Nanoscale. 2013;5(5):1789.
101. Burton LA, Whittles TJ, Hesp D, Linhart WM, Skelton JM, Hou B, Webster RF, O'Dowd G, Reece C, Cherns D, Fermin DJ, Veal TD, Dhanak VR, Walsh A. Electronic and optical properties of single crystal SnS_2: an earth-abundant disulfide photocatalyst. J Mater Chem A. 2016;4(4):1312–8.
102. Wei SH, Zunger A. Calculated natural band offsets of All II-VI and III-V semiconductors: chemical trends and the role of cation D orbitals. Appl Phys Lett. 1998;72(16):2011–3.
103. Mesa F, Gordillo G, Dittrich T, Ellmer K, Baier R, Sadewasser S. Transient surface photovoltage of P-Type Cu_3BiS_3. Appl Phys Lett. 2010;96(8):82113.
104. Kim J, Lägel B, Moons E, Johansson N. Kelvin probe and ultraviolet photoemission measurements of indium tin oxide work function: a comparison. Synth Met. 2000:311–314.
105. Bak T, Nowotny J, Rekas M, Sorrell CC. Photo-electrochemical hydrogen generation from water using solar energy. Materials-Related Aspects Int J Hydrogen Energy. 2002;27(10):991–1022.
106. Klein A. Interface properties of dielectric oxides. J Am Ceram Soc. 2016;99(2):369–87.
107. Klein A. Energy band alignment in chalcogenide thin film solar cells from photoelectron spectroscopy. J Phys: Condens Matter. 2015;27(13):134201.
108. Burton LA, Walsh A. Band alignment in SnS thin-film solar cells: possible origin of the low conversion efficiency. Appl Phys Lett. 2013;102(13):132111.

109. Hinuma Y, Oba F, Kumagai Y, Tanaka I. Ionization potentials of (112) and $(11\bar{2})$ facet surfaces of $CuInSe_2$ and $CuGaSe_2$. Phys Rev B. 2012;86(24):245433.
110. Teeter G. X-ray and ultraviolet photoelectron spectroscopy measurements of Cu-doped CdTe(111)-B: observation of temperature-reversible Cu_xTe precipitation and effect on ionization potential. J Appl Phys. 2007;102(3):34504.
111. Pookpanratana S, Repins I, Bär M, Weinhardt L, Zhang Y, Félix R, Blum M, Yang W, Heske C, Bar M, Felix R. CdS/Cu(In, Ga)Se_2 interface formation in high-efficiency thin film solar cells. Appl Phys Lett. 2010;97(7):74101.
112. Kumar SG, Rao KSRK. Physics and chemistry of CdTe/CdS thin film heterojunction photovoltaic devices: fundamental and critical aspects. Energy Environ Sci. 2014;7(1):45–102.
113. Sugiyama M, Shimizu T, Kawade D, Ramya K, Ramakrishna Reddy KT. Experimental determination of vacuum-level band alignments of SnS-based solar cells by photoelectron yield spectroscopy. J Appl Phys. 2014;115(8):83508.
114. Minemoto T, Matsui T, Takakura H, Hamakawa Y, Negami T, Hashimoto Y, Uenoyama T, Kitagawa M. Theoretical analysis of the effect of conduction band offset of window/CIS layers on performance of CIS solar cells using device simulation. Sol Energy Mater Sol Cells. 2001;67(1–4):83–8.
115. Sinsermsuksakul P, Hartman K, Bok Kim S, Heo J, Sun L, Hejin Park H, Chakraborty R, Buonassisi T, Gordon RG. Enhancing the efficiency of SnS solar cells via band-offset engineering with a zinc oxysulfide buffer layer. Appl Phys Lett. 2013;102(5):53901.
116. Sinsermsuksakul P, Sun L, Lee SW, Park HH, Kim SB, Yang C, Gordon RG. Overcoming efficiency limitations of SnS-based solar cells. Adv Energy Mater. 2014;4(15):1400496.
117. Ley L, Pollak RA, McFeely FR, Kowalczyk SP, Shirley DA. Total valence-band densities of states of III-V and II-VI compounds from X-ray photoemission spectroscopy. Phys Rev B. 1974;9(2):600–621.
118. Zhang Y, Yuan X, Sun X, Shih B-C, Zhang P, Zhang W. Comparative study of structural and electronic properties of Cu-based multinary semiconductors. Phys Rev B. 2011;84(7):75127.
119. Maeda T, Wada T. Characteristics of chemical bond and vacancy formation in chalcopyrite-type $CuInSe_2$ and related compounds. Phys Status Solidi. 2009;6(5):1312–6.
120. Lukashev P, Lambrecht WRL, Kotani T, van Schilfgaarde M. Electronic and crystal structure of $Cu_{2-x}S$: full-potential electronic structure calculations. Phys Rev B. 2007;76(19):195202.
121. Mann JB, Meek TL, Knight ET, Capitani JF, Allen LC. Configuration energies of the D-block elements. J Am Chem Soc. 2000;122(21):5132–7.
122. Mann JB, Meek TL, Allen LC. Configuration energies of the main group elements. J Am Chem Soc. 2000;122(12):2780–3.
123. Walsh A, Payne DJ, Egdell RG, Watson GW. Stereochemistry of post-transition metal oxides: revision of the classical lone pair model. Chem Soc Rev. 2011;40(9):4455–63.
124. Yu L, Kokenyesi RS, Keszler DA, Zunger A. Inverse design of high absorption thin-film photovoltaic materials. Adv Energy Mater. 2013;3(1):43–8.
125. Menéndez-Proupin E, Gutiérrez G, Palmero E, Peña JL. Electronic structure of crystalline binary and ternary Cd – Te – O compounds. Phys Rev B. 2004;70(3):35112.
126. Meijer PHE, Pecheur P, Toussaint G. Electronic structure of the Cd vacancy in CdTe. Phys status solidi. 1987;140(1):155–62.
127. Yang B, Wang L, Han J, Zhou Y, Song H, Chen S, Zhong J, Lv L, Niu D, Tang J. $CuSbS_2$ as a promising earth-abundant photovoltaic absorber material: a combined theoretical and experimental study. Chem Mater. 2014;26(10):3135–43.
128. Chen S, Yang J-H, Gong XG, Walsh A, Wei S-H. Intrinsic point defects and complexes in the quaternary kesterite semiconductor Cu_2ZnSnS_4. Phys Rev B. 2010;81(24):245204.
129. Siebentritt S, Igalson M, Persson C, Lany S. The electronic structure of chalcopyrites-bands, point defects and grain boundaries. Prog Photovoltaics Res Appl. 2010;18(6):390–410.
130. Zakutayev A, Caskey CM, Fioretti AN, Ginley DS, Vidal J, Stevanovic V, Tea E, Lany S. Defect tolerant semiconductors for solar energy conversion. J Phys Chem Lett. 2014;5(7):1117–25.

131. Sullivan I, Zoellner B, Maggard PA. Copper(I)-based P -type oxides for photoelectrochemical and photovoltaic solar energy conversion. Chem Mater. 2016;28(17):5999–6016.
132. Chen S, Gong XG, Walsh A, Wei S-H. Crystal and electronic band structure of Cu_2ZnSnX_4 (X = S and Se) photovoltaic absorbers: first-principles insights. Appl Phys Lett. 2009;94(4):41903.
133. Payne BP, Biesinger MC, McIntyre NS. X-ray photoelectron spectroscopy studies of reactions on chromium metal and chromium oxide surfaces. J Electron Spectros Relat Phenomena. 2011;184(1–2):29–37.
134. Uchida K, Ayame A. Dynamic XPS measurements on bismuth molybdate surfaces. Surf Sci. 1996;357–358:170–5.
135. Whitcher TJ, Yeoh KH, Chua CL, Woon KL, Chanlek N, Nakajima H, Saisopa T, Songsiririt-thigul P. The effect of carbon contamination and argon ion sputtering on the work function of chlorinated indium tin oxide. Curr Appl Phys. 2014;14(3):472–5.
136. Bär M, Schubert BA, Marsen B, Wilks RG, Blum M, Krause S, Pookpanratana S, Zhang Y, Unold T, Yang W, Weinhardt L, Heske C, Schock HW. Cu_2ZnSnS_4 thin-film solar cell absorbers illuminated by soft X-rays. J Mater Res. 2012;27(8):1097–104.
137. Durose K, Asher SE, Jaegermann W, Levi D, McCandless BE, Metzger W, Moutinho H, Paulson PD, Perkins CL, Sites JR, Teeter G, Terheggen M. Physical characterization of thin-film solar cells. Prog Photovoltaics. 2004;12(2–3):177–217.
138. Morasch J, Li S, Brötz J, Jaegermann W, Klein A. Reactively magnetron sputtered Bi_2O_3 thin films: analysis of structure, optoelectronic, interface, and photovoltaic properties. Phys status solidi. 2014;211(1):93–100.
139. Shuk P. Oxide ion conducting solid electrolytes based on Bi_2O_3. Solid State Ionics. 1996;89(3–4):179–96.
140. Xu Y, Schoonen MAA. The absolute energy positions of conduction and valence bands of selected semiconducting minerals. Am Miner. 2000;85(3–4):543–56.
141. Wagner CD, Davis LE, Zeller MV, Taylor JA, Raymond RH, Gale LH. Empirical atomic sensitivity factors for quantitative analysis by electron spectroscopy for chemical analysis. Surf Interface Anal. 1981;3(5):211–25.
142. Gerson AR, Bredow T. Interpretation of sulphur 2p XPS spectra in sulfide minerals by means of Ab initio calculations. Surf Interface Anal. 2000;29(2):145–50.
143. Altamura G, Vidal J. Impact of minor phases on the performances of CZTSSe thin-film solar cells. Chem Mater. 2016;28(11):3540–63.
144. Meeder A, Weinhardt L, Stresing R, Marron DF, Wurz R, Babu SM, Schedel-Niedrig T, Lux-Steiner MC, Heske C, Umbach E. Surface and bulk properties of $CuGaSe_2$ thin films. J Phys Chem Solids. 2003;64(9–10):1553–7.
145. Delbos S. Kësterite thin films for photovoltaics: a review. EPJ Photovoltaics. 2012;3:35004.
146. Schmid D, Ruckh M, Grunwald F, Schock HW. Chalcopyrite/defect chalcopyrite heterojunctions on the basis of $CuInSe_2$. J Appl Phys. 1993;73(6):2902–9.
147. Tuttle JR, Albin DS, Noufi R. Thoughts on the microstructure of polycrystalline thin film $CuInSe_2$ and its impact on material and device performance. Sol Cells. 1991;30(1–4):21–38.
148. Welch AW, Zawadzki PP, Lany S, Wolden CA, Zakutayev A. Self-regulated growth and tunable properties of $CuSbS_2$ solar absorbers. Sol Energy Mater Sol Cells. 2015;132:499–506.
149. Pfisterer F, Bloss WH. Development of Cu_2S–CdS thin film solar cells and transfer to industrial production. Sol Cells. 1984;12(1–2):155–61.
150. Popovici I, Duta A. Tailoring the composition and properties of sprayed $CuSbS_2$ thin films by using polymeric additives. Int J Photoenergy. 2012;2012:1–6.
151. Makovicky E. The Phase transformations and thermal expansion of the solid electrolyte Cu_3BiS_3 between 25 and 300 °C. J Solid State Chem. 1983;49(1):85–92.
152. Makovicky E. Polymorphism in Cu_3SbS_3 and Cu_3BiS_3: the ordering schemes for copper atoms and electron microscope observations. Neues Jahrb für Mineral Abhandlungen. 1994;168(2):185–212.
153. Makovicky E, Skinner BJ. On crystallography and structures of copper-rich sulphosalts between 25–170C. Acta Crystallogr A. 1975;31:S65.

Chapter 5
The Electronic Structures of SnS, SnS_2, and Sn_2S_3 for Use in PV

I have no doubt that we will be successful in harnessing the sun's energy... If sunbeams were weapons of war, we would have had solar energy centuries ago.

Sir George Porter, Nobel Laureate in Chemistry, 1967

This chapter involves the electronic characterisation of the single crystal tin sulphides: SnS, SnS_2, and Sn_2S_3, for PV devices and other uses. The majority of the results presented here are published as:

Whittles, T. J.; Burton, L. A.; Skelton, J. M.; Walsh, A.; Veal, T. D.; Dhanak, V. R. Band Alignments, Valence Bands, and Core Levels in the Tin Sulfides SnS, SnS_2, and Sn_2S_3: Experiment and Theory. *Chem. Mater.* **2016**, *28* (11), 3718–3726.

With other results published as:

Burton, L. A.; Whittles, T. J.; Hesp, D.; Linhart, W. M.; Skelton, J. M.; Hou, B.; Webster, R. F.; O'Dowd, G.; Reece, C.; Cherns, D.; Fermin, D. J.; Veal, T. D.; Dhanak, V. R.; Walsh,

---

The original version of this chapter was revised: Tables was reformatted and the references to the tables were cited. The correction to this chapter is available at https://doi.org/10.1007/978-3-319-91665-1_8

© Springer International Publishing AG, part of Springer Nature 2018

T. J. Whittles, *Electronic Characterisation of Earth-Abundant Sulphides for Solar Photovoltaics*, Springer Theses, https://doi.org/10.1007/978-3-319-91665-1_5

A. Electronic and Optical Properties of Single Crystal SnS_2: An Earth-Abundant Disulfide Photocatalyst. *J. Mater. Chem. A* **2016**, *4* (4), 1312–1318.

Although the material SnS has shown promising photovoltaic properties, efficiencies of solar cell devices incorporating this material are relatively poor and difficulties in identifying separate phases are known. The presentation of XPS measurements, coupled with optical absorption measurements and theoretical DoS calculations from DFT of the three phases of tin sulphide give some explanation as to why this has been the case. Differences in the core-level XPS spectra of the three phases, including a large 0.9 eV BE shift between the Sn $3d_{5/2}$ peak for SnS and SnS_2 and also a significant shift between the Sn–S and Sn–O environments allow the different oxidation states of tin to be identified and therefore make this technique useful when identifying phase-pure, mixed-phase, and even contaminated materials within the Sn–S system. A band alignment of the three phases, derived from agreement between XPS/IPES, optical absorption measurements and calculations from DFT, show that SnS and Sn_2S_3 have relatively low IP values,[1] when compared to other common thin-film absorber materials. This results in a band misalignment of SnS and CdS, showing how CdS is an inappropriate choice of window layer for this material. Further comparison between the band alignments of the three phases reveals the detrimental effect that these materials would have on solar cell performance if they were to form as impurities, and other roles which SnS_2 can play. Then, by analysing the calculated pDoS, corroborated by valence band XPS measurements, it is found that the cause of the low IP in the phases containing the Sn^{2+} oxidation state is due to a raising of the VBM, which in turn is caused by the occupied Sn 5*s* orbital. The subsequent bonding of these electrons via the revised lone-pair model accounts for the extra states at the top of the VB in these phases. This understanding, linked to the band misalignment, gives a more comprehensive explanation of the poor efficiencies. Also, the effects of the presence of impurities and contamination within these materials are discussed and found to be another possible cause for the low efficiencies.

5.1 Tin Sulphides: The Materials

The binary tin sulphides (Sn_xS_y) are of interest here and have a recent history for use in PV, compared with the length of time that the materials have been known for. Their uses extend beyond PV, and are also important regarding the growth of CZTS, further discussion of which is made in Chap. 6.

5.1.1 Structures

Tin has two oxidation states, Sn^{2+} ($[Kr]4d^{10}5s^{2}5p^{0}$) and Sn^{4+} ($[Kr]4d^{10}5s^{0}5p^{0}$). Advances in the study of the thermodynamics of the Sn–S system [1, 2] have shown there to exist three main stable phases of tin sulphide [3]. That is, tin monosulphide (SnS), tin disulphide (SnS_2) and tin sesquisulphide (Sn_2S_3).

[1] 4.71 & 4.66 eV, respectively.

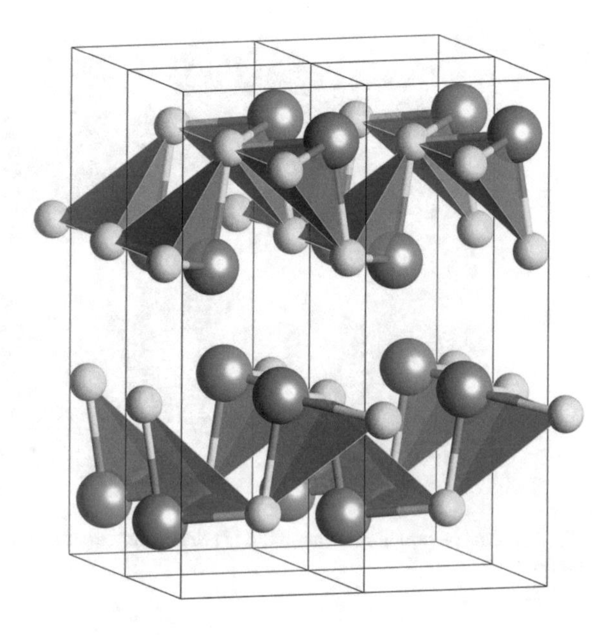

Fig. 5.1 Crystal structure of herzenbergite [4] SnS, showing the individual atoms of Sn (grey) and S (yellow)

SnS

SnS, with the Sn^{2+} cation, forms as the mineral herzenbergite [4], named after the chemist Robert Herzenberg, the crystal structure of which is shown in Fig. 5.1. The Sn atoms are coordinated tetrahedrally to sulphur atoms; however, because of the Sn 5*s* lone-pair associated with the Sn^{2+} oxidation state, one sulphur atom is missing from the final tetrahedral site and the lone-pair electron density is projected into the void [3, 5]. As a result, the structure forms tetrahedra with the tin atom at a vertex of SnS_3 structural units. Furthermore, this structure consists of double layers, which are split through the plane in which the lone-pair electron density is projected [6].

SnS_2

SnS_2, with the Sn^{4+} cation, forms as the mineral berndtite [7], named after mineralogist Fritz Berndt, the crystal structure of which is shown in Fig. 5.2. The tin atoms are coordinated 6-fold to sulphur atoms in regular octahedra, consisting of edge sharing SnS_6 units. This results in distinct single layers, which are weakly bound by VdW interactions [8, 9], similar to the structures seen for the other layered metal dichalcogenides such as TiS_2, MoS_2 or WSe_2, which have a variety of uses because of this structure, such as catalysis [10], lubricants [11], and batteries [12].

Sn_2S_3

Sn_2S_3 takes a mixed oxidation state[2] [13], and forms as the mineral ottemannite [14], named after mineralogist Joachim Ottemann, the crystal structure of which is shown in Fig. 5.3. The structure can be thought of as a 1:1 combination of the structures of SnS and SnS_2, consisting of ribbons made up of edge sharing octahedra from

[2] $Sn^{2+}Sn^{4+}S_3$.

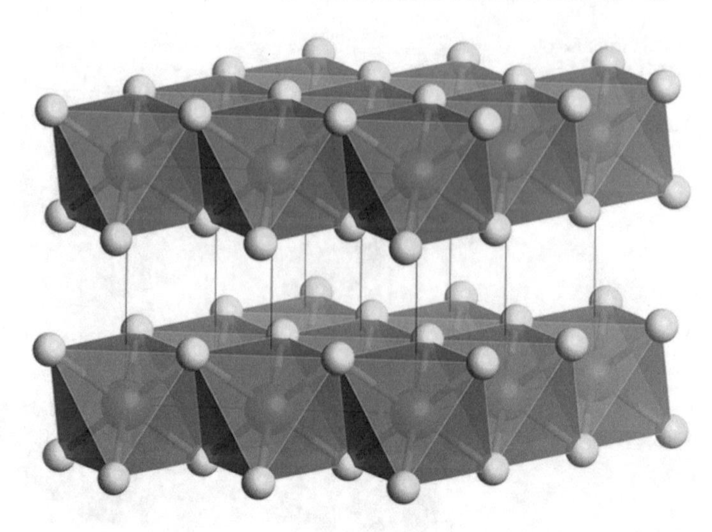

Fig. 5.2 Crystal structure of berndtite [7] SnS_2, showing the individual atoms of Sn (grey) and S (yellow)

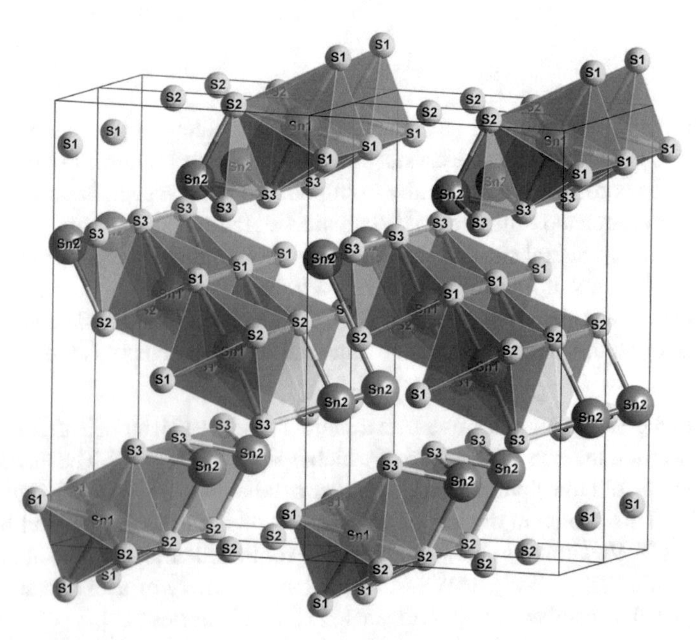

Fig. 5.3 Crystal structure of ottemannite [14] Sn_2S_3, showing the individual atoms of Sn (grey) and S (yellow). The inequivalent tin and sulphur coordination sites are marked

the Sn^{4+} ion, which are edge capped by the distorted tetrahedra from the Sn^{2+} ion [15]. Consequently, there are two inequivalent tin sites, one for each of the oxidation states, and also three inequivalent sulphur sites, marked in Fig. 5.3.

5.1.2 Properties and Growth

Such differences in the electronic structures lead to contrasting materials properties between the phases. Indeed, both SnS_2 and Sn_2S_3 display n-type conductivity [16, 17], arising from the dominant sulphur vacancy associated with the Sn^{4+} oxidation state [18]. Conversely, the p-type conductivity of SnS [19] is attributed to the dominant tin vacancy associated with the Sn^{2+} oxidation state [18].

SnS has demonstrated electronic properties which are relevant for its use as a PV solar absorber. It has a bandgap of 1.1–1.5 eV [20, 21], a high optical absorption coefficient [22–24], and intrinsic p-type conductivity [25, 26]. The 2D SnS_2 is expected to have similar electronic properties to the other 2D semiconductors [27, 28], has been shown to have a bandgap of 2.18–2.44 eV [29–31], and has potential as a photocatalyst [32, 33]. Studies of Sn_2S_3 are somewhat lacking, and there are few reports of the electronic properties; however, it could have as yet unknown applications [5].

Thin-film deposition techniques for preparing the tin sulphides are many and varied. Sputtering [34], electrochemical deposition [35], spray pyrolysis [24, 36], CBD [37], CVD [38], SILAR [31], and thermal evaporation [39, 40], have been successful, amongst others. This variety of growth methods is beneficial for device research as many different aspects of the growth method can be varied and the effects of which, studied. However, with this comes the inevitability that the measured properties of the 'material' can vary somewhat from sample to sample. Even though these values are reported to arise from the 'material' in question, this is generally accepted to be because of differing levels of impurities or contamination within the samples, and therefore the properties are in fact some average[3] of the properties of each of the materials present.

5.1.3 SnS as a Solar Absorber

SnS as a solar absorber was first envisioned in the late 1980s [37, 41], with the first cells being built and tested a few years later [42]. Progress continued and device efficiencies of 0.5% were reported by 1997 [43]. However, since then, not much progress has been made, and fabricated devices have almost exclusively shown conversion efficiencies below 4% [24, 40, 44–47], with the record being 4.36% [48]. A summary of the champion device efficiencies and their structures are detailed in Table 5.1.

It has been suggested that poor band alignments between SnS and the commonly used CdS window layer in device structures is the reason for these low efficiencies [44], with previous studies, in which the band positions of SnS were determined both theoretically [49], and experimentally [50], supporting this. Therefore, other pairings of window layer materials have been used, and as can be seen in Table 5.1, the best efficiencies have been achieved when using materials other than CdS.

[3] Dependent upon measurement technique.

Table 5.1 Performance of SnS-based solar PV devices

Absorber Growth Method	Device Architecture	Efficiency	Year	Reference
ALD	Si/SiO_2/Mo/SnS/Zn(O,S)/ZnO/ITO	4.36%	2014	Sinsermsuksakul et al. [48]
Evaporation	Si/SiO_2/Mo/SnS/Zn(O,S)/ZnO/ITO	3.88%	2014	Steinmann et al. [45]
ALD	SLG/Mo/SnS/SnO_2/Zn(O,S)/ZnO/ITO	2.9%	2015	Park et al. [46]
Pulsed-CVD	SLG/Mo/SnS/Zn(O,S)/ZnO/ITO	2.04%	2013	Sinsermsuksakul et al. [44]
Evaporation	SLG/AZO/CdS/SnS	1.42%	2013	Schneikart et al. [40]
Spray Pyrolysis	SLG/SnO_2/SnS/CdS/In	1.3%	2006	Reddy et al. [24]
Evaporation	SLG/AZO/ZnO/CdS/SnS/Au	1.17%	2013	Schneikart et al. [40]
Evaporation	SLG/AZO/ZnO/SnS/Au	0.5%	2013	Schneikart et al. [40]
Hot-wall Deposition	SLG/Mo/SnS/CdS/ZnO	0.50%	2012	Bashkirov et al. [47]
Spray Pyrolysis	Al/CdS/SnS/Ag	0.5%	1997	Reddy & Reddy [43]
Evaporation	SLG/ITO/CdS/SnS/Ag	0.29%	1994	Noguchi et al. [42]
CBD	SLG/FTO/CdS/SnS/Ag	0.2%	2008	Avellaneda et al. [122]
Evaporation	SLG/AZO/SnS/Au	0.12%	2013	Schneikart et al. [40]

5.1.4 Motivation and Scope of This Study

As a binary compound, SnS is simple from a materials point of view when compared to ternary or quaternary compounds that are used as solar absorber materials. This is one reason why interest in SnS is maintained. The promising properties that SnS continues to show, coupled with the poor efficiencies that SnS solar cells continue to demonstrate, along with the other possible uses for the other phases within this family, are the main motivations for this study.

Here, it is expected that insights into the poor efficiencies can be provided by studying the effects that the uncommon bonding environment of SnS has upon the electronic structure of the material, and in contrast to the other phases of tin sulphide, can explain the fundamental differences of SnS.

Also, as it is common for PV absorbers to be compromised on performance by the effects of secondary phase formation and contamination of the materials, confidence in the identification of the three phases is paramount. Utilising spectroscopic or diffraction techniques is challenging, due to difficulties in the assignment of different phases and in identifying mixed-phase systems [51–54]. Therefore, in order to eliminate any effects upon the underlying materials properties which may arise from imperfections in the growth, the formation of impurities, or from the contacting of the materials in a junction, it is important to study the separate phases in isolation. This will ensure that robust identification methods can be developed, the possible effects of the presence of phase impurities can be investigated, and that a benchmark can be set, against which, future (and past) studies can be compared, in order to elucidate changes or similarities observed therewith.

To this end, measurements and calculations of the electronic structures of single crystals of SnS, SnS_2, and Sn_2S_3 are presented with discussion of the differences between the spectra for the three phases and the effects that these have upon the use of tin sulphides within PV devices.

5.2 Experimental Details

Complete details of the experimental systems can be found in Chap. 2 and the specific experimental details pertinent to this study, including a breakdown of the full cleaning procedures are detailed in Appendix C.

5.2.1 Growth of Single Crystal SnS, SnS_2, and Sn_2S_3

The tin sulphide crystals studied throughout this chapter were donated by the Walsh Group in the Centre for Sustainable Chemical Technologies, Department of Chemistry, at the University of Bath, UK. Single crystals of SnS, SnS_2 and Sn_2S_3 were synthesized by iodine-assisted CVT, as described in Sect. 2.1.1, and the associated literature [18], including details of the thermodynamics of the reactions, and the amount of materials and temperature gradients used. Previous characterisation of these crystals, via XRD and EDX, showed them to be phase-pure and of the correct stoichiometry [18]. The crystals are shown in Fig. 5.4 after growth in the ampoules. The individual phases were identified by the nature of the crystals, with SnS being dark grey (a), SnS_2 forming yellow flakes (b), and Sn_2S_3 crystallising as shiny black needles (c). These observations are in agreement with other studies [54], and further confirm the phase-purity.

5.2.2 Characterisation of Single Crystal SnS, SnS_2, and Sn_2S_3

Photoemission Spectroscopy

PES was used to investigate the electronic structure of the crystals. They were affixed to UHV sample plates using double-sided carbon tape and were selected for measurement using ceramic tweezers in order to cause minimal damage to the fragile crystals. The morphology of the crystals made it difficult to provide a flat surface

Fig. 5.4 Single crystals of **a** SnS, **b** SnS_2, and **c** Sn_2S_3 contained in their ampoules post-growth. The crystals are present as the mass at the right of the ampoule for SnS and SnS_2 and the left for Sn_2S_3. Extra mass within the ampoule are remnants from the growth

in order to maximize photoemission intensity; nevertheless, a steel blade was used to cleave the top layer of the crystals in order to provide a fresh surface. Ar^+ ion sputtering at 500 eV and radiative heating up to 300 °C were further utilized to clean contamination from the sample surfaces, the full details of which are given in Appendix C, Sect. C.5.3.

XPS was used to study the core-level electronic structure, SEC, VBM and occupied VBDoS, and IPES was used to study the CBM of the crystals. Charging of the semiconducting crystals was checked by measuring the C 1*s* photoelectron region for many hours in order to obtain a significant signal from contaminant carbon. The C 1*s* photoelectron peak was found to have a BE of 284.95 eV, which was consistent across all samples and in agreement with other measurements of these materials [55–59]. It was thus concluded that no charging effects were observed in the spectra.

During cleaning, the crystals underwent several changes which are of interest to this study and are presented in Sect. 5.4. As such, presented spectra are referred to in terms of their stage in the cleaning procedure by the terminology shown in Table 5.2.

Optical Absorption Spectroscopy

Values of the direct and indirect bandgaps for the crystals of SnS and SnS_2 were determined from extrapolation of the absorption spectra derived from reflectance and transmittance measurements via FTIR spectroscopy between 4 and 350 K. The data reduction and fitting procedures are detailed in Sect. 2.5.2. Data for the Sn_2S_3 crystals were not collected as the size, shape and morphology of these crystals proved unsuitable for mounting in the spectrometer, and because an accurate thickness mea-

Table 5.2 Spectral references for the crystals throughout cleaning

SnS		SnS_2		Sn_2S_3	
Stage of Cleaning	**Reference**	**Stage of Cleaning**	**Reference**	**Stage of Cleaning**	**Reference**
As Received	*As In*	As Received	*As In*	As Received	*As In*
After Sputtering	*Sp.*	After 1 Sputter	*S1*	After Sputtering and Annealing	*Clean*
After Annealing	*Clean*	After 2 Sputters	*S2*		
		After 3 Sputters	*S3*		
		After 4 Sputters	*S4*		
		After 5 Sputters	*S5*		
		After 5 Sputters & 1 Anneal	*S5A1*		
		After 6 Sputters & 1 Anneal	*S6A1*		
		After 7 Sputters & 1 Anneal	*S7A1*		
		After 8 Sputters & 1 Anneal	*S8A1*		
		After 9 Sputters & 1 Anneal	*S9A1*		
		After 9 Sputters & 2 Anneals	*S9A2*		
		After Cleaving	*Cleaved*		

Full details of the cleaning are given in Appendix C, Sect. C.5.3

surement was not possible, this meant that the data reduction method used here was unviable for this sample.

Density Functional Theory

Electronic-structure calculations were carried out on the three tin sulphides at room temperature within the Kohn–Sham DFT [60, 61] formalism, as implemented in the VASP code [62], starting from the atomic structure of the crystals, obtained through XRD analysis. The band structures, pDoS in the valence and conduction bands,

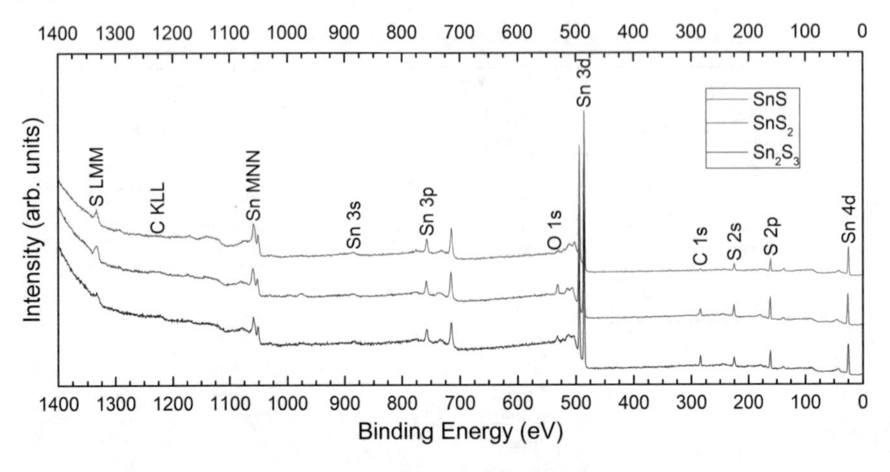

Fig. 5.5 XPS survey spectra for the tin sulphides after surface cleaning

bandgaps, and positions of the valence and conduction bands with respect to the vacuum level, were determined.

5.3 Electronic Structure Studies of Single Crystal SnS, SnS_2, and Sn_2S_3

This section presents PES measurements of the clean SnS, SnS_2, and Sn_2S_3 single crystals, along with the density of states obtained from DFT and the natural band alignments, also derived from PES and optical absorption spectroscopy. Sputtering and annealing of the crystals helped to achieve a clean surface for representative measurements. By studying the XPS spectra throughout the cleaning cycles, the crystals were deemed to be clean when the contaminant levels fell below sensitivity and the Sn 3*d* spectrum demonstrated the expected peak shapes of the pure materials. A full dissemination of the cleaning procedures is detailed in Appendix C, Sect. C.5.3, and notable effects that were observed during the cleaning procedures are discussed in Sect. 5.4.

5.3.1 XPS Core-Level Analysis

The survey spectra for the crystals, shown in Fig. 5.5, demonstrate all of the expected peaks for tin and sulphur, as well as remnant O and C after cleaning. A summary of the BE and FWHM fitted below, for the main CL features of Sn and S, is given in Table 5.3.

Table 5.3 XPS BE of the peaks of the tin sulphides after surface cleaning

Crystal	Sn-S Bonding			Sn-O Bonding		S-C Bonding
	Sn $3d_{5/2}$		S $2p_{3/2}$	Sn $3d_{5/2}$		S $2p_{3/2}$
	Sn^{2+}	Sn^{4+}		Sn^{2+}	Sn^{4+}	
SnS	485.57 (0.81)	—	161.07 (0.70)	486.45 (1.23)	—	—
SnS_2	—	486.45 (0.94)	161.45 (0.89)	—	487.16 (1.13)	162.09 (1.27)
Sn_2S_3	485.82 (0.78)	486.37 (0.91)	161.49 (0.78)	—	—	—

All values are given in eV and the peak FWHM are given in parentheses

For the Sn $3d_{5/2}$ regions, shown in Fig. 5.6, a single, high-intensity Sn $3d_{5/2}$ peak was observed for both the SnS and SnS_2 crystals, with the peak for SnS_2 at 0.9 eV higher BE than for SnS. These were assigned to the single oxidation states of Sn^{2+} (red dash) and Sn^{4+} (blue dot), respectively, confirming the single phase of both materials. Trace amounts of tin oxide remained as a contaminant on both SnS and SnS_2 and is also shown in Fig. 5.6. Although the peak attributed to Sn-oxide in the SnS spectrum (pink dot dash) has a similar BE to that of Sn^{4+} in SnS_2 (blue dot), the peak is assigned to the oxide because of the observance of its drastic reduction in area after surface cleaning.[4] The oxides forming on SnS and SnS_2 are different in BE; however, they correspond to the binding energies [63] of SnO on SnS (pink dot dash) and SnO_2 on SnS_2 (cyan dot dash), which is understandable as there is no change in the oxidation state of tin between the sulphide and the corresponding oxide. Although it is believed that the Sn retains its oxidation state upon oxidation of these materials, the peak from the oxide has a larger chemical shift compared to the sulphide because of the higher electronegativity of oxygen over sulphur [56].

For Sn_2S_3, the Sn $3d_{5/2}$ spectrum was fitted with two peaks separated by 0.6 eV, which were assigned to the Sn^{2+} (red dash) and Sn^{4+} (blue dot) oxidation states present in this mixed-valency compound. This assignment is strengthened by the fact that the area ratio was calculated to be 1:1, as expected from the distribution of tin ions in the crystal structure,[5] and that the FWHM values are comparable to those of the corresponding peaks in the single-valency materials, confirming the expected distribution of tin oxidation states, and providing strong evidence for a phase-pure material. However, the difference in BE of the peaks is reduced compared to the single-valency phases; this can be explained by the fact that the photoemitted electrons from Sn^{2+} experience a stronger bond due to the presence of Sn^{4+} ions, shifting the corresponding Sn $3d_{5/2}$ peak to a higher BE, while the converse occurs for electrons photoemitted from Sn^{4+} ions, shifting the peak to a lower BE.

[4]See Sect. 5.4.1.

[5]See Fig. 5.3.

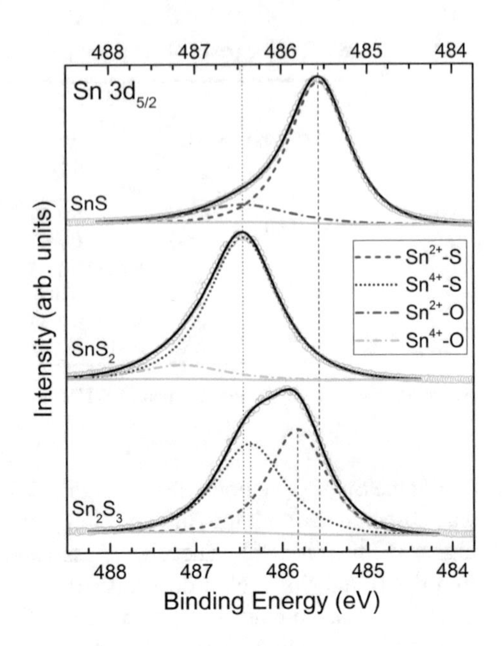

Fig. 5.6 XPS Sn $3d_{5/2}$ region of the tin sulphides after surface cleaning. Fitted peaks demonstrate the shift between the different oxidation states of Sn. SnS and SnS_2 showed trace amounts of tin oxide, shown in pink and cyan, respectively. Peak envelope shown in black

The S $2p$ doublet, shown in Fig. 5.7, was very well resolved for all three samples, and was fitted using two peaks separated by 1.18 eV with an area ratio 1:2, corresponding to the $2p_{1/2}$ and $2p_{3/2}$ features, respectively [64]. It was anticipated that the spectrum for Sn_2S_3 should show more than one S $2p$ doublet, judging by the three different coordination sites of sulphur present in the crystal structure of this compound.[6] However, within the resolution of this study, no further peaks could be resolved and the BE shifts are unknown. La Rocque et al. [59] found there to be two distinct S $2p$ doublets for Sn_2S_3; however, these were attributed to the contributions from the different oxidation states of Sn. Judging by the crystal structure, the different sulphur sites are not bonded individually to the different Sn sites, but are shared between them.

Trace amounts of sulphur-containing contamination remained on SnS_2 and is also shown in Fig. 5.7 (green dot dash). The peaks assigned to these contaminants have BE which are in agreement with similar compounds [65] and the reasons for their continued presence only on SnS_2 after cleaning are discussed in Sect. 5.4.2.

The BE for the Sn $3d_{5/2}$ and S $2p_{3/2}$ peaks found here are shown in Fig. 5.8 along with literature values of other XPS-measured samples of tin sulphides. The fitted Sn $3d_{5/2}$ BE of 485.57 eV and 486.45 eV for SnS and SnS_2, respectively, are in good agreement with other studies of these materials [48, 66], judging by the relative separation of the Sn $3d_{5/2}$ and S $2p_{3/2}$ peaks [40, 51, 54, 55, 57, 67], or once the charge referencing methods used[7] are taken into account [53, 68]. However, there are also several reports where the results are in stark contrast to those presented here,

[6] See Fig. 5.3.

[7] Shown in the top axis of Fig. 5.8.

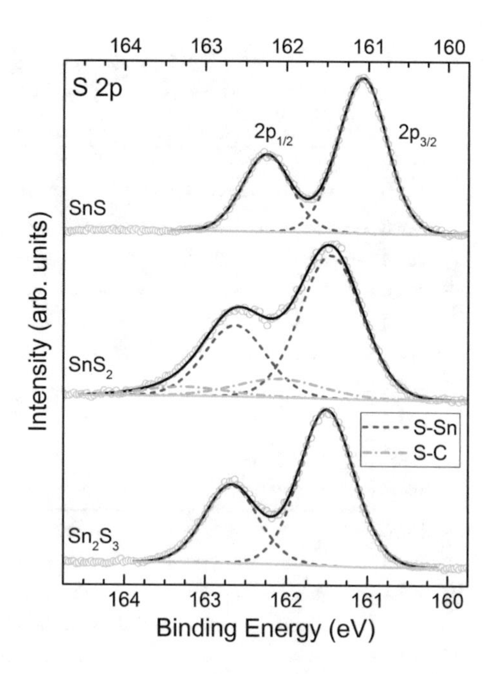

Fig. 5.7 XPS S 2*p* region of the tin sulphides after surface cleaning. SnS_2 showed trace amounts of sulphur-containing contamination, shown in green. Peak envelope shown in black

demonstrated by the scatter of the data points shown in Fig. 5.8. This is due to a number of issues, including: an unsubstantiated method of charge-referencing based on sulfur [54, 69]; a mixed-phase system, believed to be pure [56, 57]; a misprint [40, 58]; or, most commonly, the reporting of S 2*p* BE without resolution into their spin-orbit split components [54, 67, 70]. There have been few XPS studies of Sn_2S_3, and only La Rocque et al. [59] attempted to resolve the two tin oxidation states, which do not agree with the values measured here. The other studies do however, report an enveloping BE of Sn $3d_{5/2}$ in Sn_2S_3, which is intermediate of the other two phases [54] with a wider FWHM [71], suggesting that the peak was in fact a convolution of peaks from the two different oxidation states. A lack of attention to the differences in choice of charge reference may have led to the misidentification of oxidation state, and hence phase, in past studies of tin sulfides [55, 57, 71].

Regardless of these discrepancies with the literature, the oxidation states of tin, and therefore the different phases of tin sulphide, can be distinguished using properly energy-referenced XPS of a sufficiently high resolution on clean samples. The use of XPS in the identification of the different phases of tin sulphide, including when they form as secondary phases in other materials, is discussed further in Chap. 6.

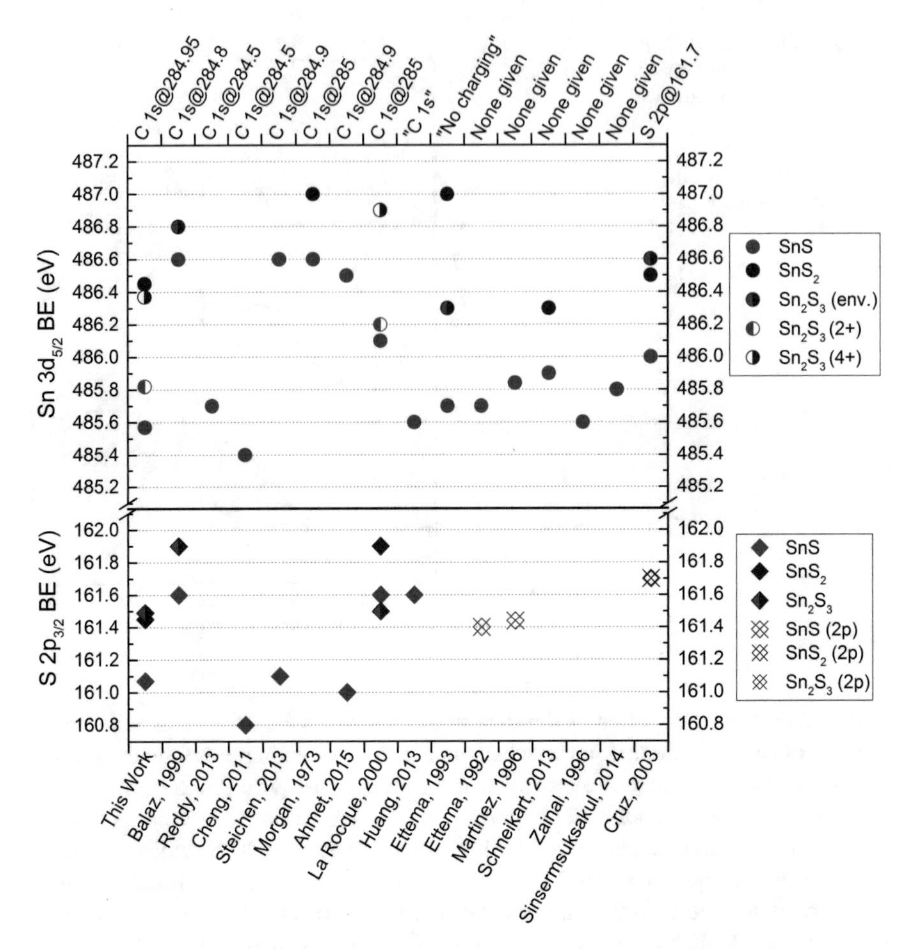

Fig. 5.8 Comparison between the XPS BE measured in this study for the clean tin sulphides, and those reported previously in the literature for SnS, SnS_2, and Sn_2S_3. Literature references are given on the bottom x-axis [40, 48, 51, 53–59, 66–68, 70, 71], and the quoted charge referencing method employed in the corresponding study are given on the top x-axis. Species assignments are as given in the corresponding reference, see text for discussion of the agreements and more details

5.3.2 *Natural Band Alignments*

As discussed in Sect. 1.6.2, the determination of the band edge positions of a material in isolation can act as a foundation from which, measurements of the band offsets within actual junctions and devices can be determined. Therefore, this section presents: measurements of the IP, EA and E_g of the crystals from PES; values for the direct and indirect bandgaps determined from optical absorption spectroscopy; equivalent values determined from DFT calculations; and a comparison of these values with relation to their effects on devices.

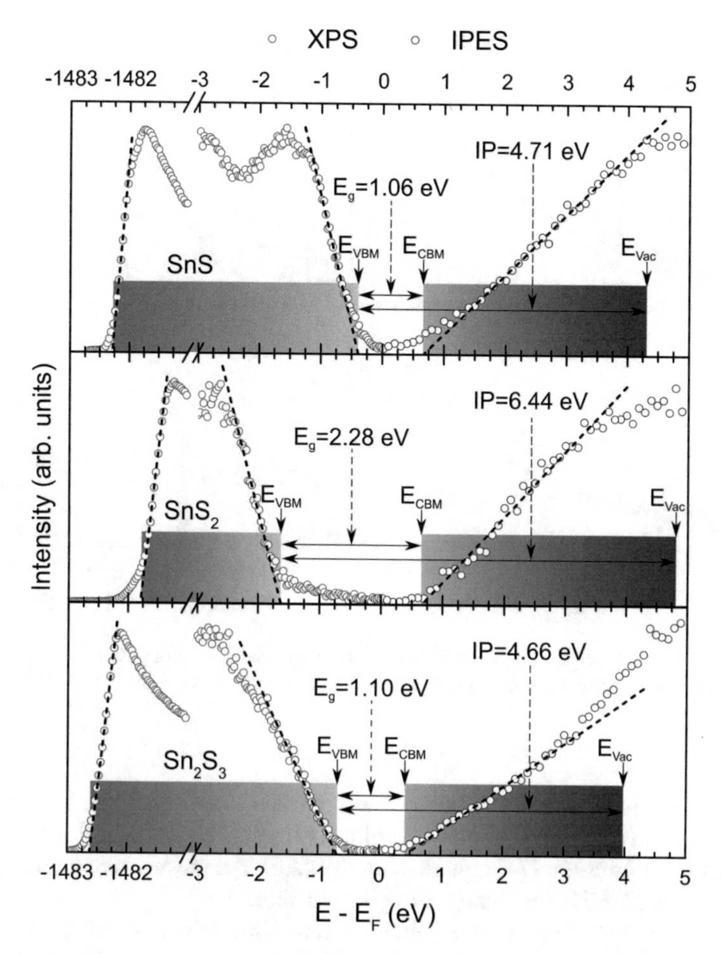

Fig. 5.9 VBM, SEC, and CBM fittings for the clean tin sulphides from XPS and IPES data

The combination of XPS and IPES measurements of the clean crystals allowed the determination of values of IP, WF, EA and E_g. The fittings of the SEC, VBM, and CBM for each of the crystals are shown in Fig. 5.9 and the values are listed in Table 5.5.

The temperature-dependent optical absorption spectra for SnS and SnS_2 are shown in Fig. 5.10a, b respectively for the indirect (top spectra) and direct (bottom spectra) bandgap extraction through the method detailed in Sect. 2.5.2. Select temperatures are shown for clarity; however, the remainder of the data sets were fitted in a similar manner. These bandgap data as a function of temperature are shown in Fig. 5.11. For comparison with other values measured in this study, the direct and indirect bandgaps at 4 and 300 K are detailed in Table 5.5. Although the linear extrapolation method of estimating direct and indirect bandgaps from absorption data should be applied with caution when these transitions are close in energy [72], support of the methodology

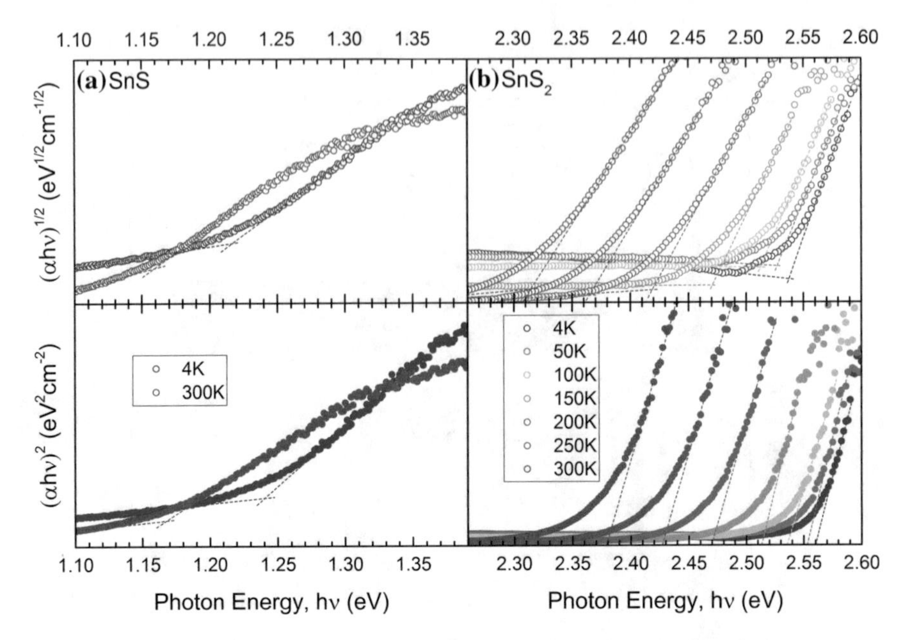

Fig. 5.10 Optical absorption data for **a** SnS and **b** SnS_2. Tauc plot fittings for the direct (bottom spectra) and indirect (top spectra) bandgaps between the temperatures 4 and 300 K. Select spectra shown for clarity

is granted by the agreement between the room temperature indirect bandgaps and those measured by PES. Also shown in Fig. 5.11 are the results of fitting the Varshni [73] and Bose–Einstein [74] equations to the data.[8] All models provided a good fit to the data, with the fitting parameters given in Table 5.4.

The data for SnS shows an unexpected feature at low temperatures. That is, below 40 K, the bandgap decreases with continued decreasing temperature. Such a trend was also observed when using different experimental systems, but was not observed with the SnS_2 data. No other study had previously observed this trend, but this is possibly because they are limited to liquid nitrogen temperatures [19, 75]. One study which did achieve liquid helium temperatures [76], shows data that may suggest evidence of such a trend. Because of this feature, which is not accounted for in the fitting models and whose origin is unknown, the temperature dependent bandgap fittings for SnS were restricted to temperatures of 40 K and above.[9]

The data and trend found for the temperature dependence of the bandgap of SnS_2 is in good agreement with a previous experimental study to liquid nitrogen temperatures [72], and the trend also matches previous calculations of the temperature dependence of the bandgap for SnS_2, albeit slightly offset in energy [77, 78].

[8]Further details can be found in Sect. 2.5.2.

[9]See Fig. 5.11a

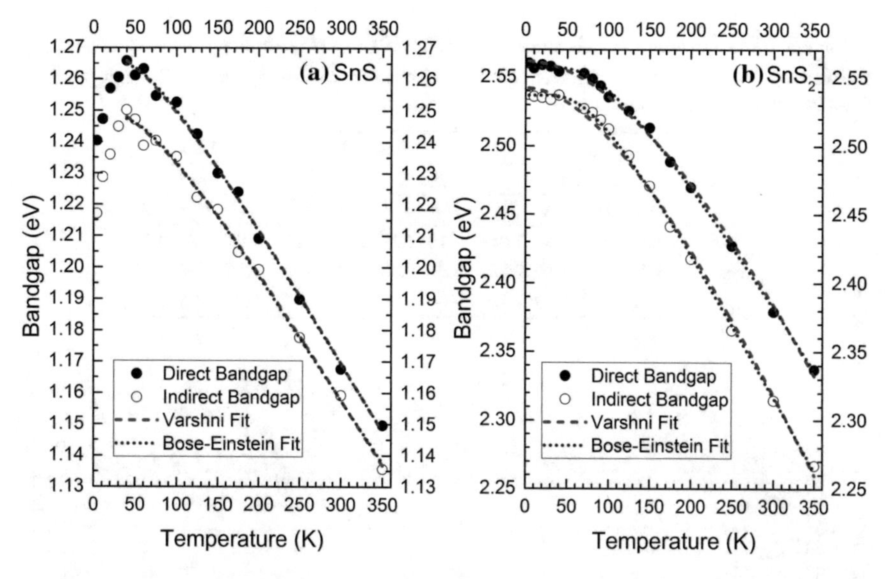

Fig. 5.11 Temperature dependence of the direct and indirect bandgaps for **a** SnS and **b** SnS_2, with fittings of the Varshni [73] and Bose–Einstein [74] equations

Table 5.4 Fitting parameters for the Varshni and Bose–Einstein equations applied to the data shown in Fig. 5.11 for the indirect E_g^i and direct E_g^d bandgaps of SnS and SnS_2

Fitting Procedure	Parameter	Value			
		SnS		SnS_2	
		E_g^i	E_g^d	E_g^i	E_g^d
Varshni [73]	E_0 (eV)	1.25	1.27	2.54	2.56
	α (meV K^{-1})	0.46	0.48	1.55	1.72
	β (K)	139	123	322	572
Bose–Einstein [74]	E_0 (eV)	1.25	1.27	2.54	2.56
	a_B (meV)	36.9	40.0	135	139
	Θ_E (K)	179	181	240	285

The room temperature values of direct and indirect bandgaps and IP for the three tin sulphides from DFT calculations are also presented in Table 5.5 for comparison. The HSE06 functional used here has been shown to be superior to other functionals with regards to predicting experimental bandgaps [79, 80]. Also, as SnS_2 is a 2D layered structure and SnS is a double-layered structure [81], long range VdW forces are important, which have been shown to be difficult to account for with DFT [82]. Nevertheless, the level of DFT utilised here has been shown previously to give good agreement with experiment for SnS [83], and it is believed to work well also for SnS_2 [32]. The agreement between the calculated and experimental values of the indirect bandgap in Table 5.5 is testament to this.

Table 5.5 Work function (WF), ionisation potential (IP), electron affinity (EA), and indirect (E_g^i) and direct (E_g^d) bandgap values as measured in this chapter for SnS, SnS_2, and Sn_2S_3, using methods from PES data, optical absorption data, and DFT calculations

Crystal	PES				Optical				DFT		
	300 K				300 K		4 K		300 K		
	WF	IP	EA	E_g	E_g^i	E_g^d	E_g^i	E_g^d	E_g^i	E_g^d	IP
SnS	4.32	4.71	3.65	1.06	1.16	1.17	1.22	1.24	1.09	1.23	4.70
SnS_2	4.81	6.44	4.16	2.28	2.31	2.38	2.54	2.56	2.19	2.74	7.30
Sn_2S_3	3.97	4.66	3.56	1.10	—	—	—	—	1.02	1.05	5.35

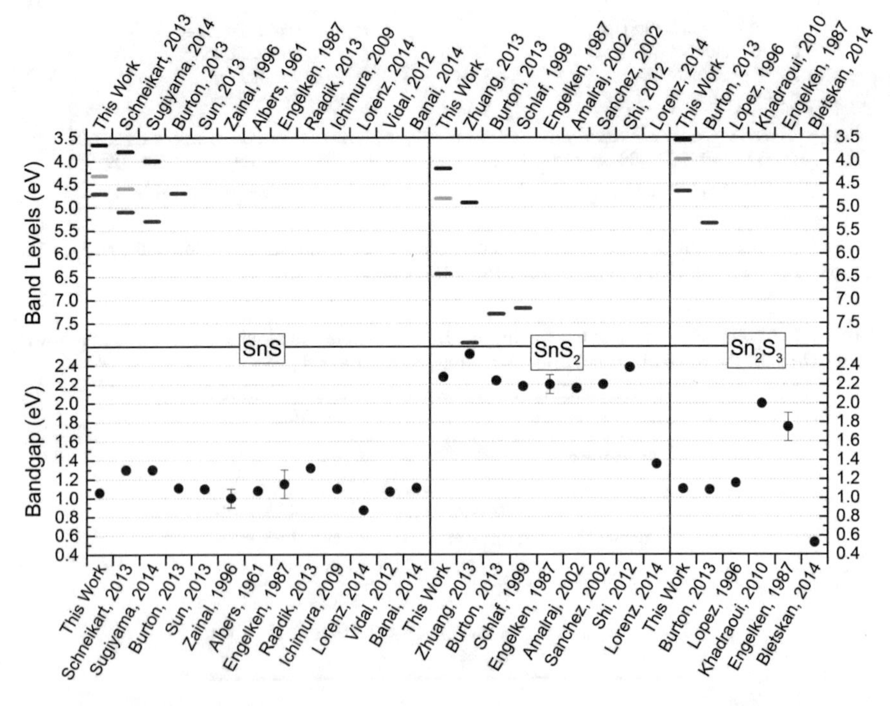

Fig. 5.12 Comparison between the electronic band levels and bandgaps measured in this study, and those reported previously in the literature for SnS, SnS_2, and Sn_2S_3. Literature references are given on the x-axis [5, 16–19, 25, 37, 40, 66, 75, 83–91, 118]. Vertical error bars on bandgap values represent a range of measurements on multiple samples

The PES bandgaps are representative of the indirect bandgaps of the materials, as PES measures the highest lying valence states and lowest lying conduction states. Therefore, from Table 5.5, the values of bandgap from PES are in agreement with the room temperature indirect bandgap extracted from optical absorption spectra and DFT. Due to this agreement, further reference to the band levels and bandgaps of these crystals shall be to those determined by PES, which are shown, along with literature reports of these values, in Fig. 5.12.

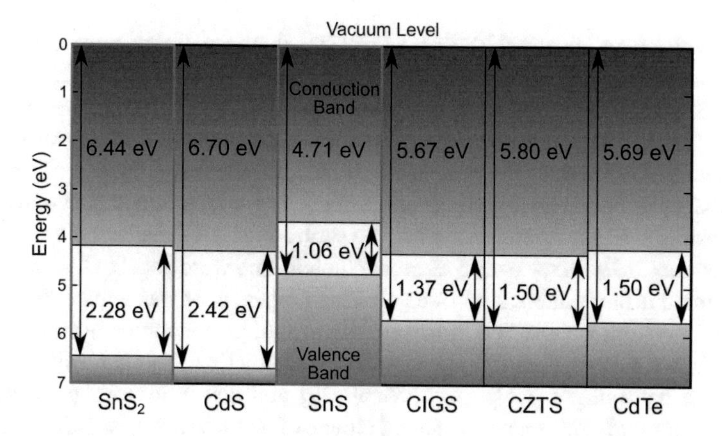

Fig. 5.13 Vacuum-aligned band diagram for SnS and SnS_2 from IP and bandgap measurements using XPS and IPES. Comparison with other common absorbers and the common n-type window layer material, CdS. Literature values of IP and bandgap are taken for CdS [49], CIGS [119], CZTS [49], and CdTe [120]

The E_g of 1.06 eV and IP of 4.71 eV for SnS are in reasonable agreement with previous studies [19, 25, 37, 40, 49, 66, 83–86], as is the E_g of 2.28 eV for SnS_2 [16, 18, 37, 87–90]. The IP of 6.44 eV for SnS_2 is; however, found to be lower than in other studies [18, 87, 88], and also lower than the calculated value from DFT. This is possibly due to band-bending effects arising from the slight contamination of the crystal as studied here.[10] It is noted that this discrepancy does not change the analysis which follows, but would serve only to exacerbate the effects described later in this section and also in Sect. 5.3.3. For Sn_2S_3, a bandgap of 1.10 eV and an IP of 4.66 eV is reported. There are few reports in the literature of the energy levels of this phase, but the bandgap determined here is in agreement with previous studies [16, 18]. The strong disagreement of some literature bandgap values for SnS_2 and Sn_2S_3 arise from either an unclear fitting method for optical data [17], a suspected mixed-phase material [37] or the notoriety of some forms of DFT to be unsuccessful with calculations of bandgap [5, 91].

By following vacuum-alignment procedures [84], band alignment diagrams were constructed between the tin sulphides and other appropriate materials for band alignment comparisons. Figure 5.13 shows the alignment of the p-type SnS with other common p-type absorber materials: CIGS, CZTS and CdTe, the n-type SnS_2 and common n-type window layer CdS.

When compared to the other common thin-film PV absorber materials, SnS was found to have a relatively low IP, leading to a band-level mismatch when using common partner materials in a solar cell,[11] confirming previous theoretical work [49], and providing some explanation to the relatively poor efficiencies seen so far.

[10] See Fig. 5.7.

[11] Such as CdS [92].

That is, with a large negative CBO[12] for CdS/SnS, there will be an increase in interface recombination [44]. This value for the CBO of CdS/SnS is in good agreement with experimental studies of the CBO of grown CdS/SnS junctions measured by XPS[13] and also PYS.[14] The agreement here between experiment and theory disputes the conclusions of a previous theoretical study [86], which found the CB of CdS to lie above that of SnS and hence give a positive CBO [86], which is in direct contradiction to the results presented here. However, the shortcomings of that study are clear: the alignment of CdS with rocksalt-SnS or zincblende-SnS were both studied, rather than the accepted orthorhombic structure [4]. This replacement was apparently justified because by taking a cubic SnS structure, this would give a commensurate interface with CdS, but nevertheless this appears to give distinctly different results.

From a band alignment point of view [40] and also a toxicity perspective, it would appear that CdS is a poorly suited choice of window layer for a SnS solar cell. Previous studies have been conducted that address this problem and explore the band alignment of alternative window layers [84, 85], proposing that suitable materials would include ZnO, $ZnIn_2Se_4$, In_2S_3 and Zn(O,S). The use of $ZnIn_2Se_4$ or In_2S_3 is also questionable from an earth-abundancy perspective, but evidence from the champion efficiencies so far[15] demonstrates the superiority of Zn(O,S) over CdS; however, it is clear that further work is required in order to further optimise cell architectures for SnS. This could also include the replacement of the back contact in order to produce an ohmic contact, because in light of the low IP of SnS, a contact metal with lower WF than the commonly used Mo may be required.

From Fig. 5.13, it can also be seen that the band positions measured here for SnS_2 are in good agreement with those of CdS and therefore, at least from an electronic perspective, the n-type SnS_2 could be considered for use as an earth-abundant replacement for CdS in solar cells where CdS is already appropriately band-aligned.

Furthermore, the alignment shown in Fig. 5.13 suggests that should SnS or SnS_2 form within a CZTS cell, they would cause detriment to the performance. The low IP of SnS means that it could act as an electron barrier, the large bandgap of SnS_2 means that it could inhibit charge transport by acting as an insulator, and because SnS_2 and Sn_2S_3 are n-type, a second diode could be formed, which would mitigate the rectifying behaviour of the cell. Further discussion of the importance of identifying secondary phases within CZTS can be found in Chap. 6.

The natural band alignments between the three phases of tin sulphide are shown in Fig. 5.14. It can be seen that SnS and Sn_2S_3 have similar band-level positions, whereas SnS_2 is offset to lower energies on the band-alignment diagram, indicating that the formation of SnS_2 within a SnS solar cell would have a detrimental effect on performance. SnS_2 has a large negative (−0.51 eV) CBO with respect to SnS and hence would act as a recombination center [44]. Again, the formation of the n-type SnS_2 or Sn_2S_3 could also act as second diodes within a SnS cell, mitigating

[12] −0.63 eV.

[13] −0.4 eV [39, 50, 93].

[14] −0.5 eV [84].

[15] See Table 5.1.

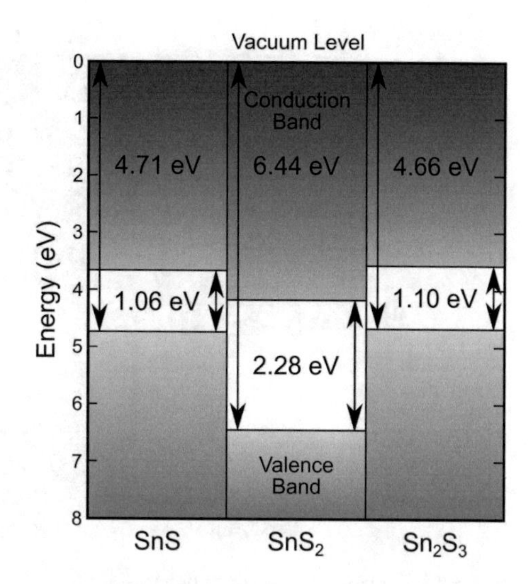

Fig. 5.14 Vacuum-aligned band diagram between SnS, SnS_2, and Sn_2S_3 from IP and bandgap measurements using XPS and IPES

its rectifying behaviour. It is possible that the effects of secondary tin sulphide phase formation could be another reason as to why SnS cells have seen such poor performances, further discussed in Sect. 5.4, especially as it has been shown that SnS_2 can form at the substrate interface during the growth of SnS [40].

An attractive proposition is to produce viable PV devices solely from tin sulphide materials, as this would simplify production and maintain earth-abundancy of the materials throughout the device. It can be seen from Fig. 5.14 that SnS/SnS_2 and SnS/Sn_2S_3 would form type II heterojunctions; however, a solar cell based on a SnS/SnS_2 junction would be unfeasible due to the aforementioned CBO between them, and indeed one study which produced a SnS/SnS_2 heterojunction demonstrated very poor diode performance [94]. A cell based on a SnS/Sn_2S_3 junction would seem electronically favourable, with a small positive CBO,[16] but producing Sn_2S_3 films of consistent quality has been shown to be problematic, depending on growth conditions [17], and by comparing to other absorber materials, it is likely that a SnS cell would favour type I heterojunction architecture [84]. The results presented here for the relatively understudied Sn_2S_3 are however, promising for the future development of this material.

Beyond potential use in PV devices, SnS_2 has also shown promise for its use in photocatalytic water splitting [32, 95]. The band levels determined for SnS_2 from Table 5.5 are shown in Fig. 5.15 with respect to the potentials required for the reduction and oxidation of water [96]. As can be seen, the bandgap of SnS_2 is wide enough to straddle both potentials, and the band levels are suitably placed so that no external bias would be required to drive the reaction. Coupled with the fact that SnS_2 is a layered material and therefore has much greater surface area available for light collection, this shows that SnS_2 is a promising candidate for water splitting, more

[16]0.09 eV.

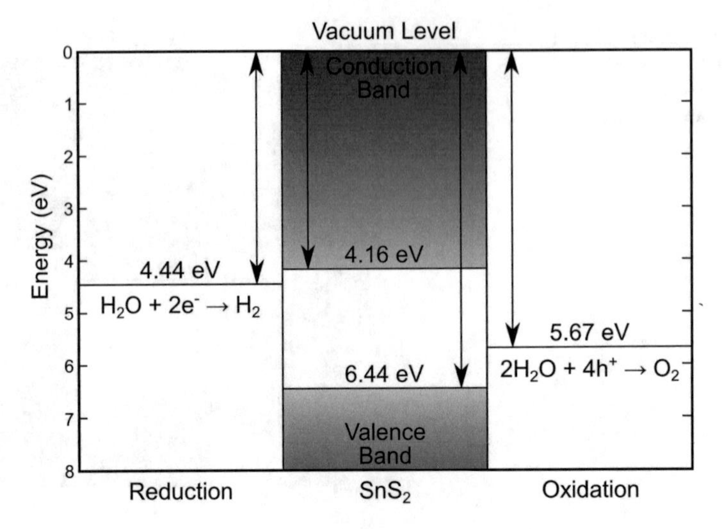

Fig. 5.15 Vacuum-aligned band diagram for SnS_2 from IP and bandgap measurements using XPS and IPES. Comparison with the reduction and oxidation potentials of water [96]

so than other materials receiving research interest in this area [97], which includes CdS.

5.3.3 Density of States Analysis

Having established the position of the band levels for the tin sulphides and the repercussions these have with regards to devices, it is important to relate the observed properties to the underlying electronic structures. By studying the DoS for these materials, one can gain insight into the nature of the bonding, reflecting the crystal structure and electronic properties.

Figure 5.16 shows the simulated pDoS curves of the three tin sulphide phases, together with the XPS VB spectra. A Shirley background was subtracted from the latter and the simulated pDoS curves were corrected using the parameters listed in Appendix C, Sect. C.3. The total DoS, obtained by summing the individual pDoS curves, are also shown for direct comparison to the XPS measurements.

The corrected DoS curves are in good agreement with the XPS data for all three phases, with all features present and only slight discrepancies in feature positions and relative intensities. The main discrepancy is found in the high BE feature, common to all spectra, which is ascribed to final-state relaxation effects, known to shift features near the bottom of the VB closer to the valence band edge [98], and which were not accounted for in the calculations.

Given the very good corroboration between theory and experiment, it is now more prudent to examine the curves before experimental corrections have been applied.

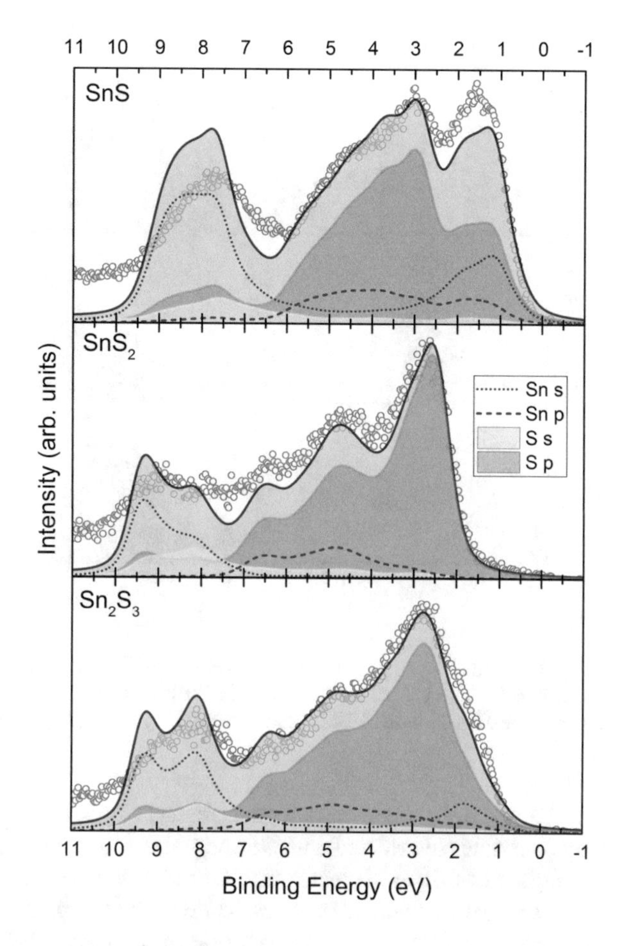

Fig. 5.16 Simulated and measured VB spectra for SnS, SnS_2 and Sn_2S_3 with respect to the Fermi-level at 0 eV. Background subtracted XPS data is compared with broadened and corrected partial DoS curves. Green data are from XPS and the black curve with grey shading is the total summed DoS

Therefore, the pDoS curves for the VB and CB before corrections are shown in Fig. 5.17.

All three phases show a VB consisting of majority S *p* states which extend from the VBM at 0 to −9 eV for SnS and Sn_2S_3, but only to −8 eV for SnS_2. All three phases also show a feature lower than the VB that consists almost entirely of S *s* states and is prominent around −14 eV for SnS and Sn_2S_3 and around −12.5 eV for SnS_2. These states, and the lack of S *s* contribution to the VB show that this orbital is somewhat localised.

The top VB of SnS is comprised of three main features. The first, at −6.5 to −10 eV, is dominated by Sn *s* states hybridized with S *p* states. The second, at −2.5 to −6.5 eV, is comprised of S *p* states, with a slight contribution from Sn *p* states. The top of the VB is formed from a hybridization of S *p*, Sn *s*, and Sn *p* states. The first CB of SnS extends from 1 to 6 eV and has majority Sn *p* states hybridised with S *p* states and a slight contribution from S *s* and also Sn *s* states near the bottom of the CB around 2 eV.

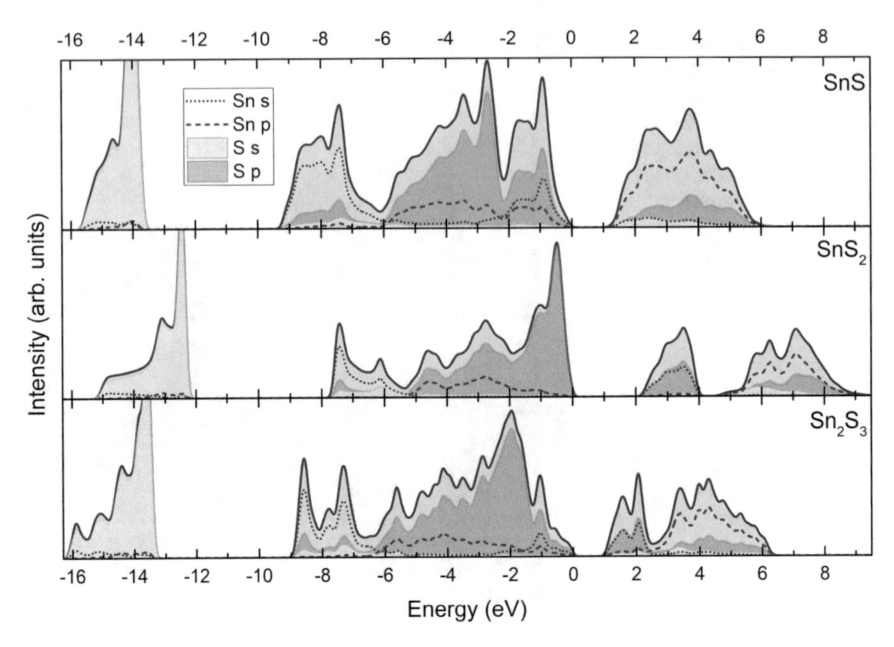

Fig. 5.17 Total and partial electronic DoS curves for SnS, SnS_2, and Sn_2S_3. Curves have been convolved with a Gaussian function (0.2 eV FWHM) in order to better distinguish features. DoS curves are aligned to the VBM. Black curve with grey shading is the total summed DoS

The top SnS_2 VB shows two main features, one at −6 to −8 eV, which is dominated by Sn *s* states hybridized with S *s* and *p* states, and a second between 0 and −5.5 eV, composed of three distinct peaks. Of these, the two at lower energy consist of S *p* states hybridised with Sn *p* states, while the largest peak at the top of the VB is due to S *p* states, with very little contribution from others. SnS_2 demonstrates two distinct CB: a lower one between 2 and 4 eV which is a shared contribution of Sn *s* states hybridised with S *p* states, and a higher one between 5 and 9 eV which comprises of Sn *p* states mixed with both S *p* and S *s* states.

The Sn_2S_3 VB is similarly composed of two main features, the first at −6.5 to −9 eV is dominated by Sn *s* states hybridized with S *s* and *p* states, while the second at 0 to −6.5 eV is similar to the corresponding feature in SnS_2, but with less well-defined peaks. Between −2 and −6 eV, it is dominated by S *p* states, with a slight contribution from Sn *p* states. In contrast to SnS_2 however, there is an extra peak at the top of the VB (−1 eV), due to a slight hybridization with Sn *s* states. In fact, by comparing the VB spectra of the three phases, it is found that the VBDoS for Sn_2S_3 appears to be an equal superposition of the VB of SnS and SnS_2, which given the crystal structure shown in Fig. 5.3, provides further evidence for the 1:1 mixing of the Sn^{2+} and Sn^{4+} cations in this material. The CB of Sn_2S_3 shows the same features as SnS_2; however, the lower CB lies between 1 and 2.5 eV and the higher CB between 2.5 and 6 eV.

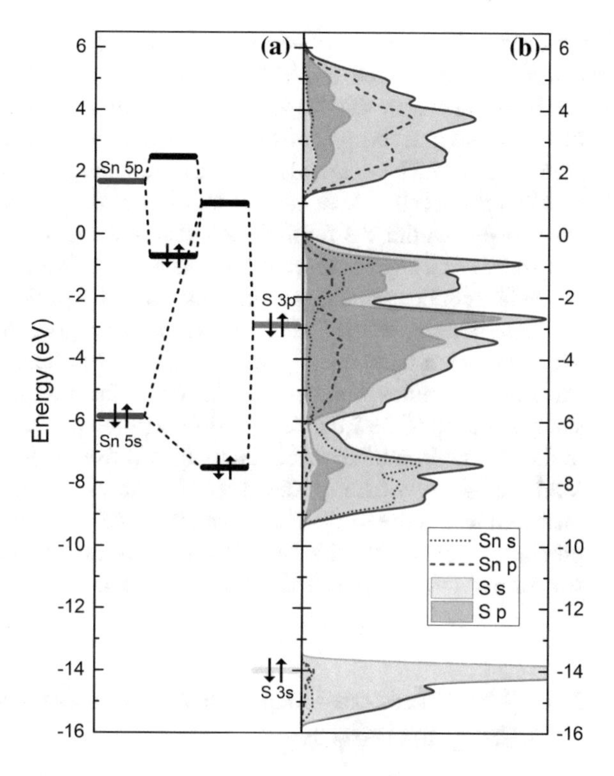

Fig. 5.18 **a** CE for the valence orbitals [121] of SnS displayed with the proposed lone-pair bonding mechanism as discussed in the text. It is noted that the CE values do not take ionisation, multi-electron occupancy or hybridisation into account and are shown only as a schematic guide. **b** DoS curves for SnS for evaluation of the orbital hybridisation contributions according to the proposed bonding mechanism

In SnS, there are Sn *s* states at the top of the VB, which are lacking in SnS_2. This can be explained by interactions through the revised lone-pair model [99, 100]. Classically, the Sn 5*s* orbital is too low lying to couple with the Sn 5*p* orbital, but mixing of these states is evident at the top of the VB for SnS. In the Sn^{2+} oxidation state, the Sn 5*s* orbital is occupied and hybridises with the full S 3*p* orbital, then the higher-lying antibonding states of this interaction further hybridise with the Sn 5*p* orbital to create the full states at the top of the VB which have Sn *s*, Sn *p*, and S *p* characteristics. This bonding mechanism is demonstrated in Fig. 5.18 with regards to the CE for SnS. There are no Sn *s* states at the top of the VB for SnS_2 because the Sn 5*s* orbital is unoccupied for the Sn^{4+} oxidation state and therefore hybridises only with the full S 3*p* orbital, creating full bonding states low in the VB,[17] and the corresponding empty antibonding states in the CB.[18] Consisting of a mixture of the Sn oxidation states, Sn_2S_3 sees some Sn *s* states at the top of the VB through the mechanism described above; however, as these only arise from the Sn^{2+} cations, they are not as prominent as in SnS.

Simulations of the electronic structures of the three phases have been carried out previously in the literature [70, 101], sometimes with support from XPS measure-

[17] −7 eV in Fig. 5.17.

[18] 3 eV in Fig. 5.17.

ments [5, 102], and agree well with the results presented here. However, no direct comparison of the DoS and XPS spectra for all three phases has been made [67], and nor have the contrasting features at the top of the VB been discussed [91]. This is therefore the first comprehensive comparison of the three phases, which employs a combination of XPS data and theoretical modelling. By examining the pDoS from the calculations, it can be seen how the hybridization of the Sn lone-pair introduces extra states higher in the VB for SnS, and also how the VB of the mixed-valency Sn_2S_3 is made up of a combination of spectral features from the single-valency materials, which has not been previously examined directly [103].

These features of the DoS for these materials give an explanation for the low IP and consequent band misalignments described in Sect. 5.3.2. SnS and Sn_2S_3 both show low IP values which are explained by the contribution of the extra lone-pair states at the top of the VB causing a shift in the level of the VBM and hence a lowering of the IP. The lack of contribution to the VB from these states for SnS_2 leads to the VBM being lower than for the other two phases. Although the phenomena of band misalignment between the phases and the lone-pair contributions in SnS have been previously noted in the literature [18, 100], an explanation in terms of the electronic structure of the materials had not been put forward.

5.4 The Effects of Contamination on Single Crystal SnS, SnS_2, and Sn_2S_3

As has been shown in Sect. 5.3.1, XPS can be used to determine the nature of contaminants present at the surface of the tin sulphides. These contaminants could have many effects on the materials, and could be important regarding their growth, as the presence of unwanted contaminants could inhibit the quality of the materials, or the performance of devices.

This section therefore presents the analysis of the crystals before they were subjected to surface cleaning, which gives some insight into the effect of atmospheric conditions on the materials. Also, throughout the cleaning, other changes took place which led to the formation of other impurities and these are also discussed.

5.4.1 Atmospheric Contamination

A comparison between the three phases of tin sulphide before and after surface cleaning is presented below in order to assess the effects of contamination. The binding energies measured in this section are presented in Tables 5.6 and 5.7.

Figure 5.19 shows the survey spectra before and after the surface cleaning for the three tin sulphides. The clean spectra are as discussed in Sect. 5.3.1 and the surveys before cleaning for SnS_2 and Sn_2S_3 appear similar but with a greater contribution

Table 5.6 XPS BE of the tin- and sulphur-relevant peaks of the tin sulphides before and after surface cleaning

Crystal	Sn-S Bonding			Sn-O Bonding			S-C Bonding
	Sn $3d_{5/2}$		S $2p_{3/2}$	Sn $3d_{5/2}$		O 1*s*	S $2p_{3/2}$
	Sn^{2+}	Sn^{4+}		Sn^{2+}	Sn^{4+}		
SnS (As In)	485.57 (0.71)	—	161.07 (0.69)	486.77 (1.31)	—	530.65 (1.31)	161.60 (1.06)
SnS (Clean)	485.57 (0.81)	—	161.07 (0.70)	486.45 (1.23)	—	530.10 (1.18)	—
SnS_2 (As In)	—	486.45 (1.33)	161.32 (0.89)	—	487.16 (1.25)	530.92 (1.71)	162.04 (0.97)
SnS_2 (Clean)	—	486.45 (0.94)	161.45 (0.89)	—	487.16 (1.13)	530.40 (1.45)	162.09 (1.27)
Sn_2S_3 (As In)	485.82 (0.77)	486.37[a] (0.91)	161.50 (0.78) 161.20 (0.60)	486.37[a] (0.91)	487.27 (1.21)	530.78 (1.38)	—
Sn_2S_3 (Clean)	485.82 (0.78)	486.37 (0.91)	161.49 (0.78)	—	—	—	—

All values are given in eV and the peak FWHM are given in parentheses
[a]These peaks are shown as one single, fitted peak in Fig. 5.20, but are probably comprised separately, as discussed in the text

from O and C features. SnS shows this trend also; however, before cleaning, iodine was also detected,[19] remnant from the synthesis [18]. Neither SnS_2 nor Sn_2S_3 showed presence of iodine, although in a similar study, iodine was found only on Sn_2S_3 [54]. This inconsistency, and the fact that iodine was removed almost completely after surface cleaning, suggests that the iodine was present only at the surface. The I $3d_{5/2}$ BE of 619.71 eV before cleaning also suggests that the iodine was present as elemental [104], and therefore unreacted, iodine.

It is noted that the surveys after cleaning appear to contain more contaminant intensity than is ideal, but it is thought that most of this intensity arises not from surface contamination, but from the carbon tape which was used to affix the crystals to the sample plate. Because of the morphology of the crystals, it proved difficult to completely cover the entire adhesive area and therefore it is possible that some signal arises from the exposed tape. Nevertheless, it is believed that this will have minimal effect on measurements, as the tape is conducting and should not interfere with the signal from the crystals, as shown in the reference spectrum in Fig. 2.17.

A comparison between the Sn $3d_{5/2}$ peaks before and after surface cleaning for the three phases is shown in Fig. 5.20. The spectra after cleaning are as described in Sect. 5.3.1. Before cleaning, all three phases showed evidence of different levels of oxidation of the tin. The oxide present on SnS is ascribed to SnO (pink dot dash) and on SnS_2 is ascribed to SnO_2 (cyan dot dash). This is the reason for the BE difference

[19]I 3*d* intensity ~620 eV.

Table 5.7 XPS BE of the contamination-relevant peaks of the tin sulphides before and after surface cleaning

Crystal	Adventitious Oxygen		Adventitious Carbon		Iodine
	O 1*s*		C 1*s*		I 3*d*$_{5/2}$
SnS (As In)	531.77 (1.91)	532.54 (2.13)	284.98 (1.33)	285.96 (1.82)	619.71 (1.71)
SnS (Clean)	531.00 (1.81)	532.54 (3.07)	284.95 (1.54)	286.73 (1.01)	619.09 (1.41)
SnS_2 (As In)	532.25 (1.83)	533.96 (2.09)	284.95 (1.81)	—	—
SnS_2 (Clean)	531.92 (2.15)	533.62 (2.02)	284.95 (2.05)	—	—
Sn_2S_3 (As In)	532.03 (1.74)	533.10(3.16)	285.03 (1.32)	286.37 (1.15)	—
Sn_2S_3 (Clean)	532.17 (1.26)	533.26 (1.81)	284.95 (1.25)	285.82 (1.85)	—

All values are given in eV and the peak FWHM are given in parentheses

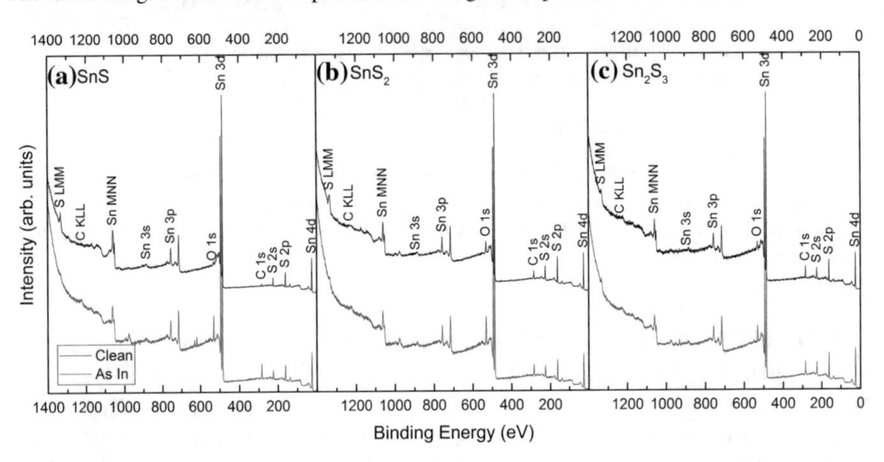

Fig. 5.19 XPS survey spectra for **a** SnS, **b** SnS_2 and **c** Sn_2S_3 before and after surface cleaning

between the oxides. On SnS, it is clear that the oxide shifts to lower BE after cleaning, suggesting that it has been reduced due to sputter damage [105]; however, as the peak from SnS did not change BE after cleaning, it is concluded that this sputter damage did not extend to the SnS.

Before cleaning, it appears that Sn_2S_3 showed presence of a small amount of SnO_2 (cyan dot dash) and that the Sn^{4+}–S peak is larger than the Sn^{2+}–S peak. This would seem surprising, as after cleaning, the Sn^{2+}:Sn^{4+}–S ratio was found to be 1:1 and a similar situation would be expected before cleaning as well. Therefore, judging by the similarity in binding energies of Sn^{4+}–S and Sn^{2+}–O,[20] it is believed that the peak labelled Sn^{4+}–S in Sn_2S_3 before cleaning (blue dot) also contains signal from Sn^{2+}–O, the ratio of Sn^{2+}:Sn^{4+}–S is indeed 1:1 and that both cations present in this material have oxidised by a similar amount, leading to a total amount of oxide which is comparable to the level of oxidation observed in SnS and SnS_2.

[20]See Table 5.3.

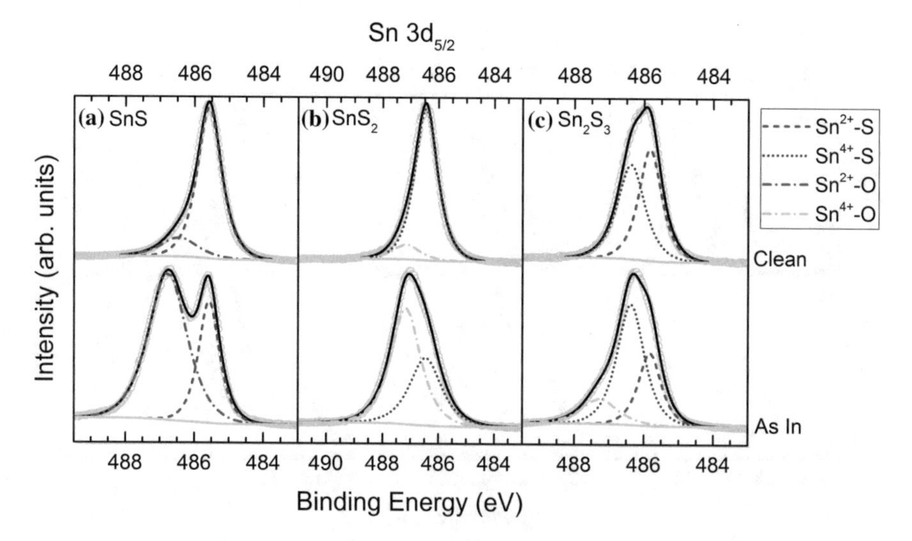

Fig. 5.20 XPS Sn $3d_{5/2}$ regions of **a** SnS, **b** SnS_2 and **c** Sn_2S_3 before and after surface cleaning. Peak envelope shown in black

The assignments of the oxides given in this study are strengthened by the agreement of the binding energy values in Table 5.6 with previous studies of tin oxide given in the literature [56, 58, 63, 66, 106, 107], both for the tin and oxygen peaks.

A comparison between the S 2*p* peaks before and after surface cleaning for the three phases is shown in Fig. 5.21. The spectra after cleaning are as described in Sect. 5.3.1. Before cleaning, on SnS and SnS_2, there is a great amount of sulphur-containing contamination (green dot dash), which is exceptionally high on SnS_2. Before cleaning, Sn_2S_3 showed a second S 2*p* doublet (blue dot) which was at lower BE than that ascribed to the sulphur in Sn_2S_3, which because of the converse shift cannot be confidently assigned to the same contamination. It is therefore believed that this extra doublet may be formed from the contributions of the different sulphur coordination environments as described in Sect. 5.3.1 and most probably there is some presence of sulphur containing contamination on this sample as well. However, because of the resolution of the study here, a fully accurate fit involving all of these features is too complex.

The acknowledgment of these contaminants, the ability to identify them and the effect that they have on the underlying materials is important. If SnO_2 or SnO were to form during or after tin sulphide growth for use in a solar cell device, it could negatively affect the performance. Both SnO_2 and SnO have been shown to have large bandgaps [106, 108, 109], meaning that they could act as insulators within a cell, inhibiting charge transport, and it has been suggested that n-type SnO_2 has unfavourable band positions to act benignly within a SnS cell with regards to CB alignments [39, 93, 108].

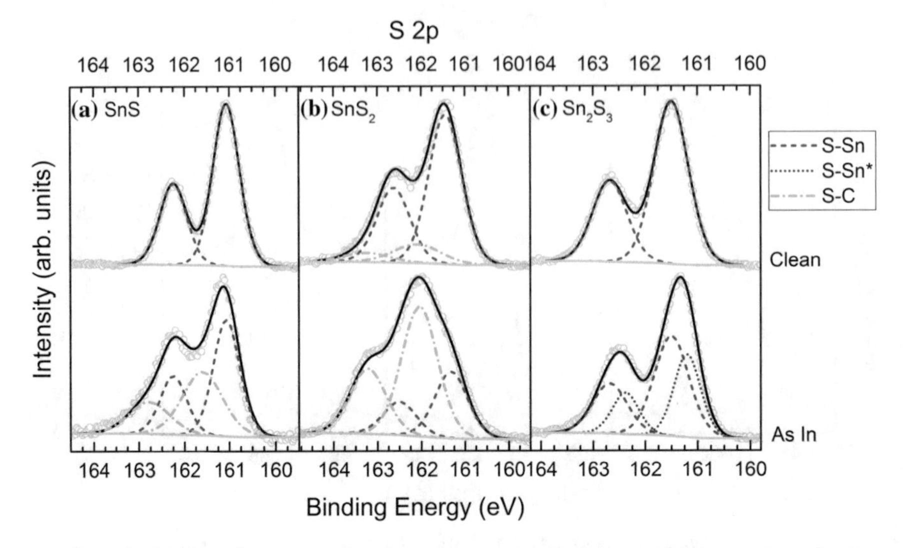

Fig. 5.21 XPS S 2*p* regions of **a** SnS, **b** SnS_2 and **c** Sn_2S_3 before and after surface cleaning. The doublet labelled S–Sn* (blue dot) is discussed in the text as a possible second coordination environment of sulphur bonded to tin in Sn_2S_3. Peak envelope shown in black

Here it is postulated that the assignments of the peaks, including those from contamination can help in future analyses of these materials. Confusion in the literature [58] has previously led to incorrect assignment of these peaks [68] because of the similarities in BE of the higher oxidation state sulphide with the lower oxidation state oxide [53]. Furthermore, it could be the case that these incorrect assignments have been a cause of the poor device performance seen by SnS so far. In cases where the oxide was mistaken for the higher oxidation state sulphide [51, 55, 66], effort may have been wasted in altering growth conditions when the cause of the oxidation should have been addressed. Furthermore, the presence of the oxide over the sulphide can be corroborated by the presence of the O 1*s* peak, which is often not addressed.

5.4.2 Cleaning-Induced Changes

SnS

The initial sputtering of the SnS crystal caused the contaminant peaks to reduce; however, this came with the formation of an extra peak in the Sn $3d_{5/2}$ spectrum (blue dot dash), shown in Fig. 5.22. This extra peak reduced on annealing of the crystal and was eradicated from the clean spectra. The BE of this peak[21] is in agreement with that for metallic tin [63, 71, 110–112], suggesting that prolonged exposure to the ion beam causes the preferential sputtering of sulphur, which has been observed

[21] 484.74 eV.

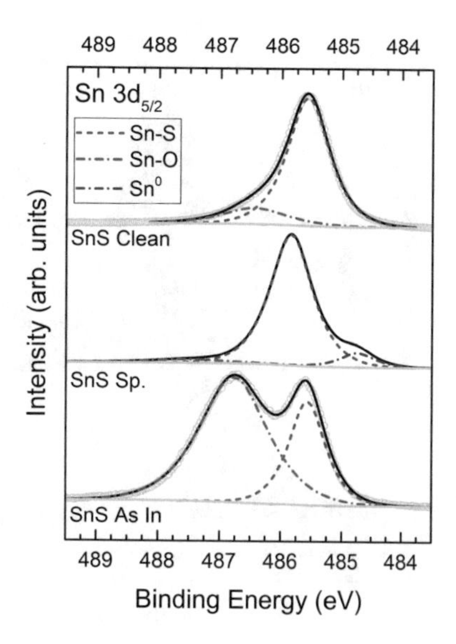

Fig. 5.22 Changes in the XPS Sn $3d_{5/2}$ spectrum of SnS during the cleaning procedures. Peak envelope shown in black. Spectral references are given in Table 5.2

in similar materials [55, 113–116]. The large shift to lower binding energies for the metallic tin peak lends ease to its identification; however, this is complicated by the small probable intensity of such a peak, if it were present. Consequently, it is believed that Steichen et al. [55] may have mistakenly reported the presence of SnS and Sn^0, when in fact, judging by the binding energies reported there, and that the survey spectrum demonstrated a large O 1*s* peak in that study, the assignment should be SnO and SnS.

SnS_2

The cleaning of the SnS_2 crystals proved difficult, because of the fragile nature and unideal morphology of these crystals for sputtering. Various cycles of sputtering and annealing were attempted, which are fully detailed in Appendix C, Sect. C.5.3. The observed changes to the Sn $3d_{5/2}$ peaks are shown in Fig. 5.23. The first sputtering cycle reduced the intensity of the oxide peak compared to the sulphide peak from the As In spectrum,[22] but after subsequent sputter cycles, very little change occurred.[23] It is also noted that after the first sputter, the BE of the oxide peak shifts closer to the sulphide peak, suggesting that the sputtering damages the natural oxide somewhat, forming a sub-oxide. Annealing was then attempted to try and either desorb the remaining oxide, else segregate it to the surface where it could then be sputtered away. After annealing,[24] the oxide was no longer present, yet a new peak at lower BE had arisen. The binding energy of this peak is in agreement with Sn^{2+} and was assigned as such. This is in agreement with the observation that Sn_2S_3 crystals were also found to have grown in the same ampoule as SnS_2 during growth [18], and

[22] See bottom spectrum in Fig. 5.23a.

[23] See S1–S5 in Fig. 5.23b.

[24] See middle spectrum of Fig. 5.23a.

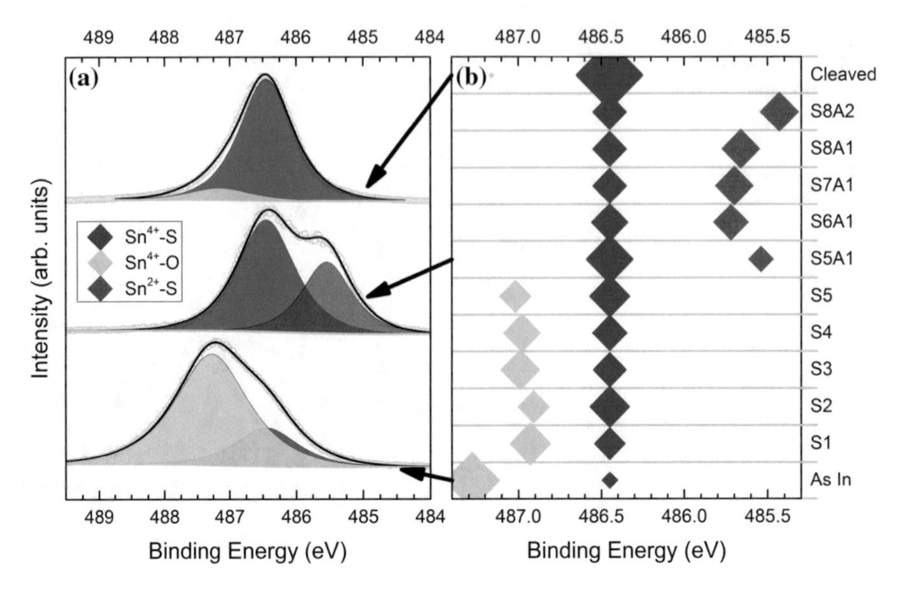

Fig. 5.23 **a** Changes in the XPS Sn $3d_{5/2}$ spectrum of SnS_2 during the cleaning procedures. **b** BE of the Sn $3d_{5/2}$ components for SnS_2 during the cleaning. The sizes of the data points represent the relative intensities of the components. Peak envelope shown in black. Spectral references are given in the Table 5.2

that SnS_2 has been shown to dissociate at higher temperatures [40, 68]. This species arose via the preferential dissociation and desorption of sulphur at elevated temperature. Subsequent sputtering after this did not remove this species, but only served to encourage its growth.[25] Again, it is noted that sputtering caused this species to shift BE closer to that of SnS_2, suggesting that the sputtering damages this species, creating a sub-sulphide that varies somewhere between SnS and SnS_2 in stoichiometry. Another annealing cycle shifted the BE of this species back to a lower value, suggesting that a purer-stoichiometry species is created through the annealing.

Finally, with no success from these sputter and anneal cycles, the crystals were removed from vacuum and cleaved again using the steel blade. Inspection of the sample after subjection to the above procedures revealed shiny black crystals forming around the edges of the yellow parent ones, confirming the formation of the Sn^{2+} species as both SnS and Sn_2S_3 demonstrated this colour. The cleaving was performed quickly and within close proximity to the load lock of the UHV system in order to minimise atmospheric contamination and retain a clean crystal surface. The spectra of this surface are presented as the clean SnS_2 data, and although there is evidence of oxide in this spectra,[26] it is very minimal compared to the As In spectrum, demonstrating the success of the cleaving. Further sputtering or annealing of this clean surface induced the same changes described above. Therefore, in future, much care

[25] See S6–S8 in Fig. 5.23b.

[26] See top spectrum of Fig. 5.23a.

must be taken when preparing the surface of SnS_2 samples, as the material is very susceptible to changes.

5.5 Summary

The electronic structures of single crystals of SnS, SnS_2, and Sn_2S_3 were studied using PES, corroborated by DoS calculations in order to gain a better understanding of the electronic structures of these materials and ultimately, how this relates to the measured electronic properties, providing explanations for the poor performance of SnS solar cells that has been seen so far. This is the first time that all three phases of tin sulphide have been studied using corroborative PES and DFT.

Prior to taking these measurements, the surface cleaning of the three phases revealed very different behaviours of the phases. SnS appears straightforward with regards to the removal of contaminants, but care must be taken not to induce the formation of metallic tin by the preferential removal of sulphur. SnS_2 suffers from a similar problem as SnS, where cleaning attempts reduced the tin within the material to its lower oxidation state; however, only through cleaving was a clean surface obtained. Although it is still relatively understudied, Sn_2S_3 appeared the most resistant to decomposition during cleaning and further work is needed to assess the thermodynamic stability of this phase [117]. These observations allowed the effects of such contamination to be assessed and it was found that oxidation would affect solar cell devices in a negative way. Because of the fragile nature of the materials, care should be taken when characterising grown material, in order to ensure that changes have not been induced between the growth and characterisation.

Once a clean surface had been obtained, it was shown that XPS measurements can provide a clear identification of the three different phases in isolation and therefore can also play an ancillary role in the identification of tin sulphide impurity phases within the PV absorbers SnS or CZTS.[27] The distinct chemical shifts between the two oxidation states of tin and between the sulphide and oxide help elucidate which bonding environments are present.

Measurements of the band levels of the three phases were shown to agree for values derived from PES, optical absorption, and DFT, giving strong merit to the accuracy of the band alignments which followed. These showed how the formation of these phases as impurities within solar cell structures can have detrimental effects on performance, which when coupled with the sometimes ambiguous measurements presented in the literature, could lend some explanation to the poor performance of SnS solar cells so far. Alignments also revealed how the n-type SnS_2 has favourable band positions for use as a possible replacement for the toxic CdS as a window layer material in solar cells and also in photocatalysis.

The alignment of SnS, when compared to other common thin-film absorbers, confirmed the low IP and higher band levels. Agreement between the calculated

[27] This role is further explored in Chap. 6.

VBDoS and VB spectra from XPS provides an explanation for these band levels, and links it back to the individual crystal structures. There is a large difference between the different oxidation states of tin depending whether the Sn 5*s* orbital is occupied (Sn^{2+}) or unoccupied (Sn^{4+}). This means, for materials containing Sn^{2+} ions (SnS and SnS_2), that Sn 5*s* electrons are available for bonding via the revised lone-pair mechanism, resulting in full bonding states at the top of the VB. These states, present in SnS and Sn_2S_3 but lacking in SnS_2, cause the upward shift of the VBM and lowering of IP, causing the band misalignment of SnS and CdS. This offers a new perspective on why SnS cells have thus far achieved relatively poor performance and suggests that in order to progress the development of this material, complete redesign of the cell architecture should be envisioned.

References

1. Sharma KC, Chang YA. The S-Zn (Sulfur-Zinc) system. J Phase Equilib. 1996;17(3):261–6.
2. Lindwall G, Shang S, Kelly NR, Anderson T, Liu Z-K. Thermodynamics of the S-Sn system: implication for synthesis of earth abundant photovoltaic absorber materials. Sol Energy. 2016;125:314–23.
3. Burton LA, Walsh A. Phase stability of the earth-abundant tin sulfides SnS, SnS_2, and Sn_2S_3. J Phys Chem C. 2012;116(45):24262–7.
4. Hofmann W. Ergebnisse Der Strukturbestimmung Komplexer Sulfide. Zeitschrift für Krist - Cryst Mater. 1935; 92 (1–6).
5. Bletskan MM, Bletskan DI. Electronic structure of Sn_2S_3 compound with the mixed valency of Tin. J Optoelectron Adv Mater. 2014;16(5–6):659–64.
6. Banai RE, Horn MW, Brownson JRS. A review of Tin (II) Monosulfide and its potential as a photovoltaic absorber. Sol Energy Mater Sol Cells. 2016;150:112–29.
7. Hazen RM, Finger LW. The crystal structures and compressibilities of layer minerals at high pressure. I. SnS_2 berndtite. Am Mineral. 1978;63:289–92.
8. Wang C, Tang K, Yang Q, Qian Y. Raman scattering, far infrared spectrum and photoluminescence of SnS_2 nanocrystallites. Chem Phys Lett. 2002;357(5–6):371–5.
9. Wilson JA, Yoffe AD. The transition metal dichalcogenides discussion and interpretation of the observed optical, electrical and structural properties. Adv Phys. 1969;18(73):193–335.
10. Zhao W, Ghorannevis Z, Chu L, Toh M, Kloc C, Tan P-H, Eda G. Evolution of electronic structure in atomically thin sheets of WS_2 and WSe_2. ACS Nano. 2013;7(1):791–7.
11. Mak KF, Lee C, Hone J, Shan J, Heinz TF. Atomically thin MoS_2: a new direct-gap semiconductor. Phys Rev Lett. 2010;105(September):2–5.
12. Whittingham MS. Lithium batteries and cathode materials. Chem Rev. 2004;104(10):4271–302.
13. Jiang T, Ozin GA. New directions in tin sulfide materials chemistry. J Mater Chem. 1998;8(5):1099–108.
14. Kniep R, Mootz D, Severin U, Wunderlich H. Structure of tin(II) tin(IV) Trisulphide, a Redetermination. Acta Crystallogr Sect B Struct Crystallogr Cryst Chem. 1982;38(7):2022–3.
15. Ichiba S, Katada M, Negita H. Mössbauer effect Of ^{119}Sn in the thermal decomposition products of Tin(IV) sulfide. Chem Lett. 1974;3(9):979–82.
16. Sanchez-Juarez A, Ortiz A. Effects of precursor concentration on the optical and electrical properties of Sn_XS_Y thin films prepared by plasma-enhanced chemical vapour deposition. Semicond Sci Technol. 2002;17(9):931–7.
17. Khadraoui M, Benramdane N, Mathieu C, Bouzidi A, Miloua R, Kebbab Z, Sahraoui K, Desfeux R. Optical and electrical properties of Sn_2S_3 thin films grown by spray pyrolysis. Solid State Commun. 2010;150(5–6):297–300.

18. Burton LA, Colombara D, Abellon RD, Grozema FC, Peter LM, Savenije TJ, Dennler G, Walsh A. Synthesis, characterization, and electronic structure of single-crystal SnS, Sn_2S_3, and SnS_2. Chem Mater. 2013;25(24):4908–16.
19. Albers W, Haas C, Vink HJ, Wasscher JD. Investigations on SnS. J Appl Phys. 1961;32(10):2220–5.
20. Devika M, Reddy NK, Ramesh K, Ganesan R, Gunasekhar KR, Gopal ESR, Reddy KTR. Thickness effect on the physical properties of evaporated SnS films. J Electrochem Soc. 2007;154(2):H67–73.
21. Tanusevski A, Poelman D. Optical and photoconductive properties of SnS thin films prepared by electron beam evaporation. Sol Energy Mater Sol Cells. 2003;80(3):297–303.
22. El-Nahass MM, Zeyada HM, Aziz MS, El-Ghamaz NA. Optical properties of thermally evaporated SnS thin films. Opt Mater (Amst). 2002;20(3):159–70.
23. Koteeswara Reddy N, Hahn YB, Devika M, Sumana HR, Gunasekhar KR. Temperature-dependent structural and optical properties of SnS films. J Appl Phys. 2007;101(9):93522.
24. Ramakrishna Reddy KT, Koteswara Reddy N, Miles RW. Photovoltaic properties of SnS based solar cells. Sol Energy Mater Sol Cells. 2006;90(18–19):3041–6.
25. Vidal J, Lany S, D'Avezac M, Zunger A, Zakutayev A, Francis J, Tate J. Band-structure, optical properties, and defect physics of the photovoltaic semiconductor SnS. Appl Phys Lett. 2012;100(3):32104.
26. Guneri E, Gode F, Ulutas C, Kirmizigul F, Altindemir G, Gumus C. Properties of P-type SnS thin films prepared by chemical bath deposition. Chalcogenide Lett. 2010;7(12):685–94.
27. Sun Y, Cheng H, Gao S, Sun Z, Liu Q, Liu Q, Lei F, Yao T, He J, Wei S, Xie Y. Freestanding tin disulfide single-layers realizing efficient visible-light water splitting. Angew Chemie Int Ed. 2012;51(35):8727–31.
28. Zhang YC, Du ZN, Li SY, Zhang M. Novel synthesis and high visible light photocatalytic activity of SnS_2 nanoflakes from $SnCl_2{\cdot}2H_2O$ and S powders. Appl Catal B Environ. 2010;95(1–2):153–9.
29. Hu X, Song G, Li W, Peng Y, Jiang L, Xue Y, Liu Q, Chen Z, Hu J. Phase-controlled synthesis and photocatalytic properties of SnS, SnS_2 and SnS/SnS_2 heterostructure nanocrystals. Mater Res Bull. 2013;48:2325–32.
30. Panda SK, Antonakos a, Liarokapis E, Bhattacharya S, Chaudhuri S. Optical properties of nanocrystalline SnS_2 thin films. Mater Res Bull. 2007;42:576–83.
31. Deshpande NG, Sagade aa, Gudage YG, Lokhande CD, Sharma R. Growth and characterization of tin disulfide (SnS_2) thin film deposited by successive ionic layer adsorption and reaction (SILAR) technique. J Alloys Compd. 2007;436:421–6.
32. Burton LA, Whittles TJ, Hesp D, Linhart WM, Skelton JM, Hou B, Webster RF, O'Dowd G, Reece C, Cherns D, Fermin DJ, Veal TD, Dhanak VR, Walsh A. Electronic and optical properties of single crystal SnS_2: an earth-abundant disulfide photocatalyst. J Mater Chem A. 2016;4(4):1312–8.
33. Li Z, Luo W, Zhang M, Feng J, Zou Z. Photoelectrochemical cells for solar hydrogen production: current state of promising photoelectrodes, methods to improve their properties, and outlook. Energy Environ Sci. 2013;6:347–70.
34. Hartman K, Johnson JL, Bertoni MI, Recht D, Aziz MJ, Scarpulla Ma, Buonassisi T. SnS thin-films by RF sputtering at room temperature. Thin Solid Films. 2011;519(21):7421–4.
35. Takeuchi K, Ichimura M, Arai E, Yamazaki Y. SnS thin films fabricated by pulsed and normal electrochemical deposition. Sol Energy Mater Sol Cells. 2003;75(3–4):427–32.
36. Thangaraju B, Kaliannan P. Spray pyrolytic deposition and characterization of SnS and SnS_2 thin films. J Phys D - Appl Phys. 2000;33:1054–9.
37. Engelken RD. Low temperature chemical precipitation and vapor deposition of Sn_xS thin films. J Electrochem Soc. 1987;134(11):2696.
38. Parkin IP, Price LS, Hibbert TG, Molloy KC. The first single source deposition of tin sulfide coatings on glass: aerosol-assisted chemical vapour deposition using $[Sn(SCH_2CH_2S)_2]$. J Mater Chem. 2001;11(5):1486–90.

39. Sugiyama M, Reddy KTR, Revathi N, Shimamoto Y, Murata Y. Band offset of SnS solar cell structure measured by X-ray photoelectron spectroscopy. Thin Solid Films. 2011;519(21):7429–31.
40. Schneikart A, Schimper H-J, Klein A, Jaegermann W. Efficiency limitations of thermally evaporated thin-film SnS solar cells. J Phys D Appl Phys. 2013;46(30):305109.
41. Ristov M, Sinadinovski G, Grozdanov I, Mitreski M. Chemical deposition of Tin(II) sulphide thin films. Thin Solid Films. 1989;173(1):53–8.
42. Noguchi H, Setiyadi A, Tanamura H, Nagatomo T, Omoto O. Characterization of vacuum-evaporated tin sulfide film for solar cell materials. Sol Energy Mater Sol Cells. 1994;35((C)):325–31.
43. Koteswara Reddy N, Ramakrishna Reddy KT. Tin Sulphide films for solar cell application. In: Conference record of the twenty sixth IEEE photovoltaic specialists conference—1997; 1997; IEEE. p. 515–8.
44. Sinsermsuksakul P, Hartman K, Bok Kim S, Heo J, Sun L, Hejin Park H, Chakraborty R, Buonassisi T, Gordon RG. Enhancing the efficiency of SnS solar cells via band-offset engineering with a zinc oxysulfide buffer layer. Appl Phys Lett. 2013;102(5):53901.
45. Steinmann V, Jaramillo R, Hartman K, Chakraborty R, Brandt RE, Poindexter JR, Lee YS, Sun L, Polizzotti A, Park HH, Gordon RG, Buonassisi T. 3.88% efficient tin sulfide solar cells using congruent thermal evaporation. Adv Mater. 2014;26(44):7488–92.
46. Park HH, Heasley R, Sun L, Steinmann V, Jaramillo R, Hartman K, Chakraborty R, Sinsermsuksakul P, Chua D, Buonassisi T, Gordon RG. Co-optimization of SnS absorber and Zn(O, S) buffer materials for improved solar cells. Prog Photovoltaics Res Appl. 2015;23(7):901–8.
47. Bashkirov SA, Gremenok VF, Ivanov VA, Shevtsova VV. Microstructure and electrical properties of SnS thin films. Phys Solid State. 2012;54(12):2497–502.
48. Sinsermsuksakul P, Sun L, Lee SW, Park HH, Kim SB, Yang C, Gordon RG. Overcoming efficiency limitations of SnS-based solar cells. Adv Energy Mater. 2014;4(15):1400496.
49. Burton LA, Walsh A. Band alignment in SnS thin-film solar cells: possible origin of the low conversion efficiency. Appl Phys Lett. 2013;102(13):132111.
50. Abdel Haleem AM, Ichimura M. Experimental determination of band offsets at the SnS/CdS and SnS/InS_xO_y heterojunctions. J Appl Phys. 2010;107(3):34507.
51. Cheng S, Conibeer G. Physical properties of very thin SnS films deposited by thermal evaporation. Thin Solid Films. 2011;520(2):837–41.
52. Ogah OE, Zoppi G, Forbes I, Miles RW. Thin films of Tin Sulphide for use in thin film solar cell devices. Thin Solid Films. 2009;517(7):2485–8.
53. Huang C-C, Lin Y-J, Chuang C-Y, Liu C-J, Yang Y-W. Conduction-type control of SnS_x films prepared by the sol–gel method for different sulfur contents. J Alloys Compd. 2013;553:208–11.
54. Cruz M, Morales J, Espinos JP, Sanz J. XRD, XPS and Sn NMR study of tin sulfides obtained by using chemical vapor transport methods. J Solid State Chem. 2003;175(2):359–65.
55. Steichen M, Djemour R, Gütay L, Guillot J, Siebentritt S, Dale PJ. Direct synthesis of single-phase P-type SnS by electrodeposition from a dicyanamide ionic liquid at high temperature for thin film solar cells. J Phys Chem C. 2013;117(9):4383–93.
56. Morgan WE, Van Wazer JR. Binding energy shifts in the X-ray photoelectron spectra of a series of related group IVa compounds. J Phys Chem. 1973;77(7):964–9.
57. Balaz P, Ohtani T, Bastl Z, Boldizarova E. Properties and reactivity of mechanochemically synthesized tin sulfides. J Solid State Chem. 1999;144(1):1–7.
58. Ahmet IY, Hill MS, Johnson AL, Peter LM. Polymorph-selective deposition of high purity SnS thin films from a single source precursor. Chem Mater. 2015;27(22):7680–8.
59. La Rocque AG, Belin-Ferré E, Fontaine MF, Senemaud C, Olivier-Fourcade J, Jumas JC. X-ray spectroscopy investigation of the electronic structure of SnS_x and $Li_{0.57}SnS_2$ compounds. Philos Mag Part B. 2000; 80(11):1933–42.
60. Hohenberg P, Kohn W. Inhomogeneous electron gas. Phys Rev. 1964;136(3B):B864–71.

61. Kohn W, Sham LJ. Self-consistent equations including exchange and correlation effects. Phys Rev. 1965;140(4A):A1133–8.
62. Kresse G, Hafner J. Ab initio molecular dynamics for liquid metals. Phys Rev B. 1993;47(1):558–61.
63. Kövér L, Kövér L, Moretti G, Kovács Z, Sanjinés R, Cserny I, Margaritondo G, Pálinkás J, Adachi H. High resolution photoemission and auger parameter studies of electronic structure of tin oxides. J Vac Sci Technol A. 1995;13(3):1382.
64. Barrie A, Drummond IW, Herd QC. Correlation of calculated and measured 2p spin-orbit splitting by electron spectroscopy using monochromatic X-radiation. J Electron Spectros Relat Phenom. 1974;5(1):217–25.
65. Lindberg BJ, Hamrin K, Johansson G, Gelius U, Fahlman A, Nordling C, Siegbahn K. Molecular spectroscopy by means of ESCA II. Sulfur compounds. Correlation of electron binding energy with structure. Phys Scr. 1970;1(5–6):286–98.
66. Zainal Z, Hussein MZ, Ghazali A. Cathodic electrodeposition of SnS thin films from aqueous solution. Sol Energy Mater Sol Cells. 1996;40(4):347–57.
67. Martinez H, Auriel C, Loudet M, Pfister-Guillouzo G. Electronic structure (XPS and Ab-Initio band structure calculation) and scanning probe microscopy images of α-Tin sulfide. Appl Surf Sci. 1996;103(2):149–58.
68. Reddy MV, Babu P, Ramakrishna Reddy KT, Miles RW. X-ray photoelectron spectroscopy and X-ray diffraction studies on tin sulfide films grown by sulfurization process. J Renew Sustain Energy. 2013;5(3):31613.
69. Hernan L, Morales J, Sanchez L, Tirado JL, Espinos JP, Gonzalez Elipe AR. Diffraction and XPS studies of misfit layer chalcogenides intercalated with cobaltocene. Chem Mater. 1995;7(8):1576–82.
70. Ettema ARHF, de Groot RA, Haas C, Turner TS. Electronic structure of SnS deduced from photoelectron spectra and band-structure calculations. Phys Rev B. 1992;46(12):7363–73.
71. Ettema ARHF, Haas C. An X-ray photoemission spectroscopy study of interlayer charge transfer in some misfit layer compounds. J Phys: Condens Matter. 1993;5(23):3817–26.
72. Powell MJ. The effect of pressure on the optical properties of 2H and 4H SnS_2. J Phys C: Solid State Phys. 1977;10(15):2967–77.
73. Varshni YP. Temperature dependence of energy gap in semiconductors. Physica. 1967;34(1):149.
74. Sarswat PK, Free ML. A study of energy band gap versus temperature for Cu_2ZnSnS_4 thin films. Phys B Condens Matter. 2012;407(1):108–11.
75. Raadik T, Grossberg M, Raudoja J, Traksmaa R, Krustok J. Temperature-dependent photoreflectance of SnS crystals. J Phys Chem Solids. 2013;74(12):1683–5.
76. Parenteau M, Carlone C. Influence of temperature and pressure on the electronic transitions in SnS and SnSe semiconductors. Phys Rev B. 1990;41(8):5227–34.
77. Reddy RR, Kumar MR, Rao TVR. Temperature and concentration dependence of energy gap and refractive index in certain mixed crystals and semiconductors. Infrared Phys. 1993;34(1):103–7.
78. Gupta VP, Agarwal P, Gupta A, Srivastava VK. Electronic properties of tin dichalcogenides (SnS_2 and $SnSe_2$). J Phys Chem Solids. 1982;43(3):291–5.
79. Le Bahers T, Rérat M, Sautet P. Semiconductors used in photovoltaic and photocatalytic devices: assessing fundamental properties from DFT. J Phys Chem C. 2014;118(12):5997–6008.
80. Perdew JP. Density functional theory and the band gap problem. Int J Quantum Chem. 2009;28(S19):497–523.
81. Makovicky E. Crystal structures of sulfides and other chalcogenides. Rev Mineral Geochem. 2006;61(1):7–125.
82. Grimme S, Ehrlich S, Goerigk L. Effect of the damping function in dispersion corrected density functional theory. J Comput Chem. 2011;32(7):1456–65.
83. Banai RE, Burton LA, Choi SG, Hofherr F, Sorgenfrei T, Walsh A, To B, Cröll A, Brownson JRS. Ellipsometric characterization and density-functional theory analysis of anisotropic optical properties of single-crystal α -SnS. J Appl Phys. 2014;116(1):13511.

84. Sugiyama M, Shimizu T, Kawade D, Ramya K, Ramakrishna Reddy KT. Experimental determination of vacuum-level band alignments of SnS-based solar cells by photoelectron yield spectroscopy. J Appl Phys. 2014;115(8):83508.
85. Sun L, Haight R, Sinsermsuksakul P, Bok Kim S, Park HH, Gordon RG. Band alignment of SnS/Zn(O,S) heterojunctions in SnS thin film solar cells. Appl Phys Lett. 2013;103(18):181904.
86. Ichimura M. Calculation of band offsets at the CdS/SnS heterojunction. Sol Energy Mater Sol Cells. 2009;93(3):375–8.
87. Zhuang HL, Hennig RG. Theoretical perspective of photocatalytic properties of single-layer SnS_2. Phys Rev B. 2013;88(11):115314.
88. Schlaf R, Pettenkofer C, Jaegermann W. Band lineup of a SnS_2/$SnSe_2$/SnS_2 semiconductor quantum well structure prepared by van Der Waals epitaxy. J Appl Phys. 1999;85(9):6550–6.
89. Amalraj L, Sanjeeviraja C, Jayachandran M. Spray pyrolysised tin disulphide thin film and characterisation. J Cryst Growth. 2002;234(4):683–9.
90. Shi C, Chen Z, Shi G, Sun R, Zhan X, Shen X. Influence of annealing on characteristics of tin disulfide thin films by vacuum thermal evaporation. Thin Solid Films. 2012;520(15):4898–901.
91. Lorenz T, Joswig J-O, Seifert G. Combined SnS@SnS_2 double layers: charge transfer and electronic structure. Semicond Sci Technol. 2014;29(6):64006.
92. Li Y-H, Walsh A, Chen S, Yin W-J, Yang J-H, Li J, Da Silva JLF, Gong XG, Wei S-H. Revised ab initio natural band offsets of all group IV, II-VI, and III-V semiconductors. Appl Phys Lett. 2009;94(21):212109.
93. Sugiyama M, Murata Y, Shimizu T, Ramya K, Venkataiah C, Sato T, Reddy KTR. Sulfurization growth of SnS thin films and experimental determination of valence band discontinuity for SnS-related solar cells. Jpn J Appl Phys 2011;50(5):05FH03.
94. Sánchez-Juárez A, Tiburcio-Silver A, Ortiz A. Fabrication of SnS_2/SnS heterojunction thin film diodes by plasma-enhanced chemical vapor deposition. Thin Solid Films. 2005;480–481:452–6.
95. Walter MG, Warren EL, McKone JR, Boettcher SW, Mi Q, Santori Ea, Lewis NS. Solar water splitting cells. Chem Rev 2010;110:6446–73.
96. Zhuang HL, Hennig RG. Single-layer group-III monochalcogenide photocatalysts for water splitting. Chem Mater. 2013;25:3232–8.
97. Kudo A, Miseki Y. Heterogeneous photocatalyst materials for water splitting. Chem Soc Rev. 2009;38(1):253–78.
98. Ley L, Pollak Ra, McFeely FR, Kowalczyk SP, Shirley Da. Total valence-band densities of states of III-V and II-VI compounds from X-ray photoemission spectroscopy. Phys Rev B. 1974;9(2):600–21.
99. Walsh A, Payne DJ, Egdell RG, Watson GW. Stereochemistry of post-transition metal oxides: revision of the classical lone pair model. Chem Soc Rev. 2011;40(9):4455–63.
100. Walsh A, Watson GW. Influence of the anion on lone pair formation in Sn(II) monochalcogenides: A DFT study. J Phys Chem B. 2005;109(40):18868–75.
101. Parker D, Singh DJ. First principles investigations of the thermoelectric behavior of tin sulfide. J Appl Phys. 2010;108(8):83712.
102. Lefebvre I, Lannoo M, Olivier-Fourcade J, Jumas JC. Tin oxidation number and the electronic structure of SnS-In_2S_3-SnS_2 systems. Phys Rev B. 1991;44(3):1004–12.
103. Lefebvre I, Lannoo M, Moubtassim ME, Fourcade JO, Jumas J-C. Lithium insertion in three-dimensional tin sulfides. Chem Mater. 1997;9(12):2805–14.
104. Sherwood PMA. X-ray photoelectron spectroscopic studies of some iodine compounds. J Chem Soc Faraday Trans. 1976;72(X):1805.
105. Lin AWC, Armstrong NR, Kuwana T. X-ray photoelectron/Auger electron spectroscopic studies of tin and indium metal foils and oxides. Anal Chem. 1977;49(8):1228–35.
106. Akgul FA, Gumus C, Er AO, Farha AH, Akgul G, Ufuktepe Y, Liu Z. Structural and electronic properties of SnO_2. J Alloys Compd. 2013;579:50–6.

107. Kwoka M, Ottaviano L, Passacantando M, Santucci S, Czempik G, Szuber J. XPS study of the surface chemistry of L-CVD SnO_2 thin films after oxidation. Thin Solid Films. 2005;490(1):36–42.
108. Scanlon DO, Watson GW. On the possibility of P-type SnO_2. J Mater Chem. 2012;22(48):25236.
109. Sivaramasubramaniam R, Muhamad MR, Radhakrishna S. Optical properties of annealed Tin(II) oxide in different ambients. Phys Status Solidi. 1993;136(1):215–22.
110. Nyholm R, Martensson N. Core level binding energies for the elements Zr-Te (Z=40-52). J Phys C: Solid State Phys. 1980;13(11):L279–84.
111. Barr TL. ESCA studies of naturally passivated metal foils. J Vac Sci Technol. 1977;14(1):660.
112. Hegde RI, Sainkar SR, Badrinarayanan S, Sinha APB. A study of dilute tin alloys by X-ray photoelectron spectroscopy. J Electron Spectros Relat Phenom. 1981;24(1):19–25.
113. Sundberg J, Lindblad R, Gorgoi M, Rensmo H, Jansson U, Lindblad A. Understanding the effects of sputter damage in W-S thin films by HAXPES. Appl Surf Sci. 2014;305:203–13.
114. Velásquez P, Ramos-Barrado JR, Leinen D. The fractured, polished and Ar^+-sputtered surfaces of natural Enargite: an XPS study. Surf Interface Anal. 2002;34(1):280–3.
115. Loeffler MJ, Dukes CA, Chang WY, McFadden LA, Baragiola RA. Laboratory simulations of sulfur depletion at Eros. Icarus. 2008;195(2):622–9.
116. Nossa A, Cavaleiro A. Chemical and physical characterization of C(N)-doped W-S sputtered films. J Mater Res. 2004;19(8):2356–65.
117. Skelton JM, Burton LA, Oba F, Walsh A. Chemical and lattice stability of the tin sulfides. J Phys Chem C. 2017;121(12):6446–54.
118. López S, Granados S, Ortíz A. Spray pyrolysis deposition of thin films. Semicond Sci Technol. 1996;11(3):433–6.
119. Hinuma Y, Oba F, Kumagai Y, Tanaka I. Ionization potentials of (112) and $(11\bar{2})$ facet surfaces of $CuInSe_2$ and $CuGaSe_2$. Phys Rev B. 2012;86(24):245433.
120. Teeter G. X-ray and ultraviolet photoelectron spectroscopy measurements of Cu-doped CdTe(111)-B: observation of temperature-reversible Cu_xTe precipitation and effect on ionization potential. J Appl Phys. 2007;102(3):34504.
121. Mann JB, Meek TL, Allen LC. Configuration energies of the main group elements. J Am Chem Soc. 2000;122(12):2780–3.
122. Avellaneda D, Nair MTS, Nair PK. Polymorphic tin sulfide thin films of zinc blende and orthorhombic structures by chemical deposition. J Electrochem Soc. 2008;155(7):D517.

Chapter 6
The Use of Photoemission Spectroscopies for the Characterisation and Identification of Cu_2ZnSnS_4 and its Secondary Phases

I'd put my money on the sun and solar energy. What a source of power! I hope we don't have to wait until oil and coal run out before we tackle that. I wish I had more years left.

Thomas Edison, Inventor, 1931

This chapter involves a comprehensive comparison and critical analysis of the use of different types of characterisation technique that are commonly used for the characterisation of various forms of the material CZTS and its use as a PV absorber.

The results of related measurements on thin-films of CZTS grown by CBD are published as:

Gupta, S.; Whittles, T.J.; Batra, Y.; Satsangi, V.; Krishnamurthy, S.; Dhanak, V. R.; Mehta, B. R. A Low-Cost, Sulfurization Free Approach to Control Optical and Electronic Properties of Cu_2ZnSnS_4 via Precursor Variation. *Sol. Energy Mater. Sol. Cells* **2016**, *157*, 820–830.

The original version of this chapter was revised: Tables was reformatted and the references to the tables were cited. The correction to this chapter is available at https://doi.org/10.1007/978-3-319-91665-1_8

© Springer International Publishing AG, part of Springer Nature 2018

T. J. Whittles, *Electronic Characterisation of Earth-Abundant Sulphides for Solar Photovoltaics*, Springer Theses, https://doi.org/10.1007/978-3-319-91665-1_6

The difficulties with both synthesising and analysing CZTS are widely acknowledged, yet it receives continued research attention because of its promise as an earth-abundant solar absorber. The growth is susceptible to the formation of secondary phases and contamination, but it is unclear as to the mechanisms of this and also the effect that such phases have on devices. Beyond this, the characterisation of CZTS itself is complicated by peak overlaps, technique insensitivity, the necessity for rigorous analysis, and a lack of standards. Here, the common characterisation techniques of XRD and RS are analysed in terms of their use for identifying secondary phases, and also for characterising mixed-phase samples of CZTS. Coupled with this, XPS is introduced as a technique that can help with these issues, but when used to its full extent, can also elucidate information about the electronic structure, such as band levels and the density of states in the VB.

In order to achieve a basic standard, from which the merits of each technique can be assessed without interference from factors such as device characteristics or growth methods, single crystal CZTS and individual samples of the possible binary secondary phases within CZTS were studied in isolation. The core-level analysis of CZTS and the secondary phases reveal complexities and nuances which are generally not considered in the literature, but which now can be used as a comparator in the future.

It was found that the use of XRD and RS combined are not unsuitable for the detection of secondary phases within CZTS, yet neither are they definitive in elucidating them, or indeed confirming phase-pure CZTS. The combined use of XPS with the two previous analysis techniques can further corroborate the presence of secondary phases, and it is shown how this technique can be used as a starting point for secondary phase analysis.

With regards to the electronic structures, corroboration of theoretical and experimental density of states curves for the VB show how the bonding nature in CZTS is different to that of CIGS, to which analogies are commonly drawn. These differences give rise to a low value of IP[1] when compared to CIGS or CdTe, supporting the idea that CdS is a poor choice of window layer for use in CZTS solar cells. Similar measurement of the possible binary secondary phases also showed the negative effects that such phases can have upon devices in terms of their electronic structures.

Further experimentation involving the single crystal CZTS gave insights in how the material behaves under vacuum at elevated temperatures, environments that are possible during growth or analysis. It is shown that instabilities of CZTS, in terms of thermodynamic processes involved with the growth and also inherent to the material itself, offer further explanation to the complexities of this material and the problems associated with the secondary phases.

[1] 5.28 eV.

6.1 Cu_2ZnSnS_4: The Material

Copper zinc tin sulphide (Cu_2ZnSnS_4) is the material of interest in this chapter and is one of the most researched materials for earth-abundant PV. Consisting of earth-abundant elements, none of which pose a significant factor economically,[2] the development of CZTS will continue, trying to drive the efficiencies beyond that of the current leaders, CdTe and CIGS. There are however, several problems which must be addressed and overcome before this material can vie for significant market share.

6.1.1 History and Uses

The family of quaternary materials to which CZTS belongs are a set of naturally occurring minerals whose structures have been determined since 1965 [1], and single crystals thereof were successfully synthesised in 1967, following the synthesis of binary and ternary compounds in the preceding years [2]. Further studies saw the refinements of the structures of these minerals [3, 4], but it was not until 1979 that the first physical characterisations of crystals of this material were made beyond the structural determinations [5]. At this point, CZTS research remained an entry in the crystallographers' catalogue until interest was renewed when thin-films were deposited and electrical and optical characterisations performed in 1988 by Ito and Nakazawa [6]. They found that CZTS had an optical absorption coefficient larger than 10^4 cm^{-1} in the visible range, intrinsic p-type conductivity and a direct bandgap of 1.45 eV. These results placed CZTS as a potential PV absorber as an alternative to CIGS. CZTS has found other semiconductor applications as well, such as the photocatalytic splitting of water [7], batteries [8], and use as the counter electrode in DSSC [9].

6.1.2 Structure

Historically, the mineral deposit $Cu_2(Fe, Zn)SnS_4$ was referred to as kesterite in the zinc-rich regime (Cu_2ZnSnS_4) and as stannite in the iron-rich regime (Cu_2FeSnS_4) [4, 10]. The latter has been known for much longer than the former [11] and only recently has it been classified under the umbrella family of $Cu_2(Fe, Zn)SnS_4$ as better determinations of these minerals were made, and from then, both terms were used interchangeably for this entire family of minerals [4].

Both of these structures are similar, and can be described as deriving from the same structure as CIGS, as was described in Sect. 1.3. From CIGS, the group III^{3+} indium and gallium are isoelectronically replaced by group II^{2+} Zn and group IV^{4+}

---

[2]See Chap. 1.

Sn, leading to the similar tetrahedral crystal structure, related to the diamond structure of silicon [12–14].

The crystal structures of kesterite CZTS, stannite CZTS, and chalcopyrite CIGS are shown in Fig. 6.1 for comparison. In kesterite, each of the cations is tetrahedrally bonded to 4 sulphur anions and likewise, each sulphur ion coordinates to two Cu, one Zn and one Sn ion. The unit cell takes the space group $I\bar{4}$, with the cations occupying FCC sites with a double axis in the c-direction, and the sulphur ions occupying opposing tetrahedral interstitial sites [4, 10]. The general structure differs from CIGS only with respect to the cation ordering. In CIGS, the cation layers are consistent 1:1 mixes of Cu:(In/Ga), whereas in kesterite they alternate between 1:1 mixes of Cu:Sn and Cu:Zn. The distortion from true tetrahedra is experienced as a result of the atomic radii mismatch between the cations, altering the bond lengths [15]. It is therefore expected that the electronic structure, and as a result, the electronic properties of CZTS will differ slightly from CIGS [13]. Despite displaying the same overall stoichiometry, it is known that kesterite and stannite adopt similar, but distinct structures [4]. This is shown in Fig. 6.1 and the difference is between the cation layers, with stannite alternating between layers consisting a 1:1 mixture of Zn:Sn and layers wholly occupied by Cu ions. This differing symmetry also gives stannite a different space group [10] of $I\bar{4}2m$.

It is now generally accepted that synthetically grown CZTS adopts the more stable kesterite structure [12, 16, 17]. This is despite historical reports which classified CZTS as the stannite structure [6] and there is much discussion and confusion in the literature [18] over which structure the synthetically grown material adopts, because there is only a small energy difference between the two phases [19]. Also, the different

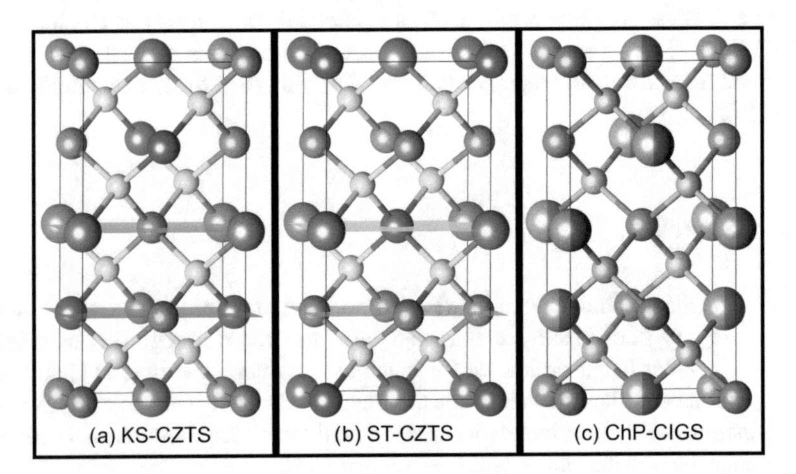

Fig. 6.1 Crystal structures of **a** kesterite (KS) CZTS [4, 20], **b** stannite (ST) CZTS [4], and **c** chalcopyrite (ChP) CIGS [21], showing individual atoms of Cu (orange), Zn (blue), Sn (grey), S (yellow), In (brown), Ga (beige), and Se (gold). Also shown are the cation planes: on the kesterite structure is the Cu/Zn (pink plane) and Cu/Sn (blue plane), and on the stannite structure is the Cu (orange plane) and Zn/Sn (green plane)

structures are thought to give differing properties, with a reduction in the bandgap being the most commonly cited [13, 15, 19].

Following the development of CZTS as a solar cell material and displacing it from the field of mineralogy, the term kesterite is now almost ubiquitously used to refer to synthetically grown instances of CZTS and as such, to avoid confusion throughout this chapter, unless explicitly stated, references to CZTS shall be to synthetically grown material in the kesterite structure.

6.1.3 CZTS as a Solar Absorber

Following the understandings gained from the crystallographic studies mentioned above, and from the initial promising electronic properties [6], the development of CZTS for its use as a solar PV absorber continued, with an initial reported efficiency of 0.66% in 1997 [22].

Initially lauded as a replacement to CIGS, the development of CZTS somewhat followed the same path, utilising similar device architectures and growth methods. As such, growth techniques for thin-films of CZTS are many and varied, following many of the same routes as CIGS [23]. Film growth has been achieved variously via: sputtering [24, 25], evaporation [26, 27], PLD [28, 29], sol-gel [30, 31], SILAR [32, 33], CBD [34, 35], and electrodeposition [36, 37].

Throughout its development, CZTS has continued to demonstrate attractive PV-relevant properties. The measured p-type conductivity is now understood to be intrinsic as a result of the Cu_{Zn} defect, which has the lowest formation energy [38]. Predictions of the bandgap [19, 39], and experimental measurements [35, 40–42], show it to be in the region of 1.5 eV, well matched to the solar spectrum. Also, it has high absorption in the visible range [43–45].

CIGS uses selenium rather than sulphur as the anion, and there also exists an analogue of CZTS using selenium, that is CZTSe. This compound has seen development alongside CZTS because it was reported to have similar properties, including a 1.5 eV bandgap [46, 47]; however, it is now accepted that although it too crystallises in the kesterite structure, CZTSe has a bandgap in the region of 1.0 eV [19, 48]; too low for effective use in PV. This finding did however, spark interest in the development of the mixed sulphide/selenide compound CZTSSe, which allows tuning of the bandgap, in order to better match the solar spectrum [17]. As such, this mixed compound has seen the highest efficiencies, with the record at 12.6% [49]. Although these higher efficiencies are promising, and the ability to tune the bandgap is beneficial, the use of selenium is questionable from an earth-abundancy point of view.[3] Therefore, compounds containing selenium are not of interest in this thesis, and henceforth, references to CZTS shall be limited to the pure sulphide material, unless explicitly stated.

[3] See Chap. 1.

Table 6.1 Selection of record-holding and historical solar cell efficiencies utilising CZTS

Absorber Growth Method	Device Architecture	Efficiency	Year	Reference
Sputtering	SLG/Mo/CZTS/$Zn_{1-x}Cd_xS$/i-ZnO/ITO	9.2%	2016	Sun et al. [52]
Evaporation	SLG/Mo/CZTS/CdS/i-ZnO/AZO	8.4%	2013	Shin et al. [27]
Sputtering	SLG/Mo/CZTS/CdS/i-ZnO/AZO	7.9%	2013	Scragg et al. [68]
Electrodeposition	SLG/Mo/CZTS/CdS/i-ZnO/ITO	7.3%	2012	Ahmed et al. [69]
Sputtering	SLG/Mo/CZTS/CdS/AZO	6.7%	2009	Katagiri et al. [70]
EBPVD	Al/ZnO/CdS/CZTS/Mo/SLG	0.66%	1997	Katagiri et al. [22]

Progress of the sulphide, selenide and mixed materials in terms of device efficiencies are detailed elsewhere [50, 51], but a historical selection of device efficiencies for the pure sulphide material are given in Table 6.1, including the current record efficiency of 9.2% reported in 2016 [52]. From this table, it can be seen that over the past few years, efficiencies have been rising, whereas for the mixed sulphide/selenide material, efficiency progress has somewhat stagnated, with the previous record set in 2014. The pure sulphide is not trailing far behind and this steady increase demonstrates the continued potential of this material.

There are several barriers which CZTS development must overcome, in order for the efficiencies to compete for market viability with CIGS and CdTe. First, there is the need to standardise the growth methods used. Currently, many methods are employed, which allows for the study of many different aspects of the material properties, such as: homogeneity [53], form factors [54, 55], defects [56], and the effects of temperature [31, 57], but in order to make industrial scale-up a possibility, a small number of growth methods must take precedence over the others; a stage which CZTS has not yet achieved.

The architecture of CZTS solar cells is also not yet fully optimised. Cell-architecture designs were taken from CIGS, but since then, issues have been faced. For example, the back contact commonly used with CIGS, molybdenum, has been shown to be detrimental to the performance of CZTS, due in part to the formation of MoS_2 at this interface [58], and investigations are in progress to replace it [59–61]. CdS is also commonly used as a window layer with CZTS, as it was with CIGS, and this combination has not led to results and efficiencies which are comparable.

The role and subsequent control of grain boundaries in solar absorber materials is also important and a current interest area for CZTS, where they have been shown to act benignly, or as recombination centres within the cell [62, 63]. The use of SLG as a substrate is prevalent throughout TFSC growth and as such, the diffusion of the sodium ions can be an issue. For CIGS, it was found that this sodium diffusion was beneficial for the cell [64], and studies on CZTS would suggest the same, with the sodium promoting large grain growth and improved conductivity [65–67].

There are many other problems and improvements that are being studied in order to progress the development of CZTS, and the reader is directed to several recent review articles, in which they are discussed in detail [23, 63, 71, 72].

Finally, there are the problems of phase-purity and the nature of the electronic structure. The importance of the accurate knowledge of the electronic structure of materials for use as solar absorbers is explained in Sect. 1.6.2 and these reasons apply also for CZTS, perhaps even more strongly due to the complex quaternary nature of this compound and the multiple valancies of the constituent elements, making the bonding regime and density of states more complex than archetypal semiconductors. There have been several studies on the electronic structure of CZTS [73], the merits of which will be discussed further in Sect. 6.7.3; however, there has been little experimental corroboration [74], most probably due to the other problem of phase-purity.

As a quaternary compound, CZTS has a complex phase diagram, which is still not complete [75, 76], and can only be represented at specific temperatures as a pseudoternary diagram of the Cu_2S–ZnS–SnS_2 system. Within this system, there is a small region at the centre where phase-pure CZTS exists, but only slight deviation from stoichiometry allows the formation of other phases within this system, and beyond it. The addressing of the combination of these two problems, along with the band alignments for the cell architecture, forms the basis of the motivation for this study. First, the possible secondary phases must be introduced.

6.2 Secondary Phases of CZTS: The Binaries

For CZTS, either during the growth, left over as precursor, post-growth, or due to degradation of the material, many different phases can occur either within, or at the surface of a grown film. These are not limited to those that make up the pseudoternary phase diagram and consist of a number of binary sulphides. Each of these materials has different properties and therefore will have different effects upon CZTS, dependent upon their quantity and location.

This section serves as an introduction to these materials, presenting their known structures, properties, and the estimated or measured effects that they have as secondary phases for CZTS.

6.2.1 Copper Sulphides

In their own right, the binary copper sulphides are interesting and useful semiconductors. They are known to occur as secondary phases within CZTS [77, 78], but, as was mentioned in Sect. 1.3.3, Cu_2S was the first material to be used as an absorber in a TFSC [79], and was one of the more successful historical materials [80, 81].

There are various phases which can exist and occur naturally in the Cu–S system [82], and include: covellite [83] CuS, chalcocite [84] Cu_2S, digenite [85] $Cu_{1.8}S$, djurleite [86] $Cu_{1.94}S$, anilite [87] $Cu_{1.75}S$, and roxbyite [88] $Cu_{1.81}S$, which are stable at room temperature, amongst others that occur through phase transitions at higher temperatures [82]. The non-integer stoichiometries are deviations from the chalcocite structure, that is $Cu_{2-x}S$, and as such have complex and vast unit cells comprising many tens of atoms. Most synthetically grown copper sulphide materials [89–91] are either CuS or Cu_2S, and the pseudoternary phase diagram of CZTS [92] is associated with Cu_2S, which is also used as a precursor for CZTS [71, 93, 94]. This means that when dealing with the possibility of copper sulphide secondary phases within CZTS, the properties of CuS and Cu_2S should represent the entire scope of this family of materials.

CuS and Cu_2S are far from simple compounds and so the identification of their formation within CZTS is a difficult task, rarely addressed with rigour, due the penchant for CZTS-synthesis within the Cu-poor regime [69]. As such, the expectation is that these phases are unlikely to form. Nevertheless, the use of copper sulphide materials as precursors for CZTS growth means that the effect of these secondary phases cannot be overlooked.

The oxidation states Cu^+ and Cu^{2+} are possible and therefore, from a simple stoichiometric point of view, the compounds Cu_2S and CuS would appear to be the cuprous (Cu^+) and cupric (Cu^{2+}) sulphides, respectively. The crystal structures however, suggest otherwise.

The crystal structure of the simpler compound, chalcocite Cu_2S, is presented in Fig. 6.2 and appears complex. However, all of the copper ions are threefold coordinated to sulphur ions, with slight deviations from a trigonal planar structure, and the hexagonal unit cell contains 24 Cu atoms and 12 S atoms [84]. This compound is correctly considered to be $Cu_2^+S^{2-}$, containing only the Cu^+ oxidation state.

The crystal structure of covellite CuS is presented in Fig. 6.3. It can be seen that this structure has distinct coordination environments for both copper and sulphur. At the first copper site (Cu_{Tri}), the copper ions are threefold coordinated to sulphur as in Cu_2S in trigonal planar polyhedra, whereas at the other copper site (Cu_{Tet}), the copper ions are fourfold coordinated to sulphur in tetrahedra. The first sulphur environment (S1) is fivefold coordinated to copper, with two of the tetrahedra aligned in the c-direction and three of the trigonal planar coppers surrounding it in the a–b plane. The final sulphur environment (S2) is coordinated to three of the copper tetrahedra surrounding the sulphur ion in the a–b plane but offset in the c–direction, and to another sulphur (S2 site) ion in the opposing c-direction [95]. This unusual structure leads to 6 copper and 6 sulphur atoms in the unit cell and is the source of confusion in the literature as to which oxidation state the elements adopt in this compound [96].

There are twice the number of Cu_{Tet} compared to Cu_{Tri}, and twice the number of sulphur ions which have the S_2 group (S2), compared with the sulphur bonded solely to copper (S1) [97]. The presence of the S–S bond means that the formal oxidation state of copper in this material cannot be described as singly ionic [83]. It has been previously described as a mixed valence compound[4] [98]; however, it was also suggested, with evidence from photoemission[5] [99–102], that the copper found in CuS is either entirely monovalent[6] [103–108], or consists of an intermediate copper valency greater than 1, but lower than 1.5[7] [109, 110]. These findings culminated in a study which suggests that Cu^{2+} is an unachievable oxidation state for copper sulfides [111], and it is now generally agreed that the structure is best represented by $Cu_3^{+}S_2^{2-}S^{2-}$, where • is a delocalised hole [112].

Cu_2S and some of the other off-stoichiometry phases have been shown to have a bandgap between 1.2 and 1.5 eV [38], and there is possibility that they can cause a shunt through a CZTS cell or act as recombination centres [113], if they were to form. Other Cu–S phases have been shown to demonstrate larger bandgaps, and because of the mobile Cu^{+} ions, can lead to junction degradation [38]. If CuS formed as a secondary phase [56], or remained after use as a precursor [114], then the large bandgap[8] [91], but high conductivity due to the hole conduction in the VB [108], could also act to shunt a CZTS cell.

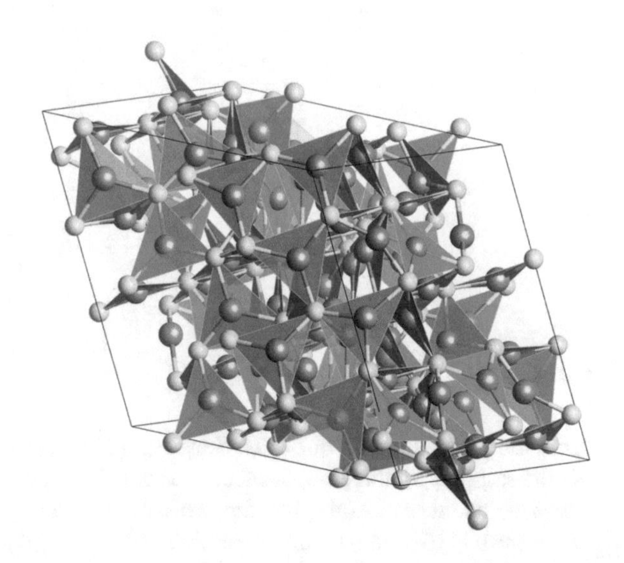

Fig. 6.2 Crystal structure of chalcocite Cu_2S [84], showing individual atoms of Cu (orange) and S (yellow)

[4] $Cu^{2+}S_2^{2-}Cu_2^{+}S^{2-}$.

[5] Further discussion can be found in Sect. 6.5.1.

[6] $Cu_3^{+}S_2^{2-}S^{-}$.

[7] $Cu_2^{3+}Cu^{+}S_2^{2-}S^{2-}$ or $Cu_3^{4/3+}S_2^{2-}S^{2-}$.

[8] 2.0 eV.

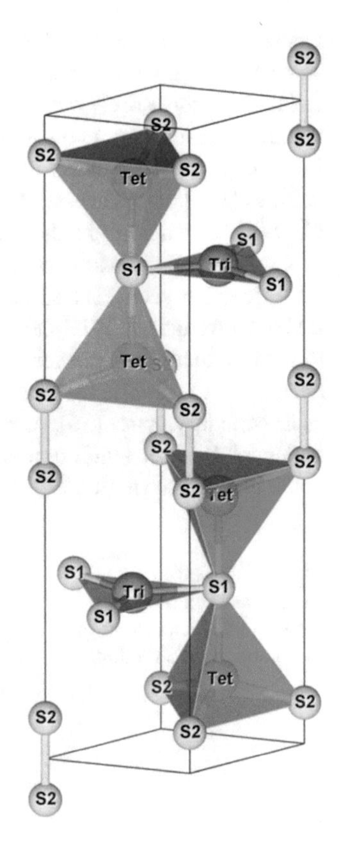

Fig. 6.3 Crystal structure of covellite CuS [83], showing individual atoms of Cu (orange) and S (yellow). The inequivalent copper and sulphur coordination sites are marked

6.2.2 *Zinc Sulphide*

From the phase diagram of the Zn-S system [115], there is only one stoichiometry of zinc sulphide, that is ZnS, because zinc only has one stable oxidation state, Zn^{2+}. ZnS occurs as the mineral sphalerite, which is the main ore of zinc found in the crust of the earth [116] and the only other phase of zinc sulphide, wurtzite occurs at high temperature[9] and need not be considered here. The crystal structure of sphalerite ZnS is presented in Fig. 6.4 and has a cubic unit cell which is based on the FCC diamond structure, with all of the zinc ions surrounded by four sulphur ions in regular tetrahedra [10].

In its own right, ZnS has several uses: it is the basis for many phosphors [117, 118], sees applications in optics [119], and is also being exploited for its semiconductor properties [120]. A typical II–VI semiconductor, ZnS has a very wide bandgap, in the region 3.5–3.7 eV [38, 121, 122], and if it was to form within CZTS, would likely act as an insulator, restricting current flow.

[9] > 1000 °C.

6.2.3 Tin Sulphides

The three main phases of tin sulphide[10] were introduced in Chap. 5, but will be discussed here as well because they are also found as secondary phases within CZTS [57, 59, 93, 94, 123–129]. These three phases are the only stable tin sulphides which form within the processing temperatures of CZTS [130], and it is expected that tin sulphides are likely to form in CZTS films because their formation is promoted in the Cu–poor regime [38].

6.2.4 Beyond Binary Secondary Phases: An Aside on Ternaries

The three binary systems described above do not exhaust the possible secondary phases of CZTS. Although this thesis is involved with the identification of these binary phases and the effects that they have upon CZTS, it becomes clear that one cannot discuss secondary phases of CZTS and disregard the ternary ones.

The several possible stoichiometries of copper tin sulphide (CTS) are the most widely recognised [132] ternary materials possible within CZTS, and it is Cu_2SnS_3 that occurs on the CZTS phase diagram [38, 76, 92, 133] and has even been studied for use as a PV absorber in its own right [134]. With a reported bandgap between

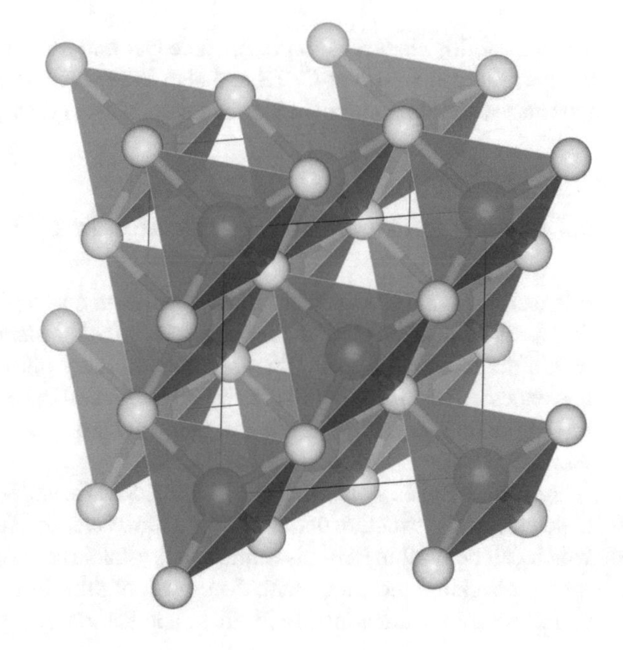

Fig. 6.4 Crystal structure of sphalerite ZnS [131], showing individual atoms of Zn (blue) and S (yellow)

[10] SnS, SnS_2, and Sn_2S_3.

0.98–1.35 eV, the formation of CTS within CZTS could have a similar effect as Cu_2S, affecting the carrier collection [76], acting as a shunt through the cell, or forming a recombination centre [135].

Whilst it is not ruled out that materials in the Cu–Zn–S (CZS) and Zn–Sn–S (ZTS) systems could form as well, the CZS system is understudied, with dubiously reported properties and growth methods [136–139], which most likely refer to doped materials rather than structurally different ones. In fact, the only reputable source of a ternary phase of the CZS system was put forward tentatively in 1970 [140] and since then, no further confirmation has been made. There are no reports of materials within the ZTS system. As such, at this stage in development, these phases can be disregarded, if they exist at all.

A Further Aside on Elemental Phases
It is also possible that elemental phases of Cu, Zn, Sn, or S could be present within the material due to their use as precursors to the growth [23, 29, 141, 142], or formed by the dissociation, rather than crystallisation of CZTS at higher temperatures. They would then act as direct shunts through the cell. However, the presence of metallic elements is unlikely due to the chalcophile elements [143], and that given the processing temperatures used, will react at least somewhat with sulphur, giving one of the materials described above.

6.3 Motivation and Scope of This Study

The aim of this chapter is to present a deeper and more complete understanding of the electronic structure of CZTS and also to establish characterisation standards in the context of helping with the identification of secondary phases in future syntheses.

6.3.1 Secondary Phase Identification in CZTS

It is well known that secondary phase formation during CZTS growth is a problem that must be addressed, by being able to accurately identify them when they form within the material. Considering the pseudoternary phase diagram for CZTS [92], it is expected that the most likely secondary phases to form are the vertex points of this phase diagram: Cu_2S, ZnS, and SnS_2, and so these materials receive the most research attention [144].

As is usual when synthesising thin-films, XRD is widely employed as a technique to determine the structure of the produced film. Overall, XRD is a powerful technique which can be used to help disseminate a crystal structure; however, when used as a ‘phase-checking’ technique with the growth of thin-films, it is usually never used to its full potential, and arguably for these applications, such use is satisfactory.

However, problems arise when XRD is not applied with rigour. As discussed in Sect. 2.6.2, this creates problems when peaks from other phases overlap with those from the expected phase and as such cause misidentification. CZTS suffers from peak overlaps, especially with sphalerite ZnS, which can be considered a parent-structure of CZTS. It is unfortunate that this phase has been shown to be the most thermodynamically stable, and therefore one of the most likely to form [145]. This specific problem is well known throughout the literature [38, 146], which has led to the standard adoption of RS measurements alongside XRD. Whilst this indeed adds more substance to claims of either phase-pure material or the identification of secondary phases, there are still associated problems with the coupled implementation of these techniques.

Other problems associated with the use of XRD and RS for secondary phase identification include the overlap of XRD peaks with CTS [38, 146] because of the similar structure. It has been suggested that the identification of copper sulphide and tin sulphide phases using XRD is straightforward due to the crystallographic differences [38, 146], yet there are other issues which complicate this analysis, further discussion of which is given in Sect. 6.8.1. Whilst Raman spectroscopy has been used to identify secondary phases [147], it is not wholly reliable for the exclusion of secondary phases. This is because the analysis depth is limited by the excitation wavelength,[11] and therefore, is unable to analyse secondary phases which may form below this depth. Also, in order to identify a phase using Raman spectroscopy, the wavelength of light used must be commensurate with the bandgap of the identifiable phases [38, 76, 146, 150], rendering the simultaneous detection of wide- and narrow-bandgap phases unfeasible.

Studies of CZTS by XRD will generally give reference to the PDF entry in the ICDD database, and while this is standardised practice, there is rarely comparison to the primary source, which could lead to a misidentification because specific measurements may not be fully comparable.

These reasons give motivation for the study of individual references for the possible secondary phases of CZTS described in Sect. 6.2, allowing the technique efficacies to be assessed.

6.3.2 *Standardising the Characterisation of CZTS*

As well as using XRD and RS to identify secondary phases within CZTS, these techniques are also used to determine the presence of the kesterite phase of CZTS, and due to the reverse of the problems described above, can lead to the misidentification of this phase. It is therefore important that other techniques are also used in order to determine the purity of the grown material.

[11] In silicon, for wavelengths less than ~450 nm, this is of the order of tens of nanometres, and extends to the order of microns for wavelengths longer than ~650 nm [148, 149]; however, it is also material dependent.

A technique that is commonly used alongside XRD and RS is XPS, usually to confirm the expected oxidation states within a grown material. Unfortunately, in the literature, from the presence of the expected oxidation states of the elements in CZTS, the conclusion is often drawn that this confirms the presence of the kesterite phase [151–155], when in fact it may mean that any number of materials with the same oxidation states are present.

XPS can also be used to assess different chemical environments, whether for different oxidation states, or different environments with the same oxidation state. Therefore, in principle, XPS can be used in the identification of secondary phases within CZTS. Such acknowledgement has been made in the literature [156], but it has very rarely been employed. In this study, it is believed that through careful measurement and characterisation of standard samples of the secondary phases, as well as a sample of phase-pure CZTS, then XPS could be used in conjunction with other techniques to determine whether these phases are present within a potentially mixed-phase sample of CZTS.

For CZTS, surface oxidation and contamination have been shown to be potential problems [157–159]. XPS has been used in the past to identify the presence of such oxidation[12] and could be used effectively for the case of CZTS as well.

Furthermore, to help determine that the correct phase of CZTS has been grown, it is important to determine the stoichiometry of the sample. This is readily achieved using EDX or ICP–MS; however, these techniques are prone to error with regards to sulphur when molybdenum is also present, and as thin-films of CZTS are commonly grown on molybdenum, this can be a problem [146]. XPS can be used for determining the surface stoichiometry of samples [164, 165], and has previously been used for CZTS thin-films [74, 114]. Thus, when coupled with other techniques, XPS could provide a more complete analysis of the stoichiometry, especially as it can be phase-selective, if properly fitted, whereas other techniques only give the overall stoichiometry of the analysed area.

It can therefore be seen that the potential of XPS for use in characterising CZTS is great, but there are standards which must be set in order to rigorously use this technique and avoid misidentifications and inaccurate analyses. This chapter aims to provide such guidelines; however, it must be kept in mind that it is not expected that this study will provide a definitive method for the identification of secondary phases in CZTS, but rather to encourage the adoption of another technique for use in this broad subject area.

6.3.3 Other Aspects of CZTS

As with all solar cell materials, the choice of partner materials, especially the window layer, is important to reduce recombination, facilitate charge extraction, and increase device efficiencies. As has been shown in Chaps. 3, 4, and 5, knowledge of the

[12]See Chaps. 3, 4, 5, and selected previous literature [160–163].

natural band levels of a solar absorber can be important for cell design. CdS is the most commonly used window layer material for CZTS, and because of the issues associated with the environmental impact of using CdS, along with allusions to CdS perhaps not being the most appropriate choice of window layer for other reasons [166, 167], it is hoped that the band levels and bonding nature of the valence and conduction bands can help justify this.

Defect analysis is an important topic for solar absorbers because they are usually the origin of the conductivity within the semiconductor, with many absorber materials being intrinsically p-type due to the presence of defects. Conversely, unwanted defects can also act in such a way to degrade the performance of the cell, through the generation of deep level trap states within the bandgap. In the case of CZTS, it is known that there exists a level of disorder between the Cu and Zn atoms and it has been suggested that this is a cause of the band tailing observed in CZTS and so acts to restrict the overall efficiency of a solar cell. Several studies have focussed on determining this level of disorder by different methods [20, 51, 168–170], and the work presented here can help give insight into the effects of it.

The complexity of CZTS is one of the reasons why the presence of secondary phases is so prevalent, and the thermodynamics involved with the annealing step of the growth are not well understood [59]. As such, because the annealing can introduce secondary phases, it is also possible that the annealing causes the breakdown of CZTS into secondary phases, which will result in them either being lost, reincorporated into the material, or forming into some other phase. Studying the thermodynamic processes involved with the growth of CZTS is important in understanding the secondary phase formation pathways, but the vast possibilities of these, which are dependent upon the growth method and choice of precursors, make this area beyond the scope of this chapter [53, 171]. However, the work presented here should give some insight into the breakdown mechanism of CZTS associated with in vacuo annealing.

6.3.4 Scope of This Chapter

XPS measurements of the binary secondary phases introduced in Sect. 6.2 will be presented in order to establish standard spectra that can be used in the future to help with the identification of these phases within CZTS samples. Following this will be the structural characterisation and XPS measurements of single crystal CZTS in order to establish standard measurements. Together, these standards for CZTS and the binary phases will be compared in a discussion of the effectiveness of the implementation of these techniques for use in identifying secondary phases within grown thin-films of CZTS.

Also presented here will be an analysis of the band levels of CZTS and the DoS with relation to the bonding mechanism and how these features relate to the electronic properties. Associated with this will be a discussion on the effects of the formation of the secondary phases within CZTS with regards to the band levels.

Finally, results will be presented from in vacuo annealing experiments conducted on the single crystal CZTS. The aims of this are threefold: to assess the effects of contaminants at the surface of CZTS films, including its susceptibility to oxidation; to assess the nature of the Cu/Zn disorder on the electronic properties and relate this to the processing temperatures used in the growth of CZTS; and to determine the mechanism through which CZTS breaks down at temperature in vacuo and assess the effects that this would have.

6.4 Experimental Details

Complete details of the experimental systems used in this chapter can be found in Chap. 2 and the specific details pertinent to this study, including a full description of the cleaning procedure used for the CZTS single crystal, are detailed in Appendix C.

6.4.1 Acquisition of Materials

Powdered samples of Cu_2S and CuS were obtained from Sigma-Aldrich Company Ltd. and were $\geq$99% pure.

Single crystal ZnS (001) was obtained from MaTeck GmbH.

Single crystals of SnS, SnS_2, and Sn_2S_3 were the same samples measured in Chap. 5, the full details of the growth can be found there, and in the appropriate literature [172].

Single crystal CZTS was donated by the Yoshino Group in the Department of Applied Physics and Electronic Engineering, at the University of Miyazaki, Japan. The crystals, grown by the travelling heater method from Sect. 2.1.1, have been shown to be phase-pure and have been characterised independently in the literature [129, 173–175] by XRD, RS, TEM, PPMS, EPMA, and Hall measurements. The crystals used in this chapter are shown in Fig. 6.5.

6.4.2 Characterisation of Binary Materials

XPS was used to study the core-level electronic structure, SEC, and VBM of the binary materials. Contamination was removed from the samples by Ar^+ ion sputtering and radiative annealing.

The powders of Cu_2S and CuS were mounted to UHV sample plates using double-sided carbon tape. Cu_2S received sputtering at 500 eV[13] for 15 min and at 1000 eV[14]

[13]Resulting in a flux ion current of 8 μA at the sample.

[14]Resulting in a flux ion current of 14 μA at the sample.

for 10 min after only small changes were observed with the lower power sputtering. CuS received sputtering at 500 eV[15] for 20 min. The BE of the C 1*s* was found to be 284.20 and 284.44 eV for Cu_2S and CuS, respectively. It was therefore concluded that no charging effects occurred in these samples because similar C 1*s* BE were observed in previous studies of these materials [100, 176].

The single crystal ZnS was attached to the UHV sample plate by spot welding tantalum straps across the edges of the surface of the crystal. This also provided an electrical connection between the crystal surface and the spectrometer. In total, the ZnS crystal received sputtering at 500 eV[16] for 5 min, at 1000 eV[17] for 30 min, at 2500 eV[18] for 20 min, and was annealed for 1 h at temperature steps of 300, 350, 400, and 450 °C. ZnS required much more aggressive sputtering and annealing because it was found that a thick layer of ZnO had formed at the surface. The large bandgap of ZnS[19] means that charging was expected [162], and it has previously been shown to differentially charge [177]. This was compensated for in the spectra by using an electron flood gun at 3.5 eV energy and correcting the spectra to the C 1*s* peak at a BE of 284.8 eV.

The tin sulphide crystals were attached to the sample plates, underwent cleaning procedures, and were charge corrected as described in Chap. 5.

All of the fitted synthetic peaks in this study were Voigt profiles, fitted after subtraction of a Shirley background, unless otherwise stated, and the constraints for peak fitting are described in Sect. 6.4.3.

Fig. 6.5 The single crystals of CZTS used throughout this work

[15]Resulting in a flux ion current of 5 μA at the sample.

[16]Resulting in a flux ion current of 9 μA at the sample.

[17]Resulting in a flux ion current of 10 μA at the sample.

[18]Resulting in a flux ion current of 13 μA at the sample.

[19]~3.6 eV.

6.4.3 Characterisation of Single Crystal CZTS

X-Ray Diffraction

For the powder XRD data, a small piece of the CZTS single crystal was ground to a powder. Measurements were then performed using a Philips X'Pert diffractometer equipped with a Cu Kα x–ray source running at 40 kV and 30 mA with an X'celerator detector.

Raman Spectroscopy

Raman spectra were acquired using a Renishaw inVia system, using exciting wavelengths of 532 nm and 633 nm.

Energy-Dispersive X-Ray Spectroscopy

EDX measurements were performed courtesy of the NiCaL within a SEM, using an accelerating voltage of 20 kV.

Photoemission Spectroscopy

PES was used to investigate the electronic structure of the CZTS crystal, which was affixed to a UHV sample plate by spot welding tantalum straps across the edges of the surface of the crystal. This provided an electrical connection between the CZTS crystal and the spectrometer. Ar^+ ion sputtering and radiative heating were utilised to clean contamination from the surface of the crystal. As the cleaning procedure forms part of the results of this experiment, details of this are given in Sect. 6.7.1 and Appendix C, Sect. C.5.4.

XPS was used to study the core-level electronic structure, combined with UPS to study the SEC, VBM and occupied VBDoS of the crystal. IPES was used to study the CBM of the crystal. Charging of the CZTS crystal was corrected by aligning the spectra to the C 1*s* peak at 284.6 eV, which has been used previously for this material [178, 179].

All of the fitted synthetic peaks in this study were Voigt profiles, fitted after subtraction of a Shirley background, unless otherwise stated. Cu 2*p* doublets were fitted with a separation of 19.80 eV [180] and an area ratio of 1:2. Zn 2*p* doublets were fitted with a separation of 23.05 eV [180], an area ratio of 1:2. Sn 3*d* doublets were fitted with a separation of 8.41 eV [181] and an area ratio of 2:3. Sn 4*d* doublets were fitted with a separation of 1.06 eV [181] and an area ratio of 2:3. Fitted S 2*p* doublets were separated by 1.20 eV [182], with an area ratio of 1:2. The FWHM between the doublet peaks for Cu 2*p* and Zn 2*p* were allowed to vary and resulted in a wider $2p_{1/2}$ peak due to Coster–Kronig effects [183, 184].

Density Functional Theory

Electronic-structure calculations were performed through the VASP code [185–188] within periodic boundary conditions. The pDoS in the valence and conduction bands were determined.

6.5 XPS Characterisation of the Binary Phases for Use as Standards

The identification of secondary phases in thin-films of CZTS is no simple task, and even with rigorous implementation of combined XRD and RS, sometimes the presence of secondary phases can still be overlooked. Utilising a wider variety of complementary techniques is key to making the identification of secondary phases within CZTS more transparent. To this end, a recent study showed that PL could be included into the secondary phase identification repertoire of techniques [189].

XPS is also a possible addition to this collection of techniques. Because of the different chemical environments between atoms within CZTS and atoms within secondary phases, there should be a representative set of peaks for each different phase. As has been shown in previous chapters, rigorous fitting models are required, with support from the literature and standard measurements, in order to support the analysis.

The main Auger features of the XPS spectra were also recorded so that the modified Auger parameter could be determined. This procedure is far more rigorous than using solely the BE position of a peak because the Auger parameter is much more sensitive to chemical changes, and also it eliminates the effect of uniform charging of samples [190].

Further to the identification of secondary phases within CZTS, the effect that they have upon the material is also important to disseminate, and for this reason, the SEC and VBM were recorded for the secondary phases in order to determine the IP and WF values for the materials.

XPS can also be used to determine the stoichiometry of CZTS; however, this is not always reliable for reasons discussed in Sect. 2.3.4. Some of these issues can be resolved by determining the stoichiometry from peaks that are close in BE, as was employed in Chap. 4, and it was suggested in the literature that for CZTS these could be Cu 3*p*, Zn 3*d*, and Sn 4*d*[20] [191]. However, the use of these peaks is unsuitable, because if oxygen were present, the O 2*s* peak overlaps the Sn 4*d* peaks, and the Zn 3*d* doublet is in the VB and susceptible to changes there. Instead, by measuring the main core-levels of CZTS and by applying stoichiometry correction factors (SCF) which are derived from known-stoichiometry materials,[21] then transmission, mean free path, and cross-section differences can be reduced.

6.5.1 Copper Sulphides

Because the samples of Cu_2S and CuS were powders, they contained a substantial amount of contamination, probably in the form of water and hydrocarbons. Sputtering

[20] ~74, 10, and 25 eV, respectively.

[21] The binary phases presented below.

the powders removed a large amount of this contamination; however, it was not possible to completely remove because of the morphology of the powder. Annealing was not performed to attempt to desorb these contaminants in order to avoid inducing changes in the materials themselves [82].

The Cu 2*p* region for the Cu_2S and CuS powders are shown in Fig. 6.6a, b, respectively. For Cu_2S, two Cu 2*p* doublets were fitted. The low-intensity, wider doublet at higher BE (purple shading) is attributed to copper bonded with oxygen (Cu–O), whose form is unknown, but given the high BE of the Cu $2p_{3/2}$ peak,[22] it is likely to be either CuO, $Cu(OH)_2$ or a mixture thereof [192–194]. The high-intensity doublet at 932.48 eV (red dash) is attributed to the copper bonded to sulphur in Cu_2S. A similar situation exists in CuS, also with a wider, lower intensity doublet (purple shading) at higher binding energy[23] being attributed to Cu–O as in Cu_2S. The copper in CuS was represented using two doublets, shown in Fig. 6.6b, with the same FWHM and an area ratio of 1:2 between the doublets. The lower intensity doublet at 932.20 eV (red dash) is attributed to the Cu_{Tri}, and the higher intensity doublet at 932.45 eV (blue dot) is attributed to Cu_{Tet}. That the higher BE doublet is twice as intense as the lower, reflects that there are twice as many instances of Cu_{Tet} than Cu_{Tri} in the crystal structure. The higher BE of the Cu_{Tet} doublet is understandable, given that there is a stronger bond felt by the copper ion here due to the presence of more surrounding anions. Such fitting of the Cu 2*p* has not previously been implemented in the literature, probably due to the small shift in BE, but suggestions alluding to it have been made [102, 109].

The S 2*p* region for the Cu_2S and CuS powders are shown in Fig. 6.7a, b, respectively. The CuS sample showed S 2*p* intensity around 169 eV (purple shading), which was not present for the Cu_2S, and is attributed to copper sulphate. This is in agreement with features seen in the Cu 2*p* region as well; the peak labelled Cu–O for CuS in Fig. 6.6b is much larger than that in Cu_2S, which is explained if copper sulphate is also present in CuS, occurring at a similar BE [195] to CuO or $Cu(OH)_2$. Common to both S 2*p* spectra is the presence of sulphur-containing contamination (green shading), which is commonly found in semiconducting materials[24] [196], and the uncertain form that this contamination takes leads to a spread of possible BE [197–200].

For Cu_2S, one S 2*p* doublet was fitted that was assigned to the sulphur in Cu_2S (red dash), as expected from the crystal structure and fitted features of the Cu 2*p* region. The spectral shape of the S 2*p* region for CuS, with three distinct peak crests, can be explained with regards to the crystal structure. The CuS S 2*p* spectrum was fitted with two doublets assigned to the different sulphur environments in CuS. The doublet at lower BE[25] is assigned to the sulphur that is bonded only to copper (red dash), whereas the higher BE doublet[26] (blue dot) is assigned to the sulphur which

[22] 933.49 eV.

[23] 933.20 eV.

[24] See Chaps. 3, 4 and 5.

[25] 161.20 eV.

[26] 161.97 eV.

is bonded to copper and another sulphur ion. This assignment is strengthened by the fact that there are twice as many instances of the sulphur bonded to copper and sulphur in the structure, reflected by the 2:1 area ratio between the two doublets, and the higher BE of this doublet is understandable as the S_2 bond is stronger than the S–Cu bond [201]. Although the crystal structure of CuS is well known in the literature, there is confusion about the expected XPS spectra. Some studies found similar spectra to those presented here [102, 103, 105] and assigned the doublets in the same manner [111, 202], whereas others assigned the doublets the other way around [101], or to non-stoichiometric material [203].

For use later in determining the stoichiometry of CZTS using XPS, SCF for Cu $2p_{3/2}$ and S $2p$ are required. The complexity of the spectra for CuS shown above makes this material unsuitable for this purpose, so Cu_2S was chosen. By normalising to a S $2p$ SCF of 1.00, taking the areas of the Cu $2p_{3/2}$ peak and S $2p$ doublet associated with Cu_2S, and by assuming an ideal stoichiometry for the Cu_2S sample, a SCF for Cu $2p_{3/2}$ of 12.85 was found. Due to the amount of contamination that was unable to be removed from the Cu_2S sample, this SCF will not be wholly accurate and caution will be applied with its use.

The binding energies found here for Cu_2S and CuS are shown, with comparison to literature values in Fig. 6.8. Goh et al. [202] and Kundu et al. [203] fitted two S $2p$ doublets for CuS, and the BE differences are in agreement with those values measured here. Although there appears to be much scatter of the BE measured throughout the literature, some general trends are commonplace. It is generally found that the Cu

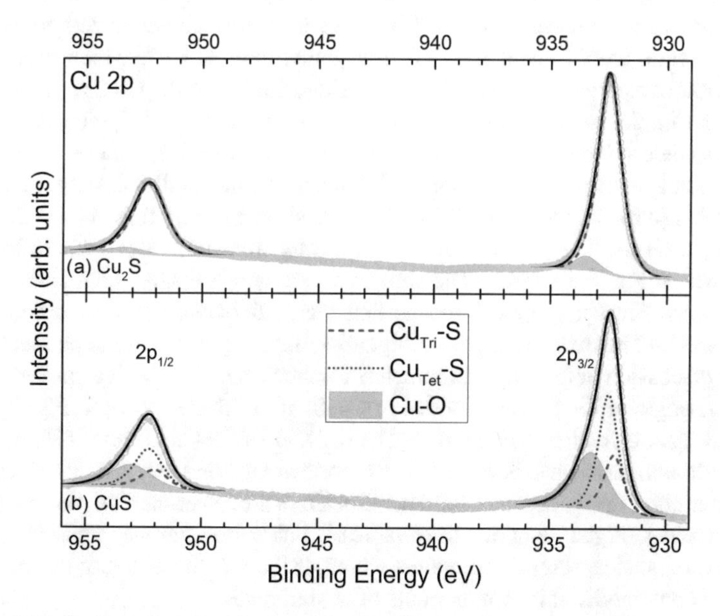

Fig. 6.6 XPS Cu $2p$ region of **a** Cu_2S, and **b** CuS. Fitted peaks reflect the differences in crystal structure between the two materials. Oxidised copper (purple shading) also shown and discussed in the text as to the origin. Peak envelope shown in black

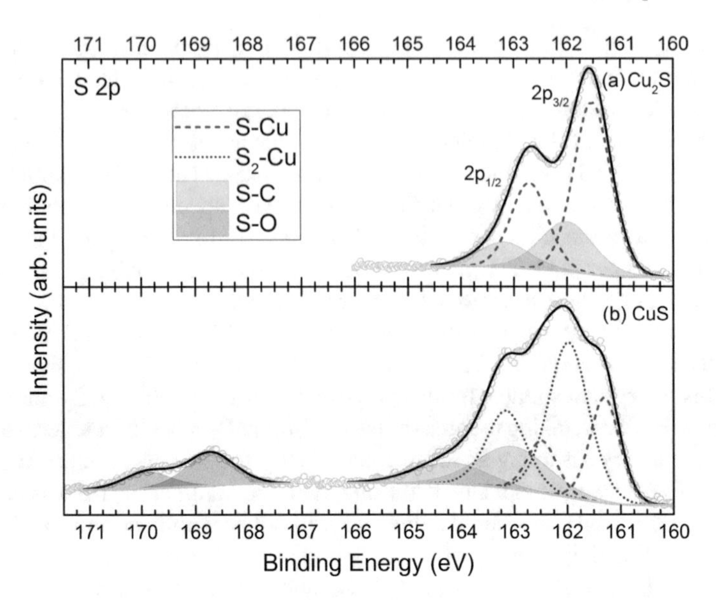

Fig. 6.7 XPS S 2*p* region of **a** Cu_2S, and **b** CuS. Fitted peaks reflect the differences in crystal structure between the two materials. Oxidised sulphur (purple shading) and contaminant sulphur (green shading) also shown and discussed in the text as to the origin. Peak envelope shown in black

$2p_{3/2}$ BE for Cu_2S is ~0.3 eV higher than that for CuS. Indeed, this finding is one of the reasons why the presence of Cu^{2+} in CuS was historically excluded, for a higher oxidation state would result in a BE higher than that of Cu^{+}, which was not the case. Without reference to both of the materials, the BE of the Cu $2p_{3/2}$ peak for an unknown sample cannot assign the phase, but the shape of the S 2*p* spectrum would be able to determine the phase of a single phase material. Nevertheless, the phases should be able to be more clearly defined by the use of the modified Auger parameter, and for this purpose, the Cu LMM regions are shown in Fig. 6.9a, b, for Cu_2S and CuS, respectively. The position of the $L_3M_{45}M_{45}$ transitions were determined and are shown on Fig. 6.9 as well. The shift between these values, which is larger than that between the Cu $2p_{3/2}$ BE means that the shift between the modified Auger parameters will be even larger, disambiguating the assignment of the phases further. Further discussion of the Auger parameters for these materials is given in Sect. 6.8.2.

The fittings for the SEC and VBM of Cu_2S and CuS are shown in Fig. 6.10a, b, respectively and yielded WF/IP values of 4.65/4.60 eV for Cu_2S and 4.74/4.67 eV for CuS. Although there have been very few reports of the band levels of these materials in the literature, the values measured here for Cu_2S are lower than literature values,[27] and for CuS are higher than literature values.[28] There are a number a possibilities for these discrepancies, foremost amongst which, is the fact that the samples measured here still contained significant amounts of contamination.

[27] ~5.3/5.4 eV [91, 204].

[28] ~3.8/4.1 eV [91, 210].

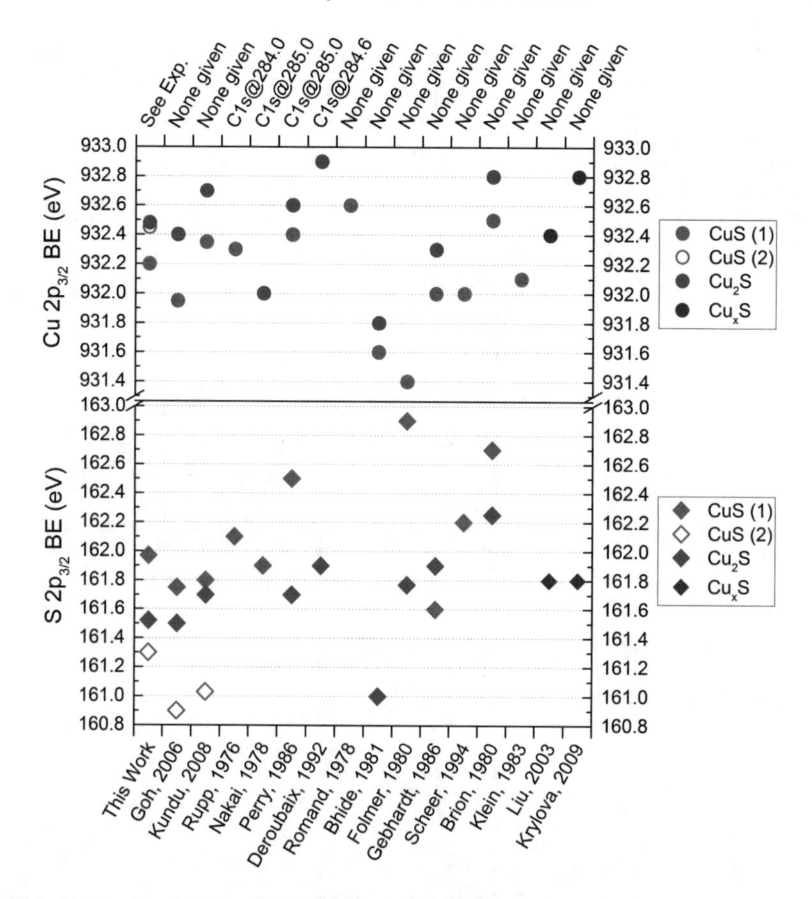

Fig. 6.8 Comparison between the XPS BE measured in this study, and those reported previously in the literature for CuS and Cu_2S. Literature references are given on the bottom x-axis [100–103, 105, 193, 195, 202–209], and the quoted charge referencing method employed in the corresponding study are given on the top x-axis. Species assignments are as given in the corresponding reference, see text for discussion of the agreements and more details

6.5.2 Zinc Sulphide

After cleaning, there was no contamination remaining on the ZnS crystal. The Zn 2*p* and S 2*p* regions are shown in Fig. 6.11a, b respectively. The single Zn 2*p* doublet (red dash) was assigned to the zinc in ZnS and the Zn $2p_{3/2}$ peak had a BE of 1022.05 eV. The S 2*p* region was fitted with two doublets. The large doublet with $2p_{3/2}$ BE of 161.79 eV (red dash) was assigned to sulphur in ZnS. The smaller doublet at lower BE (orange dash) was also assigned to sulphur in ZnS and not to contamination due to the lack of carbon and oxygen presence in the survey spectrum after cleaning. This extra peak is thought to be a result of the differential charging. This was attempted to be compensated for by using an electron flood gun which seemed to eliminate the

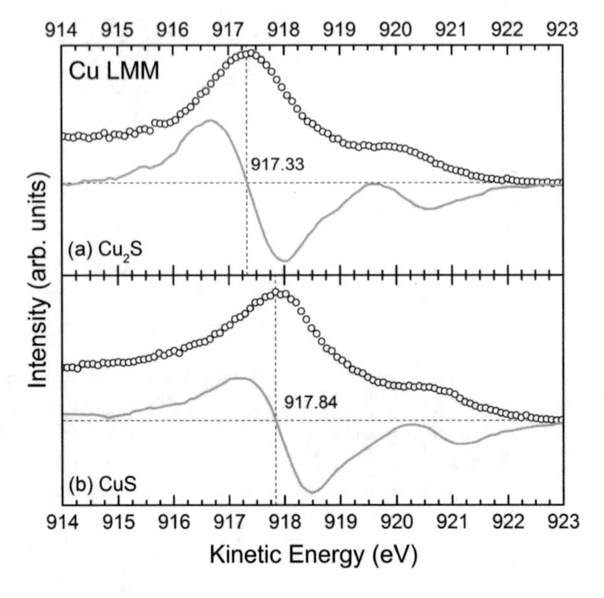

Fig. 6.9 Auger spectra for the Cu LMM region of **a** Cu_2S, and **b** CuS. Auger peak determination was performed using the x-axis intercept of the differential spectra, shown in grey

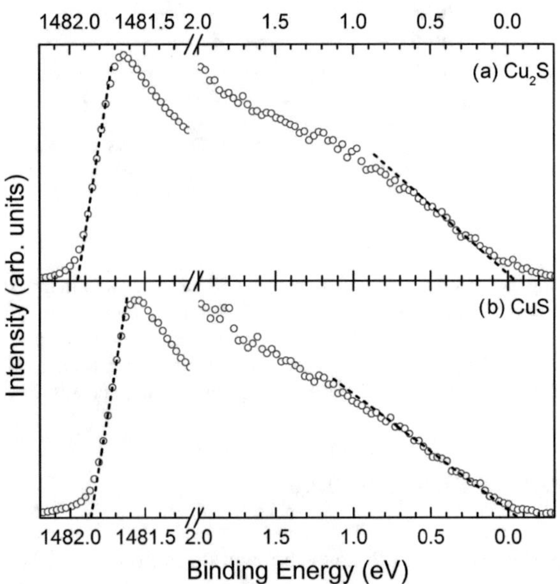

Fig. 6.10 VBM and SEC fittings for **a** Cu_2S, and **b** CuS after surface cleaning from XPS data

differential charging effects. However, it would seem that this was not completely effective and the small doublet is attributed to part of the sample that did not reach equilibrium through the charging and compensation.

For use later in determining the stoichiometry of CZTS using XPS, SCF for Zn $2p_{3/2}$ and S $2p$ are required. By normalising to a S $2p$ SCF of 1.00, taking the areas

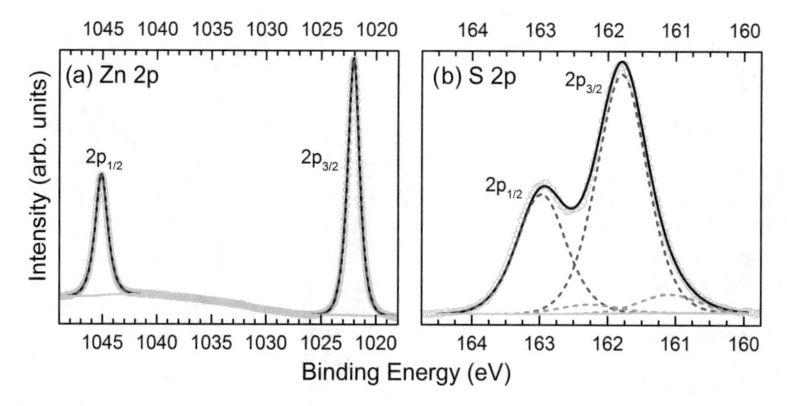

Fig. 6.11 XPS **a** Zn $2p$ region and **b** S $2p$ region of ZnS. Fitted peaks for ZnS shown in red, and secondary peaks shown in orange attributed to the differential charging, further discussed in the text. Peak envelope shown in black

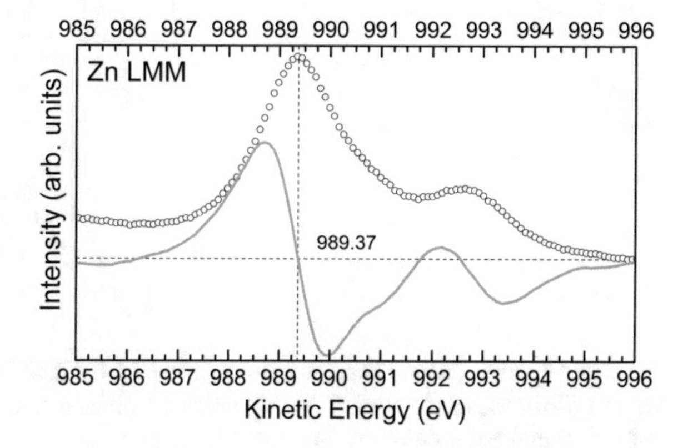

Fig. 6.12 Auger spectra for the Zn LMM region of ZnS. Auger peak determination was performed using the x-axis intercept of the differential spectrum, shown in grey

of the Zn $2p_{3/2}$ peak and S $2p$ doublet, and by assuming an ideal stoichiometry for the ZnS crystal, a SCF for Zn $2p_{3/2}$ of 15.60 was found.

Use of the flood gun means that the spectrum must be aligned to a known reference. In this case, the C $1s$ region was measured for many hours in order to acquire signal from the very low content of contaminant carbon remaining at the surface. However, because of the aforementioned differential charging, it may be the case that this carbon was compensated for differently to the ZnS and the BE measured here may not be completely reliable. For this reason, use of the Auger parameter is ideal as this measurement is independent of any charge corrections. Therefore, the Zn LMM region was also recorded and is shown in Fig. 6.12. The position of the $L_3M_{45}M_{45}$ transition was determined and is shown in Fig. 6.12 as well.

A comparison between the binding energies for ZnS measured here and those recorded previously in the literature are shown in Fig. 6.13. Although the BE are in general agreement, it is difficult to assess the practicality of using the Zn $2p_{3/2}$ BE as zinc is able to take only one oxidation state and the shifts are very small

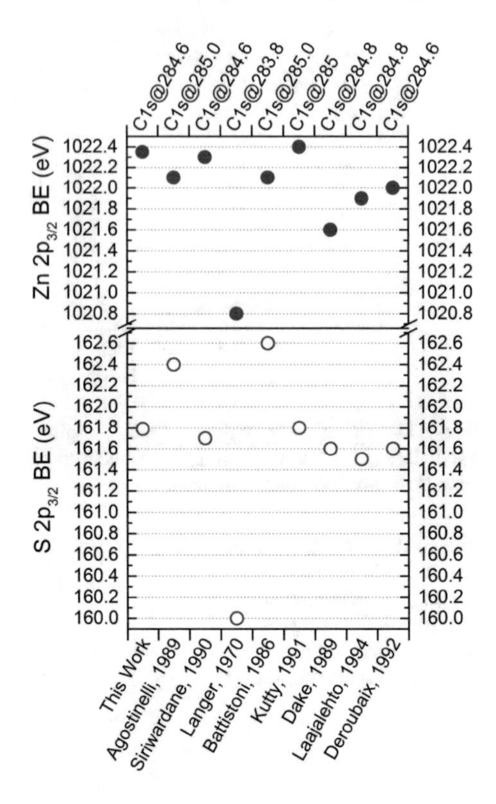

Fig. 6.13 Comparison between the XPS BE measured in this study, and those reported previously in the literature for ZnS. Literature references are given on the bottom x-axis [117, 177, 193, 211–215], and the quoted charge referencing method employed in the corresponding study are given on the top x-axis. See text for discussion of the agreements and more details

between species [216]. Variations shown in Fig. 6.13, at least to some extent, are due to the different choices of C 1*s* BE used for charge correction.[29] However, the use of the Auger parameter can render this correction unnecessary in the case of phase identification. Calculation and discussion of the Auger parameter for this material is given in Sect. 6.8.2.

The fittings for the SEC and VBM for ZnS are shown in Fig. 6.14 and yielded values of WF and IP for this material of 5.16 eV and 6.74 eV, respectively. This value of IP is in agreement with reported values in the literature [217, 218], but the WF value is lower than reported [219], probably as a result of the charging issues discussed above, because the measurement of IP should be independent of charge referencing, whereas WF is dependent upon it. Further discussion of the implications of these measurements is given in Sect. 6.8.4.

[29] Shown on the top axis.

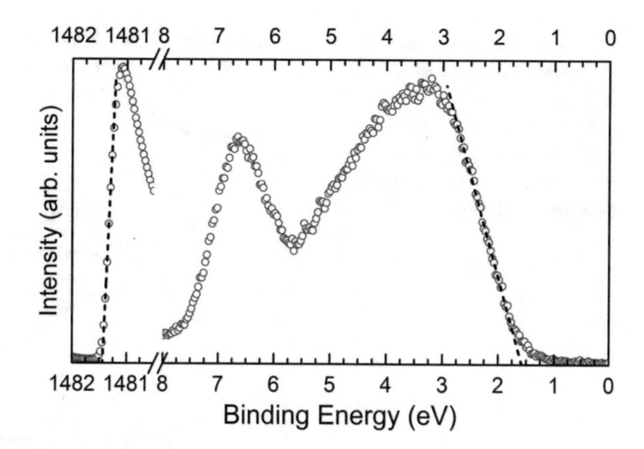

Fig. 6.14 VBM and SEC fittings for ZnS after surface cleaning from XPS data

6.5.3 Tin Sulphides

The CL fittings and band level determinations for the tin sulphides were given in Chap. 5 and were shown to be significantly different between the phases so that confident determination between the two different oxidation states of tin is possible. However, with regards to CZTS, tin is in the Sn^{4+} oxidation state, the same as in SnS_2, and therefore it may become difficult to distinguish between these two phases. For this reason, the Sn MNN Auger regions were also recorded for SnS, SnS_2, and Sn_2S_3 and are shown in Fig. 6.15a, b, c, respectively. The positions of the $M_4N_{45}N_{45}$ transitions were determined and are shown in Fig. 6.15 as well. Calculation and discussion of the Auger parameters is given in Sect. 6.8.2.

For use later in determining the stoichiometry of CZTS using XPS, SCF for Sn $3d_{5/2}$ and S $2p$ are required. By normalising to a S $2p$ SCF of 1.00, taking the areas of the Sn $3d_{5/2}$ peak and S $2p$ doublet associated with SnS, and by assuming an ideal stoichiometry for the SnS crystal, a SCF for Sn $3d_{5/2}$ of 12.69 was found.

6.6 Structural Characterisation of Single Crystal CZTS for Use as a Standard

There are few reports in the literature of standard measurements of CZTS, with most referencing to other technical samples. Therefore, in order to assess the effectiveness of the combination of characterisation techniques analysed here for use in secondary phase identification, it is prudent to make individual independent measurements upon a standard sample of CZTS, that is, a single crystal.

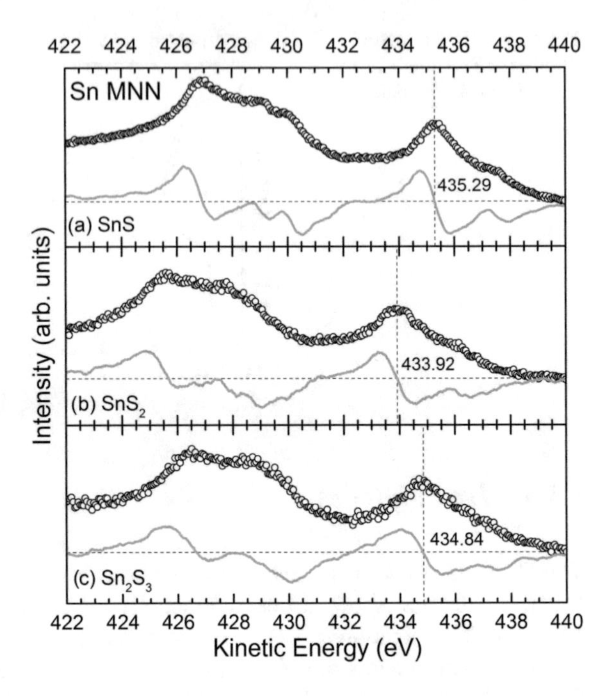

Fig. 6.15 Auger spectra for the Sn MNN region of **a** SnS, **b** SnS_2, and **c** Sn_2S_3. Auger peak determination was performed using the x-axis intercept of the differential spectra, shown in grey

6.6.1 Phase Determination

The powder XRD pattern for the CZTS crystal is shown in Fig. 6.16. The peaks are sharp, showing the high crystallinity of the powdered crystal. Also shown (red circles) is the commonly used ICDD reference for kesterite CZTS.[30] All of the indexed peaks in this entry are accounted for in the measured data, but with differing intensities for some of the peaks, specifically the (220/204), (312/116), and (224) reflections. Also shown is the theoretical diffraction pattern (green circles), calculated from the kesterite crystal structure [4] using the method in Sect. 2.6.2.[31] Previously in the literature, the use of this ICDD entry as a reference has been criticised for not including all diffraction peaks [153], leading to wrongly drawn conclusions on observations of 'extra' peaks. This criticism appears to be founded given that the theoretically calculated diffraction pattern appears to reproduce the experimental peaks more fully than this ICDD entry.

The Raman spectra for the powdered CZTS crystal are shown in Fig. 6.17 for two different excitation wavelengths. As can be seen, the different wavelengths of excitation have different sensitivities to certain modes, but the main peak at 337 cm^{-1} is common to both spectra. In order to assess the presence of secondary phases, different wavelengths must be used; however, it must also be kept in mind that the

[30] PDF# 00–026–0575.

[31] With the calculated pattern data shown in Appendix D, Sect. D.3.1.

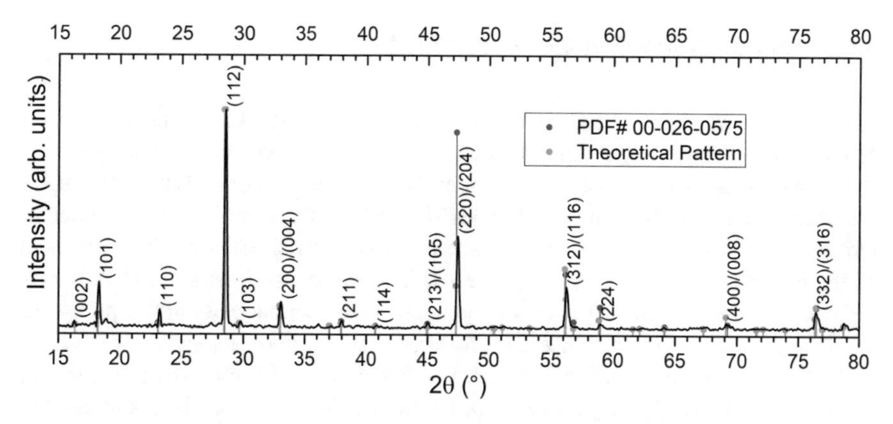

Fig. 6.16 Powder XRD pattern for the single crystal CZTS, compared with ideal patterns derived from the commonly used ICDD entry (red data), and theoretically from the crystal structure (green data) [4, 20]. Strongest reflection labels are shown (Color figure online)

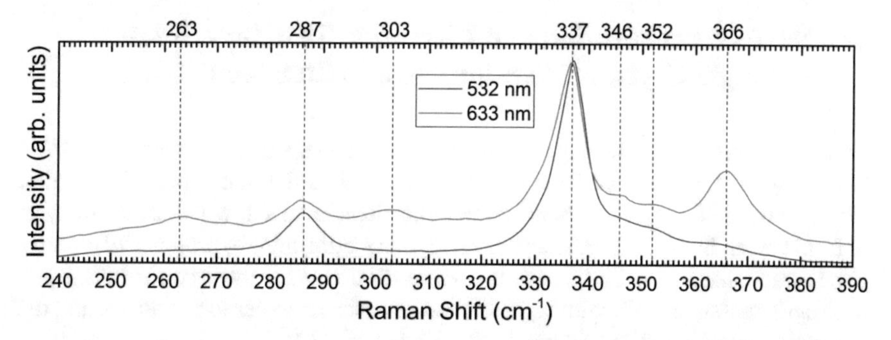

Fig. 6.17 Raman spectra for the single crystal CZTS using different excitation wavelengths. Strongest peak positions marked

spectrum from CZTS itself will change and therefore differences seen between the spectra are not necessarily proof of the presence of secondary phases.

Whilst there are many reports of RS of CZTS samples, there is little universal agreement in the literature because of the differences between the spectra of different wavelengths, and there is no ubiquitously used wavelength, like there is for XRD. Nevertheless, studies which used the same or similar experimental systems to those used here, demonstrate comparable spectra with peaks appearing at the same values [146, 220, 221].

Further discussion of the observed XRD and Raman spectra for the CZTS crystal compared to the individual secondary phases is given in Sect. 6.8.1 with regards to using these techniques to identify or exclude secondary phases in CZTS.

6.6.2 Bulk Stoichiometry Determination

The bulk stoichiometry of the crystal was determined by EDX, and the atomic percentages and compositional ratios are detailed in Table 6.2. These results suggest that the crystal is slightly Zn–rich and Sn–poor, but that the copper and sulphur content are near-stoichiometric. This result is similar to that observed in previous characterisations [175], but differs from record efficiency reports, which explain that most of the highest achieving devices are determined to be Cu–poor and Zn–rich, acting to encourage the formation of V_{Cu}, generating the conductivity and suppressing the deep level Sn_{Zn} defect [63, 222]. The stoichiometry here suggests that the Sn_{Zn} defect should also be suppressed, but the V_{Cu} defect will not be encouraged as much, possibly leading to differing transport properties in single crystal when compared to thin-film CZTS.

6.7 Surface Preparation and Electronic Structure Studies of Single Crystal CZTS for Use as a Standard

Whilst it is known how the expected XRD and Raman spectra of phase-pure CZTS should appear, there is confusion in the literature regarding the expected form that PES spectra should take and there are complications involved, which make analysis far from trivial. In the first instance therefore, this section aims to present PES spectra of the clean, phase-pure CZTS in order to help disambiguate these confusions. Also, other applications of PES with regards to solar absorber material properties and the electronic structure will be shown, in the context of CZTS.

Table 6.2 Stoichiometry values and elemental ratios of single crystal CZTS obtained from EDX measurements compared with ideal values

Parameter	Value	Ideal
Cu at%	26.28	25.00
Zn at%	13.71	12.50
Sn at%	10.57	12.50
S at%	49.43	50.00
Cu/(Zn+Sn)	1.08	1.00
Zn/Sn	1.30	1.00
S/(Cu+Zn+Sn)	0.98	1.00

6.7.1 Surface Preparation

The presence of contaminants and secondary phases can influence the PES spectra and perceived properties of a material. Therefore, it is important that standard spectra be free of such contamination, and this was achieved by in vacuo sputtering and annealing. However, it is also important to acknowledge the presence of such contamination, because it can provide information such as: the reasons why the contamination has formed; the origin of it; and the likelihood of it to form in the future. From this, possible prevention measures can be explored. It is also key to determine the effects of such contamination.

Here, the term 'contamination' refers not to the secondary phases which are possible to form in CZTS as these shall be dealt with later in Sect. 6.8, but rather to the products of the processes of surface oxidation and adsorption of water and hydrocarbons.

Expected XPS Spectra of CZTS

Subtleties in XPS spectra can make the analysis more difficult, and one must critically examine a material before measurement in order to pre-emptively acknowledge any of these features. This is especially the case when a material requires cleaning, as there must be an expected 'goal' spectra to be achieved. An example of how this applies is given by the S 2*p* spectrum of CuS, presented in Sect. 6.5.1. Here, if it was not known that there are two distinct sulphur sites in the crystal structure of CuS, which lead to the distinctive spectrum seen in Fig. 6.7b, then one may think that the second doublet was due to contamination, and if such a sample was undergoing surface cleaning procedures, then one may attempt to eliminate this second sulphur doublet which would culminate in damaging the material.

It is known that the oxidation states [14] within CZTS are Cu^+, Zn^{2+}, Sn^{4+}, and S^{2-}, which should not cause the formation of any extra features within the spectra, and the BE are expected to be in reasonable agreement with other compounds involving these oxidation states, and the expected ranges of BE are shown in Table 6.3. From the crystal structure,[32] each Cu, Zn, and Sn atom is coordinated tetrahedrally to four S atoms, and each S atom is coordinated tetrahedrally to two Cu, one Zn, and one Sn atom. These coordination environments suggest that only one peak for each element should be associated with CZTS.

Table 6.3 Expected XPS binding energy ranges for CZTS based on the formal oxidation states

Peak	Range (eV)	References
Cu^+ $2p_{3/2}$	932–933	[192, 193, 195]
Zn^{2+} $2p_{3/2}$	1021–1023	[162, 192, 193, 215, 233]
Sn^{4+} $3d_{5/2}$	486–488	[196, 234]
S^{2-} $2p_{3/2}$	160–163	[235, 236]

[32] See Fig. 6.1.

It is now possible to consider overlapping peaks within the spectra that could complicate the analysis. This is achieved by studying the features of standard spectra of the expected elements within a sample [180, 181, 223]. The main Cu 2*p*, Cu LMM, Zn 2*p*, Sn MNN, S 2*p*, S LMM, O 1*s* and C 1*s* regions should show no overlapping peaks and the only problem occurs with the Zn LMM region, which overlaps with the Sn $3d_{3/2}$ peak [224]. For these overlapping regions, there exist procedures by which these features can be fitted, compensated for, or effectively ignored, and the applied procedures used here will be discussed in Sect. 6.7.2.

Using these presuppositions about the expected spectra, sputtering and annealing could then be undertaken in order to clean the contamination from the surface of the crystal. It is well known that this can sometimes induce changes in materials that affect their properties [225–228], and for this reason, care must be taken when using these cleaning techniques. Sputtering is employed in the literature to remove surface contamination for routine thin-film XPS analysis of CZTS; however, high-energy sputtering is commonly used[33] [52, 152, 166, 224, 229–231] which, even though it is sometimes stated that no preferential sputtering was observed, can damage the film. This damage can then be overlooked if no fitting was performed or if the spectra before cleaning are not shown. Indeed, some studies recognise this problem and suggest that the ion energy should be kept as low as possible to stop the preferential sputtering of sulphur and the subsequent formation of metallic species. Bär et al. observed that ion energies of 50 and 100 eV did not cause structural changes [232], whereas with 500 eV, the formation of metallic species occurred [158].

Using these observations, a positive feedback cleaning procedure was applied, which used low energy sputtering: increasing the energy if no changes were taking place.

Cleaning Procedure

Throughout this section, spectra will be presented at different stages of the cleaning procedure and as such, are given referential labels to differentiate between them. These references are detailed, along with a complete description of the experimental procedures used in the cleaning, in Appendix C, Sect. C.5.4.

Spectra were recorded before the cleaning procedure began (As In) in order to assess the level of contamination on the crystal, and also to act as a starting point for the cleaning. The survey spectrum for the CZTS crystal before cleaning is shown in Fig. 6.18. It shows all of the expected peaks for Cu, Zn, Sn, S, O, and C. Due to the many elements present in this survey, the SNR is poorer than was observed for other materials consisting of fewer elements.

The fitted spectral regions of Cu 2*p*, Zn 2*p*, Sn $3d_{5/2}$, S 2*p*, and O 1*s* for the CZTS crystal before cleaning are shown in Fig. 6.19. Peaks attributed to CZTS (red dash) were the dominant species for all elements. The Cu 2*p* region was fitted with two doublets, with the second (blue dot) initially believed to be some form of copper oxide, but further discussion later shows that this requires further analysis.[34] The Zn 2*p* region was fitted with a single dou-

[33]>1000 eV.

[34]See Sect. 6.7.2.

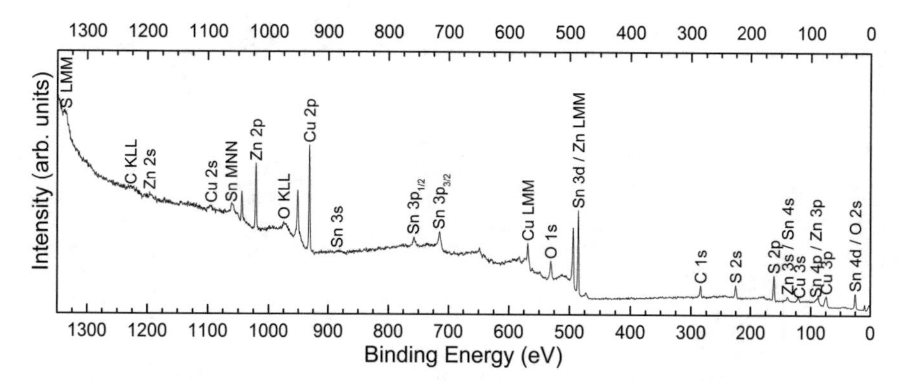

Fig. 6.18 XPS survey spectrum for the single crystal CZTS before surface cleaning

blet attributed to Zn in CZTS. The Sn $3d_{5/2}$ peak is shown rather than the whole Sn $3d$ region because of overlap with the Zn LMM Auger features. This will be addressed later, but for now, the fitting of this single component will suffice during the cleaning. Two peaks were fitted for the Sn $3d_{5/2}$ region, with the second being attributed to SnO_2 (pink dash dot), akin to the discussion in Chap. 5, where the oxide peak has a higher BE [164, 237] than the CZTS peak, and the Sn^{4+} in CZTS maintains this oxidation state when forming the oxide. The S $2p$ region was fitted with a single, very well resolved doublet, and the O $1s$ region was fitted with two peaks, a narrow peak at lower BE which corresponds to the metal oxides (pink dash dot), and a wide, higher BE peak which is attributed to adventitious (adv.) oxygen species (orange dash).

The relatively low level of contamination from the survey spectrum and apparent evidence that only tin and possibly copper have oxidised, bodes well for the surface cleaning, as there is not much to be removed. There is also no evidence of extra sulphides in the crystal, strengthened by the strong resolution seen between the splitting of the S $2p$ peaks, because even though different sulphides show very small shifts in the S $2p_{3/2}$ peak, even this would impair the resolved splitting of this doublet.[35]

Initially, the crystal was cleaned using sputtering with 200 eV ion energy; however, although this caused the reduction of the carbon and oxygen peaks in the survey spectrum, after several sputtering iterations, it became apparent that it was inducing structural changes, because of the development of a shoulder on the Sn $3d_{5/2}$ peak.[36] It was thought that this extra species could be removed by more aggressive sputtering[37]; however, the first few iterations of this aggressive sputtering caused the growth of this extra species up to a point where further sputtering caused no further changes. It was thus concluded that this extra species was indeed induced by the sputtering and that even the initial 200 eV ion energy was too aggressive. It was also noted that

[35]See, for instance Fig. 6.7.

[36]Data not shown.

[37]>1500eV.

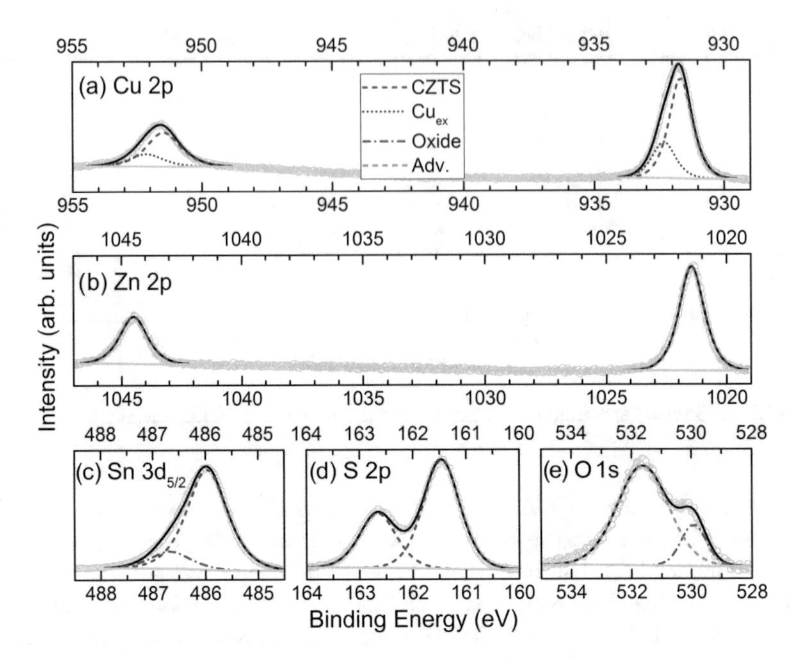

Fig. 6.19 XPS **a** Cu 2*p*, **b** Zn 2*p*, **c** Sn $3d_{5/2}$, **d** S 2*p*, and **e** O 1*s* regions of the single crystal CZTS before surface cleaning. Fitted peaks for CZTS shown in red, and other fitted peaks attributed as discussed in the text. Peak envelopes shown in black

annealing of the crystal after the formation of this extra species could not remove it, even after several days of annealing. This negative result served as motivation to study the effects of structural degradation of the CZTS crystal, the results of which are presented in Sect. 6.9.3.

The crystal was therefore re-prepared ex situ and the cleaning procedure restarted using a lower ion energy of 100 eV. This then caused the reduction of carbon and oxygen as well, but after several iterations, the same problems were not experienced. Instead, the spectra showed the reduction and elimination of the SnO_2 peak without the formation of any other species. These changes are shown in Fig. 6.20. As was mentioned previously, using the entire Sn 3*d* region for analysis at this stage is impractical due to the overlap of the Zn LMM Auger features with the Sn $3d_{3/2}$ peak and as such, only the Sn $3d_{5/2}$ peak was used for fitting.[38] This however, increases the tenuousness of the fitting, so the Sn 4*d* region was also fitted in the same manner[39] and led to the same conclusions.

The spectra shown in Fig. 6.20 are at various stages of the cleaning procedure, which consisted of sputtering with 100 eV ion energy for 5 min, followed by annealing at 230 °C for 15 min, referred to as a cleaning cycle. It can be seen, between the As In and Cl. 1 tin spectra, which corresponds to one cleaning cycle, that the SnO_2 peaks

[38]See Fig. 6.20a.

[39]See Fig. 6.20b.

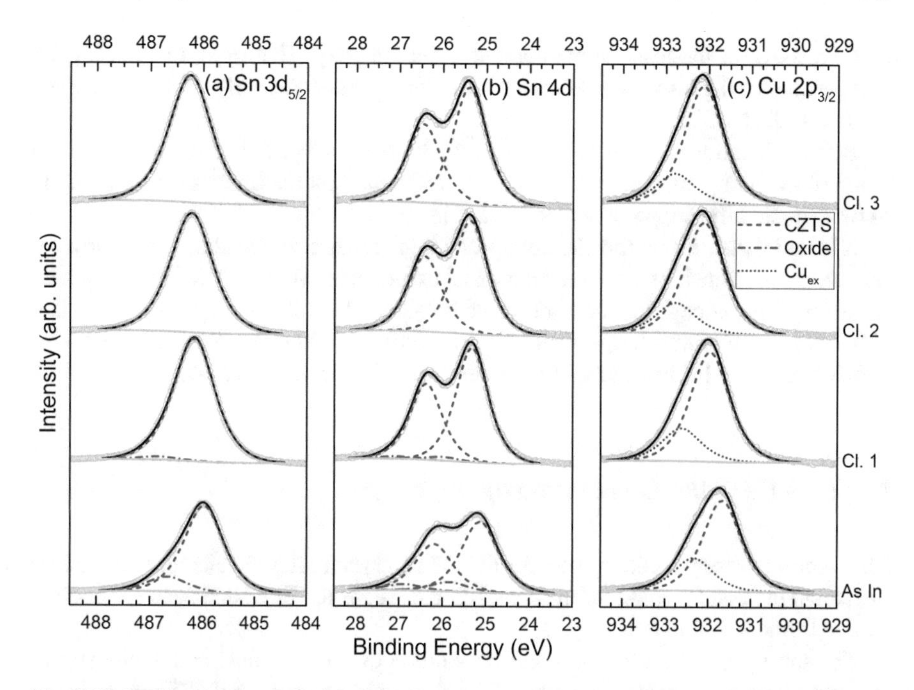

Fig. 6.20 XPS **a** Sn $3d_{5/2}$, **b** Sn $4d$, and **c** Cu $2p_{3/2}$ regions of the single crystal CZTS during the surface cleaning. Cleaning references (As In, Cl. 1–3) are detailed and explained in Appendix C, Sect. C.5.4. Peak envelopes shown in black

reduce, showing that the majority of this tin oxide is located only at the surface of the sample. Then, a subsequent six cleaning cycles (Cl. 1–Cl. 2), saw the removal of the tin oxide species, and a further six cleaning cycles brought about no further changes in the spectra (Cl. 2–Cl. 3). At this point, for fear of the structural changes that were brought about previously through extended and aggressive sputtering, four final cleaning cycles were applied, followed by a final anneal for 3 h, which also resulted in no further changes to the spectra. The crystal was then deemed to be clean enough for the spectra to be recorded for the clean crystal. It is noted that some contaminant carbon and oxygen remained present in the survey spectrum, but it was concluded that removal of these species would probably cause the structural changes seen before and as the tin oxide was removed, these other contaminants were probably contained within voids of the imperfect crystal surface,[40] and could therefore be largely ignored in the resulting analysis [238].

Shown in Fig. 6.20c is the Cu $2p_{3/2}$ region throughout cleaning. The second peak (Cu_{ex}) is present throughout the cleaning procedure and does not change with regards to relative peak intensity or peak separation. This is contradictory to the assignment of this peak as originating from copper oxide because given the removal of the tin oxide by the cleaning procedure, one may reasonably expect the removal of copper

[40]See Fig. 6.5.

oxide also, or at the least, some change in its intensity. Therefore, this extra peak was assigned to be also associated to the CZTS; the true nature of which is discussed further in Sect. 6.7.2.

Before cleaning, the finding that only the tin had oxidised is in agreement with a study of the native oxidation of CZTS, which found that the tin was the most likely to oxidise, and the copper least likely [158].

It is noted that during the cleaning procedures no discernible changes in terms of relative peak intensities or the formation of extra peaks were detected in the spectra for all regions, except for those described above. The only change was a constant shift to higher binding energies for all peaks, which arises due to the charge transfer and Fermi-level shifting caused by the removal of the contamination.

6.7.2 XPS Core-Level and Auger Analysis

This section presents XPS spectra for the CZTS crystal after the cleaning procedure described in Sect. 6.7.1. All of the peak BE and Auger KE fitted and described below can be found in Table 6.4.

The survey spectrum for the clean crystal is shown in Fig. 6.21 and demonstrates the same features as were observed in Fig. 6.18, with the only real differences being the reduction of the carbon and oxygen peaks. This is indication that with such a complex material as CZTS, the survey spectra cannot be used alone in determining the cleanliness of a sample, because the changes that take place within the regions of interest are rarely visible in the survey.

In order to more confidently identify secondary phases of CZTS in future analyses, the main Auger features were recorded as well as the CL regions. These are presented alongside their respective CL region and the calculation and use of the Auger parameters for CZTS will be discussed in Sect. 6.8.2.

The Cu 2*p* and Cu LMM regions are shown in Fig. 6.22a, b, respectively. The Cu 2*p* region was fitted with two doublets, as discussed in Sect. 6.7.1, with the main Cu $2p_{3/2}$ peak at 932.13 eV (red dash) being attributed to Cu^+ in CZTS. The

Table 6.4 XPS BE and Auger KE of the clean single crystal CZTS

Element in CZTS	Core-Level Peak	Binding Energy (eV)	Auger Feature	Kinetic Energy (eV)
Cu	$2p_{3/2}$	932.13 (1.07) 932.79 (1.07)	$L_3M_{45}M_{45}$	917.07
Zn	$2p_{3/2}$	1021.80 (1.06)	$L_3M_{45}M_{45}$	990.02
Sn	$3d_{5/2}$ $4d_{5/2}$	486.26 (0.95) 25.41 (0.93)	$M_4N_{45}N_{45}$	434.19
S	$2p_{3/2}$	161.71 (0.81)	$L_{23}M_{23}M_{23}$	149.73

Peak FWHM are given in parentheses

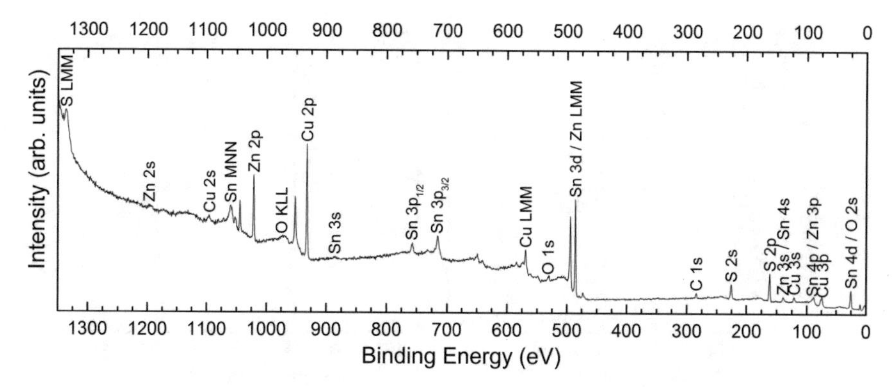

Fig. 6.21 XPS survey spectrum for the single crystal CZTS after surface cleaning

second peak at 932.79 eV (blue dot) was excluded as originating from the oxide due to the cleaning procedure and the low susceptibility of copper to oxidise in CZTS [158]. This then raises the question as to the origin of this peak, because without it, a satisfactory fit cannot be achieved for the Cu 2*p* region. Although the crystal structure would suggest that there is one coordination environment for the copper in CZTS, crystallographically, there are two different copper positions [4, 20], either in the same plane as tin or zinc. It is known that the copper in the Cu/Zn plane undergoes relatively low enthalpy of formation site-switching with the zinc atoms, and that only element-sensitive neutron diffraction can elucidate this level of switching [168, 239]. This second copper species is therefore tentatively attributed to an amount of disordered copper, occupying a site normally occupied by an element with a higher oxidation state (Zn^{2+}), which causes charge transfer around this atom, reducing the nuclear screening and resulting in a higher BE for this environment of copper [240, 241]. It is noted that a corresponding second species is not observed in the Zn 2*p* region; however, this may be a result of the small shifts seen in Zn 2*p* peaks. Further work is required in order to fully disseminate the origin of this second species, which could be more confidently assigned through the use of higher resolution SRPES. Further discussion of this is given in Sect. 6.9.2. The Cu LMM region demonstrates a sharp peak, and the calculation and discussion of the Auger parameter is given in Sect. 6.8.2.

The Sn 3*d*, Sn 4*d* and Sn MNN regions are shown in Fig. 6.23a, b, c, respectively. The Sn 4*d* region was also fitted in order to corroborate the fitting employed in the Sn 3*d* region because of the overlap with the Zn LMM features and the need for the extraction of this spectrum, as was discussed in Sect. 6.7.1. The Sn $3d_{5/2}$ region was fitted first and then the Sn $3d_{3/2}$ peaks were constrained, according to the parameters described in Sect. 6.4.3; the remaining intensity in this region is then attributed to the Zn LMM features. The single doublet (red dash) was attributed to Sn^{4+} in CZTS and was corroborated by the single doublet in the Sn 4*d* region. The Sn MNN region demonstrates two main features: a broad, high intensity feature at lower KE associated with $M_5N_{45}N_{45}$ transitions, and a sharper, low intensity feature

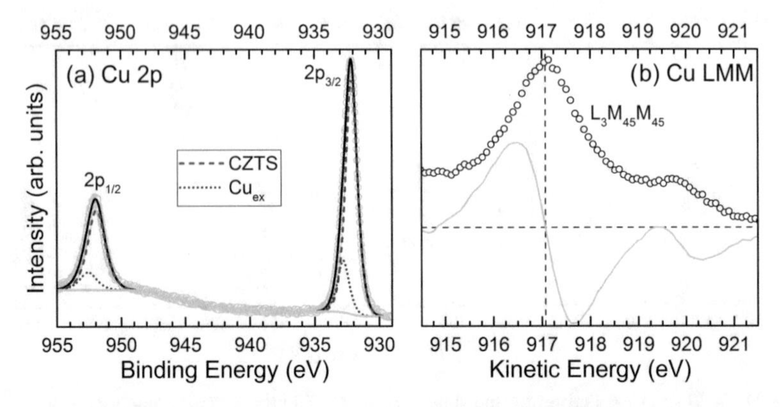

Fig. 6.22 XPS **a** Cu 2*p* and **b** Cu LMM regions of the single crystal CZTS after surface cleaning. Fitted CZTS peaks shown in red, and extra peaks (blue dot), discussed in the text. Peak envelope shown in black, and Auger peak determination was performed using the x-axis intercept of the differential spectrum, shown in grey

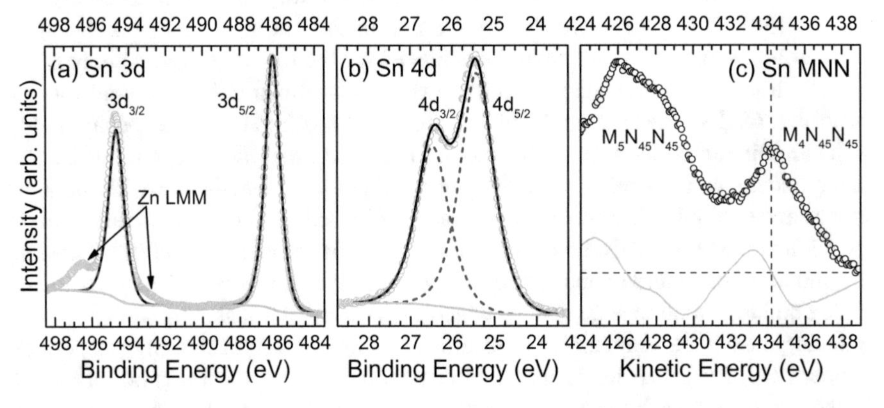

Fig. 6.23 XPS **a** Sn 3*d*, **b** Sn 4*d*, and **c** Sn MNN regions of the single crystal CZTS after surface cleaning. Features arising from Zn LMM are marked. Fitted CZTS peaks shown in red, peak envelopes shown in black, and Auger peak determination was performed using the x-axis intercept of the differential spectrum, shown in grey

at higher KE associated with $M_4N_{45}N_{45}$ transitions. Because it is sharper, this feature is taken as the main Auger peak, and the calculation and discussion of the Auger parameter is given in Sect. 6.8.2.

The Zn 2*p* and Zn LMM regions are shown in Fig. 6.24a, b, respectively. For the clean crystal, this region was fitted with a single doublet (red dash) which was attributed to Zn^{2+} in CZTS. The Zn LMM region was generated by subtracting the Sn 3*d* features from the fitted spectrum as shown in Fig. 6.23a, the position of the $L_3M_{45}M_{45}$ peak was determined, and the calculation and discussion of the Auger parameter is given in Sect. 6.8.2.

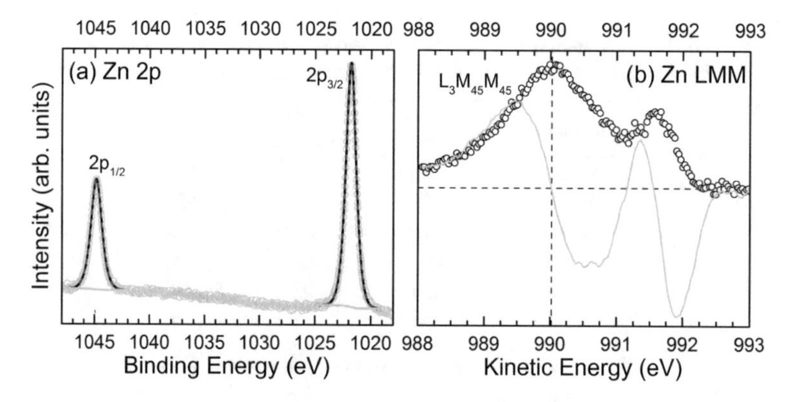

Fig. 6.24 XPS **a** Zn 2*p* and **b** Zn LMM regions of the single crystal CZTS after surface cleaning. The Zn LMM spectrum was determined by subtracting the Sn $3d_{3/2}$ peaks in Fig. 6.23a. Fitted CZTS peaks shown in red, peak envelopes shown in black, and Auger peak determination was performed using the x-axis intercept of the differential spectrum, in grey

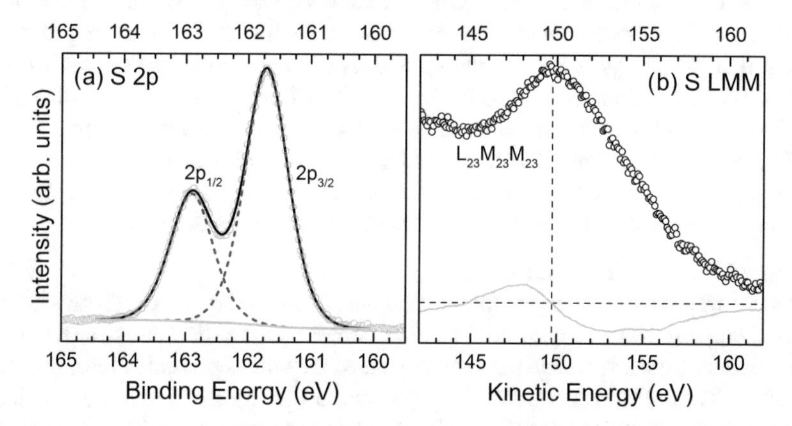

Fig. 6.25 XPS **a** S 2*p* and **b** S LMM regions of the single crystal CZTS after surface cleaning. Fitted CZTS peaks shown in red, peak envelopes shown in black, and Auger peak determination was performed using the x-axis intercept of the differential spectrum, shown in grey

The S 2*p* and S LMM regions are shown in Fig. 6.25a, b, respectively. A single, sharp, well resolved S 2*p* doublet was attributed to S^{2-} in CZTS. The S LMM region shows a broad peak with a steep background. The position of the $L_{23}M_{23}M_{23}$ peak was determined, and the calculation and discussion of the Auger parameter is given in Sect. 6.8.2.

The analysis of the CL and Auger spectra for clean CZTS can help determine the formation of secondary phases in the future, because should the spectra of a grown thin-film of CZTS differ significantly, then one could conclude that there exist secondary phases within the film which require further investigation. This matter is discussed further in Sect. 6.8.2.

Table 6.5 XPS surface stoichiometry values and elemental ratios of single crystal CZTS, using the SCF described in the text, before and after surface cleaning, compared with ideal values

Parameter	After Cleaning	Before Cleaning	Ideal
Cu at%	25.11	27.26	25.00
Zn at%	12.05	14.49	12.50
Sn at%	17.81	14.42	12.50
S at%	45.03	43.84	50.00
Cu/(Zn+Sn)	0.84	0.94	1.00
Zn/Sn	0.68	1.00	1.00
S/(Cu+Zn+Sn)	0.82	0.78	1.00

Peak FWHM are given in parentheses

Using the SCF determined from Sect. 6.5, stoichiometry values were calculated using the areas of the peaks fitted above. These results are given in Table 6.5 for the CZTS crystal both before and after the surface cleaning. Before cleaning, it would appear that the crystal is S-poor, with the metals in near stoichiometric ratios. This differs from the values calculated using EDX,[41] but is most probably explained by the fact that XPS-derived stoichiometry values are very surface sensitive, and the deviations seen here most likely result from the volatility of sulphur in vacuum. After cleaning, the surface stoichiometry remains S-poor, but is now also Zn-poor and Sn-rich, explained by the higher volatility of zinc in vacuum.

Comparison to the Literature

The measured binding energies for the elements of single crystal CZTS presented in Table 6.4 are shown in Fig. 6.26 for better comparison to the literature. As can be seen, there is quite a variation between the measured values presented throughout the literature. These discrepancies could be accounted for by the differences in charge referencing methods, but as this is often not reported, compensating for this is not always possible. Homogeneous charging of the sample, or a shift in Fermi-level, would cause all of the measured binding energies to shift by a similar amount; however, aligning the reported binding energies did not generate full agreement between the values shown in Fig. 6.26 either. This does not mean that there is no agreement between the spectra, as BE values can differ between samples for many reasons such as: a lack of, or unrepresentative peak fitting; charge transfer due to differing growth methods or post growth changes; the presence of secondary phases or contamination; and poor quality data. It is therefore prudent to analyse each report on its own merit.

XPS analyses are limited if they fail to quote the measured BE or present the spectra. Dhakal et al. [153] reported the use of XPS alongside other common characterisation techniques in order to exclude the presence of secondary phases at the interface of a thin-film of CZTS and CdS; however, with no spectra shown or binding

[41] See Table 6.2.

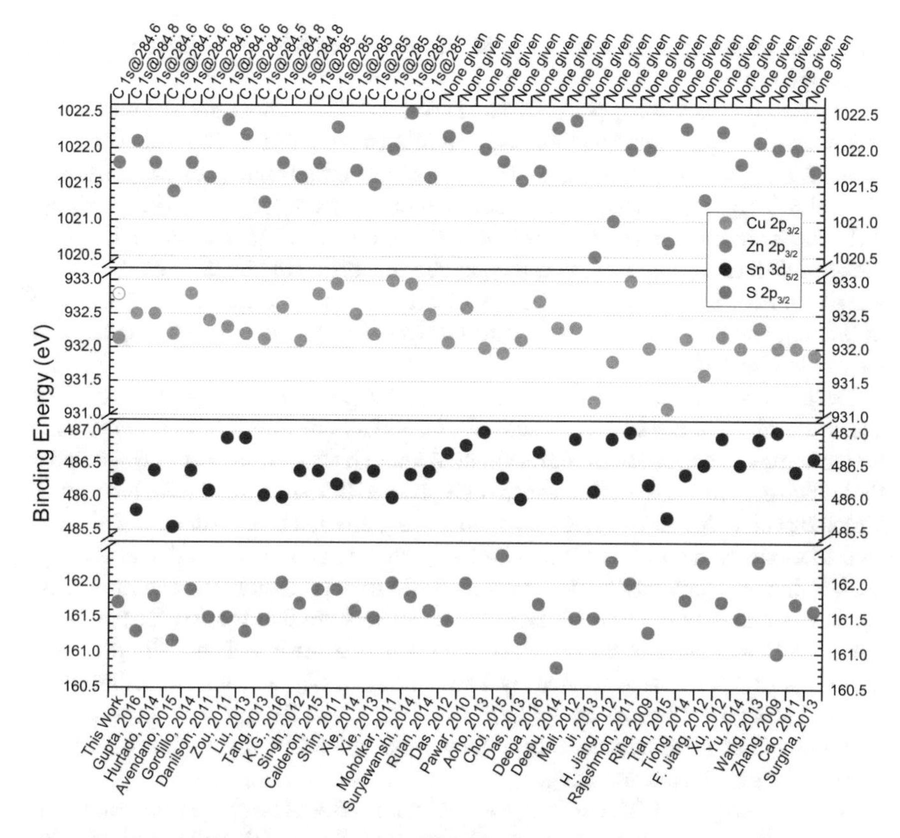

Fig. 6.26 Comparison between the XPS BE measured in this study for the clean single crystal CZTS, and those reported previously in the literature for CZTS. Literature references are given on the bottom x-axis [29, 33, 44, 93, 125, 141, 152, 154–156, 167, 178, 179, 220, 224, 231, 242–265], and the quoted charge referencing method employed in the corresponding study are given on the top x-axis. See text for discussion of the agreements and more details

energies quoted, this analysis cannot be assessed and the conclusions are therefore compromised. Ni et al. [114] reported the growth of CZTS thin-films which were sulphurised using a novel sulphur agent, and XPS was reported to be used to determine the stoichiometry, which without the spectra being presented, must be interpreted with caution, leading to the conclusions about the stoichiometry being questioned, which dictated further experiment in that study.

Survey spectra are vital for initially assessing the quality and cleanliness of a sample, but in-depth analysis can rarely be performed from them because of the sacrifice in resolution, yet conclusions have sometimes been drawn from analysis of only the survey spectra. Kim et al. [266] used XPS in analysing the surface stoichiometry of CZTS thin-films, contrasting the results to bulk stoichiometry measurements using EDX. Differences in the stoichiometry were attributed to different levels of oxidation, but because this analysis was taken from the survey spectra, and no fitting or binding energies were reported, it cannot be determined what chemical environment the oxygen is in.

The pitfall of not fitting peaks to data is common throughout the literature, and a secondary phase or contamination, which is shifted by only a small amount or is low in intensity, may be overlooked. Harel et al. [267] recorded high quality data from the CL and VB regions of samples of CZTS and discussed at length the compositional differences observed. They further discuss BE shifts and FWHM changes of the peaks between different compositions and conclude that different chemical environments are present for both sulphur and tin. However, because no fitting was performed or binding energies quoted, these extra species could not be assigned or quantified, weakening the conclusions.

Despite this, the majority of XPS studies at least quote the binding energies found, and some of these are in agreement with the values measured in this study, when taking some correction factor into account. Although peak fitting was only applied in the known overlap regions,[42] and not to the Cu 2*p* or Zn 2*p* regions, the BE reported by Hurtado et al. [178] are in agreement with those presented here. Danilson et al. [224] studied monograin powders of the mixed sulphur-selenium compound, and although peak fitting was only applied for the S 2*p*/Se 3*p* overlap region, the determined BE are in agreement with those measured here, suggesting that the BE for the mixed compound are similar to those for the pure sulphide. Electrochemically deposited thin-films of CZTS were studied by Pawar et al. [154], and polycrystalline bulk material was grown by Das et al. [220], who both found BE that were in agreement with the values measured here, once they were systematically shifted in order to account for presumable charging.

Some other quoted BE are in agreement with those presented in this study, but are not in universal agreement across the elements. This is believed to be due either to poor quality data, or to a lack of peak fitting, which would reveal the presence of extra species, whether from secondary phases or contamination. This is difficult to assess, but in some cases, the general shape of the spectra allows it. Cao and Shen [245] and Wang et al. [246] studied nanoparticles of CZTS, and although the BE agree somewhat with the values measured here, there are obvious shoulders on the Cu 2*p*, Sn 3*d*, and S 2*p* peaks, suggesting the presence of secondary phases or contamination, which is not addressed in the analysis. Similar situations exist with the spectra presented by: Riha et al. [155], Tian et al. [247], Zou et al. [248] and Jiang et al. [249], where the presented Sn 3*d* peaks are very wide and have obvious shoulders even though no fitting was performed. Choi et al. [167], Xu et al. [250], Yu et al. [251] and Liu et al. [252], who presented very noisy Cu 2*p* data, which showed the presence of shoulder peaks, as well as the satellite structure associated with copper oxide, without any acknowledgement of these features in the discussion. Jiang et al. [253], Rajeshmon et al. [44], and Zhang et al. [254], who presented noisy data with poor resolution, possibly obscuring the presence of multiple species within the spectra. Moholkar et al. [255], who presented spectra in which obvious shoulders and secondary peaks developed throughout an annealing process, but were attributed

[42] S 2*p* and Sn 3*d*/Zn LMM.

in a confused discussion of the spectra to 'broadening' of the peaks due to the annealing. Calderón et al. [256], who presented binding energies in agreement with those measured here, excepting Zn $2p_{3/2}$, on which there is a shoulder toward higher binding energies.

Because of the small spin orbit splitting for the S $2p$ doublet,[43] measurements not utilising monochromated radiation, or with poor resolution are usually unable to visibly resolve the splitting of these peaks, and as such, the enveloping S $2p$ BE is quoted over the more assessable S $2p_{3/2}$ peak BE. Such is the case with reports from Baryshev et al. [93] and Rajeshmon et al. [44], and is thought to be one of the reasons for the discrepancies in the binding energies. Other differences arise where fitting has been performed inexpertly, evidenced from the unphysical and implausible fitting constraints employed; for example Gordillo et al. [141] used a S $2p$ spin-orbit splitting value of 0.9 eV and vastly different FWHM between the doublet peaks, or Mali et al. [156], who synthesised nanoflakes of CZTS by the SILAR method and presented high quality XPS data, which was hindered by poor fitting that included the use of enveloping S $2p$ peaks rather than resolving them into their components, and also by fitting a background which did not correspond to the baseline. Other problems arise when extra peaks are sometime foisted as spectral artefacts, such as a report by Surgina et al. [29], who used SRPES to measure CZTS grown by PLD and found two Sn $3d$ doublets using a good fitting model, but then attributed the second doublet to "so-called 'final-state' effects" with no further explanation or citation.

The formation of contamination or secondary phases can be easily overlooked as is shown above, and cause potential problems either later in the characterisation, or in future growth of materials, led by these analyses. Furthermore, it is posited that several reports of 'phase-pure' CZTS synthesis are not so, which could cause further problems with citations of these reports. Nevertheless, it is promising to review several reports that acknowledge the potential influences of secondary phases or contamination, measured by XPS. Deepa et al. [257] measured the XPS depth profile of thin-films of CZTS grown by spray pyrolysis and found that signal from oxygen and carbon was present throughout the depth profile, which they acknowledged was a consequence of the growth method. Shin et al. [125] and Suryawanshi et al. [33] determined the presence of Cu_2S and SnS_2 within a thin-film of CZTS using XRD and XPS, finding the Cu $2p_{3/2}$ peak from Cu_2S to occur at higher BE than CZTS and the Sn $3d_{5/2}$ peak to also occur at higher BE than CZTS. However, this analysis is questioned somewhat because the peak fittings are not shown and the enveloping data does not fully resolve the components, with only shoulders visible. Zou et al. [248] acknowledged that ZnS could be a possible secondary phase in their CZTS nanocrystals and measured a reference sample of ZnS, comparing it to their Zn $2p$ spectrum from CZTS to rule out the formation of this secondary phase. This approach is similar to that employed here and its usefulness shall be assessed in Sect. 6.8.3.

Presumably presented in good faith, it is common to read reports that utilised XPS, yet drew conclusions which are unfounded [152]. This usually is the case where XPS is used as a secondary measurement and is not the main focus of the report; never-

---

[43] 1.20 eV.

theless, it is important to present accurate data and interpretations thereof, so that meaningful conclusions can be drawn. Sometimes, spectral regions are clipped too narrowly [151], meaning that the full peak shape cannot be visualised and assessed, or the quoted BE values do not correspond to those displayed on the spectra [258]. Other issues arise with stating experimental details copied verbatim from reports of other research groups [154, 224, 253, 268, 269], which may not represent the actual experimental details, leading to data that cannot be critically assessed later.

One common discussion matter found in analyses is the confirmation of oxidation states within a material from the agreement of the measured BE with a literature value. These citations are usually from other characterisations of CZTS, but sometimes are from completely different materials that happen to have the same elements in the same oxidation state. This is not always a justified analysis and will be further addressed in Sect. 6.8.2. Another unjustified analysis, which is related to this, is when it is stated that the oxidation state is confirmed by the spin-orbit splitting value of a doublet, and sadly, this is rife in the literature of CZTS [141, 154, 156, 179, 245, 259–261].

The binding energies reported by Xie et al. [243] are in agreement with the values presented here, despite no peak fitting being presented, and the fact that the survey showed large quantities of surface contamination in the form of carbon, oxygen and sodium, which were not addressed. Also, the resolution of this study is very poor, with very wide peaks, which could mask intensity arising from secondary phases. What is most troubling however, is that these spectra are replicated from a past article from Xie et al. [262], with no acknowledgment of the previous work, in which, different binding energies were reported. A similar situation exists with a report from Tiong et al. [263], who presented XPS data from CZTS nanocrystals that is of poor quality, and possibly obscures the identification of secondary peaks. They then state that the measured binding energies are in agreement with the literature and cite a previous report by the same authors, which replicates the same data [264].

Despite these weaknesses in the literature, the agreements are promising for the results presented here, as they show that the BE values measured from the single crystal are comparable to those measured from thin-films or other forms, and can therefore be used as standards. It is clear, given the scattered nature of the binding energies in Fig. 6.26, that it is necessary to utilise a more robust measurement for analysing XPS data that is not susceptible to charging or other effects which can shift binding energies. That is, the Auger parameter, which is analysed in the case for CZTS in Sect. 6.8.2.

6.7.3 Density of States Analysis and Natural Band Alignment

One powerful application of PES that is underused is its ability to give a semi-direct representation of the VBDoS of a material, which is beneficial for material studies. Furthermore, it can provide experimental corroboration of theoretically calculated density of states, and also help in the determination of the VBM position.

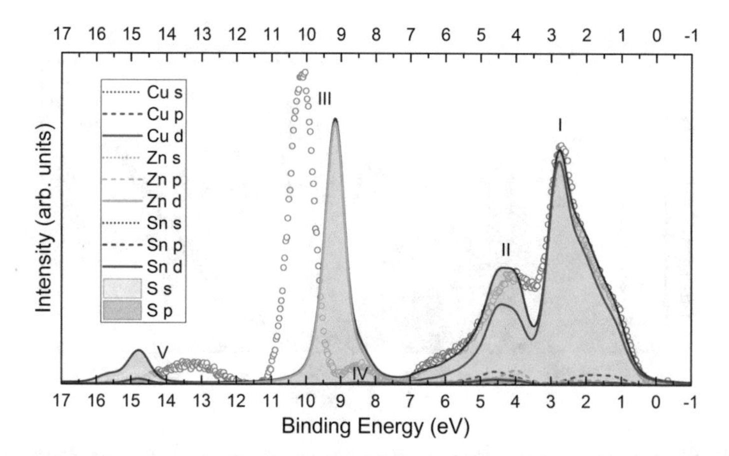

Fig. 6.27 Simulated and measured XPS VB spectra for CZTS with respect to the Fermi-level at 0 eV. Background subtracted XPS data is compared with broadened and corrected partial DoS curves. Green data are from XPS and the black curve with grey shading is the total summed DoS

PES Spectra

The VB spectrum of the clean CZTS crystal, obtained from XPS, is shown in Fig. 6.27 along with the theoretically calculated pDoS curves. A Shirley background was subtracted from the XPS spectra (green open circles) and the pDoS curves were corrected using the parameters listed in Appendix C, Sect. C.3. The total DoS curve (black line, grey shading) was obtained by summing the individual corrected pDoS curves, and is also shown in order to compare it with the XPS spectrum.

The theoretical DoS curves were aligned to the leading edge of the VB spectra and it was found that the top of the VB (I) is reproduced excellently by the calculations. Beyond this edge, the agreement is less satisfactory, but is still good, with all features present and accountable for with the correct intensity ratios. The bottom of the first VB (II, IV), and a feature at the very bottom of the VB region (V) are at lower binding energies in the XPS spectrum, because of final-state relaxation effects in XPS, which are known to shift features higher in BE closer to the valence band edge [270]. Feature III is mainly due to Zn *d* states and is found at higher energy in the XPS spectra than the DoS. This is thought to be because the level of DFT used here has incorrectly calculated the BE of the somewhat localised Zn 3*d* states. Indeed, previous studies comparing lower levels of theory show how the Zn *d* states were originally calculated with even lower BE, increasing when utilising higher levels of theory [14].

The combination of XPS/UPS and IPES allows the determinations of the values of IP, WF, EA, and E_g using the methods described in Sect. 2.3.4. The linear extrapolation fittings of the SEC, VBM and CBM from XPS and IPES are shown in Fig. 6.28. The leading edges of the SEC and VBM are clear, so the linear extrapolation of these to the baseline caused no problems with regards to finding the linear region to fit to. However, also shown in Fig. 6.28 is the broadened DoS curve (grey shading), which

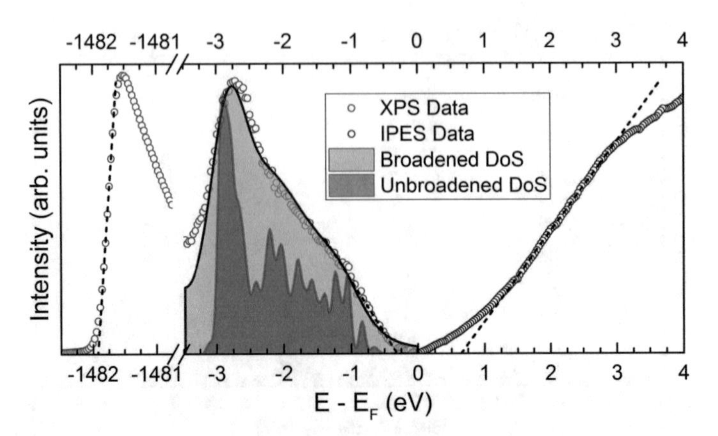

Fig. 6.28 VBM, SEC, and CBM fittings for the clean single crystal CZTS from XPS and IPES data. Band levels were determined as described in the text using linear extrapolations and fitting the DoS curve

was fitted to the leading edge of the XPS spectrum and the unbroadened DoS (green shading), which was shifted by the same amount. Historical development of the use of XPS for performing band offset measurements led to the lauding of a technique for determining the VBM that was more precise than the linear extrapolation method. Proposed by Kraut et al. [271], this involves the fitting of theoretically calculated density of states to the leading edge of the VB spectra and using the onset of the DoS as the true position of the VBM. This method then negates any effects which arise due to instrumental broadening that can shift the onset of the leading edge of experimental spectra. The necessity for accurate DoS calculations confined this technique to the realm of simple semiconductors and surface science, with the linear extrapolation technique taking precedence with regards to grown thin-film heterojunctions. The relative benefits and drawbacks for these VBM determination techniques are discussed further in Sect. 7.2.2, and how they apply to different spectra with varying degrees of accuracy. In Fig. 6.28, one can see that the linear extrapolation technique and DoS onset technique result in different values for the VBM to Fermi-level separation: 0.30 and 0.60 eV, respectively. Therefore, the value found from the onset of the aligned DoS shall be the one used in the remainder of this analysis.

IPES is a lower resolution technique than XPS and confidence in fitting the linear extrapolation of the leading edge is less certain. Nevertheless, a linear portion was assessable and this was fitted. It may be the case that it is necessary to perform a DoS fitting to the CB spectra measured by IPES in order to determine the true position of the CBM; however, this was not performed as the broadening criteria and cross sections are unknown with regards to IPES.

The VB region measured using UPS is shown in Fig. 6.29 along with the theoretically calculated pDoS curves. Here, the pDoS curves were corrected for experimental comparison using the parameters listed in Appendix C, Sect. C.3. The total DoS curve (black line, grey shading) was obtained by summing the individual corrected pDoS

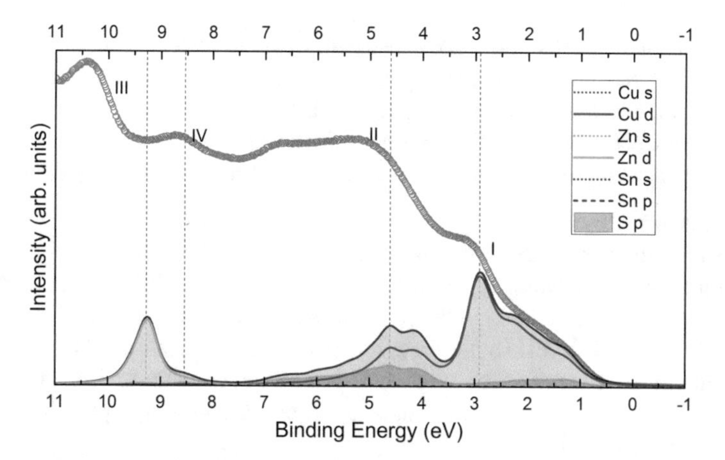

Fig. 6.29 Simulated and measured UPS VB spectra for CZTS with respect to the Fermi-level at 0 eV. UPS data is compared with broadened and corrected partial DoS curves. Green data are from UPS and the black curve with grey shading is the total summed DoS. Part labels are discussed and referred to throughout the text

curves, and is also shown in order to compare it with the UPS spectrum. Because of the correction factor that was necessary for determining the VBM position from XPS, a similar correction will be necessary here as well, so the corrected DoS curves were aligned to the leading edge of the UPS spectra and it was found that the linear extrapolation method and the DoS alignment method yielded values for the VBM to Fermi-level separation of 0.54 and 0.70 eV, respectively. Again, the difference here shows the necessity of performing this correction, and the aligned DoS value is used later.

It is noted that in Fig. 6.29, the UPS spectrum is shown without a background subtracted from it. This is due to the complexity of backgrounds for UPS spectra in general, coupled with the complex nature of the CZTS spectrum. The Shirley background type for XPS core-level peaks fails somewhat when used for UPS spectra because of the rising background and multiple features within the UPS spectrum [272]. A different background function, proposed by Li et al. [273], has seen success when applied to UPS spectra; however, this background function requires that the start and end points be free from intensity due to spectral features, and therefore cannot be used for the CZTS spectrum because there is no such region, given the photon energy used here. Nevertheless, whatever the true nature of the background for this spectrum, the leading edge and determined VBM position will be affected minimally by this choice, because the background remains low until a significant onset of states.

Regardless of these problems, one can still analyse the features of the spectra. They are much the same as those in the XPS spectrum shown in Fig. 6.27, with the differences in cross section for different photon energies causing the differences in relative intensities. The top of the VB (I), and other features occurring in the VB match in BE between the peaks of the UPS spectrum and the pDoS curves (II, IV). Similar to the XPS spectrum, feature III appears at higher BE in the UPS spectrum than the corresponding feature in the calculated DoS, showing that the Zn 3*d* orbital had an incorrectly calculated BE. Comparing the relative intensities of the DoS features with the spectral features in relation to a rising baseline, they are in agreement.

Natural Band Alignments

The extracted band positions from the fittings described above are listed in Table 6.6. First, it must be acknowledged that because of the error associated with the fitting of the IPES spectrum, discussed in the previous section, the actual separation between the CBM and Fermi-level is believed to be greater than that estimated from the linear extrapolation used here, which would result in a lower value of EA, and a larger bandgap, bringing the measured values much closer to the accepted bandgap for CZTS of ~1.5 eV [77].

Given that both sets of values in Table 6.6 use the same IPES spectrum to determine the CBM position, it is differences in the XPS and UPS spectra which generate the differences between the band position values. The differences between the XPS and UPS spectra are not vast, and given the error associated with the method used for determining the VBM position, one could argue that the VBM to Fermi-level separation between the techniques are in agreement. The value of the WF is independent of these fittings however, being based upon the fitting of the SEC and the spectrometer calibration. In this case, the WF from UPS was found to be lower than that of XPS, explainable because UPS is a much more surface sensitive technique, making it much more susceptible to the slight amount of contamination remaining on the crystal, which is known to reduce the WF [274, 275]. Therefore, for these reasons, it is concluded that XPS is a better technique to use when dealing with samples which cannot be cleaned to UHV standard, as it probes more of the bulk and is less sensitive to contamination.

The values given in Table 6.6 are shown for comparison to literature values in Fig. 6.30a. The natural band positions of CZTS are very rarely reported, and even more rarely by photoemission techniques. Much more often quoted are the CBO or VBO in relation to a constructed heterojunction, and whilst this is important, it is dependent upon the growth conditions. Instead, analysis of the natural band positions can act as a starting point, allowing predictions to be made about the behaviour of a junction before it is constructed. The bandgap values measured in this study are lower

Table 6.6 Ionisation potential (IP), work function (WF), electron affinity (EA), band-edge to Fermi-level separations, and bandgap values for the clean single crystal CZTS from XPS/UPS and IPES measurements

Technique	IP	WF	EA	VBM–E_F	CBM–E_F	E_g
XPS/IPES	5.28	4.68	4.03	0.6	0.65	1.25
UPS/IPES	5.11	4.41	3.76	0.7	0.65	1.35

Peak FWHM are given in parentheses

than other reports, and this is due to the position of the CBM being underestimated in this study, as discussed above. Despite this, the band levels measured in this study are in general agreement with the literature presented in Fig. 6.30, especially with regards to values of IP, both experimental and theoretical [33, 217, 260, 276], including previous work of the author, which measured the band levels of thin-films of CZTS grown by CBD [244]. Disagreement between the values measured here and the literature can be explained by Olopade et al. [277] using simulation parameters whose origin was not explained, or Kida et al. [278] measuring nanocrystals of CZTS, which were shown to display higher band levels than bulk CZTS. Much more often measured, the bandgap of a sample of CZTS is almost universally quoted throughout the literature,[44] and is found to have an average value of ~1.50 eV across experimental and theoretical techniques.

It is now practical to assess the natural band positions of CZTS with comparison to other common solar cell materials. This is shown in Fig. 6.31, having taken the standard value of 1.5 eV for the bandgap of CZTS. In the previous chapters of this thesis, band alignment diagrams have been presented with values of CZTS taken from the literature [281]. There, the values showed the bands for CZTS to be almost in agreement with those of CdTe and CIGS, leading to the view that in terms of band alignment, CZTS should behave similarly to these other absorbers. This is also the reason why CdS was chosen as a window layer for CZTS [22, 282], because of the success seen with CdTe/CdS and CIGS/CdS solar cells. This band alignment proliferated through theoretical studies involving CZTS [294–296] and supported the view that CZTS is similar to CIGS or CdTe. However, the results presented here, which are shown to be in agreement with experimental literature on the band levels of CZTS, suggest that the alignment of CZTS and CdS is perhaps, somewhat unfavourable.

From Fig. 6.31, it can be seen that the CB of CZTS lies above the CB of CdS, resulting in a negative CBO of −0.50 eV when considering the bandgap value of 1.50 eV, and still −0.25 eV when using the data from IPES (dash line). This is unfavourable for solar cell heterojunctions, where it forms a 'cliff-like' configuration [75], leading to increased recombination at the interface and a low V_{OC} [166, 167, 297–299]. The CBO found in this study is shown for comparison to the literature reports of band alignments of CZTS and various window layers, using a variety of techniques in Fig. 6.32. It has been suggested that an optimal solar cell heterojunction will have a positive CBO less than 0.4 eV [297, 300], and this ideal region is also shown in Fig. 6.32 (green shading). With the exception of a single experimental [301] and theoretical study [302], the CBO for CdS/CZTS is consistently determined to be negative throughout a variety of experimental and theoretical techniques (red circles).

Consequently, it has been suggested that this misalignment is one reason why CZTS has not achieved better efficiencies. This, coupled with the environmental aspects of the use of cadmium in a solar cell where the absorber was developed in part to overcome the negative aspects associated with CdTe, has led to alternative window layers being studied. ZnO and ZnS are commonly used semiconductors

[44] Shown in Fig. 6.30b.

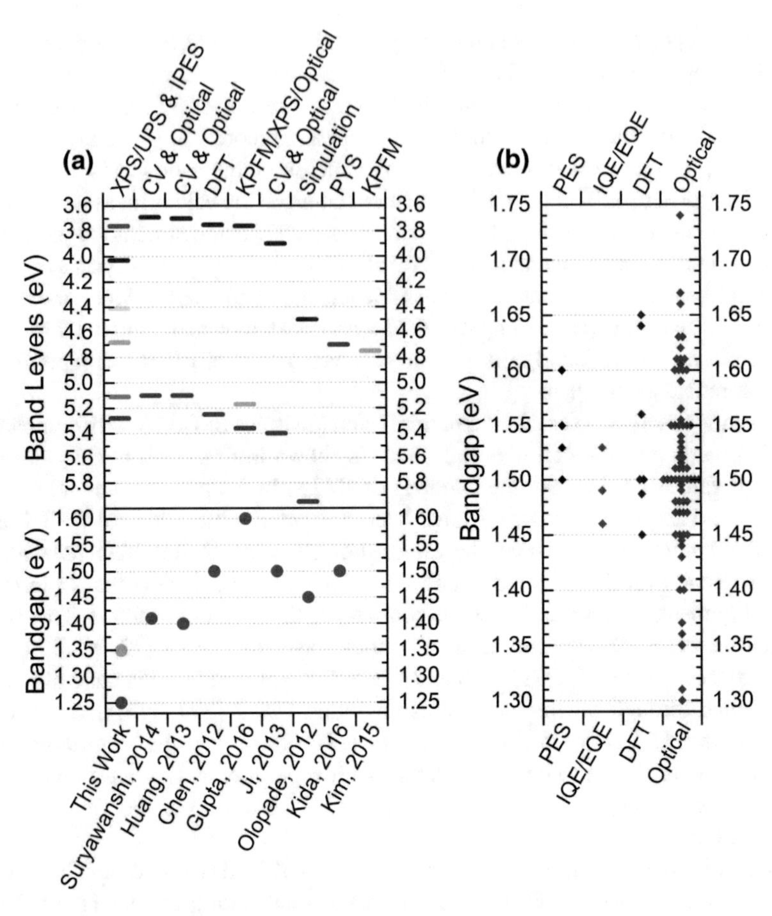

Fig. 6.30 Literature band levels and bandgaps of CZTS. **a** Comparison between the electronic band levels and bandgaps measured in this study, and those reported previously in the literature for CZTS. Literature references are given on the bottom x-axis [33, 217, 244, 260, 266, 276–278], with the corresponding experimental techniques used on the top x-axis. Semi-transparent data are from UPS. **b** Literature bandgap values for CZTS, by different methods, shown on the x-axis [13–15, 18, 19, 24, 31, 34, 35, 39, 40, 42, 44, 45, 48, 94, 125, 141, 153, 155, 156, 179, 222, 243, 246, 248, 252–255, 257–259, 261, 279–293]. Measurements of the same value are displaced on the x-axis

for other applications and were shown to give a positive CBO when interfaced with CZTS (blue and teal circles, respectively); however, the CBO is judged to be too large, forming a barrier against electron extraction [297, 299, 302, 303]. In_2S_3 has been praised as a potential window layer [304], and has been found to be successful with CIGS cells [305]. Band alignment studies of In_2S_3/CZTS shown in Fig. 6.32 (brown circles) would suggest that success could be found here too; however, this brings about the problems associated with the scarcity of indium.[45] With ZnS producing

[45]See Sect. 1.4.

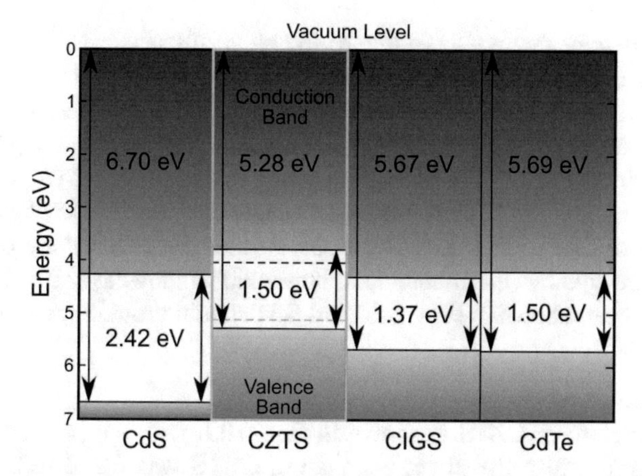

Fig. 6.31 Vacuum-aligned band diagram for CZTS from IP and bandgap measurements using XPS/UPS and IPES, and taking a bandgap value of 1.50 eV. Values from combined XPS/IPES measurements are shown as dashed lines and from UPS/IPES are shown as grey dashed lines. Comparison with other common absorbers and the common n-type window layer material, CdS. Literature values of IP and bandgap are taken for CdS [281], CIGS [307], and CdTe [226] (Color figure online)

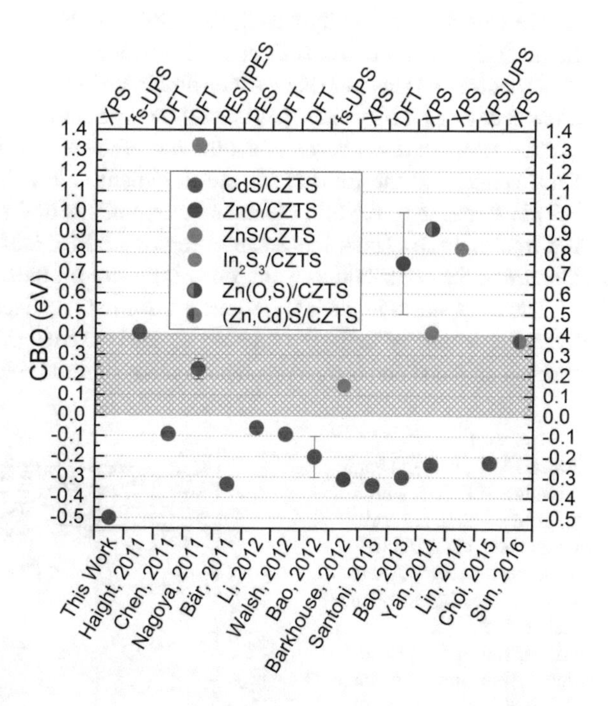

Fig. 6.32 Literature CBO for CZTS. Comparison between the CBO for CdS/CZTS taken from Fig. 6.31 in this study, and literature values of the CBO between various window layers and CZTS. Literature references are given on the bottom x-axis [52, 166, 167, 229, 230, 232, 294, 299, 301–303, 306, 308–310], and the method used in the corresponding study are given on the top x-axis. Error bars represent the spread of data measured in a single study

too large a positive CBO, and CdS producing too large a negative CBO with CZTS, it was thought that by mixing these two compounds, band level tuning could result in a favourable CBO using a (Zn,Cd)S/CZTS junction. Although this was shown experimentally to be favourable (red/teal circles), it still has the associated problems with utilising cadmium. Indeed, the record efficiency for CZTS solar cells is held by a device utilising (Zn,Cd)S as a window layer [52], but the record efficiency device using the mixed sulphur/selenium compound, higher than the pure sulphide, uses a CdS window layer [49], again suggesting that the CdS/CZTS junction is less favourable than the CdS/CZTSSe junction, because the CB of CZTSe has been shown to be lower than CZTS [283, 294, 301, 306]. Further research is therefore needed to identify a more suitable, earth-abundant, non-toxic window layer for CZTS, but the natural band levels shown in Figs. 6.30 and 6.31 should provide good grounding for this.

Density of States Analysis

The band structure of CZTS is shown in Fig. 6.33, with the projected total DoS also shown. From this, the direct bandgap of CZTS was found to be 1.47 eV at the gamma point, in agreement with previous studies of the band structure [14, 39], and the commonly used value of 1.50 eV. The direct nature of the bandgap means that this should be representative of the photoemission measurements, as they were performed under normal emission, and the band edge measurements should therefore reproduce the states at the gamma point.

The element- and orbital-separated pDoS curves for CZTS are shown in Fig. 6.34. In the DoS shown in Figs. 6.27 and 6.29, which were corrected for comparison with PES spectra, the Cu *d* and Zn *d* states dominate because of the larger photoionisation cross sections for *d*-orbitals [311], with contributions from the other orbitals hardly visible. Without these corrections, the top of the VB (I, II, III) is still dominated by Cu *d* states, and the Zn *d* states are dominant below the VB (IV), located around −9 eV. Because of the four elemental components of CZTS, it is important to study the role of each. Thus, the zoomed region of Fig. 6.34 is also shown in order to determine the contribution from lower intensity orbitals.

The Zn *d* states in the VB (IV) are somewhat localised and contribute only slight hybridised states with the S *p* orbitals, around −3.5 eV (V). Most of the top of the VB consists of Cu *d* and S *p* states, which are split into three distinct regions (I, II, III),

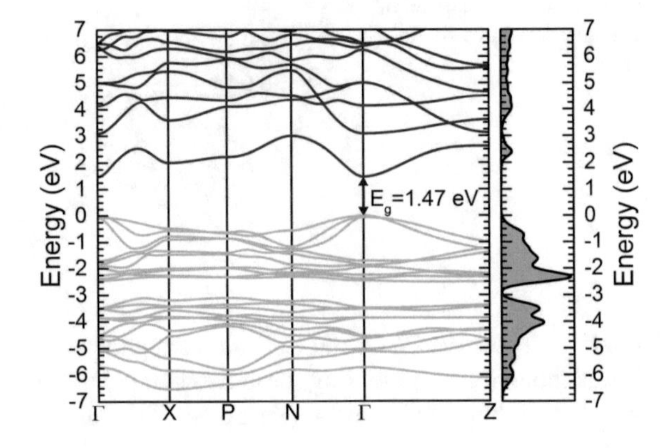

Fig. 6.33 Calculated band structure of CZTS, with valence states in red and conduction states in blue. The structure is aligned to coincide the highest valence state to 0 eV. The corresponding integrated density of states is shown to the right, and the location of the direct band gap is marked (Color figure online)

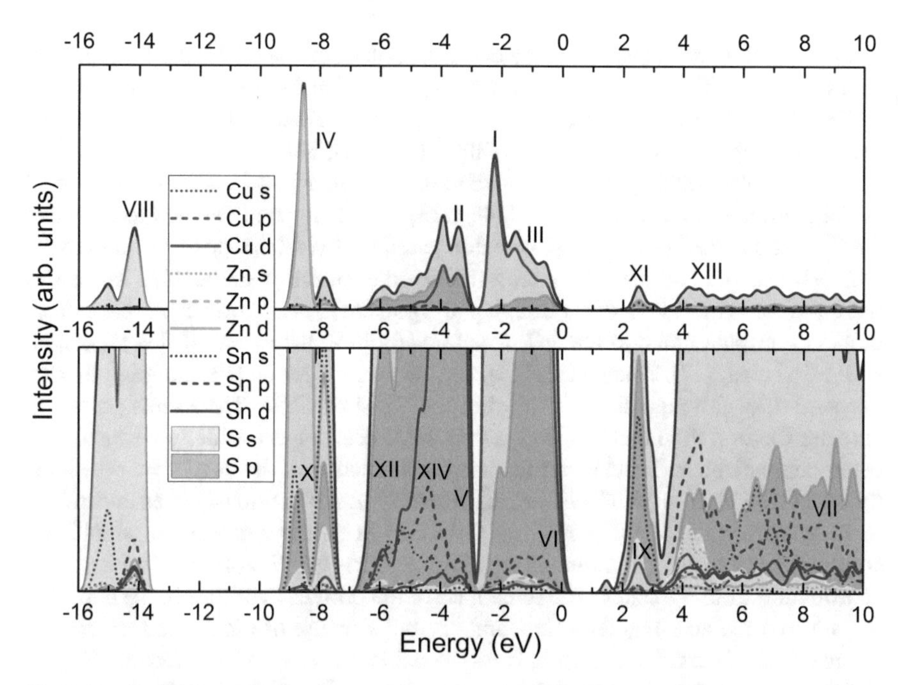

Fig. 6.34 Total and partial electronic DoS curves for CZTS. Bottom region is a y-axis zoom of the top region. Curves have been convolved with a Gaussian function (0.3 eV FWHM) in order to better distinguish features and are aligned to the VBM. Black curve with grey shading is the total summed DoS. Part labels are discussed and referred to throughout the text

and arise from the tetrahedral Cu–S bonding units in CZTS. Here, the crystal field splits the Cu *d* states into a non-bonding e_g doublet, and a t_{2g} triplet, which is able to bond with the S *p* states [13, 14, 39]. This is reflected in the pDoS, where there is a large central feature (I) consisting of the non-bonding e_g Cu *d* states, and on either side are two weaker features (II, III), representing the Cu *d* t_{2g} orbitals hybridised with S *p* orbitals. Beneath the majority Cu *d*/S *p* states is a slight contribution to the VBM from Cu *p*/S *p* bonding orbitals (VI), with the corresponding antibonding orbitals situated high in the CB (VII). The final region of semi-localised states is that from S *s* around −14.5 eV (VIII), too tightly bound to interact, except with Sn *s* states, which also have a high BE, and being unoccupied in the Sn^{4+} oxidation state, result in a small amount of intensity from S *s* states in the CB (IX) because of the Sn *s*/S *s* antibonding states found there. The remainder of Sn *s* states are found to be hybridised with S *p* orbitals, with the bonding states overlapping the Zn *d* states (X), and the antibonding states constituting most of the CBM (XI). The lower tail of the first VB is comprised of a significant portion of Zn *s*/S *p* bonding states (XII), with the corresponding antibonding states found in the CB (XIII) around 4 eV.

The unoccupied states are split into two distinct bands. The first (XI), consisting mostly of antibonding Sn *s*/S *p* states as described above, is separated from the rest of the CB (XIII), which has a mixed contribution from antibonding Sn *p*/S *p*, Cu *s*/S *p*, and Zn *p*/S *p* states, but also some contributions from other orbitals. The bonding states from Sn *p*/S *p*, Cu *s*/S *p*, and Zn *p*/S *p* are located deeper in the VB (XIV, V).

The DoS for CZTS shown here is in general agreement with previous calculations in the literature, including the nature of the VBM and CBM [13, 17, 18, 43, 144, 306, 312], where the bonding here is likened to that in CIGS [14, 19, 39]. Experimental corroboration exists with Bär et al. [144, 313], who compared the DoS to XES and XAS spectra and Xiao et al. [74], who compared the overall shape of the DoS to the XPS VB spectra, and there they reported that the Zn *d*/S *p* antibonding states (V) were a cause of the VBM level through p–d repulsion [314]. However, as these states do not contribute much to the VB, it is not certain to the extent of this repulsion, with Mitzi et al. [17], Zhang et al. [13], and Zhao and Persson [43] claiming that Zn plays no role with regards to the level of the VBM or CBM. The results presented here are the first time that the DoS curves have been experimentally corrected for better comparison and corroboration, and the extraction of the band level positions, followed by a discussion of the bonding regime, has not previously been addressed.

In summary, the VBM is majorly Cu *d*/S *p* in nature with no real significant contribution from other orbitals and the CBM is mostly Sn *s*/S *p* in nature, with a small contribution from S *s* states. The remainder of the orbitals hybridise in the expected ways, given the near-regular tetrahedral crystal structure of CZTS, and the formal occupancies of the orbitals arising from the oxidation states of the elements[46] [14]. The apparent similarity between the electronic structures of CZTS and CIGS suggest that the materials should exhibit similar band levels. As this is not the case, evidenced by the results presented above, it is believed that this is due to the differences in the electronic structures. First, CIGS contains selenium rather than sulphur, forming a different VB, and the use of tetravalent tin over trivalent indium/gallium, which forms the CB, also results in a different bonding environment. It has also been suggested previously that the reduction in symmetry of the CZTS crystal structure over that of CIGS, is the cause for the separated CB in CZTS, which is not present in CIGS [39]. It is then believed that these differences are the reason why CZTS has higher band levels than CIGS [15].

6.8 A Comparison of Technique Efficacies for the Characterisation and Identification of CZTS and Its Secondary Phases

Having analysed the measurements of the single crystal CZTS and the possible binary secondary phases, it is now necessary to compare the spectra in order to assess the possibility of using each technique to determine each secondary phase and also whether there exist any remaining ambiguities in the analysis.

[46] Cu^{+}, Zn^{2+}, Sn^{4+}, S^{2-}.

6.8.1 Structural Analysis Techniques

As discussed in Sect. 6.3.1, the problems associated with using XRD to unambiguously determine phase-pure CZTS is not possible and RS should be used in conjunction to help determine this.

With regards to XRD, studies of CZTS are usually compared against previous results without assessing how the possibility of secondary phase formation would alter the observed spectrum, if at all. Theoretically calculated diffraction patterns for CZTS and the possible secondary phases are shown in Fig. 6.35.[47] The kesterite CZTS structure is derived from the sphalerite structure of ZnS, and for this reason, the main peaks associated with CZTS overlap with those of ZnS, which makes the identification of ZnS in CZTS near-impossible with lab-based XRD. A similar situation exists with CTS, which is also believed to be sphalerite-derived, meaning the same situation exists. Conversely, CZTS may be identified using XRD, as there are many more identifiable peaks from CZTS, due to the broken symmetry, because of the extra elements. These extra peaks however, have small structure factors and are therefore not very intense in the spectrum. This makes their detection difficult in a sample displaying widened peaks because of poorer crystallinity, and therefore means that they are not significant above the background.

The other possible secondary phases shown in Fig. 6.35 are identifiable by XRD within CZTS because of at least one characteristic peak which is not overlapped by those from CZTS. This is particularly the case for the tin sulphides, which all have strong peaks at lower angles. The copper sulphides also have peaks that are not overlapped by those from CZTS; however, because the main peak of Cu_2S is close to one of CZTS, and the other peaks are significantly lower in intensity, this could make identification of Cu_2S difficult by XRD if it existed in low quantities or the crystallinity was poor; likely with a secondary phase. Similarly, the main peak of CuS is between those that nearly overlap with peaks from CZTS, and as such, may not be distinguishable from the background.

In summary, the use of XRD alone could identify the presence of CZTS and tin sulphide secondary phases; however, the determination of phase-pure CZTS is not possible because of the direct overlap of peaks from ZnS and CTS, also rendering their identification not possible, and the multiple peaks of the copper sulphides, some of which overlapping with those of CZTS, makes their detection not impossible, but difficult.

The Raman peaks from the measurements of the CZTS crystal presented in Fig. 6.17 are summarised and compared to standard literature values for the spectra of the secondary phases in Fig. 6.36. The most obvious differences between the spectra presented here are that several different wavelengths are used for the excitation. This selection is sometimes dictated by the use of quasi-resonant Raman scattering, whereby the peaks become more intense when an excitation wavelength is used that corresponds to the bandgap of the material under study [150]; however, it is some-

[47]With the calculated pattern data shown in Appendix D, Sect. D.3.1.

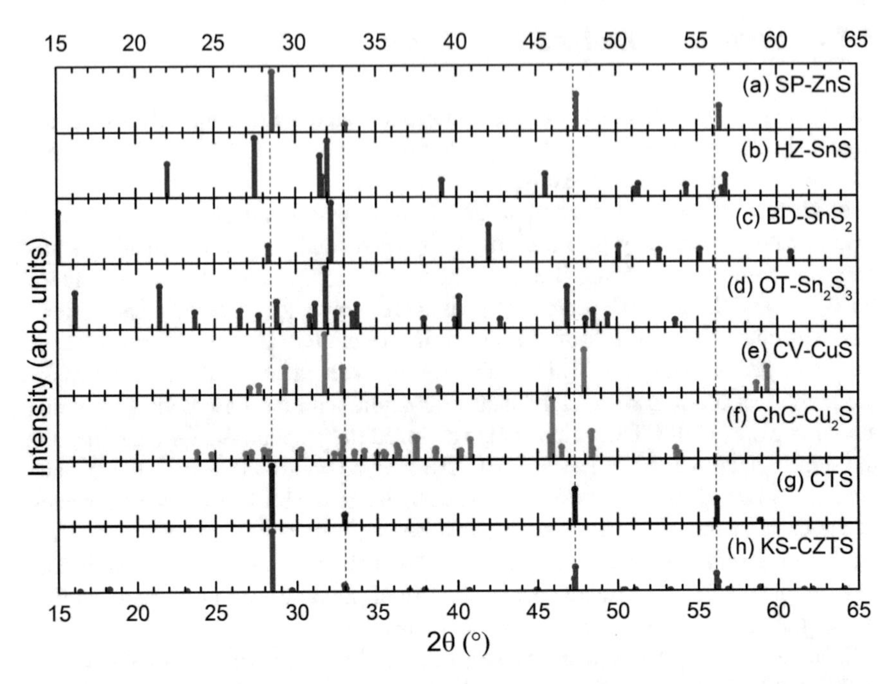

Fig. 6.35 Comparison of theoretically calculated ideal XRD patterns from the crystal structures of **a** sphalerite [131] (SP) ZnS, **b** herzenbergite [315] (HZ) SnS, **c** berndtite [316] (BD) SnS_2, **d** ottemannite [317] (OT) Sn_2S_3, **e** covellite [83] (CV) CuS, **f** chalcocite [84] (ChC) Cu_2S, **g** CTS [318], and **h** kesterite [4, 20] (KS) CZTS

times the case that the selection of excitation wavelength is restricted by available equipment.

No peaks are shown for Cu_2S because it is believed that none of the modes are Raman active, judging by the broad, featureless spectrum observed by Wang et al. [319], and as such, this secondary phase is undetectable by RS. In terms of peak locations, Fig. 6.36 would suggest that the identification of CuS or the tin sulphides would be possible using Raman spectroscopy, but that the identification of ZnS and CTS would be difficult. Whilst the identification of CTS would indeed appear to be problematic, the identification of ZnS using RS is relieved somewhat if one utilises quasi-resonant Raman scattering, because the bandgaps of CZTS and ZnS are significantly different.[48] On account of this, the signal from ZnS should appear resonant with the use of shorter wavelength excitations and with a comparison to longer wavelength measurements, one should be able to determine the presence of ZnS within a CZTS sample.

Therefore, the use of a single excitation wavelength of Raman spectroscopy is not sufficient to determine phase-pure CZTS because secondary phases may produce much less intense signal, depending on their amount and also the excitation wave-

[48]~1.5 and 3.8 eV, respectively.

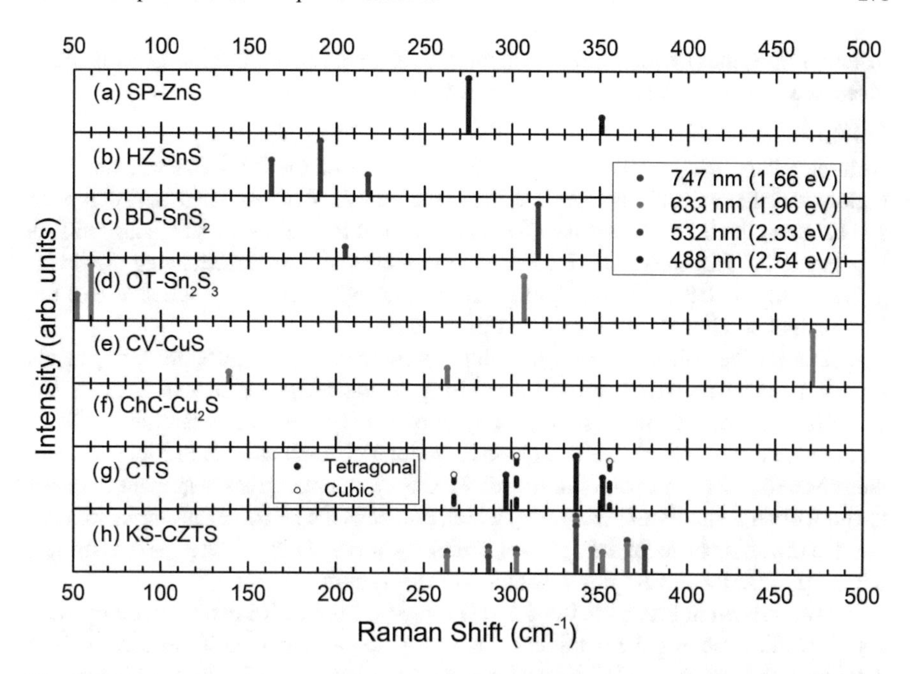

Fig. 6.36 Comparison of the peaks from Raman spectra of **a** sphalerite [146] (SP) ZnS, **b** herzenbergite [320] (HZ) SnS, **c** berndtite [321] (BD) SnS_2, **d** ottemannite [322] (OT) Sn_2S_3, **e** covellite [89] (CV) CuS, **f** chalcocite [319] (ChC) Cu_2S, **g** CTS [146], and **h** kesterite [45] (KS) CZTS. The colours determine the excitation wavelength used in that study

length used. Whilst it may not be practical, the use of multiple excitation wavelengths is necessary in order to draw the conclusion that phase-pure CZTS is present [323].

Furthermore, these techniques are limited for secondary phase identification because of the inherent limits of the techniques themselves. First, there is the detection limit associated with them, and the effective detection limit may be lessened by the presence of CZTS, which should constitute the majority of the intensity [324]. Also, physical requirements of the techniques dictate that samples must be crystalline, in order to result in peaks which are narrow enough to be detectable, and with the formation of secondary phases, it is possible that they occur in an amorphous state, rather than being incorporated with the crystal structure [38, 76, 276, 284].

6.8.2 Core-Level and Auger Photoelectron Spectroscopy

The standard spectra from the CZTS crystal presented in Sect. 6.7.2 can act as a starting point for determining whether this technique is suitable for secondary phase identification in CZTS. Therefore, these spectra, and the standard spectra of the possible secondary phases will be compared here and assessed for the possibility of identifying them within a real sample. All of the energies and fitting parameters discussed in this section are detailed in Table 6.7.

Sulphur

Although it is usually the most overlooked aspect of the XPS spectra, the S 2*p* region can give valuable insight into the presence of secondary phases. A comparison between the S 2*p* regions for CZTS and each of the measured secondary phases is presented in Fig. 6.37. All of the compounds show similar spectral shapes, distinguished only by BE shifts, excepting that of CuS, which is different because of the two distinct sulphur environments in this structure.

Although the shifts between the binding energies of the different compounds are small, it is the small spin-orbit splitting of the S 2*p* region that benefits the analysis of secondary phases. A good-quality, highly crystalline sample of CZTS will demonstrate a sharp, well resolved S 2*p* doublet, such as seen in Fig. 6.37, and therefore, the measurement of an ill-resolved S 2*p* doublet gives testament to either the poor crystal quality of the sample, which broadens the peaks of the single doublet, else the presence of secondary phases, which even when present in small quantities, will jeopardise the enveloping resolution of this region.

As was shown in Fig. 6.25, the S LMM region for the CZTS crystal demonstrated a wide peak, lessening the confidence in its position and making it more difficult to determine the presence of additional phases. Coupled with this, the S LMM region occurs at relatively low KE, and therefore is affected more by the presence of contaminant overlayers, and indeed is rarely visible in survey spectra of samples with contamination.[49] For these reasons, the use of the S LMM region and the calculated

Table 6.7 Ionisation potential (IP), work function (WF), electron affinity (EA), band-edge to Fermi-level separations, and bandgap values for the clean single crystal CZTS from XPS/UPS and IPES measurements

Sample	Cation CL		Cation	α'	S $2p_{3/2}$	
	BE	FWHM	Auger KE		BE	FWHM
CuS	932.20 932.45	1.00 1.00	917.84	1850.04 1850.29	161.97 161.30	0.89 0.65
Cu_2S	932.48	1.16	917.33	1849.81	161.52	0.82
ZnS	1022.05	1.27	989.37	2011.42	161.79	0.86
SnS	485.57	0.81	435.29	920.86	161.07	0.70
SnS_2	486.45	0.94	433.92	920.37	161.45	0.89
Sn_2S_3	485.82 486.37	0.78 0.91	434.84	920.66 921.21	161.50	0.78
CZTS	932.13 932.79 1021.80 486.26	1.07 1.07 1.06 0.95	917.07 990.02 434.19	1849.20 1849.86 2011.82 920.45	161.71	0.81

All values given in eV

[49] See, for instance Figs. 3.3, 3.15, 4.4, and 5.19.

Fig. 6.37 Comparison of the XPS S 2p region of CZTS, CuS, Cu_2S, ZnS, SnS, SnS_2, and Sn_2S_3

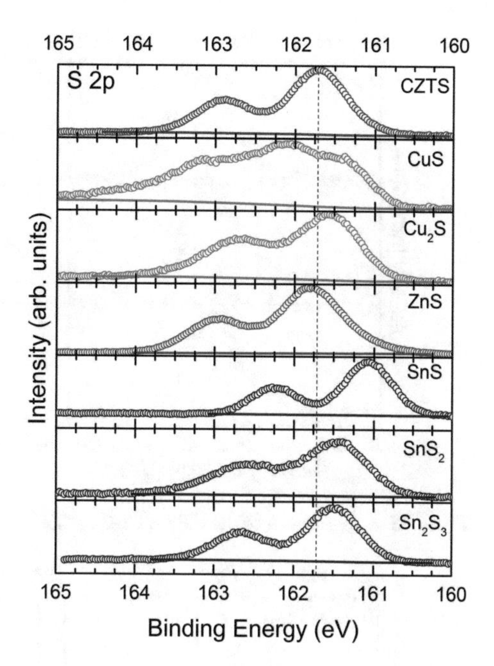

Auger parameter is not recommended for helping with secondary phase identification.

Cations

A comparison of the Cu 2$p_{3/2}$ and Cu LMM regions between CZTS and the copper sulphides is shown in Fig. 6.38a, b, respectively. Despite each of these compounds containing Cu^+, the copper sulphides have higher binding energies than CZTS and could be identified by the presence of a shoulder peak. Then, the difference between the Cu LMM regions of the two copper sulphides could allow distinction to be made between the phases. It is noted however, that the presence of oxidised copper could lead to the same result, but this could be checked using the O 1s region.

A comparison of the Zn 2$p_{3/2}$ and Zn LMM regions between CZTS and ZnS is shown in Fig. 6.39a, b, respectively. The BE for ZnS is higher than for CZTS and could therefore be identified by the presence of a shoulder peak. The Zn LMM spectra for ZnS and CZTS are also different; however, this does not arise completely from differences between the materials, but also from artefacts from the Sn 3$d_{5/2}$ peaks, which needed to be subtracted from this region.

A comparison of the Sn 3$d_{5/2}$ and Sn MNN regions between CZTS and the tin sulphides is shown in Fig. 6.40a, b, respectively. The differences between the spectra for the tin sulphides are discussed at length in Chap. 5. Regarding the CZTS Sn 3$d_{5/2}$ spectrum, it most closely resembles that of SnS_2, unsurprising as they both contain only the Sn^{4+} cation. This allows the presence of a compound containing Sn^{2+} to be addressed on BE alone, as the shift is quite large, and will result in a shoulder toward the lower BE side of CZTS. The use of the Sn MNN region will aid in this

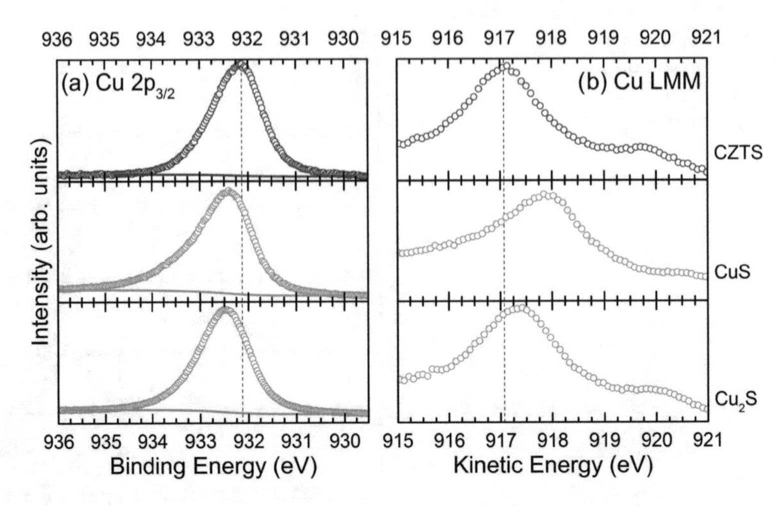

Fig. 6.38 Comparison of the XPS **a** Cu $2p_{3/2}$ and **b** Cu LMM regions for CZTS, CuS, and Cu_2S

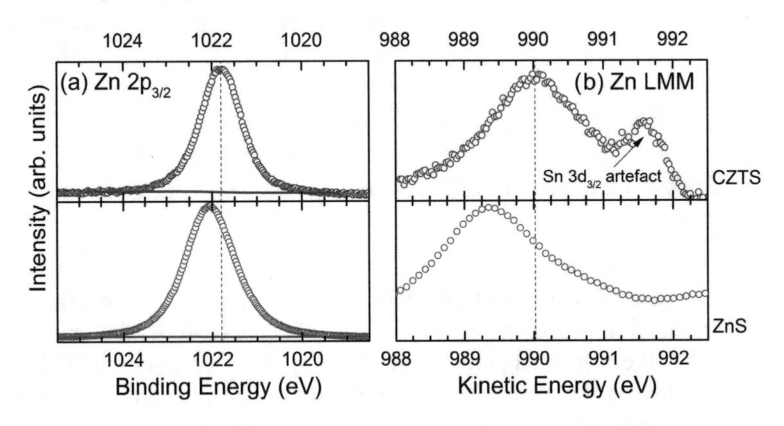

Fig. 6.39 Comparison of the XPS **a** Zn $2p_{3/2}$ and **b** Zn LMM regions for CZTS and ZnS

identification as the four compounds show different Sn MNN spectral shapes, not only shifts in KE. Again, the presence of oxidised tin could complicate this analysis, but this could also be checked using the O 1*s* region.

Auger Parameter

One of the biggest difficulties involved with determining secondary peaks within an XPS spectrum is the possibility for variation in the BE of a measured species from the standard value derived from a clean sample. This means that the use of absolute binding energies is often impractical because the value could change because of charging, Fermi-level shifting, or charge transfer effects because of the contacting of a secondary phase with the host material. This can be overcome by using the Auger parameter; introduced in Sect. 2.3.4, it combines the positions of the main CL of an element with the corresponding Auger line in such a way as to allow any

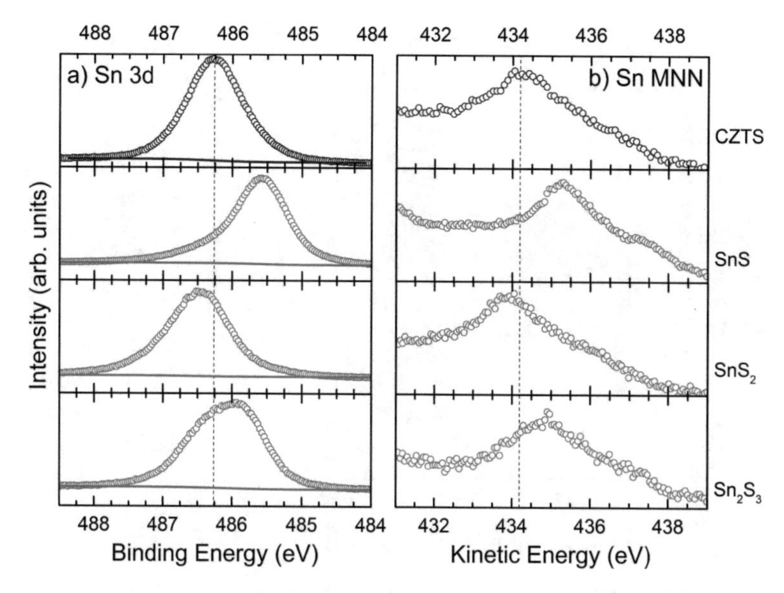

Fig. 6.40 Comparison of the XPS **a** Sn $3d_{5/2}$ and **b** Sn MNN regions for CZTS, SnS, SnS_2, and Sn_2S_3

effects requiring the homogeneous correction of the BE scale to be neglected, so that its value changes only because of chemical state changes. Furthermore, it is more sensitive to chemical state changes than the position of the individual peaks alone, assuming no final state effects are present.

Using the values calculated in the data presented in previous sections, displayed in Table 6.7, along with values taken from the appropriate literature, Wagner plots were drawn for each of the cations of CZTS. Such a figure displays the BE of the main CL, KE of the main Auger line, and the Auger parameter for different compounds, allowing easier identification of chemical species, and trends between them [190].

The Wagner plot for copper is shown in Fig. 6.41, containing data measured from CZTS and the copper sulphides, along with literature reports of metallic copper, oxides, sulphides, hydroxides and sulphates. The measured values of the copper Auger parameter for CuS and Cu_2S are in agreement with those from the literature. Three distinct regions are seen in this figure, which are labelled and associated with the presence of the different oxidation states of copper. Compounds containing Cu^{2+} or Cu^0 demonstrate a high Auger parameter value greater than 1851 eV and the Cu^{2+} compounds are distinguished from the metallic copper by the higher BE Cu $2p_{3/2}$ peak, and also the associated shake-up structure seen in the spectra, present only with the Cu^{2+} oxidation state [111].

Compounds that demonstrate an Auger parameter value of less than ~1850.5 eV contain the Cu^+ oxidation state, delineating this group of compounds for more effective determination. Furthermore, between these such compounds, although the binding energies of the CL are very similar, the Auger line KE is variable enough so that

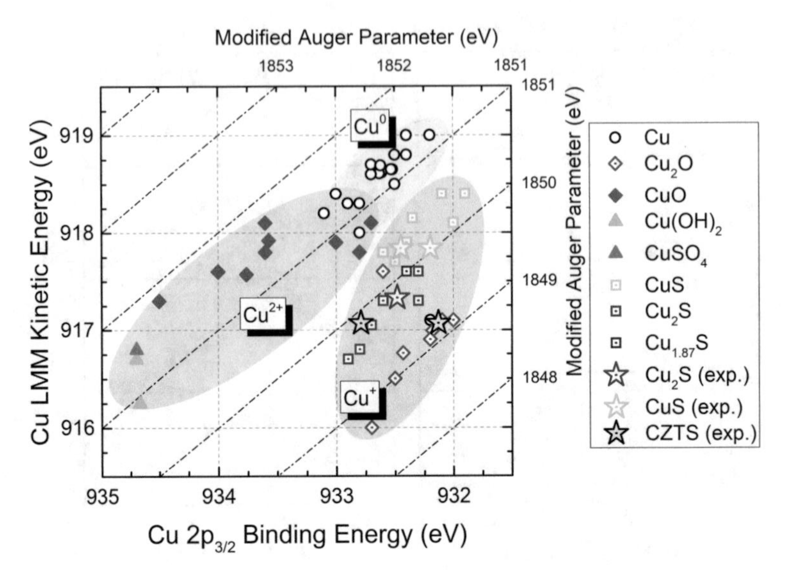

Fig. 6.41 Wagner plot for copper, showing experimentally determined values for CZTS, CuS and Cu_2S, and values from the literature for other copper compounds [102, 105, 177, 192–195, 203, 205, 207, 208, 325–336]. Regions of compounds with different oxidation states highlighted

the Auger parameter of CuS, Cu_2S, and Cu_2O are discernible. CZTS is represented as two points in Fig. 6.41 because of the two distinct copper peaks that were fitted in Sect. 6.7.2 and therefore appear to be in conflict with the Auger parameters for Cu_2S and Cu_2O. However, it is thought that in thin-films of CZTS, this level of resolution will not be accessible between the two copper species and therefore spectra found there will be well represented by a single peak, which will have a BE intermediary of the two measured here. This would then result in a copper Auger parameter which would lie in between the two shown in Fig. 6.41, therefore allowing the unambiguous measurement of the Auger parameter for each compound, including CZTS.

The Wagner plot for zinc is shown in Fig. 6.42, containing data measured from CZTS, ZnS and ZnO, along with literature reports of metallic zinc, oxides, sulphides, hydroxides, carbonates and sulphates. Although the binding energies measured in this study for ZnS and ZnO are higher than those found in the literature, probably because of charging effects in these insulating materials, the measured Auger parameters are in agreement, demonstrating the ability of the Auger parameter to determine chemical states, regardless of charging effects. There is a large spread of binding energies throughout the literature for the same materials, which makes the determination of chemical state using BE alone difficult. However, by assessing the Auger parameters, it can be seen that there is distinction between the different compounds. Separated by almost 2 eV, metallic zinc shows a high Auger parameter around 2014 eV, making its identification straightforward. As zinc is only able to take one cationic oxidation state, Zn^{2+}, there is only one other region shown in Fig. 6.42. Those compounds with generally high binding energies and a resulting Auger parameter of less than

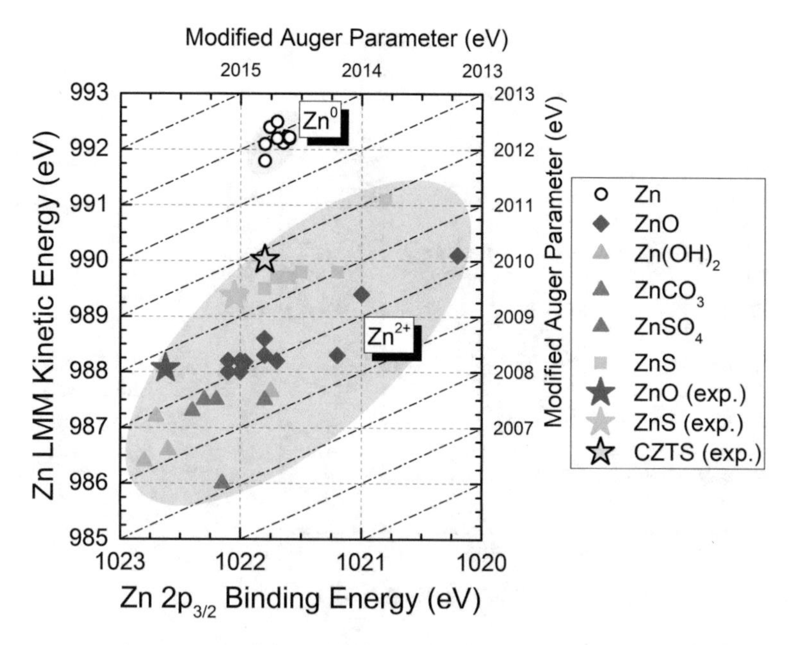

Fig. 6.42 Wagner plot for zinc, showing experimentally determined values for CZTS, ZnS and ZnO, and values from the literature for other zinc compounds [162, 193, 208, 213, 215, 216, 325, 331, 332, 337–339]. Regions of compounds with different oxidation states highlighted

2010 eV are from zinc compounds with polyatomic anions, values between 2010 and 2011 eV from ZnO, and compounds with an Auger parameter between 2011 and 2012 eV are from ZnS. Being most closely related to ZnS, CZTS demonstrates an Auger parameter close to this, but being higher, it is also able to be distinguished.

The Wagner plot for tin is shown in Fig. 6.43, containing data measured from CZTS and the tin sulphides, along with literature reports of metallic tin, oxides and sulphides. The measured values of the tin Auger parameter for SnS, SnS_2, and Sn_2S_3 are in general agreement with those from the literature. However, given the large shifts in BE between the oxidation states, and also the other spectral differences which were explored in Chap. 5, it would seem necessary to use the Auger parameter only in order to distinguish between sulphide and oxide, although the O 1*s* peak can help with this also. CZTS demonstrates an Auger parameter close to that of SnS_2, its most closely related tin compound, and without careful analysis of the spectra, could be mistaken.

Judging by the expected and measured Auger parameters presented above, it would appear that determination of secondary phases within CZTS using XPS would be somewhat straightforward. Unfortunately, there are aspects of the spectra which complicate the analysis further. First, the determination of the position of the Auger line is clear for a single-phase, clean compound, but for a mixed-phase sample, one of two approaches must be taken. Either, the enveloping Auger line energy must be determined, which will give some indication of the presence of secondary phases,

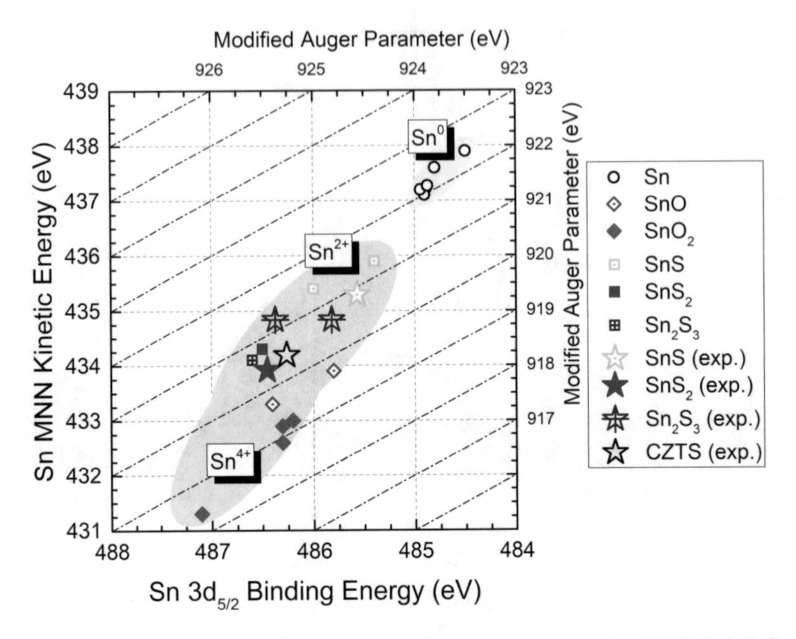

Fig. 6.43 Wagner plot for tin, showing experimentally determined values for CZTS, SnS, SnS_2, and Sn_2S_3, and values from the literature for other tin compounds [164, 234, 237, 325, 337, 340–343]. Regions of compounds with different oxidation states highlighted

as it will skew the Auger parameter from the expected values. Otherwise, one must attempt to fit the Auger region, by either using standard spectra in a qualitative way, or by using theoretical lineshapes for the Auger regions. This latter method is complex, especially because of the overlap of the Zn LMM region with the Sn $3d_{3/2}$ peak, which, as the determination of the zinc chemical environments is already one of the most difficult, only compounds the problem.

There are several approaches which are used to overcome the problem of overlapping peaks of interest in XPS spectra; however, they have been shown to be unsuitable for use with CZTS. First, a different CL is chosen to be analysed rather than the main peak, and while this is effective as the same chemical information is contained in each orbital, it is impractical for CZTS because, as can be seen in Fig. 6.21, the secondary peaks for each element either have their own overlapping problems,[50] or are significantly lower in intensity,[51] meaning that data acquisition times will become excessive. Another method is to use x-rays with a different photon energy, which shifts the Auger features by the difference in photon energy on the BE scale. However, for CZTS, this again is impractical because the use of twin anode sources removes the benefits gained by monochromated x-rays, resulting in poorer resolution and the presence of satellites, which further exacerbates the complexities.

[50]Zn 3*p*.

[51]Cu 3*p* and Sn 3*p*.

Much work has been conducted on oxides, more so than sulphides, and in the same manner of determining secondary sulphide phases within CZTS, the presence of oxide phases could also be assessed using XPS. Such a method was employed in a study by Bär et al. [158], who used standard Sn MNN spectra of oxides and sulphides in order to assess their presence in CZTS. Beyond this, other studies have mentioned the possible use of the Auger parameter to aid with secondary phase or oxide identification [151, 283], but none have presented data using it.

Despite these problems, it is still believed that XPS can play a vital role in helping to determine the presence of secondary phases within CZTS, provided that: the spectra are of sufficiently high quality and resolution, it is critically analysed, and also it is used in conjunction with other techniques.

An Aside on CTS

Although samples of phase-pure CTS were unavailable for measurement in this study, it is important, as will become clear in Sect. 6.9.3, for this possible secondary phase to be analysed to the best ability using reports from the literature. This is difficult as CTS is a newly studied material, and as such, reports of XPS measurements are limited and the growth is non-standardised and inconsistent.

Despite this, because of the similarly derived structure of CTS to CZTS, it is expected that the binding energies from XPS will be very similar, but because CTS contains no zinc, which is arguably the most difficult element to characterise in CZTS, then it is possible that the presence of secondary peaks in the copper and tin spectra could lead to a conclusion that this secondary phase is present.

6.8.3 Stoichiometry Determination

Although the use of XPS for stoichiometry determination was presented in Sect. 6.7.2, it is necessary to use correction factors, which may not be available in all instances. Also, it remains that stoichiometry from XPS is very surface sensitive and that while surface stoichiometry measurements are sometimes important, it is probably more useful to determine the bulk stoichiometry when growing CZTS and therefore XPS should only be used as a secondary technique, for comparison to results from either EDX or ICP–MS.

6.8.4 The Effects of Secondary Phase Formation

Having determined the possibility of secondary phase identification using the above techniques, it is practical to assess the effects of the secondary phases, should they form within CZTS. The electronic band levels determined using XPS coupled with the known properties of the secondary phases from the literature can help elucidate this.

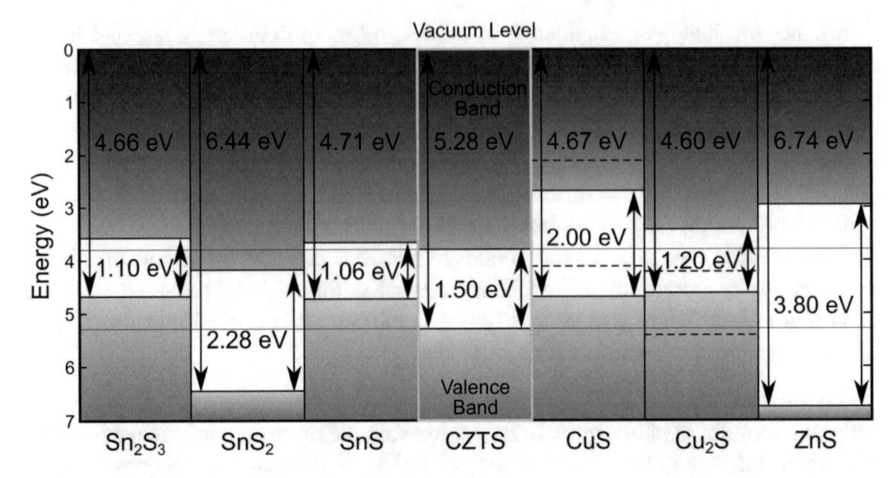

Fig. 6.44 Vacuum-aligned band diagram for CZTS in comparison with the secondary phases measured in this study. IP values are from XPS. Bandgap values are taken from Chap. 5 for the tin sulphides, the accepted value of 1.50 eV for CZTS, and the literature for CuS [91, 344], Cu_2S [204, 345], and ZnS [217, 218, 346, 347]. Differing literature values for the IP of CuS [91, 210], and Cu_2S [91, 204], are shown as dashed lines

The values for IP measured for the secondary phases and CZTS are shown, with either literature, or measured values for the bandgaps, in the vacuum-aligned band diagram in Fig. 6.44. Also, the main effects of the formation of each phase within CZTS are summarised in Table 6.8, with their detection feasibility using the techniques discussed above.

Altamura and Vidal [38] composed a similar diagram to Fig. 6.44 and the results were found to be in agreement. The natural band offsets shown give some indication to the effect of the secondary phase on the carrier properties of the sample; however, depending on the nature and formation mechanism of the secondary phase, these could give differing alignments. Nevertheless, it would seem that the formation of any and all of these secondary phases would be detrimental to a CZTS cell in some way. The wide-bandgap, insulating nature of ZnS or SnS_2 would hinder carrier extraction, whilst the small bandgap and metallic nature of SnS, CuS, and Cu_2S could lead to shunting through the cell or unfavourable offsets, which could form recombination centres, or electron blocking layers. Finally, the n-type conductivity of ZnS, SnS_2, and Sn_2S_3 could form a second diode in the cell, mitigating its rectifying behaviour.

In summary, the problems associated with secondary phase formation within CZTS are known, and the results presented here are in agreement with what has been observed in the synthesis of this material, and therefore more sophisticated analysis techniques are required to help determine the presence of them. It has been shown that the difficulties associated with XRD and RS can be somewhat remedied when

Table 6.8 Summary of the properties and effects of secondary phase formation in CZTS, with the efficacy of the analysis techniques to determine that phase

Phase	Bandgap	Conductivity	Likelihood of Formation	Effective Detection Methods[a]	Ineffective Detection Methods	Impact on Cell
CZTS	1.5	p-type	—	XRD RS XPS	—	—
Cu_2S	1.2[b]	p-type metallic	Phase diagram endpoint Suppressed by Cu-poor	—	RS	Shunting
CuS	2.0[c]	hole conduction	Used as a precursor[d] Suppressed by Cu-poor	XPS	XRD	Electron barrier
ZnS	3.8[e]	n-type insulator	Phase diagram endpoint	XPS RS[f]	XRD	Insulating
SnS	1.0	p-type	Oxidation state change Degradation[g] Used as a precursor[d]	XPS RS	—	Electron barrier
SnS_2	2.3	n-type	Phase diagram endpoint	RS	—	Diode Recombination
Sn_2S_3	1.1	n-type	Most stable tin sulphide[h]	XPS RS	—	Electron barrier Diode

All values derived experimentally in this study unless otherwise stated
[a]Effectiveness of XRD and RS limited by crystallinity. Use of RS assumes multiple wavelengths. Use of XPS assumes Auger parameter analysis and fitting model applied
[b]Liu et al. [204] and Marshal and Mitra [345]
[c]Kar et al. [91] and Kalanur and Seo [344]
[d]Ni et al. [114], Fella et al. [50], Thimsen et al. [78]
[e]Chen and Wang [217], Van de Walle and Neugebauer [218], Klein [346] and Chen et al. [347]
[f]Quasi-resonant wavelength required
[g]Weber et al. [127]
[h]Altamura and Vidal [38]

coupled with the use of XPS, because the inherent differences between the techniques can complement and contrast the findings of each. This combination of XRD, RS, and XPS is powerful, but determining the spatial location of such secondary phases, and sub-micron detection still remains a challenge [76].

6.9 Induced Changes in CZTS by Means of in Vacuo Annealing

This section now explores the changes that occur to CZTS as a result of surface cleaning for measurement, and also the effect that vacuum and elevated temperatures have upon the material. Both of these are environments which samples of CZTS may be exposed to, as a result of the growth, or during characterisation. Whilst it is assumed that no changes occur during the characterisation of a sample, dynamic processes are real and must be addressed if they are affecting the measured properties.

Several studies have explored the thermodynamics of the reaction processes involved with the growth of CZTS, and because these are not the focus of the present study, the reader is directed to them for more comprehensive analyses [46, 59, 171].

6.9.1 *The Effects of Contamination*

From the spectra presented in Sect. 6.7.1, it was shown that surface contamination was present on the CZTS crystal. Such contamination is inevitable for any sample that has been exposed to atmospheric conditions, and can only be avoided by growing and characterising the sample in situ. Such hydrocarbon contamination is somewhat benign when compared to the effect of carbon present from growth conditions, such as precursors [159, 348], but the presence of oxides on a sample of CZTS can cause significant problems to device performance [38, 158].

The hydrocarbon contamination was readily removed after minimal cleaning cycles, but it was also observed that some oxidation of the tin had also occurred, which took further cleaning cycles to remove. The lack of oxidation of the copper, zinc, or sulphur of the CZTS crystal shows that the native oxidation of CZTS is mostly centred on the tin, as has been previously concluded [158]. This observation has ramifications for future growth of CZTS thin-films, because as phase-pure CZTS should show only the native oxidation of tin, if the other elements show significant levels of oxidation, then it could be indicative that there is some deviation from phase-purity, possibly due to remnant precursor, the formation of secondary phases,[52] or that a processing problem has allowed exposure to oxygen during the growth.

6.9.2 *Changes in CZTS at 25 °C ≤ T ≤ 300 °C*

Whilst it is not the main focus of this study, the Cu/Zn disorder transition in CZTS is the main focus of several current studies using DFT [124], optical spectroscopy

[52] Which are more susceptible to oxidation.

[169], RS [168] and neutron scattering [170, 239], with the hope to elucidate the cause and effect of this disorder upon device performance [51]. Here, the changes in XPS spectra of the CZTS crystal are presented at elevated temperatures, in the hope of gaining some insight into this disorder transition.

In CZTS, there is a low enthalpy of formation for the site-switching of the copper and zinc atoms in plane, which does not exist in the Cu/Sn plane or indeed in CIGS because of the larger chemical mismatch between copper and tin/indium/gallium [51], compared with the comparative size of the copper and zinc atoms [19]. It was then found that the level of disorder present within CZTS is temperature dependent, with a critical temperature around 250 °C, below which, the Cu/Zn plane is mostly ordered, and above which, this plane rapidly becomes disordered [168, 170]. Given this low transition temperature, there are implications for the growth of CZTS, which usually takes place above it, suggesting that the rate and method of cooling for the material is important regarding the reversibility of this transition [168]. Although it has been shown that this transition results in bandgap changes of up to ~0.3 eV [51, 169], and that it has been suggested that these antisite defects then result in shallow states which increase the band tailing [239], the true effect of this disorder upon cell performance still requires determination [51].

In order to induce this transition within the CZTS crystal, PES measurements were performed whilst heating the crystal to a temperature above the known critical temperature, in this case, 300 °C, and then allowing the crystal to cool slowly back to room temperature. Full details of the annealing temperatures and times are detailed in Appendix C, Sect. C.5.4. Throughout the anneal, the crystal was monitored by XPS in order to determine any changes which may occur, and the crystal was kept at 300 °C for a sufficiently long time,[53] in order to ensure any changes would be observed.

Compared to the clean crystal, no extra species were fitted during the annealing, and the fittings used were the same as described in Sect. 6.7.2. This demonstrates that the annealing did not cause any oxidation or reduction of the material. Binding energies and FWHM of the main peaks of CZTS, as well as the position of the bands, at different stages of the annealing process, are shown in Fig. 6.45a, b and c, respectively. The binding energies of all of the peaks and the VBM shift by the same amount between each stage of annealing, suggesting that this is due only to Fermi-level shifting within the bandgap. This, and the relative constancy of the FWHM of the peaks, suggests that no changes have occurred in the level of disorder, because with such a transition, one could expect contrasting shifts, or widening of the peaks from copper and zinc. The shifts in the band levels also do not give evidence to the observation of the transition, because at temperature, the bandgap narrows slightly, as expected [349], and after cooling back to room temperature, the bands return to a similar position, albeit slightly higher than before, in agreement with the shift in Fermi-level. The disagreement between the band positions derived from XPS and UPS are because of the problems associated with fitting the band edges of this material, which were discussed in Sect. 6.7.3.

[53] >30 h.

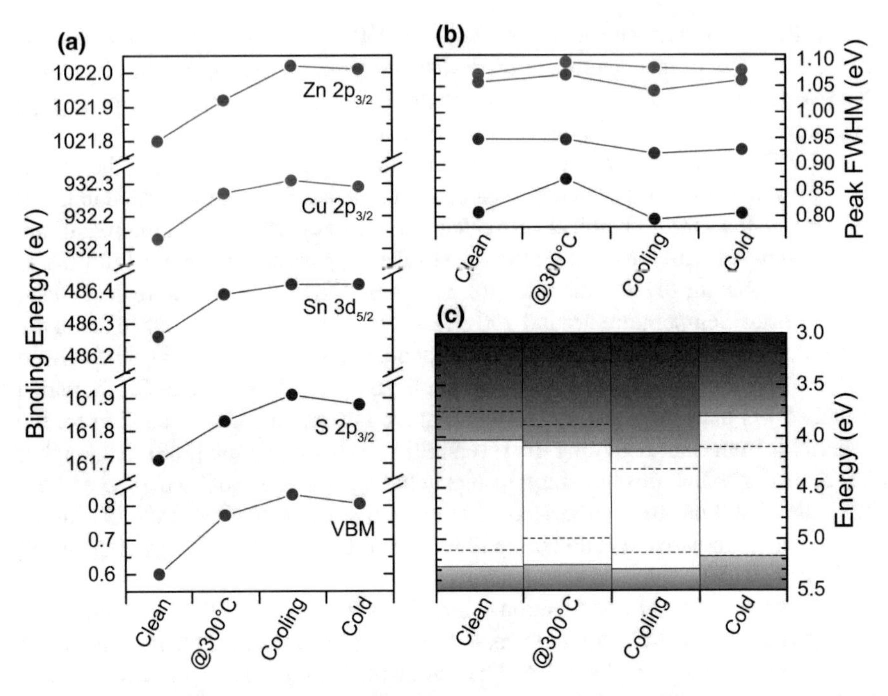

Fig. 6.45 XPS changes throughout annealing the single crystal CZTS to 300 °C. **a** Main core-level peak BE and VBM positions. **b** Main core-level peak FWHM. **c** Band-level positions from XPS (solid lines) and UPS (dashed lines), showing the top of the valence band (red), and bottom of the conduction band (blue) (Color figure online)

Figure 6.46a, b shows the XPS stoichiometry values through the annealing, in the form of atomic percentages and the commonly cited ratios of elements within CZTS, respectively. Although the surface stoichiometry from XPS was criticised in Sect. 6.8.3, the relative shifts between the different stages of the annealing will be representative of the changes occurring at the surface. From the atomic percentages, it can be seen that the sulphur and tin content remain relatively constant, the copper sees a general increase, and the zinc sees a general decrease, and these trends are reflected in the ratios as well, the crystal surface becomes more copper-rich, and more zinc-poor after the anneal.

It has been reported that the disorder transition is reversible, and the results seen here would suggest that this is not the case, because the observed changes do not revert to the same values as before the annealing took place. It is therefore concluded that although changes to the crystal were induced by the annealing, it cannot be fully assessed as to whether they occurred because of the disorder transition. More probably, the changes observed were caused as a result of the dissociation of zinc from the surface of the crystal, which explains the stoichiometry changes observed and also why the values did not revert to their original values after cooling, as would be expected with observations of the disorder transition. This result further suggests

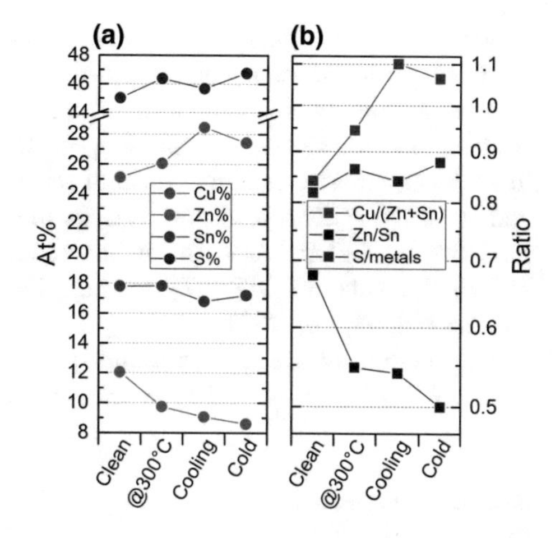

Fig. 6.46 XPS changes throughout annealing the single crystal CZTS to 300 °C. **a** Atomic percentages of the constituent elements. **b** Elemental ratios, shown logarithmically

that the reversibility of this disorder transition holds true for atmospheric annealing, but not in vacuo.

6.9.3 Degradation of CZTS

The growth of CZTS is performed at relatively high temperatures in vacuum for some methods, and as such, it is worth determining the prolonged effects of temperature under vacuum, in order to gain insights about the degradation experienced by CZTS under these conditions. In order to simulate these conditions on the CZTS crystal, PES measurements were performed whilst heating the crystal to temperatures up to 475 °C, and then allowing the crystal to cool slowly back to room temperature. This temperature was chosen because it is known that full decomposition of CZTS occurs at temperatures above 500 °C [53] and this was not desired. Full details of the annealing temperatures and times are detailed in Appendix C, Sect. C.5.4.

Several changes took place throughout the annealing, in terms of peak positions and areas. The stoichiometric changes, in terms of atomic percentages of the elements and the common elemental ratios used for CZTS, are shown throughout the annealing in Fig. 6.47a, b, respectively, and the corresponding changes in terms of peak positions and FWHM are shown in Fig. 6.48a, b, respectively.

Between the clean spectra and that at 360 °C, a similar trend is observed to that in Sect. 6.9.2: that the crystal experiences a loss of zinc, resulting in a Cu–rich and S–rich surface at 360 °C and that the binding energies of all the peaks, including the position of the VBM are shifted higher, demonstrating the continued Fermi-level shifting.

Between 360 and 475 °C, the values are relatively constant, suggesting that at this temperature, some form of equilibrium is reached. On prolonged exposure at 450 °C, the surface loses zinc, tin, and sulphur, with a large increase in copper. After cooling, the stoichiometry values continue with these trends, but the BE and FWHM return closer to the values before the annealing. This shows that the stoichiometry changes are due to elemental loss, whereas the widening and shifting of the peaks is reversible. These observations suggest that zinc is lost from CZTS at relatively low temperatures in vacuum, probably because of the high vapour pressure of zinc, and at 450 °C, CZTS begins to degrade through the loss of zinc, tin, and sulphur, probably via ZnS and SnS. This is in agreement with previous studies which found ZnS to segregate in CZTS [27], and that at temperatures above 400 °C, tin was lost via SnS [38, 53, 127, 171].

Degradation of CZTS was found during the initial cleaning of the crystal in Sect. 6.7.1, where a secondary species in the tin spectra formed with high-energy

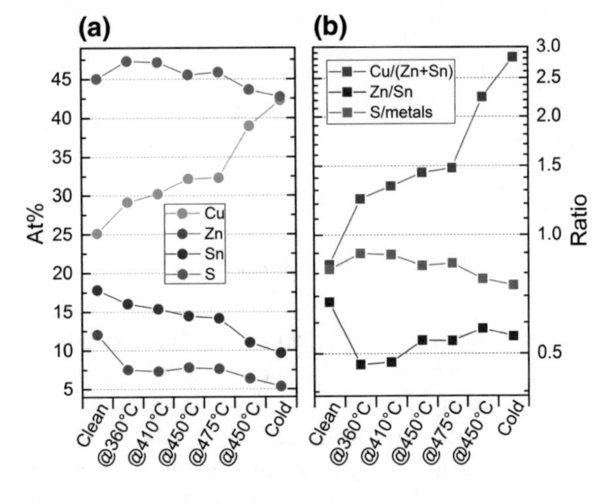

Fig. 6.47 XPS changes throughout annealing the single crystal CZTS to 475 °C. **a** Atomic percentages of the constituent elements. **b** Elemental ratios, shown logarithmically

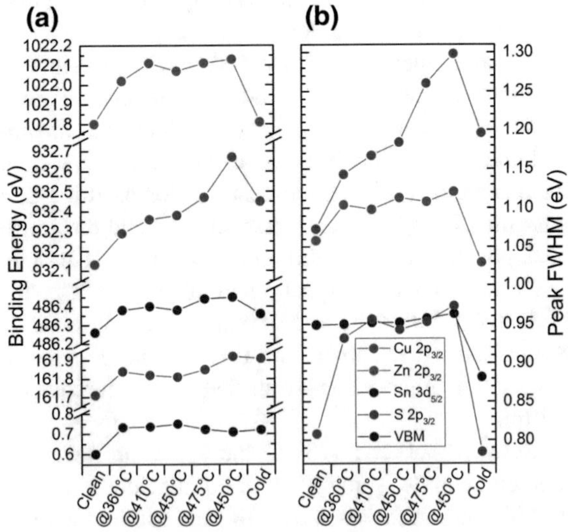

Fig. 6.48 XPS changes throughout annealing the single crystal CZTS to 475 °C. **a** Main core-level peak BE and VBM positions. **b** Main core-level peak FWHM

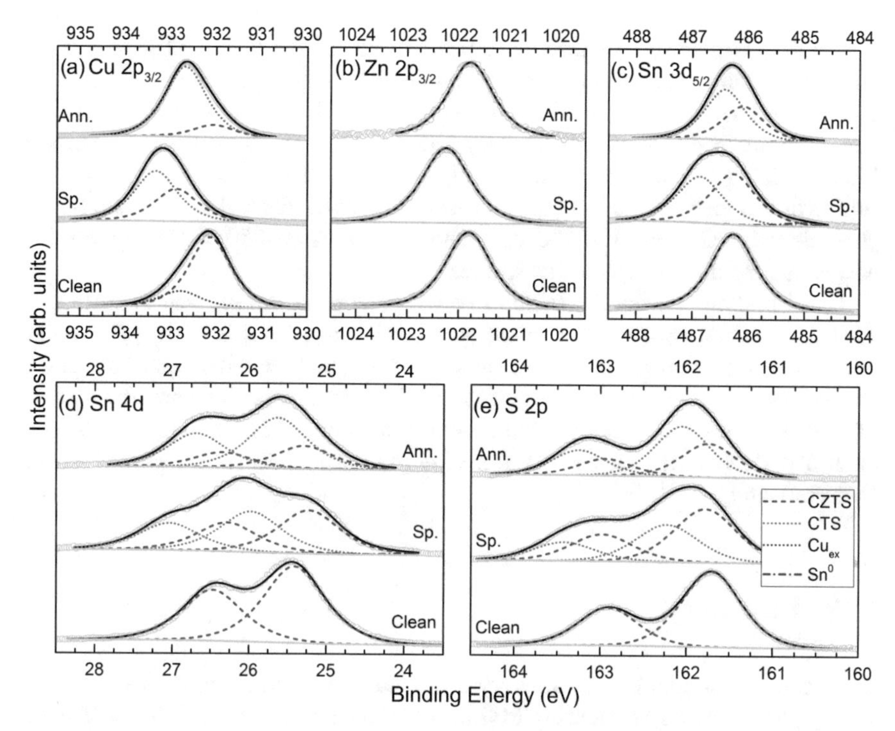

Fig. 6.49 XPS **a** Cu $2p_{3/2}$, **b** Zn $2p_{3/2}$, **c** Sn $3d_{5/2}$, **d** Sn $4d$, and **e** S $2p$ regions of CZTS for the clean crystal (Clean), after aggressive sputtering (Sp.), and subsequent annealing (Ann.). Fitted peaks for CZTS shown in red, and other fitted peaks attributed as discussed in the text. Peak envelopes shown in black. Spectral references are detailed and explained in Appendix C, Sect. C.5.4

sputtering. Therefore, in order to study the nature of this species, the crystal was deliberately sputtered aggressively, followed by annealing at 450 °C. Full details of this can be found in Appendix C, Sect. C.5.4.

The aggressive sputtering caused the immediate removal of all surface contamination, showing that the remnant carbon and oxygen in Fig. 6.21 is indeed only present at the surface. XPS regions for Cu $2p_{3/2}$, Zn $2p_{3/2}$, Sn $3d_{5/2}$, Sn $4d$, and S $2p$ are shown in Fig. 6.49a, b, c, d and e, respectively. In the spectra after sputtering, and after annealing, an extra species appears in the Cu $2p$ and S $2p$ spectra at higher BE (green dots), which increased after the annealing. A corresponding extra species was found in the Sn $3d_{5/2}$ spectra at higher BE as well, but there was also evidence of intensity at lower BE (blue dash dot). These extra peaks were corroborated by the Sn $4d$ spectra, which demonstrate the same peak ratios. The Sn peaks at lower BE are attributed to metallic tin, which is in agreement with the literature [181, 234, 350–352].

The extra peak at higher BE was originally thought be from SnS, as the degradation mechanism explored above would suggest. However, the formation of SnS from CZTS would require a change in oxidation state of the tin, from Sn^{4+} to Sn^{2+}, and as the extra species of tin now appears at higher BE than CZTS, this would seem implausible. So, given the extra peaks also observed in the copper and sulphur spectra, they were attributed to the formation of CTS. Annealing the crystal after this sputtering had formed the CTS only caused this species to increase. The higher BE than CZTS for the copper, tin and sulphur in CTS is in agreement with the few studies of this material [353–357]. Also, studies showed that in the CTS–ZnS binary phase diagram, CTS and CZTS coexist under a solidus at ~850 °C [92], explaining why annealing did not reduce this species.

These findings on the degradation of CZTS after sputtering and annealing in vacuum have consequences for future characterisations. That is, there exists a fine line between the removal of surface contamination, and the induction of structural changes that damage the material, making further characterisation unrepresentative. The results shown here can therefore help in avoiding these problems with thin-films of CZTS, which may suffer from the presence of secondary phases before characterisation.

6.10 Summary

The electronic structures of single crystal, phase-pure CZTS and its possible secondary phases were studied using PES in order to better understand the underlying electronic properties of these materials, and also in order to assess the viability of using XPS as a technique to help characterise samples of CZTS, which may contain these secondary phases. Also, the use of XRD and RS for this matter was discussed by comparing and contrasting expected spectra of the secondary phases with measurements of phase-pure CZTS. XPS was then further used to study the behaviour of CZTS in vacuum and at elevated temperatures in order to determine the processes that can occur.

The core-level and Auger spectra for the possible binary secondary phases Cu_2S, CuS, ZnS, SnS, SnS_2, and Sn_2S_3 showed large and interesting differences, especially for CuS and the tin sulphides,[54] which allow the spectra of the different phases to be distinguished.

The standard XRD and RS measurements of the single crystal CZTS showed that the commonly used XRD reference spectra lacks the full range of peaks and therefore can lead to misidentification, and the wavelength-dependence of the RS spectra means that peaks which are present at one wavelength but not another, could have led to misidentification in previous analyses, where spectra were compared to a reference that used a different wavelength.

An analysis of the core-level XPS spectra for the single crystal CZTS demonstrated the complexities of these spectra and allowed a critique of the current literature, finding it severely wanting both in terms of quality of data and quality of analysis. However, with careful and pre-informed approaches, the analysis of CZTS using

---

[54]Explored in Chap. 5.

XPS was found to be useful for its ability to determine oxidation states, elemental presence, contaminant effects, surface stoichiometry for comparison to bulk values, and the possibility to help reveal secondary phases.

Corroboration of the theoretical DoS for CZTS with both XPS and UPS valence band measurements allowed a deeper analysis of the band structure and bonding regime in this material, which found that the main contributions to the VB by Cu *d*/S *p* states are also influenced by states from the other cations, and that the split CB, with the CBM comprised of Sn *s*/S *p* states, differs somewhat from the often-compared material CIGS. These differences can be attributed to the differing structures of the two materials, involving the use of sulphur over selenium, but more importantly, the use of group II, Zn^{2+} and group IV, Sn^{4+} elements over the group III Ga/In^{3+} in CIGS, and how the different occupancies of the formal oxidation states of these elements affect the bonding nature of the valence and conduction bands.

The fitting of the theoretical DoS to the XPS VB spectra also showed the weaknesses of the linear extrapolation method of VBM determination, at least in the case of CZTS, but then allowed more accurate determination of the VBM. The band level positions determined here, revealed a lower value of IP than was previously assumed for CZTS. This finding, in agreement with other measurements of CZTS thin-films, gives some explanation as to why lower CZTS solar cell efficiencies are found, when using CdS as a window layer. A band misalignment is formed, and this, from both a design and environmental perspective, supports research which envisions to replace CdS with a more suitable candidate.

Assessments of the combined use of XRD, RS, and XPS, in terms of determining either phase-pure CZTS or secondary phases within it, showed the matter to be complex, even with high-quality data and in–depth analysis. Peak overlaps in XRD renders the detection of ZnS and CTS problematic, and the conclusion of phase-pure CZTS a difficult one to draw. RS was shown to be effective in determining the presence of secondary phases with differing bandgaps, through the use of quasi-resonant RS and multiple excitation wavelengths. However, this required prior knowledge of the expected secondary phases within a film, which may not be clear.

The use of XPS was shown to be able to provide some relief from these problems by offering another avenue by which secondary phases can be assessed, but that some caveats are required for the analysis. First, high-quality data must be recorded and a fitting model employed. Second, as the absolute binding energies can be influenced by external factors, the use of the Auger parameter is recommended for the assignment of peaks. Finally, if the measurements warrant surface cleaning, then extreme caution must be applied with regards to the cleaning procedure, because it was shown that the surface of CZTS is delicate in terms of the possibility for induction of unwanted phases.

Measurement of the band levels of CZTS and the secondary phases, along with their known properties, show that formation of any of these phases within CZTS, without exception, can act detrimentally to the performance of a cell, whether through the creation of second diodes, acting to block electron transport, or the formation of recombination centres. Furthermore, the use of XPS was shown to be effective in determining the presence of oxidation and contamination at the surface of CZTS, and analysing the effects they have upon the material.

Whilst it was found that the onset of the disorder transition in CZTS could not be established through the use of XPS and in vacuo annealing, the breakdown mechanism of CZTS through the loss of zinc at lower temperatures, followed by the subsequent loss of tin and sulphur at higher temperatures, show how films are expected to behave in such environments in future growth and analyses.

Overall, with regards to the detection and analysis of secondary phases within CZTS, there are many factors which must be acknowledged, in order to address the issues involved. Here, the merits of three powerful analysis tools have been shown to cooperate, in order to provide a more complete analysis; however, this does not completely address all of the issues. It is therefore vital, considering these problems, that as many techniques as possible are properly implemented, in order to elucidate the presence of secondary phases within CZTS, because it is believed that this is one of the main reasons why this promising, earth-abundant absorber is not achieving the efficiencies needed to progress it along the road to market viability.

References

1. Hahn H, Schulze H. Über Quaternäre Chalkogenide Des Germaniums Und Zinns. Naturwissenschaften. 1965;52(14):426.
2. Nitsche R, Sargent DF, Wild P. Crystal growth of quaternary 1_2246_4 chalcogenides by iodine vapor transport. J Cryst Growth. 1967;1(1):52–3.
3. Schäfer W, Nitsche R. Tetrahedral quaternary chalcogenides of the type Cu_2-II-IV-$S_4(Se_4)$. Mater Res Bull. 1974;9(5):645–54.
4. Hall SR, Szymanski JT, Stewart JM. Kesterite, $Cu_2(Zn, Fe)SnS_4$ and stannite $Cu_2(Fe, Zn)SnS_4$, structurally similar but distinct minerals. Can Mineral. 1978;16(2):131–7.
5. Guen L, Glaunsinger WS, Wold A. Physical properties of the quarternary chalcogenides $Cu_2{}^{I}B^{II}C^{IV}X_4$ (B^{II} = Zn, Mn, Fe, Co; C^{IV} = Si, Ge, Sn; X = S, Se). Mater Res Bull. 1979;14(4):463–7.
6. Ito K, Nakazawa T. Electrical and optical properties of stannite-type quaternary semiconductor thin films. Jpn J Appl Phys. 1988;27:2094–7.
7. Yokoyama D, Minegishi T, Jimbo K, Hisatomi T, Ma G, Katayama M, Kubota J, Katagiri H, Domen K. H_2 evolution from water on modified Cu_2ZnSnS_4 photoelectrode under solar light. Appl Phys Express. 2010;3(10):101202.
8. Jiang Q, Chen X, Gao H, Feng C, Guo Z. Synthesis of Cu_2ZnSnS_4 as novel anode material for lithium-ion battery. Electrochim Acta. 2016;190:703–12.
9. Xin X, He M, Han W, Jung J, Lin Z. Low-cost copper zinc tin sulfide counter electrodes for high-efficiency dye-sensitized solar cells. Angew Chemie Int Ed. 2011;50(49):11739–42.
10. Makovicky E. Crystal structures of sulfides and other chalcogenides. Rev Mineral Geochemistry. 2006;61(1):7–125.
11. Klaproth MH. Chemische Untersuchung Des Zinnkieses. In: Beiträge zur chemischen Kenntniss der Mineralkörper, Zweiter Band; Berlin; 1797. p. 257–264.
12. Schorr S. Structural aspects of adamantine like multinary chalcogenides. Thin Solid Films. 2007;515(15):5985–91.
13. Zhang Y, Yuan X, Sun X, Shih B-C, Zhang P, Zhang W. Comparative study of structural and electronic properties of Cu-based multinary semiconductors. Phys Rev B. 2011;84(7):75127.

14. Paier J, Asahi R, Nagoya A, Kresse G. Cu_2ZnSnS_4 as a potential photovoltaic material: a hybrid hartree-fock density functional theory study. Phys Rev B. 2009;79(11):115126.
15. Botti S, Kammerlander D, Marques MAL. Band structures of Cu_2ZnSnS_4 and Cu_2ZnSnS_4 from many-body methods. Appl Phys Lett. 2011;98(24):241915.
16. Kumar M, Persson C. Cu_2ZnSnS_4 and Cu_2ZnSnS_4 as potential earth-abundant thin-film absorber materials: a density functional theory study. Int J Theor Appl Sci. 2013;5(1):1–8.
17. Mitzi DB, Gunawan O, Todorov TK, Wang K, Guha S. The path towards a high-performance solution-processed kesterite solar cell. Sol Energy Mater Sol Cells. 2011;95(6):1421–36.
18. Khare A, Himmetoglu B, Johnson M, Norris DJ, Cococcioni M, Aydil ES. Calculation of the lattice dynamics and raman spectra of copper zinc tin chalcogenides and comparison to experiments. *J Appl Phys.* 2012;*111* (8).
19. Chen S, Gong XG, Walsh A, Wei S-H. Crystal and electronic band structure of Cu_2ZnSnX_4 (X = S and Se) photovoltaic absorbers: first-principles insights. Appl Phys Lett. 2009;94(4):41903.
20. Lafond A, Choubrac L, Guillot-Deudon C, Fertey P, Evain M, Jobic S. X-ray resonant single-crystal diffraction technique, a powerful tool to investigate the kesterite structure of the photovoltaic Cu_2ZnSnS_4 compound. Acta Crystallogr Sect B Struct Sci Cryst Eng Mater. 2014;70(2):390–4.
21. Hall SR, Stewart JM. The crystal structure refinement of chalcopyrite, $CuFeS_2$. Acta Crystallogr Sect B: Struct Crystallogr Cryst Chem. 1973;29(3):579–85.
22. Katagiri H, Sasaguchi N, Hando S, Hoshino S, Ohashi J, Yokota T. Preparation and evaluation of Cu_2ZnSnS_4 thin films by sulfurization of E-B evaporated precursors. Sol Energy Mater Sol Cells. 1997;49(1–4):407–14.
23. Delbos S. Kësterite thin films for photovoltaics: a review. EPJ Photovoltaics. 2012;3:35004.
24. Seol J-S, Lee S-Y, Lee J-C, Nam H-D, Kim K-H. Electrical and optical properties of Cu_2ZnSnS_4 thin films prepared by rf magnetron sputtering process. Sol Energy Mater Sol Cells. 2003;75(1–2):155–62.
25. Inamdar AI, Lee S, Jeon K-Y, Lee CH, Pawar SM, Kalubarme RS, Park CJ, Im H, Jung W, Kim H. Optimized fabrication of sputter deposited Cu_2ZnSnS_4 (CZTS) thin films. Sol Energy. 2013;91:196–203.
26. Shi C, Shi G, Chen Z, Yang P, Yao M. Deposition of Cu_2ZnSnS_4 thin films by vacuum thermal evaporation from single quaternary compound source. Mater Lett. 2012;73:89–91.
27. Shin B, Gunawan O, Zhu Y, Bojarczuk NA, Chey SJ, Guha S. Thin film solar cell with 8.4% power conversion efficiency using an earth-abundant Cu_2ZnSnS_4 absorber. Prog Photovoltaics Res Appl. 2013;21(1):72–6.
28. Sun L, He J, Kong H, Yue F, Yang P, Chu J. Structure, composition and optical properties of Cu_2ZnSnS_4 thin films deposited by pulsed laser deposition method. Sol Energy Mater Sol Cells. 2011;95(10):2907–13.
29. Surgina GD, Zenkevich AV, Sipaylo IP, Nevolin VN, Drube W, Teterin PE, Minnekaev MN. Reactive pulsed laser deposition of Cu_2ZnSnS_4 thin films in H_2S. Thin Solid Films. 2013;535(1):44–7.
30. Tanaka K, Moritake N, Uchiki H. Preparation of Cu_2ZnSnS_4 thin films by sulfurizing sol–gel deposited precursors. Sol Energy Mater Sol Cells. 2007;91(13):1199–201.
31. Maeda K, Tanaka K, Nakano Y, Uchiki H. Annealing temperature dependence of properties of Cu_2ZnSnS_4 thin films prepared by sol-gel sulfurization method. Jpn J Appl Phys. 2011;50(5):05FB08.
32. Mali SS, Shinde PS, Betty CA, Bhosale PN, Oh YW, Patil PS. Synthesis and characterization of Cu_2ZnSnS_4 thin films by silar method. J Phys Chem Solids. 2012;73(6):735–40.
33. Suryawanshi MP, Shin SW, Ghorpade UV, Gurav KV, Hong CW, Agawane GL, Vanalakar SA, Moon JH, Yun JH, Patil PS, Kim JH, Moholkar AV. Improved photoelectrochemical performance of Cu_2ZnSnS_4 (CZTS) thin films prepared using modified successive ionic layer adsorption and reaction (SILAR) sequence. Electrochim Acta. 2014;150:136–45.
34. Subramaniam EP, Rajesh G, Muthukumarasamy N, Thambidurai M, Asokan V, Velauthapillai D. Solar cells of Cu_2ZnSnS_4 thin films prepared by chemical bath deposition method. Indian J Pure Appl Phys. 2014;52:620–4.

35. Kumar AV, Park N-K, Kim E-T. A simple chemical approach for the deposition of Cu_2ZnSnS_4 thin films. Phys Status Solidi. 2014;211(8):1857–9.
36. Scragg JJ, Dale PJ, Peter LM. Synthesis and characterization of Cu_2ZnSnS_4 absorber layers by an electrodeposition-annealing route. Thin Solid Films. 2009;517(7):2481–4.
37. Chan CP, Lam H, Surya C. Preparation of Cu_2ZnSnS_4 films by electrodeposition using ionic liquids. Sol Energy Mater Sol Cells. 2010;94(2):207–11.
38. Altamura G, Vidal J. Impact of minor phases on the performances of CZTSSe thin-film solar cells. Chem Mater. 2016;28(11):3540–63.
39. Persson C. Electronic and optical properties of Cu_2ZnSnS_4 and $Cu_2ZnSnSe_4$. J Appl Phys. 2010;*107* (5).
40. Emrani A, Vasekar P, Westgate CR. Effects of sulfurization temperature on CZTS thin film solar cell performances. Sol Energy. 2013;98(PC):335–40.
41. He J, Sun L, Chen S, Chen Y, Yang P, Chu J. Composition dependence of structure and optical properties of $Cu_2ZnSn(S,Se)_4$ solid solutions: an experimental study. J Alloys Compd. 2012;511(1):129–32.
42. Feng J, Huang X, Chen W, Wu J, Lin H, Cheng Q, Yun D, Zhang F. Fabrication and characterization of Cu_2ZnSnS_4 thin films for photovoltaic application by low-cost single target sputtering process. Vacuum. 2016;126:84–90.
43. Zhao H, Persson C. Optical properties of $Cu(In,Ga)Se_2$ and $Cu_2ZnSn(S,Se)_4$. Thin Solid Films. 2011;519(21):7508–12.
44. Rajeshmon VG, Kartha CS, Vijayakumar KP, Sanjeeviraja C, Abe T, Kashiwaba Y. Role of precursor solution in controlling the opto-electronic properties of spray pyrolysed Cu_2ZnSnS_4 thin films. Sol Energy. 2011;85(2):249–55.
45. Lin X, Kavalakkatt J, Kornhuber K, Levcenko S, Lux-Steiner MC, Ennaoui A. Structural and optical properties of Cu_2ZnSnS_4 thin film absorbers from ZnS and Cu_3SnS_4 nanoparticle precursors. Thin Solid Films. 2013;535:10–3.
46. Matsushita H, Maeda T, Katsui A, Takizawa T. Thermal analysis and synthesis from the melts of cu-based quaternary compounds $Cu–III–IV–VI_4$ and $Cu_2–II–IV–VI_4$ (II = Zn, Cd; III = Ga, In; IV = Ge, Sn; VI = Se). J Cryst Growth. 2000;208(1–4):416–22.
47. Wibowo RA, Lee ES, Munir B, Kim KH. Pulsed laser deposition of quaternary $Cu_2ZnSnSe_4$ thin films. Phys Status Solidi Appl Mater Sci. 2007;204(10):3373–9.
48. Ahn S, Jung S, Gwak J, Cho A, Shin K, Yoon K, Park D, Cheong H, Yun JH. Determination of band gap energy (E_g) of $Cu_2ZnSnSe_4$ thin films: on the discrepancies of reported band gap values. Appl Phys Lett. 2010;97(2):21905.
49. Wang W, Winkler MT, Gunawan O, Gokmen T, Todorov TK, Zhu Y, Mitzi DB. Device characteristics of CZTSSe thin-film solar cells with 12.6% efficiency. Adv Energy Mater. 2014;4(7):1301465.
50. Fella CM, Romanyuk YE, Tiwari AN. Technological status of $Cu_2ZnSn(S,Se)_4$ thin film solar cells. Sol Energy Mater Sol Cells. 2013;119:276–7.
51. Bourdais S, Choné C, Delatouche B, Jacob A, Larramona G, Moisan C, Lafond A, Donatini F, Rey G, Siebentritt S, Walsh A, Dennler G. Is the Cu/Zn disorder the main culprit for the voltage deficit in kesterite solar cells? Adv Energy Mater. 2016;6(12):1502276.
52. Sun K, Yan C, Liu F, Huang J, Zhou F, Stride JA, Green M, Hao X. Over 9% efficient kesterite Cu_2ZnSnS_4 solar cell fabricated by using $Zn_{1-x}Cd_xS$ Buffer Layer. Adv Energy Mater. 2016;6(12):1600046.
53. Redinger A, Berg DM, Dale PJ, Siebentritt S. The consequences of kesterite equilibria for efficient solar cells. J Am Chem Soc. 2011;133(10):3320–3.
54. Abermann S. Non-vacuum processed next generation thin film photovoltaics: towards marketable efficiency and production of CZTS based solar cells. Sol Energy. 2013;94:37–70.
55. Tian Q, Xu X, Han L, Tang M, Zou R, Chen Z, Yu M, Yang J, Hu J. Hydrophilic Cu_2ZnSnS_4 nanocrystals for printing flexible, low-cost and environmentally friendly solar cells. CrystEngComm. 2012;14(11):3847.
56. Chen S, Yang J-H, Gong XG, Walsh A, Wei S-H. Intrinsic point defects and complexes in the quaternary kesterite semiconductor Cu_2ZnSnS_4. Phys Rev B. 2010;81(24):245204.

57. Zhang J, Long B, Cheng S, Zhang W. Effects of sulfurization temperature on properties of CZTS films by vacuum evaporation and sulfurization method. Int J Photoenergy. 2013;2013:1–6.
58. Cui H, Liu X, Liu F, Hao X, Song N, Yan C. Boosting Cu_2ZnSnS_4 solar cells efficiency by a thin Ag intermediate layer between absorber and back contact. Appl Phys Lett. 2014;104(4):41115.
59. Scragg JJ, Dale PJ, Colombara D, Peter LM. Thermodynamic aspects of the synthesis of thin-film materials for solar cells. ChemPhysChem. 2012;13(12):3035–46.
60. Huang TJ, Yin X, Qi G, Gong H. CZTS-based materials and interfaces and their effects on the performance of thin film solar cells. Phys status solidi—Rapid Res Lett. 2014;8(9):735–62.
61. Scragg JJ, Watjen JT, Edoff M, Ericson T, Kubart T, Platzer-Bjorkman C. A detrimental reaction at the molybdenum back contact in $Cu_2ZnSn(S,Se)_4$ thin-film solar cells. J Am Chem Soc. 2012;134(47):19330–3.
62. Haight R, Shao X, Wang W, Mitzi DB. Electronic and elemental properties of the $Cu_2ZnSn(S,Se)_4$ surface and grain boundaries. Appl Phys Lett. 2014;104(3):33902.
63. Liu X, Feng Y, Cui H, Liu F, Hao X, Conibeer G, Mitzi DB, Green M. The current status and future prospects of kesterite solar cells: a brief review. Prog Photovoltaics Res Appl 2016;15, n/a-n/a.
64. Rudmann D, Bilger G, Kaelin M, Haug FJ, Zogg H, Tiwari AN. Effects of NaF coevaporation on structural properties of $Cu(In,Ga)Se_2$ thin films. Thin Solid Films. 2003;431–432(3):37–40.
65. Prabhakar T, Jampana N. Effect of sodium diffusion on the structural and electrical properties of Cu_2ZnSnS_4 thin films. Sol Energy Mater Sol Cells. 2011;95(3):1001–4.
66. Hlaing Oo WM, Johnson JL, Bhatia A, Lund EA, Nowell MM, Scarpulla MA. Grain size and texture of Cu_2ZnSnS_4 thin films synthesized by cosputtering binary sulfides and annealing: effects of processing conditions and sodium. J Electron Mater. 2011;40(11):2214–21.
67. Li JV, Kuciauskas D, Young MR, Repins IL. Effects of sodium incorporation in co-evaporated $Cu_2ZnSnSe_4$ thin-film solar cells. Appl Phys Lett. 2013;102(16):163905.
68. Scragg JJ, Kubart T, Wa JT, Ericson T, Linnarsson MK, Platzer-bjo C. Effects of back contact instability on Cu_2ZnSnS_4 devices and processes. Chem Mater. 2013;25:3162–71.
69. Ahmed S, Reuter KB, Gunawan O, Guo L, Romankiw LT, Deligianni H. A high efficiency electrodeposited Cu_2ZnSnS_4 solar cell. Adv Energy Mater. 2012;2(2):253–9.
70. Katagiri H, Jimbo K, Maw WS, Oishi K, Yamazaki M, Araki H, Takeuchi A. Development of CZTS-Based thin film solar cells. Thin Solid Films. 2009;517(7):2455–60.
71. Song X, Ji X, Li M, Lin W, Luo X, Zhang H. a review on development prospect of CZTS based thin film solar cells. Int J Photoenergy. 2014;2014:1–11.
72. Ji S, Ye C. Cu_2ZnSnS_4 as a new solar cell material: the history and the future. Rev Adv Sci Eng. 2012;1(1):42–58.
73. Siebentritt S, Schorr S. Kesterites-a challenging material for solar cells. Prog Photovoltaics Res Appl. 2012;20(5):512–9.
74. Xiao Z-Y, Li Y-F, Yao B, Deng R, Ding Z, Wu T, Yang G, Li C-R, Dong Z-Y, Liu L, Zhang L, Zhao H-F. Bandgap engineering of $Cu_2Cd_xZn_{1-x}SnS_4$ alloy for photovoltaic applications: a complementary experimental and first-principles study. J Appl Phys. 2013;114(18):183506.
75. Siebentritt S. Why are kesterite solar cells not 20% efficient? Thin Solid Films. 2013;535(1):1–4.
76. Kumar M, Dubey A, Adhikari N, Venkatesan S, Qiao Q. Strategic review of secondary phases, defects and defect-Complexes in kesterite CZTS–Se solar cells. Energy Environ Sci. 2015;8(11):3134–59.
77. Suryawanshi MP, Agawane GL, Bhosale SM, Shin SW, Patil PS, Kim JH, Moholkar AV. CZTS based thin film solar cells: a status review. Mater Technol. 2013;28(1–2):98–109.
78. Thimsen E, Riha SC, Baryshev SV, Martinson ABF, Elam JW, Pellin MJ. Atomic layer deposition of the quaternary chalcogenide Cu_2ZnSnS_4. Chem Mater. 2012;24(16):3188–96.
79. Chopra KL, Paulson PD, Dutta V. Thin-film solar cells: an overview. Prog Photovoltaics Res Appl. 2004;12(23):69–92.

80. Hall RB, Birkmire RW, Phillips JE, Meakin JD. Thin-film polycrystalline $Cu_2S/Cd_{1-x}Zn_xS$ solar cells of 10% efficiency. Appl Phys Lett. 1981;38(11):925–6.
81. Bhat PK, Das SR, Pandya DK, Chopra KL. Back Illuminated high efficiency thin film Cu_2S/CdS solar cells. Sol Energy Mater. 1979;1(3–4):215–9.
82. Chakrabarti DJ, Laughlin DE. The Cu-S (Copper-Sulfur) system. Bull Alloy Phase Diagrams. 1983;4(3):254–71.
83. Evans HT Jr, Konnert JA. Crystal structure refinement of covellite. Am Mineral. 1976;61:996–1000.
84. Evans HT. Crystal structure of low chalcocite. Nat Phys Sci. 1971;232(29):69–70.
85. Donnay G, Donnay JDH, Kullerud G. Crystal and twin structure of digenite, Cu_9S_5. Am Mineral. 1957;1958(43):228–42.
86. Evans HT Jr. Djurleite ($Cu_{1.94}S$) and low chalcocite (Cu_2S): new crystal structure studies. Science. 1979;203(4378):356–8.
87. Koto K, Morimoto N. The crystal structure of anilite. Acta Crystallogr Sect B: Struct Crystallogr Cryst Chem. 1970;26(7):915–24.
88. Mumme WG, Gable RW, Petricek V. The crystal structure of roxbyite, $Cu_{58}S_{32}$. Can Mineral. 2012;50(2):423–30.
89. Munce CG, Parker GK, Holt SA, Hope GA. A raman spectroelectrochemical investigation of chemical bath deposited Cu_xS thin films and their modification. Colloids Surfaces A Physicochem. Eng Asp. 2007;295(1–3):152–8.
90. He YB, Polity A, Österreicher I, Pfisterer D, Gregor R, Meyer BK, Hardt M. Hall effect and surface characterization of Cu_2S and CuS films deposited by RF reactive sputtering. Phys B Condens Matter 2001;308–310, 1069–1073.
91. Kar P, Farsinezhad S, Zhang X, Shankar K. Anodic Cu_2S and CuS nanorod and nanowall arrays: preparation, properties and application in CO_2 photoreduction. Nanoscale. 2014;6:14305.
92. Olekseyuk ID, Dudchak IV, Piskach LV. Phase equilibria in the Cu_2S–ZnS–SnS_2 system. J Alloys Compd. 2004;368(1–2):135–43.
93. Baryshev SV, Riha SC, Zinovev AV. Solar absorber Cu_2ZnSnS_4 and its parent multilayers $ZnS/SnS_2/Cu_2S$ synthesized by atomic layer deposition and analyzed by x-ray photoelectron spectroscopy. Surf Sci Spectra. 2015;22(1):81–99.
94. Moriya K, Watabe J, Tanaka K, Uchiki H. Characterization of Cu_2ZnSnS_4 thin films prepared by photo-chemical deposition. Phys Status Solidi Curr Top Solid State Phys. 2006;3(8):2848–52.
95. Gotsis HJ, Barnes AC, Strange P. Experimental and theoretical investigation of the crystal structure of CuS. J Phys: Condens Matter. 1992;4(50):10461–8.
96. Roberts HS, Ksanda CJ. The crystal structure of covellite. Am J Sci. 1929;s5–17(102):489–503.
97. Ohmasa M, Suzuki M, Takéuchi Y. A refinement of the crystal structure of covellite CuS. Mineral J. 1977;8(6):311–9.
98. Pattrick RAD, Mosselmans JFW, Charnock JM, England KER, Helz GR, Garner CD, Vaughan DJ. The structure of amorphous copper sulfide precipitates: an x-ray absorption study. Geochim Cosmochim Acta. 1997;61(10):2023–36.
99. van der Laan G, Pattrick RAD, Henderson CMB, Vaughan DJ. Oxidation state variations in copper minerals studied with Cu 2p x-ray absorption spectroscopy. J Phys Chem Solids. 1992;53(9):1185–90.
100. Rupp H, Weser U. X-ray photoelectron spectroscopy of copper(II), copper(I), and mixed valence systems. Bioinorg Chem. 1976;6(1):45–59.
101. Nakai I, Sugitani Y, Nagashima K, Niwa Y. X-ray photoelectron spectroscopic study of copper minerals. J Inorg Nucl Chem. 1978;40(5):789–91.
102. Gebhardt JE, McCarron JJ, Richardson PE, Buckley AN. The effect of cathodic treatment on the anodic polarization of copper sulfides. Hydrometallurgy. 1986;17(1):27–38.
103. Folmer JCW, Jellinek F. The valence of copper in sulphides and selenides: an x-ray photoelectron spectroscopy study. J Less Common Met. 1980;76(1–2):153–62.

104. Takéuchi Y, Kudoh Y, Sato G. The crystal structure of covellite CuS under high pressure up to 33 Kbar. Zeitschrift für Krist. 1985;173(1–2):119–28.
105. Perry DL, Taylor JA. X-ray photoelectron and auger spectroscopic studies of Cu_2S and CuS. J Mater Sci Lett. 1986;5(4):384–6.
106. Fjellvag H, Gronvold F, Stolen S, Andresen AF, Muller-Kafer R, Simon A. Low-temperature structural distortion in CuS. Zeitschrift fur Krist—New Cryst Struct. 1988;184(1–2):111–21.
107. Liang W, Whangbo M-H. Conductivity anisotropy and structural phase transition in covellite CuS. Solid State Commun. 1993;85(5):405–8.
108. Nozaki H, Shibata K, Ohhashi N. Metallic hole conduction in CuS. J Solid State Chem. 1991;91(2):306–11.
109. Kumar P, Nagarajan R, Sarangi R. Quantitative X-ray absorption and emission spectroscopies: electronic structure elucidation of Cu_2S and CuS. J Mater Chem C Mater Opt Electron Devices. 2013;1(13):2448–2454.
110. Mazin II. Structural and electronic properties of the two-dimensional superconductor CuS with 1[1/3]-valent copper. Phys Rev B—Condens Matter Mater Phys. 2012;85(11):1–5.
111. Goh SW, Buckley AN, Lamb RN. Copper(II) sulfide? Miner Eng. 2006;19(2):204–8.
112. Di Benedetto F, Borgheresi M, Caneschi A, Chastanet G, Cipriani C, Gatteschi D, Pratesi G, Romanelli M, Sessoli R. First evidence of natural superconductivity: covellite. Eur J Mineral. 2006;18(3):283–7.
113. Wagner R, Wiemhöfer H-D. Hall effect and conductivity in thin films of low temperature chalcocite Cu_2S at 20 °C as a function of stoichiometry. J Phys Chem Solids. 1983;44(8):801–5.
114. Ni H-C, Lin C-H, Lo K-Y, Tsai C-H, Chen T-P, Tsai J-L, Gong J-R. Properties of Cu_2ZnSnS_4 films by sulfurization of Cu_2ZnSnS_4 precursors using ditert-butylsulfide at atmospheric pressure. ECS J Solid State Sci Technol. 2015;4(8):Q72–4.
115. Sharma KC, Chang YA. The S-Zn (Sulfur-Zinc) system. J Phase Equilibria. 1996;17(3):261–6.
116. Schwab B, Ruh A, Manthey J, Drosik M. Zinc. In Ullmann's Encyclopedia of Industrial Chemistry. Weinheim, Germany: Wiley-VCH Verlag GmbH and Co. KGaA; 2015, p. 1–25.
117. Kutty TRN. A Controlled copper-coating method for the preparation of ZnS: Mn DC electroluminescent powder phosphors. Mater Res Bull. 1991;26(5):399–406.
118. Prener JS, Williams FE. Activator Systems in Zinc Sulfide Phosphors. J Electrochem Soc. 1956;103(6):342.
119. Shadfan A, Pawlowski M, Wang Y, Subramanian K, Gabay I, Ben-Yakar A, Tkaczyk T. Design and fabrication of a miniature objective consisting of high refractive index zinc sulfide lenses for laser surgery. Opt Eng. 2016;55(2):25107.
120. Khamala B, Franklin L, Malozovsky Y, Stewart A, Saleem H, Bagayoko D. Calculated electronic, transport, and bulk properties of zinc-blende zinc sulphide (Zb-ZnS). Comput Condens Matter. 2016;6:18–23.
121. Yu F, Ou S-L, Yao P-C, Wu B-R, Wuu D. Structural, surface morphology and optical properties of ZnS films by chemical bath deposition at various Zn/S molar ratios. J Nanomater. 2014;2014(6):1–7.
122. Hwang D, Ahn J, Hui K, Hui K, Son Y. Structural and optical properties of ZnS thin films deposited by RF magnetron sputtering. Nanoscale Res Lett. 2012;7(1):26.
123. Schneikart A, Schimper H-J, Klein A, Jaegermann W. Efficiency limitations of thermally evaporated thin-film SnS solar cells. J Phys D Appl Phys. 2013;46(30):305109.
124. Nagoya A, Asahi R, Wahl R, Kresse G. Defect formation and phase stability of Cu_2ZnSnS_4 photovoltaic material. Phys Rev B. 2010;81(11):113202.
125. Shin SW, Pawar SM, Park CY, Yun JH, Moon JH, Kim JH, Lee JY. Studies on Cu_2ZnSnS_4 (CZTS) Absorber layer using different stacking orders in precursor thin films. Sol Energy Mater Sol Cells. 2011;95(12):3202–6.
126. Ansari MZ, Khare N. Structural and optical properties of CZTS thin films deposited by ultrasonically assisted chemical vapour deposition. J Phys D Appl Phys. 2014;47(18):185101.
127. Weber A, Mainz R, Schock HW. On the Sn loss from thin films of the material system Cu–Zn–Sn–S in high vacuum. J Appl Phys. 2010;107(1):13516.

128. Muska K, Kauk M, Altosaar M, Pilvet M, Grossberg M, Volobujeva O. Synthesis of Cu_2ZnSnS_4 monograin powders with different compositions. Energy Procedia. 2011;10:203–7.
129. Nagaoka A, Yoshino K, Taniguchi H, Taniyama T, Kakimoto K, Miyake H. Growth and characterization of Cu_2ZnSnS_4 single crystals. Phys Status Solidi. 2013;210(7):1328–31.
130. Sharma RC, Chang YA. The S − Sn (sulfur-tin) system. Bull Alloy Phase Diagrams. 1986;7(3):269–73.
131. Skinner BJ. Unit cell edges of natural and synthetic sphalerites. Am Mineral. 1961;46(11/12):1388–411.
132. Lokhande AC, Chalapathy RBV, He M, Jo E, Gang M, Pawar SA, Lokhande CD, Kim JH. Development of Cu_2SnS_3 (CTS) thin film solar cells by physical techniques: a status review. Sol Energy Mater Sol Cells. 2016;153:84–107.
133. Lund EA, Du H, Hlaing OO, M W, Teeter G, Scarpulla MA. Investigation of combinatorial coevaporated thin film Cu_2ZnSnS_4 (II): beneficial cation arrangement in Cu-Rich growth. J Appl Phys. 2014;115(17):173503.
134. Zakutayev A, Baranowski LL, Welch AW, Wolden CA. Toberer ES. Comparison of Cu_2SnS_3 and $CuSbS_2$ as potential solar cell absorbers. In: 2014 IEEE 40th Photovoltaic Specialist Conference (PVSC); IEEE, vol. 3, p. 2436–2438; 2014.
135. Wu C, Hu Z, Wang C, Sheng H, Yang J, Xie Y. Hexagonal Cu_2SnS_3 with metallic character: another category of conducting sulfides. Appl Phys Lett. 2007;91(14):143104.
136. Kitagawa N, Ito S, Nguyen D-C, Nishino H. Copper zinc sulfur compound solar cells fabricated by spray pyrolysis deposition for solar cells. Nat Resour. 2013;4(1):142–5.
137. Uhuegbu, CC, Babatunde EB. Spectral analysis of copper zinc sulphide ternary thin film grown by solution growth technique. Am J Sci Ind Res. 2010;1(3):397–400.
138. Jose E, Kumar MS. Room-temperature wide-range luminescence and structural, optical, and electrical properties of silar deposited Cu-Zn-S nano-structured thin films. In: Lakhtakia A, Mackay TG, Suzuki M, editors. Proceedings of the SPIE; 2016. vol. 9929, p. 992917.
139. Sreejith MS, Deepu DR, Sudhakartha C, Rajeevkumar K, Vijayakumar KP. Improvement of sprayed $CuZnS/In_2S_3$ solar cell efficiency by making multiple band gap nature more prominent. J Renew Sustain Energy. 2016;8(2).
140. Clark AH, Sillitoe RH. Cuprian sphalerite and a probable copper-zinc sulfide, cachiyuyo de llampos, copiapó. Chile Am Mineral. 1021;1970:55.
141. Gordillo G, Calderón C, Bartolo-Pérez P. XPS analysis and structural and morphological characterization of Cu_2ZnSnS_4 thin films grown by sequential evaporation. Appl Surf Sci. 2014;305:506–14.
142. Scragg JJ, Dale PJ, Peter LM, Zoppi G, Forbes I. New routes to sustainable photovoltaics: evaluation of Cu_2ZnSnS_4 as an alternative absorber material. Phys Status Solidi Basic Res. 2008;245(9):1772–8.
143. Goldschmidt VM, The principles of distribution of chemical elements in minerals and rocks. the seventh hugo müller lecture, delivered before the chemical society on March 17th. J Chem Soc. 1937;1937:655–73.
144. Bär M, Schubert BA, Marsen B, Wilks RG, Blum M, Krause S, Pookpanratana S, Zhang Y, Unold T, Yang W, Weinhardt L, Heske C, Schock HW. Cu_2ZnSnS_4 thin-film solar cell absorbers illuminated by soft x-rays. J Mater Res. 2012;27(8):1097–104.
145. Just J, Lützenkirchen-Hecht D, Frahm R, Schorr S, Unold T. Determination of secondary phases in kesterite Cu_2ZnSnS_4 thin films by x-ray absorption near edge structure analysis. Appl Phys Lett. 2011;2011(99):262105.
146. Fernandes PA, Salomé PMP, da Cunha AF. Study of polycrystalline Cu_2ZnSnS_4 films by raman scattering. J Alloys Compd. 2011;509(28):7600–6.
147. Fontané X, Calvo-Barrio L, Izquierdo-Roca V, Saucedo E, Pérez-Rodriguez A, Morante JR, Berg DM, Dale PJ, Siebentritt S. In-depth resolved raman scattering analysis for the identification of secondary phases: characterization of Cu_2ZnSnS_4 layers for solar cell applications. Appl Phys Lett. 2011;98(18):181905.

148. Song J, Yang C, Hu H, Dai X, Wang C, Zhang H. Penetration depth at various raman excitation wavelengths and stress model for raman spectrum in biaxially-strained Si. Sci China Phys Mech Astron 2013;56(11):2065–2070.
149. Holtz M, Duncan WM, Zollner S, Liu R. Visible and ultraviolet raman scattering studies of $Si_{1-x}Ge_x$ alloys. J Appl Phys. 2000;88(5):2523–8.
150. Nguyen VT, Nam D, Gansukh M, Park S-N, Sung S-J, Kim D-H, Kang J-K, Sai CD, Tran TH, Cheong H. Influence of sulfate residue on Cu_2ZnSnS_4 thin films prepared by direct solution method. Sol Energy Mater Sol Cells. 2015;136:113–9.
151. Vaccarello D, Tapley A, Ding Z. Optimization of the Cu_2ZnSnS_4 nanocrystal recipe by means of photoelectrochemical measurements. Rsc Adv. 2013;3(11):3512–5.
152. Aono M, Yoshitake K, Miyazaki H. XPS depth profile study of CZTS thin films prepared by spray pyrolysis. Phys Status Solidi. 2013;10(7–8):1058–61.
153. Dhakal TP, Peng C, Reid Tobias R, Dasharathy R, Westgate CR. Characterization of a CZTS thin film solar cell grown by sputtering method. Sol Energy. 2014;100:23–30.
154. Pawar BS, Pawar SM, Shin SW, Choi DS, Park CJ, Kolekar SS, Kim JH. Effect of complexing agent on the properties of electrochemically deposited Cu_2ZnSnS_4 (CZTS) thin films. Appl Surf Sci. 2010;257(5):1786–91.
155. Riha SC, Parkinson BA, Prieto AL. Solution-based synthesis and characterization of Cu_2ZnSnS_4 nanocrystals. J Am Chem Soc. 2009;131(34):12054–5.
156. Mali SS, Patil BM, Betty CA, Bhosale PN, Oh YW, Jadkar SR, Devan RS, Ma Y-R, Patil PS. Novel synthesis of kesterite Cu_2ZnSnS_4 nanoflakes by successive ionic layer adsorption and reaction technique: characterization and application. Electrochim Acta. 2012;66:216–21.
157. Park S-N, Sung S-J, Son D-H, Kim D-H, Gansukh M, Cheong H, Kang J-K. Solution-processed Cu_2ZnSnS_4 absorbers prepared by appropriate inclusion and removal of thiourea for thin film solar cells. Rsc Adv. 2014;4(18):9118–25.
158. Bär M, Schubert B-A, Marsen B, Krause S, Pookpanratana S, Unold T, Weinhardt L, Heske C, Schock H-W. Native oxidation and Cu-Poor surface structure of thin film Cu_2ZnSnS_4 solar cell absorbers. Appl Phys Lett. 2011;99(11):112103.
159. Tanaka K, Kato M, Uchiki H. Effects of chlorine and carbon on Cu_2ZnSnS_4 thin film solar cells prepared by spray pyrolysis deposition. J Alloys Compd. 2014;616:492–7.
160. Durose K, Asher SE, Jaegermann W, Levi D, McCandless BE, Metzger W, Moutinho H, Paulson PD, Perkins CL, Sites JR, Teeter G, Terheggen M. Physical characterization of thin-film solar cells. Prog Photovoltaics. 2004;12(2–3):177–217.
161. Würz R, Rusu M, Schedel-Niedrig T, Lux-Steiner MC, Bluhm H, Hävecker M, Kleimenov E, Knop-Gericke A, Schlögl R. In situ x-ray photoelectron spectroscopy study of the oxidation of $CuGaSe_2$. Surf Sci. 2005;580(1–3):80–94.
162. Buckley AN, Wouterlood HJ, Woods R. The surface composition of natural sphalerites under oxidative leaching conditions. Hydrometallurgy. 1989;22(1–2):39–56.
163. Fritsche J, Gunst S, Golusda E, Lejard MC, Thißen A, Mayer T, Klein A, Wendt R, Gegenwart R, Bonnet D, Jaegermann W. Surface analysis of CdTe thin film solar cells. Thin Solid Films. 2001;387(1–2):161–4.
164. Lin AWC, Armstrong NR, Kuwana T. X-ray photoelectron/auger electron spectroscopic studies of tin and indium metal foils and oxides. Anal Chem. 1977;49(8):1228–35.
165. Delgass WN, Haller GL, Kellerman R, Lunsford JH. X-ray photoelectron spectroscopy. In: Delgass W, editor. Spectroscopy in heterogeneous catalysis. New York: Elsevier; 1979. p. 267–322.
166. Yan C, Liu F, Song N, Ng BK, Stride JA, Tadich A, Hao X. Band alignments of different buffer layers (CdS, Zn(O, S), and In_2S_3) on Cu_2ZnSnS_4. Appl Phys Lett. 2014;104(17):173901.
167. Choi Y, Baek M, Zhang Z, Dao V-D, Choi H-S, Yong K. A two-storey structured photoanode of a 3D Cu_2ZnSnS_4/CdS/ZnO@steel composite nanostructure for efficient photoelectrochemical hydrogen generation. Nanoscale. 2015;7(37):15291–9.
168. Scragg JJS, Choubrac L, Lafond A, Ericson T, Platzer-Björkman C. A low-temperature order-disorder transition in Cu_2ZnSnS_4 thin films. Appl Phys Lett. 2014;104(4):041911–4.

169. Valentini M, Malerba C, Menchini F, Tedeschi D, Polimeni A, Capizzi M, Mittiga A. Effect of the order-disorder transition on the optical properties of Cu_2ZnSnS_4. Appl Phys Lett. 2016;108(21):2–7.
170. Ritscher A, Hoelzel M, Lerch M. The order-disorder transition in Cu_2ZnSnS_4—a neutron scattering investigation. J Solid State Chem. 2016;238:68–73.
171. Scragg JJ, Ericson T, Kubart T, Edoff M, Platzer-Björkman C. Chemical insights into the instability of Cu_2ZnSnS_4 films during annealing. Chem Mater. 2011;23(20):4625–33.
172. Burton LA, Colombara D, Abellon RD, Grozema FC, Peter LM, Savenije TJ, Dennler G, Walsh A. Synthesis, characterization, and electronic structure of single-crystal SnS, Sn_2S_3, and SnS_2. Chem Mater. 2013;25(24):4908–16.
173. Nagaoka A, Yoshino K, Taniguchi H, Taniyama T, Miyake H. Growth of Cu_2ZnSnS_4 single crystal by traveling heater method. Jpn J Appl Phys. 2011;50:128001.
174. Nagaoka A, Yoshino K, Aoyagi K, Minemoto T, Nose Y, Taniyama T, Kakimoto K, Miyake H. Thermo-physical properties of Cu_2ZnSnS_4 single crystal. J Cryst Growth. 2014;393:167–70.
175. Nagaoka A, Yoshino K, Taniguchi H, Taniyama T, Miyake H. Preparation of Cu_2ZnSnS_4 single crystals from Sn solutions. J Cryst Growth. 2012;341(1):38–41.
176. Partain LD, Schneider RA,Donaghey, LF, McLeod PS. Surface chemistry of Cu_xS and Cu_xS/CdS determined from x-ray photoelectron spectroscopy. J Appl Phys. 1985;57(11):5056.
177. Laajalehto K, Kartio I, Nowak P. XPS study of clean metal sulfide surfaces. Appl Surf Sci. 1994;81(1):11–5.
178. Hurtado M, Cruz SD, Becerra RA, Calderon C, Bartolo-Perez P, Gordillo G. XPS analysis and structural characterization of CZTS thin films prepared using solution and vacuum based deposition techniques. In: 2014 IEEE 40th Photovoltaic Specialist Conference (PVSC); IEEE; 2014, p. 0368–0372.
179. Avendano CAM, Mathews NR, Pal M, Delgado FP, Mathew X. Structural evolution of multilayer SnS/Cu/ZnS stack to phase-pure Cu_2ZnSnS_4 thin films by thermal processing. ECS J Solid State Sci Technol. 2015;4(3):P91–P96.
180. Lebugle A, Axelsson U, Nyholm R, Mårtensson N. Experimental L and M core level binding energies for the metals ^{22}Ti to ^{30}Zn. Phys Scr. 1981;23(5A):825–7.
181. Nyholm R, Martensson N. Core level binding energies for the elements Zr-Te ($Z = 40$–52). J Phys C: Solid State Phys. 1980;13(11):L279–84.
182. Barrie A, Drummond IW, Herd QC. Correlation of calculated and measured 2p spin-orbit splitting by electron spectroscopy using monochromatic x-radiation. J Electron Spectros Relat Phenomena. 1974;5(1):217–25.
183. Coster D, Kronig LRD. New type of auger effect and its influence on the x-ray spectrum. Physica. 1935;2(1–12):13–24.
184. Nyholm R, Martensson N, Lebugle A, Axelsson U. Auger and coster-kronig broadening effects in the 2p and 3p photoelectron spectra from the metals ^{22}Ti-^{30}Zn. J Phys F: Met Phys. 1981;11(8):1727–33.
185. Kresse G, Hafner J. Ab initio molecular dynamics for liquid metals. Phys Rev B. 1993;47(1):558–61.
186. Kresse G, Hafner J. Ab initio molecular-dynamics simulation of the liquid-metal–amorphous-semiconductor transition in germanium. Phys Rev B. 1994;49(20):14251–69.
187. Kresse G, Furthmüller J. Efficiency of ab-initio total energy calculations for metals and semiconductors using a plane-wave basis set. Comput Mater Sci. 1996;6(1):15–50.
188. Kresse G, Furthmüller J. Efficient iterative schemes for ab initio total-energy calculations using a plane-wave basis set. Phys Rev B. 1996;54(16):11169–86.
189. Salomé PMP, Fernandes PA, Leitão JP, Sousa MG, Teixeira JP, da Cunha AF. Secondary crystalline phases identification in $Cu_2ZnSnSe_4$ thin films: contributions from raman scattering and photoluminescence. J Mater Sci. 2014;49(21):7425–36.
190. Wagner CD, Joshi A. The auger parameter, its utility and advantages: a review. J Electron Spectros Relat Phenomena. 1988;47:283–313.

191. Bär M, Schubert B-A, Marsen B, Krause S, Pookpanratana S, Unold T, Weinhardt L, Heske C, Schock H-W. Impact of KCN etching on the chemical and electronic surface structure of Cu_2ZnSnS_4 thin-film solar cell absorbers. Appl Phys Lett. 2011;99(15):152111.
192. Biesinger MC, Lau LWM, Gerson AR, Smart RSC. Resolving surface chemical states in XPS analysis of first row transition metals, oxides and hydroxides: Sc, Ti, V, Cu and Zn. Appl Surf Sci. 2010;257(3):887–98.
193. Deroubaix G, Marcus P. X-ray photoelectron spectroscopy analysis of copper and zinc oxides and sulphides. Surf Interface Anal. 1992;18(1):39–46.
194. Capece FM, Di Castro V, Furlani C, Mattogno G, Fragale C, Gargano M, Rossi M. "Copper chromite" catalysts: XPS structure elucidation and correlation with catalytic activity. J Electron Spectros Relat Phenomena. 1982;27(2):119–28.
195. Klein JC, Proctor A, Hercules DM, Black JF. X-ray excited auger intensity ratios for differentiating copper compounds. Anal Chem. 1983;55(13):2055–9.
196. Whittles TJ, Burton LA, Skelton JM, Walsh A, Veal TD, Dhanak VR. Band alignments, valence bands, and core levels in the tin sulfides SnS, SnS_2, and Sn_2S_3: experiment and theory. Chem Mater. 2016;28(11):3718–26.
197. Lindberg BJ, Hamrin K, Johansson G, Gelius U, Fahlman A, Nordling C, Siegbahn K. Molecular spectroscopy by means of ESCA II. sulfur compounds. correlation of electron binding energy with structure. Phys Scr. 1970;1(5–6):286–98.
198. Paraknowitsch JP, Thomas A, Schmidt J. Microporous sulfur-doped carbon from thienyl-based polymer network precursors. Chem Commun. 2011;47(29):8283.
199. Zhang L, Li Z, Yang Y, Zhou Y, Li J, Si L, Kong B. Research on the composition and distribution of organic sulfur in coal. Molecules. 2016;21(5):630.
200. Marsh H, Sherwood PMA, Augustyn D. An XPS study of binding energies of sulphur in carbon. Fuel. 1976;55(2):97–8.
201. Lide DR. CRC Handbook of chemistry and physics 84th Edition; 2003.
202. Goh SW, Buckley AN, Lamb RN, Rosenberg RA, Moran D. The oxidation states of copper and iron in mineral sulfides, and the oxides formed on initial exposure of chalcopyrite and bornite to air. Geochim Cosmochim Acta. 2006;70(9):2210–28.
203. Kundu M, Hasegawa T, Terabe K, Yamamoto K, Aono M. Structural studies of copper sulfide films: effect of ambient atmosphere. Sci Technol Adv Mater. 2008;9(3):35011.
204. Liu GM, Schulmeyer T, Brotz J, Klein A, Jaegermann W. Interface properties and band alignment of Cu_2S/CdS thin film solar cells. Thin Solid Films. 2003;431:477–82.
205. Romand M, Roubin M, Deloume JP. ESCA studies of some copper and silver selenides. J Electron Spectros Relat Phenomena. 1978;13(3):229–42.
206. Bhide VG, Salkalachen S, Rastog AC, Rao CNR, Hegde MS. Depth profile composition studies of thin film CdS:Cu_2S solar cells using XPS and AES. J Phys D Appl Phys. 1981;14(9):1647–56.
207. Scheer R. Photoemission study of evaporated $CuInS_2$ thin films. II. Electronic surface structure. J Vac Sci Technol A Vacuum Surfaces Film, 1994;12(1):56.
208. Brion D. Etude Par Spectroscopie de photoelectrons de la degradation superficielle de FeS_2, $CuFeS_2$, ZnS et PbS a L'air et Dans L'eau. Appl Surf Sci. 1980;5(2):133–52.
209. Krylova V, Andrulevičius M, Optical, XPS and XRD studies of semiconducting copper sulfide layers on a polyamide film. Int J Photoenergy 2009, 2009 (Ii).
210. Panigrahi S, Basak D. Solution-processed novel Core–shell N–p heterojunction and its ultrafast UV photodetection properties. RSC Adv. 2012;2(31):11963.
211. Agostinelli E, Battistoni C, Fiorani D, Mattogno G, Nogues M. An XPS study of the electronic structure of the $Zn_xCd_{1-x}Cr_2X$=S, Se) spinel system. J Phys Chem Solids. 1989;50(3):269–72.
212. Siriwardane RV, Poston JA. Interaction of H_2S with Zinc titanate in the presence of H_2 and CO. Appl Surf Sci. 1990;45(2):131–9.
213. Langer DW, Vesely CJ. Electronic core levels of zinc chalcogenides. Phys Rev B. 1970;2(12):4885–92.

214. Battistoni C, Gastaldi L, Lapiccirella A, Mattogno G, Viticoli S. Octahedral versus tetrahedral coordination of the Co(II) Ion in layer compounds: $Co_xZn_{1-x}In_2S_4(O \leqslant x \leqslant 0.46)$ solid solution. J Phys Chem Solids. 1986;47(9):899–903.
215. Dake LS, Baer DR, Zachara JM. Auger Parameter Measurements of zinc compounds relevant to zinc transport in the environment. Surf Interface Anal. 1989;14(1–2):71–5.
216. Schön G. Auger and direct electron spectra in x-ray photoelectron studies of zinc, zinc oxide, gallium and gallium oxide. J Electron Spectros Relat Phenomena. 1973;2(1):75–86.
217. Chen S, Wang L. Thermodynamic oxidation and reduction potentials of photocatalytic semiconductors in aqueous solution. Chem Mater. 2012;24(18):3659–66.
218. Van de Walle CG, Neugebauer J. Universal alignment of hydrogen levels in semiconductors. Insulators and Solutions Nature. 2003;423(6940):626–8.
219. Swank RK. Surface properties of II–VI compounds. Phys Rev. 1967;153(3):844–9.
220. Das S, Krishna RM, Ma S, Mandal KC. Single phase polycrystalline Cu_2ZnSnS_4 grown by vertical gradient freeze technique. J Cryst Growth. 2013;381:148–52.
221. Guc M, Levcenko S, Bodnar IV, Izquierdo-Roca V, Fontane X, Volkova LV, Arushanov E, Pérez-Rodríguez A. Polarized raman scattering study of kesterite type Cu_2ZnSnS_4 single crystals. Sci Rep. 2016;6(1):19414.
222. Scragg JJ, Dale PJ, Peter LM. towards sustainable materials for solar energy conversion: preparation and photoelectrochemical characterization of Cu_2ZnSnS_4. Electrochem Commun. 2008;10(4):639–42.
223. Moulder JF, Stickle WF, Sobol PE, Bomben KD. Handbook of x-ray photoelectron spectroscopy; 1995.
224. Danilson M, Altosaar M, Kauk M, Katerski A, Krustok J, Raudoja J. XPS study of CZTSSe monograin powders. Thin Solid Films. 2011;519(21):7407–11.
225. Sundberg J, Lindblad R, Gorgoi M, Rensmo H, Jansson U, Lindblad A. Understanding the effects of sputter damage in W-S thin films by HAXPES. Appl Surf Sci. 2014;305:203–13.
226. Teeter G. X-ray and ultraviolet photoelectron spectroscopy measurements of Cu-Doped CdTe(111)-B: observation of temperature-reversible Cu_xTe precipitation and effect on ionization potential. J Appl Phys. 2007;102(3):34504.
227. Hoye RLZ, Schulz P, Schelhas LT, Holder AM, Stone KH, Perkins JD, Vigil-Fowler D, Siol S, Scanlon DO, Zakutayev A, Walsh A, Smith IC, Melot BC, Kurchin RC, Wang Y, Shi J, Marques FC, Berry JJ, Tumas W, Lany S, Stevanović V, Toney MF, Buonassisi T. Perovskite-inspired photovoltaic materials: toward best practices in materials characterization and calculations. Chem Mater. 2017;29(5):1964–88.
228. Vasquez RP, Grunthaner FJ. Chemical composition of the SiO_2/InSb interface as determined by x-ray photoelectron spectroscopy. J Appl Phys. 1981;52(5):3509–14.
229. Santoni A, Biccari F, Malerba C, Valentini M, Chierchia R, Mittiga A. Valence band offset at the Cds/Cu_2ZnSnS_4 interface probed by x-ray photoelectron spectroscopy. J Phys D Appl Phys. 2013;46(17):175101.
230. Lin L, Yu J, Cheng S, Lu P, Lai Y, Lin S, Zhao P. Band alignment at the In_2S_3/Cu_2ZnSnS_4 heterojunction interface investigated by x-ray photoemission spectroscopy. Appl Phys A. 2014;116(4):2173–7.
231. Ruan C-H, Huang C-C, Lin Y-J, He G-R, Chang H-C, Chen Y-H. Electrical properties of $Cu_xZn_ySnS_4$ films with different Cu/Zn ratios. Thin Solid Films. 2014;550:525–9.
232. Bär M, Schubert B-A, Marsen B, Wilks RG, Pookpanratana S, Blum M, Krause S, Unold T, Yang W, Weinhardt L, Heske C, Schock H-W. Cliff-like conduction band offset and KCN-induced recombination barrier enhancement at the CdS/Cu_2ZnSnS_4 thin-film solar cell heterojunction. Appl Phys Lett. 2011;99(22):222105.
233. Buckley AN, Woods R, Wouterlood HJ. An XPS investigation of the surface of natural sphalerites under flotation-related conditions. Int J Miner Process. 1989;26(1–2):29–49.
234. Kövér L, Kövér L, Moretti G, Kovács Z, Sanjinés R, Cserny I, Margaritondo G, Pálinkás J, Adachi H. High resolution photoemission and auger parameter studies of electronic structure of tin oxides. J Vac Sci Technol, A. 1995;13(3):1382.

235. Gopalakrishnan J, Murugesan T, Hegde MS, Rao CNR. Study of transition-metal monosulphides by photoelectron spectroscopy. J Phys C: Solid State Phys. 1979;12(23):5255–61.
236. Yu X-R, Liu F, Wang Z-Y, Chen Y. Auger parameters for sulfur-containing compounds using a mixed aluminum-silver excitation source. J Electron Spectros Relat Phenomena. 1990;50(2):159–66.
237. Stranick MA. SnO_2 by XPS. Surf Sci Spectra; 1993, 2(1):50.
238. Cumpson PJ, Zalm PC. Thickogram: a method for easy film thickness measurement in XPS. Surf Interface Anal. 2000;29(6):403–6.
239. Bosson CJ, Birch MT, Halliday DP, Knight KS, Tang CC, Kleppe AK, Hatton PD. Crystal structure and cation disorder in bulk Cu_2ZnSnS_4 using neutron diffraction and x-ray anomalous scattering. In: 2016 IEEE 43rd Photovoltaic Specialists Conference (PVSC); IEEE, 2016; p. 0405–10.
240. Shaw A, Wrench JS, Jin JD, Whittles TJ, Mitrovic IZ, Raja M, Dhanak VR, Chalker PR, Hall S. Atomic layer deposition of Nb-Doped ZnO for thin film transistors. Appl Phys Lett. 2016;109(22):222103.
241. Druska P, Steinike U, Šepelák V. Surface structure of mechanically activated and of mechanosynthesized zinc ferrite. J Solid State Chem. 1999;146(1):13–21.
242. Das S, Mandal K, Low-Cost Cu_2ZnSnS_4 thin films for large-area high-efficiency heterojunction solar cells. In: 2012 38th IEEE photovoltaic specialists conference; 2012; vol. 4, p. 2668–73.
243. Xie M, Zhuang D, Zhao M, Li B, Cao M, Song J. Fabrication of Cu_2ZnSnS_4 thin films using a ceramic quaternary target. Vacuum. 2014;101:146–50.
244. Gupta S, Whittles TJ, Batra Y, Satsangi V, Krishnamurthy S, Dhanak VR, Mehta BR. A low-cost, sulfurization free approach to control optical and electronic properties of Cu_2ZnSnS_4 via precursor variation. Sol Energy Mater Sol Cells. 2016;157:820–30.
245. Cao M, Shen Y. A mild solvothermal route to kesterite quaternary Cu_2ZnSnS_4 nanoparticles. J Cryst Growth. 2011;318(1):1117–20.
246. Wang W, Shen H, Jiang F, He X, Yue Z. Low-cost chemical fabrication of Cu_2ZnSnS_4 microparticles and film. J Mater Sci: Mater Electron. 2013;24(6):1813–7.
247. Tian Q, Huang L, Zhao W, Yang Y, Wang G, Pan D. Metal sulfide precursor aqueous solutions for fabrication of $Cu_2ZnSn(S,Se)_4$ thin film solar cells. Green Chem. 2015;17(2):1269–75.
248. Zou C, Zhang L, Lin D, Yang Y, Li Q, Xu X, Chen X, Huang S. Facile synthesis of Cu_2ZnSnS_4 nanocrystals. CrystEngComm. 2011;13(10):3310.
249. Jiang H, Dai P, Feng Z, Fan W, Zhan J. Phase selective synthesis of metastable orthorhombic Cu_2ZnSnS_4. J Mater Chem. 2012;22(15):7502.
250. Xu J, Yang X, Yang Q, Wong T, Lee C. Cu_2ZnSnS_4 hierarchical microspheres as an effective counter electrode material for quantum dot sensitized solar cells. J Phys Chem C. 2012;116(37):19718.
251. Yu X, Ren A, Wang F, Wang C, Zhang J, Wang W, Wu L, Li W, Zeng G, Feng L. Synthesis and characterization of CZTS thin films by sol-gel method without sulfurization. Int J Photoenergy. 2014;2014:1–6.
252. Liu W, Guo B, Mak C, Li A, Wu X, Zhang F. Facile synthesis of ultrafine Cu_2ZnSnS_4 nanocrystals by hydrothermal method for use in solar cells. Thin Solid Films. 2013;535(1):39–43.
253. Jiang F, Shen H, Wang W. Optical and electrical properties of Cu_2ZnSnS_4 film prepared by sulfurization method. J Electron Mater. 2012;41(8):2204–9.
254. Zhang X, Shi X, Ye W, Ma C, Wang C. Electrochemical deposition of quaternary Cu_2ZnSnS_4 thin films as potential solar cell material. Appl Phys A. 2009;94(2):381–6.
255. Moholkar AV, Shinde SS, Babar AR, Sim KU, Lee HK, Rajpure KY, Patil PS, Bhosale CH, Kim JH. Synthesis and characterization of Cu_2ZnSnS_4 thin films grown by PLD: solar cells. J Alloys Compd. 2011;509(27):7439–46.
256. Calderón C, Gordillo G, Becerra R, Bartolo-Pérez P. XPS analysis and characterization of thin films Cu_2ZnSnS_4 grown using a novel solution based route. Mater Sci Semicond Process. 2015;39:492–8.

257. Deepa KG, Chandrabose G, Jampana N. Tailoring the properties of Cu_2ZnSnS_4 thin films grown by ultrasonic spray pyrolysis. J Anal Appl Pyrolysis; 2016, 120:356–360.
258. Deepa KG, Jampana N. Development of an automated ultrasonic spray pyrolysis system and the growth of Cu_2ZnSnS_4 thin films. J Anal Appl Pyrolysis. 2016;117:141–6.
259. Tang D, Wang Q, Liu F, Zhao L, Han Z, Sun K, Lai Y, Li J, Liu Y. An alternative route towards low-cost Cu_2ZnSnS_4 thin film solar cells. Surf Coatings Technol. 2013;232:53–9.
260. Ji S, Shi T, Qiu X, Zhang J, Xu G, Chen C, Jiang Z, Ye C. A route to phase controllable $Cu_2ZnSn(S_{1-x}Se_x)_4$ nanocrystals with tunable energy bands. Sci Rep. 2013;3:2733.
261. Singh A, Geaney H, Laffir F, Ryan KM. Colloidal synthesis of wurtzite Cu_2ZnSnS_4 nanorods and their perpendicular assembly. J Am Chem Soc. 2012;134(6):2910–3.
262. Xie M, Zhuang D, Zhao M, Zhuang Z, Ouyang L, Li X, Song J. Preparation and characterization of Cu_2ZnSnS_4 thin films and solar cells fabricated from quaternary Cu-Zn-Sn-S target. Int J Photoenergy. 2013;2013:1–9.
263. Tiong VT, Bell J, Wang H. One-step synthesis of high quality kesterite Cu_2ZnSnS_4 nanocrystals—a hydrothermal approach. Beilstein J Nanotechnol. 2014;5:438–46.
264. Tiong VT, Zhang Y, Bell J, Wang H. Phase-selective hydrothermal synthesis of Cu_2ZnSnS_4 nanocrystals: the effect of the sulphur precursor. CrystEngComm. 2014;16(20):4306–13.
265. Deepu DR, Rajeshmon VG, Kartha CS, Vijayakumar KP. XPS depth profile study of sprayed CZTS thin films. In: AIP conference proceedings; 2014; vol. 1591, p. 1666–68.
266. Kim GY, Jo W, Lee KD, Choi HS, Kim JY, Shin HY, Nguyen TTT, Yoon S, Joo BS, Gu M, Han M. Optical and surface probe investigation of secondary phases in Cu_2ZnSnS_4 films grown by electrochemical deposition. Sol Energy Mater Sol Cells. 2015;139:10–8.
267. Harel S, Guillot-Deudon C, Choubrac L, Hamon J, Lafond A. Surface composition deviation of Cu_2ZnSnS_4 derivative powdered samples. Appl Surf Sci. 2014;303:107–10.
268. Zhong J, Xiang W, Cai Q, Liang X. Synthesis, characterization and optical properties of flower-like Cu_3BiS_3 nanorods. Mater Lett. 2012;70:63–6.
269. Chen D, Shen G, Tang K, Liu X, Qian Y, Zhou G. The synthesis of Cu_3BiS_3 nanorods via a simple ethanol-thermal route. J Cryst Growth. 2003;253(1–4):512–6.
270. Ley L, Pollak RA, McFeely FR, Kowalczyk SP, Shirley DA. Total valence-band densities of states of III–V and II–VI compounds from x-ray photoemission spectroscopy. Phys Rev B. 1974;9(2):600–621.
271. Kraut EA, Grant RW, Waldrop JR, Kowalczyk SP. Precise determination of the valence-band edge in x-ray photoemission spectra: application to measurement of semiconductor interface potentials. Phys Rev Lett. 1980;44(24):1620–3.
272. Végh J. The shirley background revised. J Electron Spectros Relat Phenomena. 2006;151(3):159–64.
273. Li X, Zhang Z, Henrich VE. Inelastic electron background function for ultraviolet photoelectron spectra. J Electron Spectros Relat Phenomena. 1993;63(3):253–65.
274. Whitcher TJ, Yeoh KH, Chua CL, Woon KL, Chanlek N, Nakajima H, Saisopa T, Songsiririthigul P. The effect of carbon contamination and argon ion sputtering on the work function of chlorinated indium tin oxide. Curr Appl Phys. 2014;14(3):472–5.
275. Helander MG, Greiner MT, Wang ZB, Tang WM, Lu ZH. Work function of fluorine doped tin oxide. J Vac Sci Technol A Vacuum Surfaces, Film. 2011;29(1).
276. Huang S, Luo W, Zou Z. Band positions and photoelectrochemical properties of Cu_2ZnSnS_4 thin films by the ultrasonic spray pyrolysis method. J Phys D Appl Phys. 2013;46(23):235108.
277. Olopade MA, Oyebola OO, Adeleke BS. Investigation of some materials as buffer layer in copper zinc tin sulphide (Cu_2ZnSnS_4) solar cells by SCAPS-1D. Adv Appl Sci Res. 2012;3(6):3396–400.
278. Kida T, Horita K, Suehiro S, Yuasa M, Quitain AT, Tanaka T, Fujita K, Ishiwata Y, Shimanoe K. Influence of processing conditions on the performance of Cu_2ZnSnS_4 nanocrystal solar cells. ChemistrySelect. 2016;1(1):86–93.
279. Bär M, Weinhardt L, Heske C. Soft X-ray and electron spectroscopy: A Unique "Tool Chest" to characterize the chemical and electronic properties of surfaces and interfaces. In: Advanced Characterization Techniques for Thin Film Solar Cells. Wiley-VCH Verlag GmbH and Co. KGaA: Weinheim, Germany, 2011; p. 387–409.

280. Ki W, Hillhouse HW. Earth-abundant element photovoltaics directly from soluble precursors with high yield using a non-toxic solvent. Adv Energy Mater. 2011;1(5):732–5.
281. Burton LA, Walsh A. Band alignment in SnS thin-film solar cells: possible origin of the low conversion efficiency. Appl Phys Lett. 2013;102(13):132111.
282. Katagiri H, Saitoh K, Washio T, Shinohara H, Kurumadani T, Miyajima S. Development of thin film solar cell based on Cu_2ZnSnS_4 thin films. Sol Energy Mater Sol Cells. 2001;65(1–4):141–8.
283. Terada N, Yoshimoto S, Chochi K, Fukuyama T, Mitsunaga M, Tampo H, Shibata H, Matsubara K, Niki S, Sakai N, Katou T, Sugimoto H. Characterization of electronic structure of $Cu_2ZnSn(S_xSe_{1-x})_4$ absorber layer and $CdS/Cu_2ZnSn(S_xSe_{1-x})_4$Interfaces by in-Situ photoemission and inverse photoemission spectroscopies. Thin Solid Films. 2015;582:166–70.
284. Fairbrother A, Fontané X, Izquierdo-Roca V, Espíndola-Rodríguez M, López-Marino S, Placidi M, Calvo-Barrio L, Pérez-Rodríguez A, Saucedo E. On the formation mechanisms of Zn-Rich Cu_2ZnSnS_4 films prepared by sulfurization of metallic stacks. Sol Energy Mater Sol Cells. 2013;112:97–105.
285. Malerba C, Azanza Ricardo CL, Valentini M, Biccari F, Müller M, Rebuffi L, Esposito E, Mangiapane P, Scardi P, Mittiga A. Stoichiometry effect on Cu_2ZnSnS_4 thin films morphological and optical properties. J Renew Sustain Energy. 2014;6(1):11404.
286. Fan D, Zhang R, Zhu Y, Peng H, Zhang J. Structural development and dynamic process in sulfurizing precursors to prepare Cu_2ZnSnS_4 absorber layer. J Alloys Compd. 2014;583:566–73.
287. Malerba C, Biccari F, Azanza Ricardo CL, Valentini M, Chierchia R, Müller M, Santoni A, Esposito E, Mangiapane P, Scardi P, Mittiga A. CZTS stoichiometry effects on the band gap energy. J Alloys Compd. 2014;582:528–34.
288. Rana TR, Shinde NM, Kim J. Novel chemical route for chemical bath deposition of Cu_2ZnSnS_4 (CZTS) thin films with stacked precursor thin films. Mater Lett. 2016;162:40–3.
289. Sarswat PK, Free ML. Demonstration of a sol-gel synthesized bifacial CZTS photoelectrochemical cell. Phys Status Solidi. 2011;208(12):2861–4.
290. Shinde NM, Deshmukh PR, Patil SV, Lokhande CD. Aqueous chemical growth of Cu_2ZnSnS_4 (CZTS) thin films: air annealing and photoelectrochemical properties. Mater Res Bull. 2013;48(5):1760–6.
291. Shinde NM, Lokhande CD, Kim JH, Moon JH. Low cost and large area novel chemical synthesis of Cu_2ZnSnS_4 (CZTS) thin films. J Photochem Photobiol A Chem. 2012;235:14–20.
292. Tanaka T, Nagatomo T, Kawasaki D, Nishio M, Guo Q, Wakahara A, Yoshida A, Ogawa H. Preparation of Cu_2ZnSnS_4 thin films by hybrid sputtering. J Phys Chem Solids. 2005;66(11):1978–81.
293. Woo K, Kim Y, Moon J. A Non-toxic, solution-processed, earth abundant absorbing layer for thin-film solar cells. Energy Environ Sci. 2012;5(1):5340.
294. Walsh A, Chen S, Wei S-H, Gong X-G. Kesterite thin-film solar cells: advances in materials modelling of Cu_2ZnSnS_4. Adv Energy Mater. 2012;2(4):400–9.
295. Savory CN, Ganose AM, Travis W, Atri RS, Palgrave RG, Scanlon DO. An assessment of silver copper sulfides for photovoltaic applications: theoretical and experimental insights. J Mater Chem A. 2016;4(32):12648–57.
296. Yuan ZK, Chen S, Xiang H, Gong XG, Walsh A, Park JS, Repins I, Wei SH. Engineering solar cell absorbers by exploring the band alignment and defect disparity: the case of Cu- and Ag-based kesterite compounds. Adv Funct Mater. 2015;25(43):6733–43.
297. Minemoto T, Matsui T, Takakura H, Hamakawa Y, Negami T, Hashimoto Y, Uenoyama T, Kitagawa M. Theoretical analysis of the effect of conduction band offset of window/CIS layers on performance of CIS solar cells using device simulation. Sol Energy Mater Sol Cells. 2001;67(1–4):83–8.
298. Sinsermsuksakul P, Hartman K, Bok Kim S, Heo J, Sun L, Hejin Park H, Chakraborty R, Buonassisi T, Gordon RG. Enhancing the efficiency of SnS solar cells via band-offset engineering with a zinc oxysulfide buffer layer. Appl Phys Lett. 2013;102(5):53901.
299. Bao W, Ichimura M. Band offsets at the ZnO/Cu_2ZnSnS_4 interface based on the first principles calculation. Jpn J Appl Phys. 2013;52(6R):61203.

300. Sugiyama M, Shimizu T, Kawade D, Ramya K, Ramakrishna Reddy KT. Experimental determination of vacuum-level band alignments of SnS-based solar cells by photoelectron yield spectroscopy. J Appl Phys. 2014;115(8):83508.
301. Haight R, Barkhouse A, Gunawan O, Shin B, Copel M, Hopstaken M, Mitzi DB. Band alignment at the $Cu_2ZnSn(S_xSe_{1-x})_4$/CdS Interface. Appl Phys Lett. 2011;98(25):253502.
302. Nagoya A, Asahi R, Kresse G. First-principles study of Cu_2ZnSnS_4 and the related band offsets for photovoltaic applications. J Phys: Condens Matter. 2011;23(40):404203.
303. Barkhouse DAR, Haight R, Sakai N, Hiroi H, Sugimoto H, Mitzi DB. Cd-free buffer layer materials on $Cu_2ZnSn(S_xSe_{1-x})_4$: band alignments with ZnO, ZnS, and In_2S_3. Appl Phys Lett. 2012;100(19):193904.
304. Calixto-Rodriguez M, Tiburcio-Silver A, Ortiz A, Sanchez-Juarez A. Optoelectronical properties of indium sulfide thin films prepared by spray pyrolysis for photovoltaic applications. Thin Solid Films. 2005;480–481:133–7.
305. Mughal MA, Engelken R, Sharma R. Progress in indium (III) sulfide (In_2S_3) buffer layer deposition techniques for CIS, CIGS, and CdTe-Based thin film solar cells. Sol Energy. 2015;120:131–46.
306. Chen S, Walsh A, Yang J-H, Gong XG, Sun L, Yang P-X, Chu J-H, Wei S-H. Compositional dependence of structural and electronic properties of $Cu_2ZnSn(S,Se)_4$ alloys for thin film solar cells. Phys Rev B. 2011;83(12):125201.
307. Hinuma Y, Oba F, Kumagai Y, Tanaka I. Ionization potentials of (112) and $(11\bar{2})$ facet surfaces of $CuInSe_2$ and $CuGaSe_2$. Phys Rev B. 2012;86(24):245433.
308. Li J, Du Q, Liu W, Jiang G, Feng X, Zhang W, Zhu J, Zhu C. The band offset at CdS/Cu_2ZnSnS_4 heterojunction interface. Electron Mater Lett. 2012;8(4):365–7.
309. Bao W, Ichimura M. Prediction of the band offsets at the CdS/Cu_2ZnSnS_4 interface based on the first-principles calculation. Jpn J Appl Phys. 2012;51:10NC31.
310. Bao W, Ichimura M. First-principles study on influences of crystal structure and orientation on band offsets at the CdS/Cu_2ZnSnS_4 interface. Int J Photoenergy. 2012;2012(1):1–5.
311. Yeh JJ, Lindau I. Atomic subshell photoionization cross sections and asymmetry parameters: $1 \leqslant Z \leqslant 103$. At Data Nucl Data Tables. 1985;32(1):1–155.
312. Ichimura M, Nakashima Y. Analysis of atomic and electronic structures of Cu_2ZnSnS_4 based on first-principle calculation. Jpn J Appl Phys. 2009;48(9):90202.
313. Bär M, Schubert B-A, Marsen B, Schorr S, Wilks RG, Weinhardt L, Pookpanratana S, Blum M, Krause S, Zhang Y, Yang W, Unold T, Heske C, Schock H-W. Electronic structure of Cu_2ZnSnS_4 probed by soft x-ray emission and absorption spectroscopy. Phys Rev B. 2011;84(3):35308.
314. Wei SH, Zunger A. Calculated natural band offsets of All II–VI and III–V semiconductors: chemical trends and the role of cation D orbitals. Appl Phys Lett. 1998;72(16):2011–3.
315. Hofmann, W. Ergebnisse der strukturbestimmung komplexer sulfide. Zeitschrift für Krist.—Cryst. Mater. 1935;92(1–6).
316. Hazen RM, Finger LW. The crystal structures and compressibilities of layer minerals at high pressure. I. SnS_2 berndtite. Am Mineral. 1978;63:289–92.
317. Kniep R, Mootz D, Severin U, Wunderlich H. Structure of tin(II) tin(IV) trisulphide, a redetermination. Acta Crystallogr Sect B Struct Crystallogr Cryst Chem. 1982;38(7):2022–3.
318. Chen X, Wada H, Sato A, Mieno M. Synthesis, electrical conductivity, and crystal structure of $Cu_4Sn_7S_{16}$ and structure refinement of Cu_2SnS_3. J Solid State Chem. 1998;139(1):144–51.
319. Wang S, Huang Q, Wen X, Li X, Yang S. Thermal oxidation of Cu_2S nanowires: a template method for the fabrication of mesoscopic Cu_xO ($X=1,2$) wires. Phys Chem Chem Phys. 2002;4(14):3425–9.
320. Raadik T, Grossberg M, Raudoja J, Traksmaa R, Krustok J. Temperature-dependent photoreflectance of SnS crystals. J Phys Chem Solids. 2013;74(12):1683–5.
321. Burton LA, Whittles TJ, Hesp D, Linhart WM, Skelton JM, Hou B, Webster RF, O'Dowd G, Reece C, Cherns D, Fermin DJ, Veal TD, Dhanak VR, Walsh A. Electronic and optical properties of single crystal SnS_2: an earth-abundant disulfide photocatalyst. J Mater Chem A. 2016;4(4):1312–8.

322. Price LS, Parkin IP, Hardy AME, Clark RJH, Hibbert TG, Molloy KC. Atmospheric pressure chemical vapor deposition of tin sulfides (SnS, Sn_2S_3, and SnS_2) on glass. Chem Mater. 1999;11(7):1792–9.
323. Dimitrievska M, Xie H, Fairbrother A, Fontané X, Gurieva G, Saucedo E, Pérez-Rodríguez A, Schorr S, Izquierdo-Roca V. Multiwavelength excitation raman scattering of $Cu_2ZnSn(S_xSe_{1-x})_4$ ($0 \leq X \leq 1$) polycrystalline thin films: vibrational properties of sulfoselenide solid solutions. Appl Phys Lett. 2014;105(3):31913.
324. Berg DM, Arasimowicz M, Djemour R, Gütay L, Siebentritt S, Schorr S, Fontané X, Izquierdo-Roca V, Pérez-Rodriguez A, Dale PJ. Discrimination and detection limits of secondary phases in Cu_2ZnSnS_4 using x-ray diffraction and raman spectroscopy. Thin Solid Films. 2014;569(C):113–123.
325. Wagner CD. Chemical shifts of auger lines, and the auger parameter. Faraday Discuss. Chem Soc. 1975;60:291.
326. Asami K. A precisely consistent energy calibration method for x-ray photoelectron spectroscopy. J Electron Spectros Relat Phenomena. 1976;9(5):469–78.
327. Battistoni C, Mattogno G, Paparazzo E, Naldini L. An XPS and auger study of some polynuclear copper compounds. Inorganica Chim Acta. 1985;102(1):1–3.
328. Fuggle JC, Källne E, Watson LM, Fabian DJ. Electronic structure of aluminum and aluminum-noble-metal alloys studied by soft-x-ray and x-ray photoelectron spectroscopies. Phys Rev B. 1977;16(2):750–61.
329. Haber J, Machej T, Ungier L, Ziółkowski J. ESCA studies of copper oxides and copper molybdates. J Solid State Chem. 1978;25(3):207–18.
330. Hussain Z, Salim MA, Khan MA, Khawaja EE. X-ray photoelectron and auger spectroscopy study of copper-sodium-germanate glasses. J Non Cryst Solids. 1989;110(1):44–52.
331. Kowalczyk SP, Pollak RA, McFeely FR, Ley L, Shirley DA. $L_{2,3}M_{45}M_{45}$ auger spectra of metallic copper and zinc: theory and experiment. Phys Rev B. 1973;8(6):2387–91.
332. Moretti G, Fierro G, Lo Jacono M, Porta P. Characterization of CuO-ZnO catalysts by x-ray photoelectron spectroscopy: precursors, calcined and reduced samples. Surf Interface Anal. 1989;14(6–7):325–36.
333. Powell CJ. Summary abstract: accurate determination of the energies of Auger electrons and photoelectrons from nickel, copper, and gold. J Vac Sci Technol. 1982;20(3):625.
334. Powell CJ. Recommended Auger-electron kinetic energies for 42 elemental solids. J Electron Spectros Relat Phenomena. 2010;182(1–2):11–8.
335. Schön G. High Resolution Auger electron spectroscopy of metallic copper. J Electron Spectros Relat Phenomena. 1972;1(4):377–87.
336. Schön, G. ESCA Studies of Cu, Cu_2O and CuO. Surf Sci. 1973;35(C): 96–108.
337. Powell CJ. Recommended Auger parameters for 42 elemental solids. J Electron Spectros Relat Phenomena. 2012;185(1–2):1–3.
338. Biesinger MC, Payne BP, Grosvenor AP, Lau LWM, Gerson AR, Smart RSC. Resolving surface chemical states in XPS analysis of first row transition metals, oxides and hydroxides: Cr, Mn, Fe. Co and Ni Appl Surf Sci. 2011;257(7):2717–30.
339. Darrah Thomas T, Weightman P. Valence electronic structure of AuZn and AuMg alloys derived from a new way of analyzing auger-parameter shifts. Phys Rev B. 1986;33(8):5406–13.
340. Cruz M, Morales J, Espinos JP, Sanz J. XRD, XPS and Sn NMR study of tin sulfides obtained by using chemical vapor transport methods. J Solid State Chem. 2003;175(2):359–65.
341. Pessa M, Vuoristo A, Vulli M, Aksela S, Väyrynen J, Rantala T, Aksela H. Solid-state effects in $M_{4,5}N_{4,5}N_{4,5}$ Auger spectra of elements from ^{49}In to ^{52}Te. Phys Rev. 1979;20(8):3115–23.
342. Stranick M.A. SnO by XPS. Surf Sci Spectra. 1993;2(1):45.
343. Venezia AM, Cavell RG. Molecular MNN Auger spectra of gaseous tetramethyltin and hexamethylditin. J Electron Spectros Relat Phenomena. 1985;36(3):281–96.
344. Kalanur SS, Seo H. Tuning plasmonic properties of cus thin films via valence band filling. RSC Adv. 2017;7(18):11118–22.
345. Marshall R, Mitra SS. Optical properties of cuprous sulfide. J Appl Phys. 1965;36(12):3882–3.

346. Klein A. Energy band alignment in chalcogenide thin film solar cells from photoelectron spectroscopy. J Phys: Condens Matter. 2015;27(13):134201.
347. Chen S, Gong XG, Walsh A, Wei S-H. Electronic structure and stability of quaternary chalcogenide semiconductors derived from cation cross-substitution of II–VI and I–III–VI_2 compounds. Phys Rev B. 2009;79(16):165211.
348. Tiong VT, Zhang Y, Bell J, Wang H. Carbon concentration dependent grain growth of Cu_2ZnSnS_4 thin films. RSC Adv. 2015;5(26):20178–85.
349. Sarswat PK, Free ML. A study of energy band gap versus temperature for Cu_2ZnSnS_4 thin films. Phys B Condens Matter. 2012;407(1):108–11.
350. Ettema ARHF, Haas C. An x-ray photoemission spectroscopy study of interlayer charge transfer in some misfit layer compounds. J Phys: Condens Matter. 1993;5(23):3817–26.
351. Barr TL. ESCA studies of naturally passivated metal foils. J Vac Sci Technol. 1977;14(1):660.
352. Hegde RI, Sainkar SR, Badrinarayanan S, Sinha APB. A study of dilute tin alloys by x-ray photoelectron spectroscopy. J Electron Spectros Relat Phenomena. 1981;24(1):19–25.
353. Dias S, Krupanidhi SB. Study of band offsets at the Cu_2SnS_3/In_2O_3: sn interface using x-ray photoelectron spectroscopy. Mater Res Express. 2015;2(6):65901.
354. Ghorpade UV, Suryawanshi MP, Shin SW, Kim I, Ahn SK, Yun JH, Jeong C, Kolekar SS, Kim JH. Colloidal wurtzite Cu_2SnS_3 (CTS) nanocrystals and their applications in solar cells. Chem Mater. 2016;28(10):3308–17.
355. Li J, Huang J, Zhang Y, Wang Y, Xue C, Jiang G, Liu W, Zhu C. Solution-processed Cu_2SnS_3 thin film solar cells. RSC Adv. 2016;6(63):58786–95.
356. Tiwari D, Chaudhuri TK, Shripathi T, Deshpande U. Synthesis of earth-abundant Cu_2SnS_3 powder using solid state reaction. J Phys Chem Solids. 2014;75(3):410–5.
357. Tiwari D, Chaudhuri TK, Shripathi T, Deshpande U, Sathe VG. Microwave-assisted rapid synthesis of tetragonal Cu_2SnS_3 nanoparticles for solar photovoltaics. Appl Phys A Mater Sci Process. 2014;117(3):1139–46.

Chapter 7
Conclusions and Recommendations for the Future

Scientists will eventually stop flailing around with solar power and focus their efforts on harnessing the only truly unlimited source of energy on the planet: stupidity. I predict that in the future, scientists will learn how to convert stupidity into clean fuel. Energy companies will place huge hamster wheels outside of convenience stores and offer free lottery tickets to people who spend five minutes running inside the wheels, which will be connected to power generators.

Scott Adams, 'Dilbert' Cartoonist, 2008

In this chapter, the findings throughout this thesis are collated, drawing comparisons and contrasts between not only the materials, but the methods and analysis techniques that were used. Finally, reference is drawn back to the contents of Chap. 1, and suggestions are made for the progression of this work.

7.1 Chapter Summaries

Throughout the previous chapters, the electronic structures of the materials: CAS, CBS, SnS, and CZTS have been studied with regards to their suitability for use as PV absorber materials. Unusual crystal structures and bonding regimes have been

© Springer International Publishing AG, part of Springer Nature 2018

T. J. Whittles, *Electronic Characterisation of Earth-Abundant Sulphides for Solar Photovoltaics*, Springer Theses, https://doi.org/10.1007/978-3-319-91665-1_7

shown to give rise to properties that may not have been expected for these materials, but have been clarified through the use of photoemission spectroscopy, corroborated by electronic-structure calculations from DFT.

7.1.1 Chapter 3

In Chap. 3, the state of the VB and bandgaps of CAS were compared experimentally and theoretically. In the VB, it was found that the strong Cu *d* states obscured Sb *s* and *p* states at the top of the VB, which is different to the case of CIGS. These extra features cause the low IP, resulting in a band misalignment with CdS. This has been part of the cause of the poor efficiencies so far, where CdS is used because of the analogies drawn between CAS and CIGS, which have now been shown to be unfounded.

A core-level XPS analysis allowed the determination of the two sulphur environments in the crystal structure and has shown the detrimental effect that contamination can have on the material. Previously in the literature, it is believed that misinterpretations have resulted in contaminant species being overlooked. A thorough fitting procedure was shown to allow the proper interpretation of these spectra, which can reveal these extra species.

A thermal treatment applied to the thin-film of CAS, which produces better quality films, had little effect on the electronic structure, but was found to inhibit the formation of oxidation at the surface of the film.

7.1.2 Chapter 4

In Chap. 4, the measured IP of CBS shows that CdS would be unsuitable as a window layer. A comparison between the VB spectrum from XPS and the theoretically calculated DoS gives an explanation for this low IP. Beneath the majority Cu *d* states in the VB, bismuth contributes *p* and *s* states as well, raising the level of the VBM. These arise from the difference in orbital energies and occupancies of CBS when compared to other absorber materials.

The application of a rigorous fitting model for the core-level XPS spectra has shown how the complexities of the spectra for CBS can be overcome, and also has revealed how the previous literature suffered by these complexities. XPS has therefore shown the detrimental effect that contamination can have on CBS films.

7.1.3 Chapter 5

In Chap. 5, the electronic structures of SnS, SnS_2, and Sn_2S_3, derived from XPS measurements and calculations from DFT, were compared, showing agreement between the nature of the DoS and also the positions of the band levels. These show how the other phases of tin sulphide can act detrimentally to solar cell performance, if they were to form in SnS. SnS_2 was shown to have favourable band positions for its use as a possible replacement to CdS, but the measured IP of SnS shows that part of the reason why SnS cells have achieved poor efficiencies is because of a band misalignment with CdS, due to the low IP. Compared to SnS_2, the different oxidation state of tin in SnS is the reason why the IP is low and extra states are formed by the lone-pair tin electrons, which raise the level of the VBM.

The effects of the surface cleaning and contaminants of the tin sulphides showed different behaviours, demonstrating how it is important to distinguish between them, because of their possible effects on SnS cells. This differentiation can be achieved with XPS, which can be used in future characterisations of these materials, and clarifies some previous confusion in the literature.

7.1.4 Chapter 6

In Chap. 6, measurements were made on single crystal CZTS and samples of the possible binary secondary phases, so that they can act as standards, and the efficacy of such characterisations were assessed, with the focus on identifying secondary phases.

Differences in the XPS core-level and Auger spectra between the binary phases and CZTS show that, with complete analysis, this technique can be used to help identify the secondary phases. The complexities of the XPS spectra from CZTS were also addressed, with reference to the current literature and the problems associated therewith. Ultimately, it was shown how XPS can be used with CZTS to determine: oxidation states, the presence and effects of contaminants, surface stoichiometry estimates, and the possible presence of secondary phases. However, the necessity for rigorous analysis and critical interpretation was also highlighted.

Complications in the XRD and RS spectra of CZTS and the binary phases explain the care required when using these techniques to determine phase-purity and identify secondary phases. The XPS analyses of the secondary phases showed that the combination of all of these techniques makes the matter of secondary phase identification in CZTS more accessible. In all cases, the need for high-quality data and analysis is paramount, with some extra aspects also required, which are not always undertaken in standard analyses: measurement of the Auger parameter, and application of a thorough fitting model.

The cleaning of the CZTS crystal for XPS measurements was shown not to be simple, with there being much scope for unintended damage to the material: another factor to be considered in future analyses. However, the presence of oxidation and contaminants were shown to be assessable using XPS, and it was also shown how this complex material can undergo changes as a result of the vacuum environment, high temperatures, or sputtering.

The DoS for CZTS was compared experimentally and theoretically to determine the bonding nature of this material, which is different from the related material CIGS, due to the presence of states from the non-copper cations, producing differences in bonding. This was then related to the measured band levels, showing how CdS is not a suitable material for use with CZTS, because of the low IP. The band levels of the binary phases were also shown to act detrimentally to a CZTS cell.

7.2 Analysis Summaries

Throughout the data presented in this thesis, it is clear that the use of PES, with a focus on XPS, is powerfully beneficial when applied to PV absorber materials. It should not however, be concluded that the use of these techniques can provide a replacement to the more standard characterisation techniques that are used. Instead, PES can provide complementary information, and also, through the analysis of the DoS and band levels shown throughout, give a better understanding to the properties of these materials.

7.2.1 XPS Core-Level Summaries

It is clear from the literature reviews presented in each individual chapter that XPS-related problems have been encountered in the past with each of these materials, which have then led to misinterpretations, possibly causing detriment to the materials. Therefore, these problems cannot be ignored, but must be acknowledged and analysed in order to drive forward the versatility of this technique.

In Chap. 3, the main problem associated with the CAS spectra was the overlap of the Sb 3*d* and O 1*s* regions. This was shown to be accounted for by the fitting procedure employed in that chapter, which allowed the determination of all of the antimony and oxygen species that could be present within a sample of CAS. The two fitted S 2*p* doublets, attributed to each of the sulphur sites in the crystal structure of CAS, was unexpected, but the reasoning given shows how confidence in this is founded and should be acknowledged for future measurements of CAS.

In Chap. 4, the main problem associated with the CBS spectra was the overlap of the Bi 4*f* and S 2*p* regions. This proved more difficult to fit than the overlapping case for CAS, because the overlapping regions both originated from the material in question and therefore could not be removed. However, reasonable success was achieved by using the fitting procedure described in that chapter. The XPS literature of CBS was then shown to suffer from a lack of fitting in this region and it is stressed how important this is, so that all present species can be identified and accounted for.

In Chap. 5, the oxidation state of the tin was identifiable for the different tin sulphides, through a large shift in BE. However, including the possibility of oxidation of the tin increases the complexity of the spectra because the oxide peak of the Sn^{2+} was shown to overlap with the sulphide peak of the Sn^{4+}. All of these species are possible within a technical sample due to the reduction or oxidation of the tin during either cleaning or growth. Therefore, regarding the tin sulphides, it is necessary to analyse the spectra with these considerations in mind, utilising the complementary anion peaks for identification, where necessary.

In Chap. 6, the XPS spectra of CZTS were shown to be very complex, even for the clean single crystal. The overlap of the Sn 3*d* and Zn LMM regions is unavoidable, given that monochromated XPS is generally required to provide a resolution which is fully analysable. The XPS spectra of the binary phases were sometimes shown to result in small BE shifts between them and CZTS, but with the inclusion of the calculation of the Auger parameter, this became achievable, given a properly fitted set of spectra. Also, the 4–component nature of the material means that the intensities of the peaks are suppressed, which is further compounded by the inclusion of any contamination. This means that long acquisition times are necessary in order to generate spectra that are of high enough quality to analyse, especially if the Auger features also need to be collected.

Throughout the results chapters, the S 2*p* spectra were sometimes fitted with a doublet that was assigned to sulphur-containing contamination (S-C). Such a species is rarely found in the literature, but then again, considering the literature reviews from the individual chapters, fitting the S 2*p* region is rare. Still, the fitting of this species, given its small size, may appear somewhat tenuous. Its inclusion is believed to be accurate because, without exception, it reduces after surface cleaning, suggesting that it is present only at the surface.

Despite these complexities, the use of properly analysed XPS spectra in the characterisation of PV absorbers is encouraged. It can be used to determine the oxidation states, help identify phase-pure material, identify and characterise contaminants, and also determine the effects of various modifications on the electronic structure. In fact, XPS is the sole technique that can effectively acknowledge specific features, such as certain oxides and other surface-specific effects, which may be missed by other characterisation techniques, because they are not sensitive to such things.

Previously in the literature, several recurrent features of analyses are reported that have been shown to limit the effectiveness of an XPS characterisation. These include: presenting data whose quality is too low for effective analysis; a reliance upon literature agreement of BE, when no analysis has been performed; a lack of fitting procedures, or unfounded ones, which lead to the exclusion of certain species; or the conclusion of BE agreement between two materials that are not necessarily related. It is therefore clear that these should be avoided in future characterisations, and in order to sustain the true capabilities of XPS, standard procedures should be followed, high-quality data should be presented, and a complete, critical analysis of the spectra should be performed with the explicit reporting of all relevant details.

7.2.2 *Band Alignments, DoS, and Bonding Analysis Summaries*

Although XPS was used in the earlier literature for studying the VB structures of materials [1–7], it is rare to find its use in contemporary reports, especially for novel, functional materials such as the new generation of PV absorbers. It has been shown in this thesis, how XPS can be coupled with advanced theoretical calculations of such materials in order to elucidate the DoS in the valence and conduction bands.

For such measurements, it is essential that the material be as clean as possible in order that a representative spectrum is measured, and also it is recommended that monochromated XPS is used in order to improve the resolution, and to remove the bremsstrahlung background and satellite features that can mar the spectrum [8]. UPS is sometimes thought to be better suited for measurements of the VB of materials, but following the results presented in Chap. 6, it seems that monochromated XPS is the better choice for VB studies, which is in agreement with earlier literature [4, 6]. XPS has several advantages over UPS for theoretically compared VB spectra: the chemical states and cleanliness of a sample are able to be checked using the core-levels; XPS spectra are not as susceptible to final-state distortion effects as UPS spectra [4]; and surface cleanliness is less important for XPS than for UPS [6].

In Chap. 6, it was discussed how the leading edge extrapolation method was unsuitable for determining the VBM of CZTS, due to the nature of the DoS at the VBM. Instead, a method which involves the fitting of the DoS curves to the experimental spectra was used. Chambers et al. [9] studied the application of these methods, based on a value termed the effective resolution (R_{eff}), which is the ratio of the energy over which the leading edge rises, to the resolution of the spectrometer used. They found that for $R_{eff} > 1$, the linear extrapolation method works because the onset of states is shallow enough that the experimental resolution will not affect it, but for $R_{eff} < 1$, the experimental resolution broadens the slope of the true edge, and as such, the linear extrapolation method will underestimate the onset of the VBM, as was the case in Chap. 6. Where possible, the linear extrapolation method is preferred, as it is much easier to implement; however, this criteria should be kept in mind when performing this determination. Table 7.1 shows estimates of the R_{eff} criterion for the materials studied in this thesis, showing how the linear extrapolation method was applicable for all of the materials, excepting CZTS.

For the band alignments shown throughout this thesis, the IP was always measured using XPS, but the EA value was calculated either from subtracting the bandgap value[1] from the IP, or from IPES spectra. The direct measurement of EA from IPES is the preferred method; however, it is not always possible or accurate due to the weaknesses associated with the IPES measurements as implemented here.[2] For this reason, the use of EA values from IPES measurements are recommended only when they are corroborated by other methods of measuring the bandgap.

---

[1]Either the calculated or literature value.

[2]Discussed in Sect. 6.7.3.

Table 7.1 Effective resolution (R_{eff}) estimates, according to Chambers et al. [9] for the materials studied in this thesis

Sample	R_{eff}
CAS	1.24
CBS	1.17
SnS	1.12
SnS_2	1.17
Sn_2S_3	1.39
CZTS	0.73
Cu_2S	1.02
CuS	1.63
ZnS	2.44

$R_{eff} < 1$ indicates that the linear extrapolation method for the determination of the VBM will not be accurate

7.3 Materials Prospects

Whilst it has been shown how the techniques used throughout this thesis are applicable for use during future characterisations of PV absorber materials, it is also important to discuss the materials themselves, in order to assess their potential as solar absorber materials. Comparisons between them will be given in Sect. 7.3.5.

7.3.1 $CuSbS_2$

As a relatively new absorber material, the poor efficiencies demonstrated so far are not completely disheartening. With regards to the measurements and results presented in Chap. 3, it would appear that one of the main reasons for these poor efficiencies is the continued reliance in drawing analogies from CIGS. Indeed, the bonding regime in CAS revealed differences that lead to the difference in band positions. Still, further research of this material is required, especially with regards to designing and developing different cell architectures to determine the optimal partner materials for CAS. Also, during the growth of CAS, more reproducible methods are required, and the use of XPS, as implemented in Chap. 3, can aid with this by identifying phase-pure material and contaminants.

7.3.2 Cu_3BiS_3

Of all the materials studied here, CBS is the youngest in terms of technological and research advancement. Theoretical studies and experimentally grown samples have demonstrated attractive PV-relevant properties; however, there is confusion as to the nature of the bandgap [10–13], and a lack of literature studies for the rigorous characterisation of the material. Nevertheless, the measurements presented in Chap. 4 show promise for future characterisations of this material, and the insights into the band positions from the bonding nature have revealed properties which can aid in cell designs for the future.

7.3.3 Tin Sulphides

SnS is arguably the most mature of the materials studied here, in terms of PV absorber applications. As a binary material, it has the advantage of simplicity over the other materials, whilst still demonstrating attractive properties for PV. Nevertheless, it was feared that SnS suffers from the impacts of secondary phases and improper cell design [14, 15]. XPS showed not only that secondary phases can hinder cell performance, but also that they can be identified using this technique, as well as acknowledging that with a different window layer, SnS cells may be able to achieve higher efficiencies. From the properties of SnS_2 that were also measured in Chap. 5, this material was shown to have potential as a window layer for PV cells, or for use in photocatalysis.

7.3.4 CZTS

CZTS is the most advanced PV absorber that was studied here, in terms of the highest-achieved efficiency [16], but is also probably the most complex. Of the many identified problems that affect the performance of CZTS solar cells [17–23], the impact of secondary phases have been alleviated somewhat by the results presented in Chap. 6. In order to reduce the impact of secondary phases, they must first be identified. Measurements of the single crystal CZTS, which are unique in terms of XPS analyses, and the systematic approach to secondary phase identification applied in Chap. 6, set standard measurements that can be compared to in future analyses, leading to the more accurate identification of these phases, but also showing the impact of them. The combined application of XPS, RS, and XRD can provide a commanding analysis, with each technique bringing different advantages. Then, after identification of these secondary phases, research can be conducted into preventing their formation, or removing them. Post-growth surface etches have been shown to be effective at removing some types of secondary phase that segregate at the surface [24], but if some were to form in the bulk of the material [25], then the growth methods would need revising.

7.3.5 Comparison

Individually, the materials studied in this thesis show promise as solar absorbers, but similar problems were common between them, which mostly involve the band misalignment with CdS, due to the low IP, shown independently to be a result of the bonding regime. Here, a direct comparison is made between these materials and CIGS.

Figure 7.1 shows the natural band alignments of the materials studied throughout this thesis, along with CIGS, with the contributions to the DoS labelled at their approximate positions. The band alignment between CIGS and CdS has been shown to be beneficial for solar cells [26–28], and it is common to find this architecture present in the high-efficiency devices [29, 30]; even modules [31]. The materials studied throughout this thesis all have band levels that are higher than CIGS, which led to the individual conclusions that CdS is a poor choice of window layer for these types of solar cell, due to the cliff-like CBO, which can generate a recombination centre or lead to voltage losses.[3] In terms of absolute values, the offsets between the materials shown in Fig. 7.1 appear small. However, when considering solar cell structures, where the bandgap is of the order 1.5 eV, an offset of 0.5 eV can impact the performance, leading to a voltage deficit, limited to two thirds of its maximum. In each individual chapter, explanations were given as to why the IP of each material was 'unusually low', in terms of the bonding regimes, apparent from the DoS presented there.

Figure 7.2 shows the schematic CE diagram for the materials studied throughout this thesis, along with CIGS, highlighting the main contributions to the VBM and CBM, and also shows the positions of these for each material, relative to the vacuum level. For binary semiconductors involving valence cation *d* states, it has been shown using a simple tight-binding model [32, 33] that, given a common-cation, the larger the anion *p*/cation *d* orbital separation, the higher the resulting VBM. CIGS can be compared to CAS, CBS, and CZTS in terms of these materials sharing the common-cation copper, but with CIGS containing selenium anions and the others containing sulphur anions. From the orbital energies of Se 4*p* and S 3*p*, compared with Cu 3*d* in Fig. 7.2, this chemical trend would suggest that CIGS should have a higher VBM than the others. As this is not the case, there must be other factors that result in the band positions observed here, and these arise from the other cations, adding another layer of complexity [34]. As such, these chemical trends cannot be simplistically applied here, as they were for binary semiconductors [33, 35].

In CIGS [36, 37], the *s* orbitals of indium and gallium are unoccupied, and close in energy to the full selenium *p* orbitals. This results in bonding states deep in the VB and antibonding states at the CBM. Also, the unoccupied *p* orbitals of indium and gallium are much higher than the selenium *p* orbitals. The interaction of these result in bonding states low in the VB and high in the CB. Importantly, the only contribution to the VBM is from Cu *d*/Se *p* interactions.

[3] See Sect. 1.2.2.

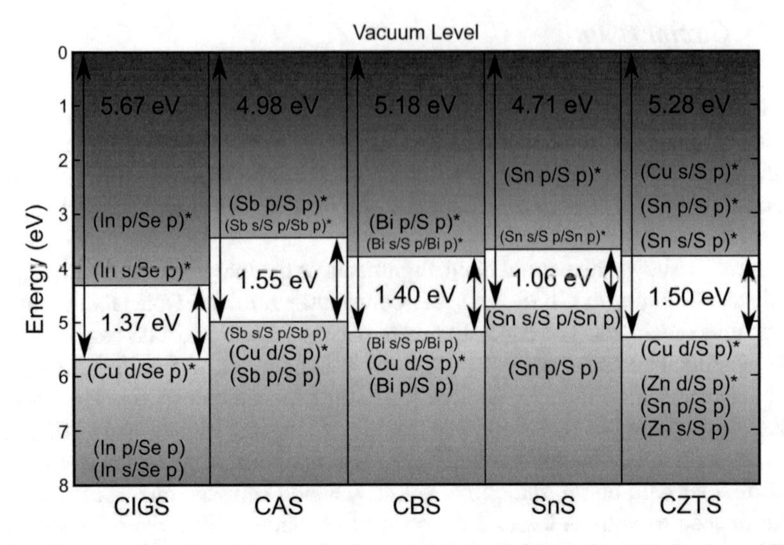

Fig. 7.1 Vacuum-aligned band diagram for the materials studied throughout this thesis, and CIGS. Values for the IP and E_g are taken from the relevant chapters for the materials studied here, and from the literature for CIGS [93]. Also marked are hybridisation contributions to the valence and conduction bands. These are taken from the calculated DoS for each of the materials studied here, and a literature report for CIGS [36]. For CIGS, 'In'-bonding also represents Ga-bonding for conciseness. Asterisks represent antibonding states, and smaller hybridisation labels represent minority contributions

For SnS, there is no copper cation, and so the bonding regime is dependent on a cation with no valence *d* orbital influence. As shown in Chap. 5, the state of the VBM in SnS is dictated strongly by lone-pair interactions, via the revised lone-pair model [38]. These interactions create full states high in the VB, shifting the VBM upwards when compared with SnS_2,[4] which has low band positions, similar to CdS. This demonstrates that states generated through lone-pair interactions via the revised lone-pair model act to raise the position of the VBM.

CAS and CBS are similar materials with regards to their electronic structures. Comparable to CIGS, they have VBM which consist mainly of Cu *d*/S *p* interactions. However, the structure of the non-copper cation is different. The antimony and bismuth *s* orbitals are full, and so it is the empty *p* orbitals which hybridise with the full S *p* orbitals; the antibonding states of which form the CBM. Also, because the *p* orbitals of antimony and bismuth are much closer in energy to the S *p* orbitals, compared with In/Ga *p* and Se *p* in CIGS, these interactions also contribute states near the top of the VB in CAS and CBS, as opposed to deeper in the VB, like in CIGS. Furthermore, the lone-pair antimony and bismuth electrons also interact weakly via the revised lone-pair model, generating still extra states at the top of the VB. Overall, it is believed that this combination of states is responsible for the higher VBM levels than for CIGS. It is possible that CAS has a higher VBM than CBS because of the

[4]Where these interactions are forbidden because of the unoccupied Sn 5*s* orbitals.

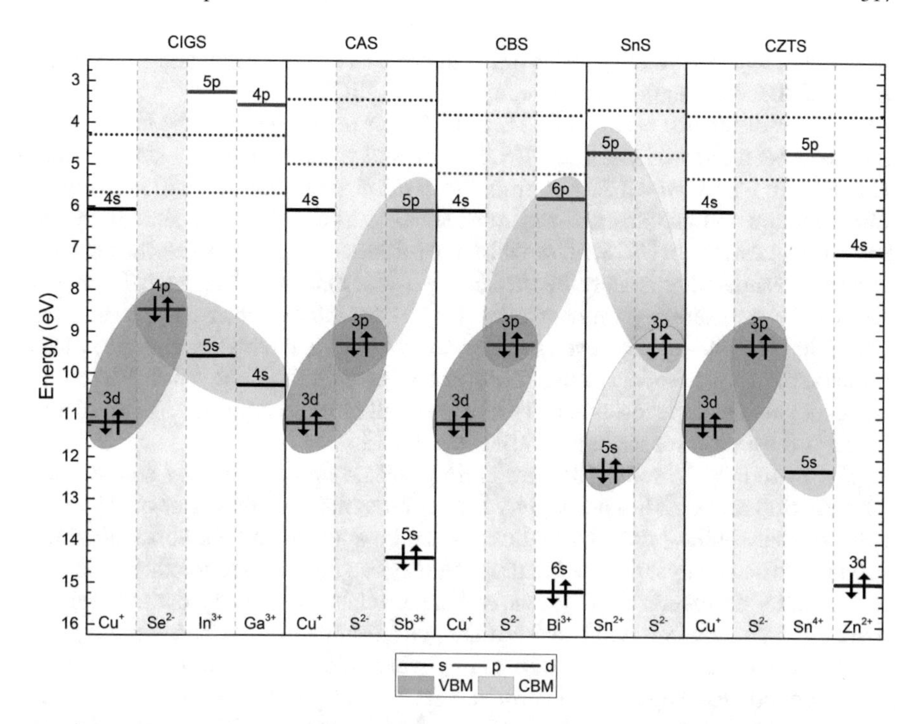

Fig. 7.2 CE [94, 95] for the valence orbitals of CIGS and the materials studied in this thesis. The formal ionic occupancies of the orbitals within the materials are marked and the main contributions to the VBM and CBM are highlighted. The gradient-shading for SnS represents that the hybridisation of all three of the highlighted orbitals contribute to both the VBM and CBM via the revised lone-pair model (see Fig. 5.18). Also shown are the VBM positions (red dot) and CBM positions (blue dot) for the individual materials relative to the vacuum level. The CE have been arbitrarily shifted to act as a guide to the eye for the bonding situations, are not absolute values, and do not take ionisation, multi-electron occupancy or hybridisation into account

stronger interactions between the non-Cu cation and sulphur, due to the stoichiometry[5], but also because of the more energetically distant orbitals of bismuth over antimony, which altogether results in a slightly higher VBM for CAS.

In Chap. 4, it was discussed how the bonding similarities between CIGS and CZTS were different to CBS. But then, in Chap. 6, it was shown how the measurement of the IP of single crystal CZTS was in agreement with other measurements [39–43], and therefore rejected the previous idea that CZTS had band levels that were comparable to CIGS[6] [15]. At first, the bonding regimes of the VBM and CBM of CIGS and CZTS do appear similar: strong Cu *d*/anion *p* VBM, and an unoccupied non-Cu cation *s* orbital[7] lower than the anion *p* orbitals, which hybridise to produce the antibonding CBM. However, in CZTS, the Sn also has a lower *p* orbital than indium

[5] Antimony in $CuSbS_2$ is more prevalent than bismuth in Cu_3BiS_3.

[6] As was initially shown in Figs. 3.10, 4.9, and 5.13, and rectified in Sect. 6.7.3.

[7] Sn 5*s* in CZTS.

or gallium, with the bonding states of the interaction of this orbital and S *p* residing higher in the VB than the corresponding states in CIGS.

Another difference between CZTS and CIGS is the presence of the third cation. As CIGS is an alloy of CIS and CGS, indium and gallium play very similar roles. However, in CZTS, which derived from CIGS through the isoelectronic substitution of the indium and gallium, one can consider the tetravalent tin in a similar vein to the trivalent cations in CIGS; all of which are *p*-block elements, but the divalent zinc must be considered separately, because it is a *d*-block element. In Sect. 6.7.3, it was discussed how several reports concluded that no Zn *d* interactions were present at the top of the VB [44–46]. However, in other reports [47], and here, the presence of Zn *d* interactions are observed in the VB, overlapping with the Cu *d* states. Therefore, it is concluded that the location and presence of all of these extra states in the VB of CZTS is the reason for the higher VBM than CIGS.

The materials CAS and CBS are closely related and are directly comparable, whereas SnS and CZTS have slightly different bonding regimes. But overall, it is believed that similar principles and trends can be applied, and the positions of the band levels and analyses of the bonding regimes can be extended to other materials for use as PV absorbers. That is, if the cation *d* orbitals are located close to the anion *p* orbitals,[8] then the VB will consist of the hybridisation of these states and the CB will most likely consist of the antibonding hybridisation of anion *p* with the most tightly bound, unoccupied cation orbital, yet the final position of the VBM will be decided by the location of the states of any other cations present in the material.

Throughout the chapters, each one of these materials was found to have a low IP compared with CIGS, and also CdTe, and were an anomaly in terms of the expected band positions. Given that they all demonstrate a similarly low IP, it could appear that CIGS is the anomaly of PV absorbers, and that a low IP is characteristic of these materials. This rationale is discouraged however, because CIGS and CdTe crystallise in the expected tetrahedral structures, whereas for the materials studied here, the distortion present in the crystal structures is part of the reason for the low IP. Therefore, it could be said that for copper-based, earth-abundant PV absorber materials,[9] a low IP is expected, and so CdS will not be a suitable window layer. CZTS would appear to be an exception to this, because the IP is still low, despite crystallising in a tetrahedral structure. This is partly because of the presence of two *d*-block cations in this material and although the effects of *p*-*d* repulsion are well known for binary semiconductors [32, 34, 48, 49], there is still scope for the study of the repulsion from two different cations, as this situation is not common in materials, especially for solar absorbers.

[8]For example with Cu^{+} and S^{2-}.

[9]Or those with strong lone-pair presence.

7.4 Final Remarks

The implementation of PES, especially XPS, has been explored, with application to characterising PV absorbers. Too regularly, it was found that XPS analyses of PV absorbers in the literature were weakened by misinterpretations of the spectra, arising because of the complexities that can sometimes be present. These problems were shown to be overcome by applying sound fitting procedures to the spectra, developed individually for each material, which then enabled the use of this technique in their characterisation. It can determine the oxidation states and chemical species present within a sample, which allows the phase-purity and presence of contaminants to be assessed. Beyond this, the wider power of this technique was demonstrated by using it to measure both the DoS in the VB, for corroboration with theoretical calculations, and the position of the VBM with respect to the vacuum level, to complement other band level measurements and perform band alignments: two aspects which are rarely applied with functional materials.

It was the intention of the work presented throughout this thesis, to determine whether CAS, CBS, SnS, and CZTS were viable contenders as earth-abundant PV absorbers, and to offer explanations as to why the efficiencies are not as high as is desirable for a market-ready material, suggesting ways in which they could be improved. Thorough exploration of the electronic structure of these materials has revealed how the positions of the band levels are not suited to CdS and other materials which commonly partner thin-film PV absorbers. Apart from this technological aspect, the use of cadmium is also a concern for devices [19, 50–54], as its use could jeopardise the goal of earth-abundancy.[10]

Overall, it is thought that with continued research into their development, all of these materials still have the potential to perform well, given the new understandings regarding the electronic structures gleaned here. Although CZTS is the most well established, it is perhaps more prudent to allow this research to step aside for the new, simpler materials, which show greater promise than CZTS did at the equivalent stage in development. This is especially the case, as novel methods of improving CZTS usually add to the complexity, through the inclusion of extra elements, rather than reducing it [22].

With relation to the energy sector in general, as presented in Sect. 1.4, these materials fulfil the criteria of earth-abundancy well, and it may be possible to sacrifice some ultimate efficiency in favour of cheaper production, both economically and environmentally speaking. Nevertheless, it is clear from the size of the PV sector, within the size of the current renewables market,[11] and within that, the amount of thin-film deployment,[12] that one technology must outshine the others in terms of cost of production and efficiency. This would generate considerate attraction to impose a significant shift toward the more widespread adoption of energy generation from PV. More likely, given the current economic climate [55], earth-abundant thin-film

[10]As discussed in Sects. 1.3.3 and 1.4.

[11]See Figs. 1.1 and 1.2.

[12]See Figs. 1.15 and 1.16.

PV can continue to establish itself within its attractive niche and gradually, greater deployment could be realised [56].

7.5 Suggestions for Future Research

The necessity for continued research into these materials was acknowledged in Sect. 7.4, and here, potential avenues for investigation are presented.

7.5.1 Best Practices

From the limitations of various literature reports presented throughout this thesis, it is important to establish best practices for the characterisation and reporting of the findings for PV absorbers. Hoye et al. [8] explored many of these, with respect to perovskite-like materials, but which are applicable to most PV absorbers. First, it is vital that any data collected is of high-quality, as analyses can never be performed accurately with poor data. For absorber characterisations, XRD, RS and XPS spectra should be collected and corroborated in order to characterise the material as fully as possible. Where possible, RS measurements should use multiple wavelengths of excitation and monochromated XPS should be used.

For XPS, the fitting procedures, analysis techniques, and measured values, used and found throughout this thesis, can be taken forward, for use in future analyses, as standards. Pre-measurement assessments of a sample should be conducted in order to gain an expectation for the number of peaks present within a sample, including any expected contamination, which then allows a fitting procedure to be implemented. Fitting should always be performed, even in the case where one peak is expected, because small shifts can cause imperceptible changes to the data, elucidated only by peak fitting. Following this, all BE should be reported, in order that they can be assessed together, and references to appropriate previous literature should be given in corroboration. At the expense of extra measurement time, the SEC and VBM of a sample can be collected during the CL acquisition as well, which has the added benefit of allowing the IP of materials to be calculated. During the same experiment, the VB spectra can also be collected, allowing analysis of the bonding nature and comparison to theoretical calculations for the DoS. Alternatively, changes in the VB spectra can reveal differences between materials, for example if an impurity is introduced.

7.5.2 Furthering the Measurements Performed Here

Whilst the analyses presented here are believed to be comprehensive and representative of the pure, clean materials, there are improvements that can be made in order to provide more confidence in the results.

Ideally, these measurements would be repeated on single crystal samples of all of the materials, but this was not the case for CAS or CBS because single crystals were not available. As research progresses with these materials, it is hoped that large single crystals can be grown and measured, which would allow in vacuo bulk cleaving: ensuring a fresh, clean surface, free of contamination. Alternatively, to remove any concern of the contamination affecting the band level measurements, samples could be grown in situ whilst performing IP measurements.

A benefit of the measurements performed here is that they can be performed with standard laboratory systems, and require no specialist equipment. However, a deeper understanding could be gained from using specialist analysis techniques. The use of synchrotron radiation allows PES to be performed with a tunable photon source of very high brightness, and benefits from very high resolution, allowing individual species to be more easily identified [57], or more accurate determination of band alignments [58] and the VB [59]. Other techniques also benefit from synchrotron radiation, such as: XRD and XRF, which allow spectra to be recorded very quickly and with high resolution [60–64]; NEXAFS, which facilitates CBM measurements via the absorption onset [42, 65], or an alternative to determining chemical environments [66]; and XAS and XES, which provide an alternative for measuring the band edge positions [67–69].

7.5.3 Advancement of the Materials

Discussions of the effects of defects on these materials were found throughout the chapters, as defects are very important to the properties and performance of absorber materials, providing the inherent conductivity, but also possibly acting detrimentally to cell performance. CZTS has received much research interest in the study of its defects [70–75], and those of SnS are well understood [76–79], given its simplicity. CAS [80, 81] and CBS on the other hand, have not received much research attention regarding their defects and as such, this is warranted. This could be achieved in a similar manner to the theoretical and experimental corroboration between the VBDoS presented here: PL could be used to experimentally determine the presence and formation of defects within samples [68, 82], whilst being corroborated by a theoretical defect analysis.

In Chap. 6, the binary secondary phases were studied, with the knowledge that these are likely to form within CZTS. It was also stated that ternary phases were a possibility, but were not fully explored, as they are currently understudied.[13] Further

[13]See Sect. 6.2.4.

investigation into these possible ternary phases is therefore necessary, especially as it was concluded that a ternary phase was found during the degradation of CZTS.[14]

The intention of the experiments performed here was to provide measurements of purely the absorber layer, in order to set standards that were not affected by external factors. This is why the materials underwent surface cleaning, and the reason why junctions were not analysed. Given that CdS was shown to be unsuitable for all of the materials studied here, it is now important to select and grow new materials to be the window layer and back contact in these solar cells, and to experimentally determine the band alignments. This research is already underway for CZTS [51, 65, 83–86] and SnS [87–91], is limited for CAS [92], but is lacking for CBS. Nevertheless, further studies of this need to continue in order to further the progress of these materials, as the results here show that this is a significant limit upon their efficiencies.

References

1. Shirley DA. High-resolution x-ray photoemission spectrum of the valence bands of gold. Phys Rev B. 1972;5(12):4709–14.
2. Shirley DA, Fadley CS. X-ray photoelectron spectroscopy in North America—the early years. J Electron Spectros Relat Phenom. 2004;137–140:43–58.
3. Ley L, Kowalczyk S, Pollak R, Shirley DA. X-ray photoemission spectra of crystalline and amorphous Si and Ge valence bands. Phys Rev Lett. 1972;29(16):1088–92.
4. Ley L, Pollak RA, McFeely FR, Kowalczyk SP, Shirley DA. Total valence-band densities of states of III-V and II-VI compounds from x-ray photoemission spectroscopy. Phys Rev B. 1974;9(2):600–21.
5. Pollak RA, Ley L, Kowalczyk S, Shirley DA, Joannopoulos JD, Chadi DJ, Cohen ML. X-ray photoemission valence-band spectra and theoretical valence-band densities of states for Ge, GaAs, and ZnSe. Phys Rev Lett. 1972;29(16):1103–5.
6. Fadley CS, Shirley DA. Electronic densities of states from x-ray photoelectron spectroscopy. J Res Natl Bur Stand Sect A Phys Chem. 1970;74A(4):543.
7. Kraut EA, Grant RW, Waldrop JR, Kowalczyk SP. Precise determination of the valence-band edge in x-ray photoemission spectra: application to measurement of semiconductor interface potentials. Phys Rev Lett. 1980;44(24):1620–3.
8. Hoye RLZ, Schulz P, Schelhas LT, Holder AM, Stone KH, Perkins JD, Vigil-Fowler D, Siol S, Scanlon DO, Zakutayev A, Walsh A, Smith IC, Melot BC, Kurchin RC, Wang Y, Shi J, Marques FC, Berry JJ, Tumas W, Lany S, Stevanović V, Toney MF, Buonassisi T. Perovskite-inspired photovoltaic materials: toward best practices in materials characterization and calculations. Chem Mater. 2017;29(5):1964–88.
9. Chambers SA, Droubay T, Kaspar TC, Gutowski M. Experimental determination of valence band maxima for $SrTiO_3$, TiO_2, and SrO and the associated valence band offsets with Si(001). J Vac Sci Technol B Microelectron Nanom Struct. 2004;22(4):2205.
10. Zeng Y, Li H, Qu B, Xiang B, Wang L, Zhang Q, Li Q, Wang T, Wang Y. Facile synthesis of flower-like Cu_3BiS_3 hierarchical nanostructures and their electrochemical properties for lithium-ion batteries. CrystEngComm. 2012;14(2):550–4.
11. Gerein NJ, Haber JA. One-step synthesis and optical and electrical properties of thin film Cu_3BiS_3 for use as a solar absorber in photovoltaic devices. Chem Mater. 2006;18(26):6297–302.

[14]See Sect. 6.9.3

12. Estrella V, Nair MTS, Nair PK. Semiconducting Cu_3BiS_3 thin films formed by the solid-state reaction of CuS and bismuth thin films. Semicond Sci Technol. 2003;18(2):190–4.
13. Yakushev MV, Maiello P, Raadik T, Shaw MJ, Edwards PR, Krustok J, Mudryi AV, Forbes I, Martin RW. Electronic and structural characterisation of Cu_3BiS_3 thin films for the absorber layer of sustainable photovoltaics. Thin Solid Films. 2014;562:195–9.
14. Burton LA, Colombara D, Abellon RD, Grozema FC, Peter LM, Savenije TJ, Dennler G, Walsh A. Synthesis, characterization, and electronic structure of single-crystal SnS, Sn_2S_3, and SnS_2. Chem Mater. 2013;25(24):4908–16.
15. Burton LA, Walsh A. Band alignment in SnS thin-film solar cells: possible origin of the low conversion efficiency. Appl Phys Lett. 2013;102(13):132111.
16. Sun K, Yan C, Liu F, Huang J, Zhou F, Stride JA, Green M, Hao X. Over 9% efficient kesterite Cu_2ZnSnS_4 solar cell fabricated by using $Zn_{1-x}Cd_xS$ buffer layer. Adv Energy Mater. 2016;6(12):1600046.
17. Delbos S. Kësterite thin films for photovoltaics: a review. EPJ Photovoltaics. 2012;3:35004.
18. Abermann S. Non-vacuum processed next generation thin film photovoltaics: towards marketable efficiency and production of CZTS based solar cells. Sol Energy. 2013;94:37–70.
19. Suryawanshi MP, Agawane GL, Bhosale SM, Shin SW, Patil PS, Kim JH, Moholkar AV. CZTS based thin film solar cells: a status review. Mater Technol. 2013;28(1–2):98–109.
20. Bourdais S, Choné C, Delatouche B, Jacob A, Larramona G, Moisan C, Lafond A, Donatini F, Rey G, Siebentritt S, Walsh A, Dennler G. Is the Cu/Zn disorder the main culprit for the voltage deficit in kesterite solar cells? Adv Energy Mater. 2016;6(12):1502276.
21. Song X, Ji X, Li M, Lin W, Luo X, Zhang H. A review on development prospect of CZTS based thin film solar cells. Int J Photoenergy. 2014;2014:1–11.
22. Ji S, Ye C. Cu_2ZnSnS_4 as a new solar cell material: the history and the future. Rev Adv Sci Eng. 2012;1(1):42–58.
23. Wang H. Progress in thin film solar cells based on Cu_2ZnSnS_4. Int J Photoenergy. 2011;2011:1–10.
24. Altamura G, Vidal J. Impact of minor phases on the performances of CZTSSe thin-film solar cells. Chem Mater. 2016;28(11):3540–63.
25. Su C-Y, Yen Chiu C, Ting J-M. Cu_2ZnSnS_4 absorption layers with controlled phase purity. Sci Rep. 2015;5:9291.
26. Klein A. Energy band alignment in chalcogenide thin film solar cells from photoelectron spectroscopy. J Phys: Condens Matter. 2015;27(13):134201.
27. Nishimura T, Hirai Y, Kurokawa Y, Yamada A. Control of valence band offset at CdS/Cu(In,Ga)Se_2 interface by inserting wide-bandgap materials for suppression of interfacial recombination in Cu(In,Ga)Se_2 solar cells CdS side Mo side Se flux CdS side. Jpn J Appl Phys. 2015;54:08KC08.
28. Morkel M, Weinhardt L, Lohmüller B, Heske C, Umbach E, Riedl W, Zweigart S, Karg F. Flat conduction-band alignment at the CdS/$CuInSe_2$ thin-film solar-cell heterojunction. Appl Phys Lett. 2001;79(27):4482–4.
29. Pookpanratana S, Repins I, Bär M, Weinhardt L, Zhang Y, Félix R, Blum M, Yang W, Heske C, Bar M, Felix R. CdS/Cu(In,Ga)Se_2 interface formation in high-efficiency thin film solar cells. Appl Phys Lett. 2010;97(7):74101.
30. Jackson P, Hariskos D, Wuerz R, Kiowski O, Bauer A, Friedlmeier TM, Powalla M. Properties of Cu(In, Ga)Se_2 solar cells with new record efficiencies up to 21.7%. Phys Status Solid Rapid Res Lett. 2015;9(1):28–31.
31. Wallin E, Malm U, Jarmar T, Edoff OLM, Stolt L. World-record Cu(In,Ga)Se_2-based thin-film sub-module with 17.4% efficiency. Prog Photovoltaics Res Appl. 2012;20(7):851–4.
32. Wei SH, Zunger A. Calculated natural band offsets of all II-VI and III-V semiconductors: chemical trends and the role of cation D orbitals. Appl Phys Lett. 1998;72(16):2011–3.
33. King PDC, Veal TD, Jefferson PH, Hatfield SA, Piper LFJ, McConville CF, Fuchs F, Furthmüller J, Bechstedt F, Lu H, Schaff WJ. Determination of the branch-point energy of InN: chemical trends in common-cation and common-anion semiconductors. Phys Rev B. 2008;77(4):45316.

34. Kilday DG, Margaritondo G, Ciszek TF, Deb SK, Wei S-H, Zunger A. Common-anion rule and its limits: photoemission studies of $CuIn_xGa_{1-x}Se_2$-Ge and $Cu_xAg_{1-x}InSe_2$-Ge interfaces. Phys Rev B. 1987;36(17):9388–91.
35. Li Y-H, Walsh A, Chen S, Yin W-J, Yang J-H, Li J, Da Silva JLF, Gong XG, Wei S-H. Revised Ab initio natural band offsets of all group IV, II-VI, and III-V semiconductors. Appl Phys Lett. 2009;94(21):212109.
36. Maeda T, Wada T. Characteristics of chemical bond and vacancy formation in chalcopyrite-type $CuInSe_2$ and related compounds. Phys Status Solid. 2009;6(5):1312–6.
37. Chen X-D, Chen L, Sun Q-Q, Zhou P, Zhang DW. Hybrid density functional theory study of $Cu(In_{1-x}Ga_x)Se_2$ band structure for solar cell application. AIP Adv. 2014;4(8):87118.
38. Walsh A, Payne DJ, Egdell RG, Watson GW. Stereochemistry of post-transition metal oxides: revision of the classical lone pair model. Chem Soc Rev. 2011;40(9):4455–63.
39. Gupta S, Whittles TJ, Batra Y, Satsangi V, Krishnamurthy S, Dhanak VR, Mehta BR. A low-cost, sulfurization free approach to control optical and electronic properties of Cu_2ZnSnS_4 via precursor variation. Sol Energy Mater Sol Cells. 2016;157:820–30.
40. Suryawanshi MP, Shin SW, Ghorpade UV, Gurav KV, Hong CW, Agawane GL, Vanalakar SA, Moon JH, Yun JH, Patil PS, Kim JH, Moholkar AV. Improved photoelectrochemical performance of Cu_2ZnSnS_4 (CZTS) thin films prepared using modified successive ionic layer adsorption and reaction (SILAR) sequence. Electrochim Acta. 2014;150:136–45.
41. Chen S, Wang L. Thermodynamic oxidation and reduction potentials of photocatalytic semiconductors in aqueous solution. Chem Mater. 2012;24(18):3659–66.
42. Ji S, Shi T, Qiu X, Zhang J, Xu G, Chen C, Jiang Z, Ye C. A route to phase controllable $Cu_2ZnSn(S_{1-x}Se_x)_4$ nanocrystals with tunable energy bands. Sci Rep. 2013;3:2733.
43. Huang S, Luo W, Zou Z. Band positions and photoelectrochemical properties of Cu_2ZnSnS_4 thin films by the ultrasonic spray pyrolysis method. J Phys D Appl Phys. 2013;46(23):235108.
44. Zhang Y, Yuan X, Sun X, Shih B-C, Zhang P, Zhang W. Comparative study of structural and electronic properties of Cu-based multinary semiconductors. Phys Rev B. 2011;84(7):75127.
45. Mitzi DB, Gunawan O, Todorov TK, Wang K, Guha S. The path towards a high-performance solution-processed kesterite solar cell. Sol Energy Mater Sol Cells. 2011;95(6):1421–36.
46. Zhao H, Persson C. Optical properties of Cu(In, Ga)Se_2 and Cu2ZnSn(S, Se)$_4$. Thin Solid Films. 2011;519(21):7508–12.
47. Xiao Z-Y, Li Y-F, Yao B, Deng R, Ding Z, Wu T, Yang G, Li C-R, Dong Z-Y, Liu L, Zhang L, Zhao H-F. Bandgap engineering of $Cu_2Cd_xZn_{1-x}SnS_4$ alloy for photovoltaic applications: a complementary experimental and first-principles study. J Appl Phys. 2013;114(18):183506.
48. Wei S-H, Zunger A. Role of metal D states in II-VI semiconductors. Phys Rev B. 1988;37(15):8958–81.
49. Wei S-H, Zunger A. Role of D orbitals in valence-band offsets of common-anion semiconductors. Phys Rev Lett. 1987;59(1):144–7.
50. Pfisterer F. Photovoltaic cells. In: Ullmann's Encyclopedia of Industrial Chemistry. Wiley-VCH Verlag GmbH & Co. KGaA: Weinheim, Germany, 2000; pp 35–154.
51. Barkhouse DAR, Haight R, Sakai N, Hiroi H, Sugimoto H, Mitzi DB. Cd-free buffer layer materials on $Cu_2ZnSn(S_xSe_{1-x})_4$: band alignments with ZnO, ZnS, and In_2S_3. Appl Phys Lett. 2012;100(19):193904.
52. Kim K, Larina L, Yun JH, Yoon KH, Kwon H, Ahn BT. Cd-free CIGS solar cells with buffer layer based on the In_2S_3 derivatives. Phys Chem Chem Phys. 2013;15(23):9239.
53. Heske C, Groh U, Fuchs O, Umbach E, Franco N, Bostedt C, Terminello LJ, Perera RCC, Hallmeier KH, Preobrajenski A, Szargan R, Zweigart S, Riedl W, Karg F. X-ray emission spectroscopy of Cu(In,Ga)(S,Se)$_2$-based thin film solar cells: electronic structure, surface oxidation, and buried interfaces. Phys Status Solid A-Appl Res. 2001;187(1):13–24.
54. Barreau N, Marsillac S, Bernède JC, Assmann L. Evolution of the band structure of β-$In_2S_{3-3x}O_{3x}$ buffer layer with its oxygen content. J Appl Phys. 2003;93(9):5456–9.
55. International Energy Agency (IEA). World Energy Outlook 2012; 2012.
56. Wallace SK, Mitzi DB, Walsh A. The steady rise of kesterite solar cells. ACS Energy Lett. 2017;2(4):776–9.

57. Surgina GD, Zenkevich AV, Sipaylo IP, Nevolin VN, Drube W, Teterin PE, Minnekaev MN. Reactive pulsed laser deposition of Cu_2ZnSnS_4 thin films in H_2S. Thin Solid Films. 2013;535(1):44–7.
58. Li J, Du Q, Liu W, Jiang G, Feng X, Zhang W, Zhu J, Zhu C. The band offset at CdS/Cu_2ZnSnS_4 heterojunction interface. Electron Mater Lett. 2012;8(4):365–7.
59. Ettema ARHF, de Groot RA, Haas C, Turner TS. Electronic structure of SnS deduced from photoelectron spectra and band-structure calculations. Phys Rev B. 1992;46(12):7363–73.
60. Weber A, Mainz R, Schock HW. On the Sn loss from thin films of the material system Cu–Zn–Sn–S in high vacuum. J Appl Phys. 2010;107(1):13516.
61. Lafond A, Choubrac L, Guillot-Deudon C, Fertey P, Evain M, Jobic S. X-ray resonant single-crystal diffraction technique, a powerful tool to investigate the kesterite structure of the photovoltaic Cu_2ZnSnS_4 compound. Acta Crystallogr Sect B Struct Sci Cryst Eng Mater. 2014;70(2):390–4.
62. Bosson CJ, Birch MT, Halliday DP, Knight KS, Tang CC, Kleppe AK, Hatton PD. Crystal structure and cation disorder in bulk Cu_2ZnSnS_4 using neutron diffraction and x-ray anomalous scattering. In: 2016 IEEE 43rd Photovoltaic Specialists Conference (PVSC), IEEE, 2016; pp 0405–0410.
63. Malerba C, Azanza Ricardo CL, Valentini M, Biccari F, Müller M, Rebuffi L, Esposito E, Mangiapane P, Scardi P, Mittiga A. Stoichiometry effect on Cu_2ZnSnS_4 thin films morphological and optical properties. J Renew Sustain Energy. 2014;6(1):11404.
64. Razmara MF. The crystal chemistry of the solid solution series between chalcostibite ($CuSbS_2$) and emplectite ($CuBiS_2$). Mineral Mag. 1997;61(404):79–88.
65. Yan C, Liu F, Song N, Ng BK, Stride JA, Tadich A, Hao X. Band alignments of different buffer layers (CdS, Zn(O, S), and In_2S_3) on Cu_2ZnSnS_4. Appl Phys Lett. 2014;104(17):173901.
66. Just J, Lützenkirchen-Hecht D, Frahm R, Schorr S, Unold T. Determination of secondary phases in kesterite Cu_2ZnSnS_4 thin films by x-ray absorption near edge structure analysis. Appl Phys Lett. 2011;2011(99):262105.
67. Bär M, Schubert BA, Marsen B, Wilks RG, Blum M, Krause S, Pookpanratana S, Zhang Y, Unold T, Yang W, Weinhardt L, Heske C, Schock HW. Cu_2ZnSnS_4 thin-film solar cell absorbers illuminated by soft x-rays. J Mater Res. 2012;27(8):1097–104.
68. Durose K, Asher SE, Jaegermann W, Levi D, McCandless BE, Metzger W, Moutinho H, Paulson PD, Perkins CL, Sites JR, Teeter G, Terheggen M. Physical characterization of thin-film solar cells. Prog Photovoltaics. 2004;12(2–3):177–217.
69. Bär M, Nishiwaki S, Weinhardt L, Pookpanratana S, Fuchs O, Blum M, Yang W, Denlinger JD, Shafarman WN, Heske C. Depth-resolved band gap in $Cu(In,Ga)(S,Se)_2$ thin films. Appl Phys Lett 2008;93(24).
70. Chen S, Yang J-H, Gong XG, Walsh A, Wei S-H. Intrinsic point defects and complexes in the quaternary kesterite semiconductor Cu_2ZnSnS_4. Phys Rev B. 2010;81(24):245204.
71. Scragg JJ, Watjen JT, Edoff M, Ericson T, Kubart T, Platzer-Bjorkman C. A Detrimental Reaction at the Molybdenum Back Contact in $Cu_2ZnSn(S,Se)_4$ thin-film solar cells. J Am Chem Soc. 2012;134(47):19330–3.
72. Kumar M, Dubey A, Adhikari N, Venkatesan S, Qiao Q. Strategic review of secondary phases, defects and defect-complexes in kesterite CZTS–Se solar cells. Energy Environ Sci. 2015;8(11):3134–59.
73. Yuan ZK, Chen S, Xiang H, Gong XG, Walsh A, Park JS, Repins I, Wei SH. Engineering solar cell absorbers by exploring the band alignment and defect disparity: the case of Cu- and Ag-based kesterite compounds. Adv Funct Mater. 2015;25(43):6733–43.
74. Yin L, Cheng G, Feng Y, Li Z, Yang C, Xiao X. Limitation factors for the performance of kesterite Cu_2ZnSnS_4 thin film solar cells studied by defect characterization. RSC Adv. 2015;5(50):40369–74.
75. Biswas K, Lany S, Zunger A. The electronic consequences of multivalent elements in inorganic solar absorbers: multivalency of Sn in Cu_2ZnSnS_4. Appl Phys Lett. 2010;96(20):94–7.
76. Vidal J, Lany S, D'Avezac M, Zunger A, Zakutayev A, Francis J, Tate J. Band-structure, optical properties, and defect physics of the photovoltaic semiconductor SnS. Appl Phys Lett. 2012;100(3):32104.

77. Banai RE, Horn MW, Brownson JRS. A review of Tin (II) monosulfide and its potential as a photovoltaic absorber. Sol Energy Mater Sol Cells. 2016;150:112–29.
78. Huang C-C, Lin Y-J, Chuang C-Y, Liu C-J, Yang Y-W. Conduction-type control of SnS_x films prepared by the sol–gel method for different sulfur contents. J Alloys Compd. 2013;553:208–11.
79. Kumagai Y, Burton LA, Walsh A, Oba F. Electronic structure and defect physics of tin sulfides: SnS, Sn_2S_3, and SnS_2. Phys Rev Appl. 2016;6(1):14009.
80. Krishnan B, Shaji S, Ernesto Ornelas R. Progress in development of copper antimony sulfide thin films as an alternative material for solar energy harvesting. J Mater Sci: Mater Electron. 2015;26(7):4770–81.
81. Yang B, Wang L, Han J, Zhou Y, Song H, Chen S, Zhong J, Lv L, Niu D, Tang J. $CuSbS_2$ as a promising earth-abundant photovoltaic absorber material: a combined theoretical and experimental study. Chem Mater. 2014;26(10):3135–43.
82. Unold T, Gütay L. Photoluminescence analysis of thin-film solar cells. In: Advanced characterization techniques for thin film solar cells. Wiley-VCH Verlag GmbH & Co. KGaA: Weinheim, Germany, 2011; pp 151–175.
83. Huang TJ, Yin X, Qi G, Gong H. CZTS-based materials and interfaces and their effects on the performance of thin film solar cells. Phys Status Solid Rapid Res Lett. 2014;8(9):735–62.
84. Lin L, Yu J, Cheng S, Lu P, Lai Y, Lin S, Zhao P. Band alignment at the In_2S_3/Cu_2ZnSnS_4 heterojunction interface investigated by x-ray photoemission spectroscopy. Appl Phys A. 2014;116(4):2173–7.
85. Bao W, Ichimura M. Band offsets at the ZnO/Cu_2ZnSnS_4 interface based on the first principles calculation. Jpn J Appl Phys. 2013;52(6R):61203.
86. Nagoya A, Asahi R, Kresse G. First-principles study of Cu_2ZnSnS_4 and the related band offsets for photovoltaic applications. J Phys: Condens Matter. 2011;23(40):404203.
87. Sugiyama M, Shimizu T, Kawade D, Ramya K, Ramakrishna Reddy KT. Experimental determination of vacuum-level band alignments of SnS-based solar cells by photoelectron yield spectroscopy. J Appl Phys. 2014;115(8):83508.
88. Park HH, Heasley R, Sun L, Steinmann V, Jaramillo R, Hartman K, Chakraborty R, Sinsermsuksakul P, Chua D, Buonassisi T, Gordon RG. Co-optimization of SnS absorber and Zn(O, S) buffer materials for improved solar cells. Prog Photovoltaics Res Appl. 2015;23(7):901–8.
89. Abdel Haleem AM, Ichimura M. Experimental determination of band offsets at the SnS/CdS and SnS/InS_xO_y heterojunctions. J Appl Phys. 2010;107(3):34507.
90. Sun L, Haight R, Sinsermsuksakul P, Bok Kim S, Park HH, Gordon RG. Band alignment of SnS/Zn(O, S) heterojunctions in SnS thin film solar cells. Appl Phys Lett. 2013;103(18):181904.
91. Devika M, Reddy NK, Patolsky F, Gunasekhar KR. Ohmic contacts to SnS films: selection and estimation of thermal stability. J Appl Phys 2008;104(12).
92. Baranowski LL, Christensen S, Welch AW, Lany S, Young M, Toberer ES, Zakutayev A. Conduction band position tuning and Ga-doping in (Cd,Zn)S alloy thin films. Mater Chem Front. 2017.
93. Hinuma Y, Oba F, Kumagai Y, Tanaka I. Ionization potentials of (112) and $(11\bar{2})$ facet surfaces of $CuInSe_2$ and $CuGaSe_2$. Phys Rev B. 2012;86(24):245433.
94. Mann JB, Meek TL, Knight ET, Capitani JF, Allen LC. Configuration energies of the D-block elements. J Am Chem Soc. 2000;122(21):5132–7.
95. Mann JB, Meek TL, Allen LC. Configuration energies of the main group elements. J Am Chem Soc. 2000;122(12):2780–3.

Correction to: Electronic Characterisation of Earth-Abundant Sulphides for Solar Photovoltaics

Correction to: T. J. Whittles, *Electronic Characterisation of Earth-Abundant Sulphides for Solar Photovoltaics*, Springer Theses, https://doi.org/10.1007/978-3-319-91665-1

In the original version of the book, the tables are to be reformatted and the references to the tables are to be cited properly throughout. The correction book has been updated with the changes.

The updated online version of this book can be found at https://doi.org/10.1007/978-3-319-91665-1

© Springer Nature Switzerland AG 2018

T. J. Whittles, *Electronic Characterisation of Earth-Abundant Sulphides for Solar Photovoltaics*, Springer Theses, https://doi.org/10.1007/978-3-319-91665-1_8

Appendix A
Extra Information for Chap. 1

A.1 Determining the Energy of the Sun

In Sect. 1.1.3, it was stated that the solar power reaching the surface of the earth is 4 orders of magnitude greater than the current global electricity demand. First, it is assumed that the world has an annual electricity demand of 20 PWh, from the 2008 value in Fig. 1.1 [1]. This equates to a required average power of

$$20\ \text{PWh} \div 8760\ \text{h} \approx 2.3\ \text{TW}\ . \tag{A.1}$$

The sun delivers 174 PW to the upper atmosphere of earth with a density of 1366 Wm^{-2}. Through the effects of the atmosphere, obliqueness, seasonal and diurnal variations, cloud cover, and prevailing conditions [2], this reduces to an average power density of 188 Wm^{-2} at the surface. Assuming a negligible change in effective surface area due to the thickness of the atmosphere, this relates to the surface of the earth receiving an average of

$$188\ \text{W}\,\text{m}^{-2} \times \frac{174\ \text{PW}}{1366\ \text{W}\ \text{m}^{-2}} \approx 24\ \text{PW}, \tag{A.2}$$

which then equates to a potential supply of energy greater than demand by

$$24\ \text{PW} \div 2.3\ \text{TW} \approx 10{,}000 = 4\ \text{OoM}. \tag{A.3}$$

A.2 Calculations for the Limiting Factors of PV Technologies

In Sect. 1.5.3, estimates were given to the cost of GW modules of various PV technologies and their ability to be deployed on a terawatt scale, given the limiting

© Springer International Publishing AG, part of Springer Nature 2018

T. J. Whittles, *Electronic Characterisation of Earth-Abundant Sulphides for Solar Photovoltaics*, Springer Theses, https://doi.org/10.1007/978-3-319-91665-1

Table A.1 Limiting element for selected solar absorber materials based on world reserve estimates [7]

Absorber Material	Limiting Element
c-Si	Silicon
CdTe	Tellurium
CIGS	Indium
CZTS	Tin
CAS	Antimony
CBS	Bismuth
SnS	Tin

element. First, the PV technologies were assumed to either have the record efficiency taken from the literature[1] [3–6] or 20%. For silicon, it was assumed that the absorber layer had a thickness of 200 μm, and the thin-film absorber layers had a thickness of 2.5 μm. In reality, the different technologies probably favour different thicknesses, but for the sake of argument, these should be representative.

For determining the required amount of absorber material required to produce a module capable of 1 GW, the power density from the sun was taken to be 1000 Wm^{-2}, the same for AM1.5G [8]. In determining the price of a module of each technology, prices for the elements were taken to be the raw materials costs and did not include any consideration of manufacturing, processing, or installation costs, and dealt only with the materials cost for the absorber layer: none of the other parts of the device. In determining the amount of possible 1 GW modules available, it was assumed that the limiting factor was the estimated reserve of the rarest elemental component of the absorber layer, shown in Table A.1, and that the entire contents of this reserve could go into producing cells. Although these assumptions mean that the price of a cell could be much higher, and the number of producible modules could be much lower, it allows some level of comparison between the technologies.

The mass of absorber layer material required to generate 1 GW of power, was calculated using

$$m_{\mathrm{GW}}(x) = \frac{\rho_x d_x}{\eta_x I_{\mathrm{sol}}}, \tag{A.4}$$

where x is the absorber in question, ρ_x is the density of the absorber material, d_x is the required thickness of the absorber material, η_x is the efficiency of the solar cell, and I_{sol} is the power density of the sun. By considering each element of an absorber material separately, the mass of each element for a 1 GW absorber was determined by

[1] 1% was assumed for CBS which has not yet had any efficiencies reported.

$$m_{\mathrm{GW}(i)}(x) = m_{\mathrm{GW}}(x) \cdot \left(\frac{M_i \cdot n_i(x)}{\sum_j (M_j \cdot n_j(x))} \right), \tag{A.5}$$

where *i* is the element in question, M_i is the molar mass of the element in question, n_i is the number of moles of the element in question in the absorber, and the sum over *j*, each of the elements in the absorber. By taking the current price of each element [7], the total cost of the absorber layer material for a 1 GW module was calculated using

$$£_{\mathrm{GW}}(x) = \sum_i \left(£_i \cdot m_{\mathrm{GW}(i)}(x) \right), \tag{A.6}$$

where $£_i$ is the cost per mass for each element, summed over the elements in an absorber material. Further, the number of producible 1 GW modules from the limiting elemental reserve for an absorber material was calculated using

$$\#_{\mathrm{GW}}(x) = \frac{m_{\mathrm{res}(k)}(x)}{m_{\mathrm{GW}(k)}(x)}, \tag{A.7}$$

where $m_{res(k)}$ is the mass of the estimated reserve [7] for the limiting element, *k* in the absorber material. The limiting element for each of the considered absorbers is given in Table A.1.

Appendix B
Extra Information for Chap. 2

B.1 Achieving UHV: Further Details

Advanced development in stainless steel production and welding techniques means that this material is now primarily used for vacuum components, designed in the form of interchangeable, flanged components, which, with the use of metal gaskets, make systems very modular and adaptable to different situations [9].

In the viscous flow regime, positive displacement pumps are used in order to bring the pressure from atmosphere to around 1×10^{-3} mbar [10]. These work by increasing and decreasing the volume of the system in such a way that as a pressure gradient is formed, flow occurs out of the system, hence decreasing the pressure. This type of pump comes in many forms but the most common is the rotary vane pump, the schematic for which is shown in Fig. B.1a. An oil-sealed rotating cam with two spring-loaded vanes creates a changing volume which is alternately connected to, and isolated from, the vacuum system. During each rotation of the cam, gas from the vacuum system is swept into the volume between the vanes, isolated from the system, compressed within the pump, and exhausted from the pump. Another pump which works on a similar principle is the scroll pump, shown in Fig. B.1b. In this type of pump, the changing volume is created by two interleaved scrolls, one of which rotates, shifting a volume of gas, compressing it at the centre of the scrolls. The action is similar to the rotary vane pump, but does not require sealing with oil: beneficial for UHV as there is no risk of oil contamination.

Once this initial pumping has achieved a level of medium vacuum, another pump is used to achieve pressures of high vacuum and below, in the molecular flow regime. The pumps used in this regime can be classified as either momentum transfer compression pumps, or entrapment pumps, where the former compresses the gas molecules into the back of the pump where they can be removed using another type of pump, and the latter serves to contain the gas molecules in such a way as they cannot contribute to the total pressure in the system [10].

© Springer International Publishing AG, part of Springer Nature 2018

T. J. Whittles, *Electronic Characterisation of Earth-Abundant Sulphides for Solar Photovoltaics*, Springer Theses, https://doi.org/10.1007/978-3-319-91665-1

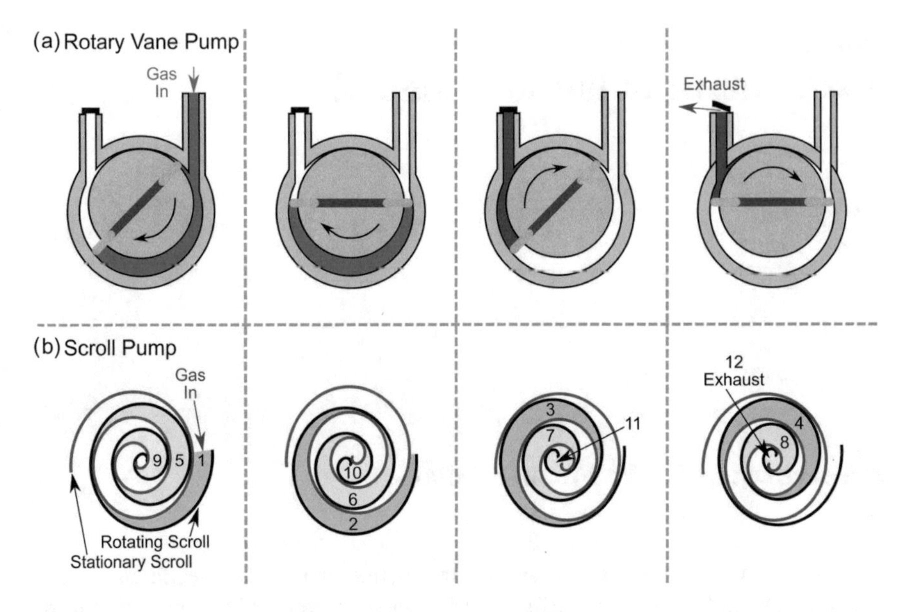

Fig. B.1 Schematic diagrams for **a** a rotary vane pump, and **b** a scroll pump. Panes show the cycling stages of the pumping mechanism, and both work by continually isolating, compressing and exhausting gas from the chamber

The most commonly used compression pumps that allow the attainment of UHV conditions are the diffusion pump and the turbomolecular pump, shown in Fig. B.2a, b, respectively. The diffusion pump entrains gas molecules in a jet of fluid, facilitated by the continual cycling of oil through the pump. This type of pump has been superseded by the turbomolecular pump, due to the high risk of oil backflow and high associated maintenance [11]. The turbomolecular pump contains a series of stators and rotors that spin at high speeds. The blades are angled in such a way as to maximise the probability of compression should a gas molecule impinge on the top rotor. This type of pump reduces the oil problems associated with diffusion pumps and by using magnetically levitated bearings instead of greased ball bearings, the use of oil can be removed completely, with this also serving to reduce vibrations.

Entrapment pumps are almost defined by the ion pumps [10]. Boasting no moving parts, high pumping speeds, and simple operation and maintenance, they are used to maintain UHV conditions on the majority of vacuum systems. The basic schematic for an ion pump is shown in Fig. B.3 but there are many variants [12]. A strong magnetic field within the central cylinders trap electrons, which then collide with gas molecules in the system and ionise them. Plates of titanium are held at a high negative potential, causing the ions to be accelerated into them. The pumping action is twofold: first, the ions bombard the plates with such force that

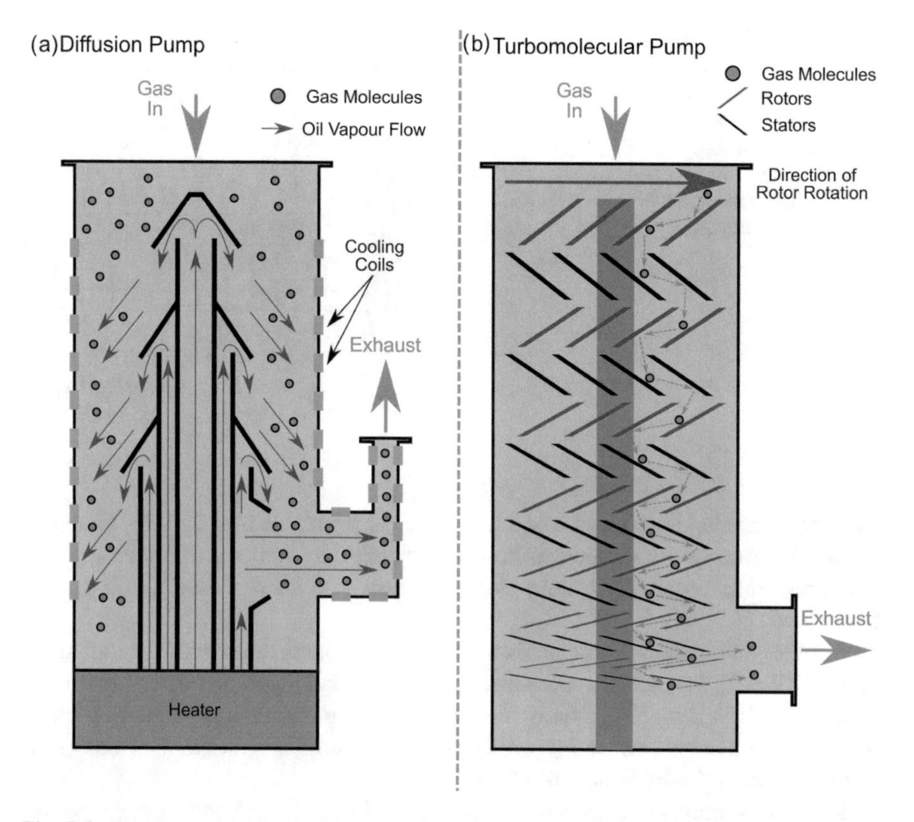

Fig. B.2 Schematic diagrams for **a** a diffusion pump, and **b** a turbomolecular pump, showing how gas molecules are compressed at the bottom of the pumps. A further type of pump (backing pump) is required at the exhaust of these pumps because they cannot compress the gas to atmospheric pressures by these mechanisms

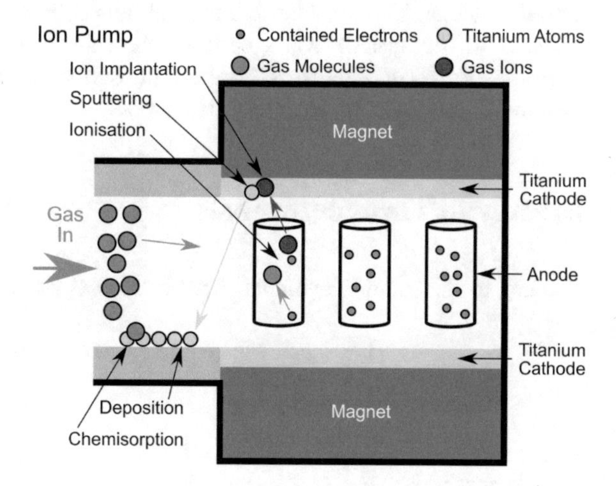

Fig. B.3 Schematic diagram for an ion pump, showing how electrons are contained within the anodes and ionise gas molecules, causing them to accelerate into the titanium cathodes, sputtering material away

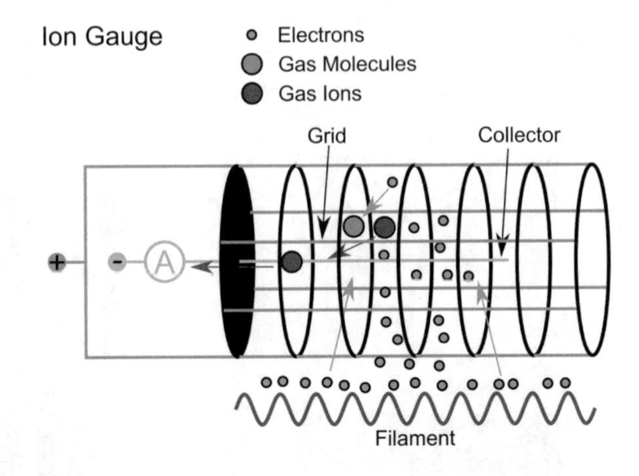

Fig. B.4 Schematic diagram of an ion gauge, showing how a hot filament produces electrons which ionise gas molecules. These are then accelerated towards a negatively charge collector, where the current is proportional to the pressure

they are implanted into it, and second, the force at which the ions impinge the plates causes titanium to be sputtered from them, which then forms a film and acts as a getter material, whereby gas molecules chemisorb with the titanium, forming stable compounds.

Another entrapment pump that is used, often in combination with the ion pump, is the TSP [13]. Instead of continually sputtering small amounts of titanium on the chamber wall to form the gettering film like the ion pump does, this pump consists of a titanium filament, which by resistive heating is sublimated onto the walls of the chamber creating a fresh titanium film.

In order to measure the pressure of the system in these conditions, ion gauges are used [13]. A schematic of an ion gauge is shown in Fig. B.4. Electrons are thermionically emitted from a hot filament and are accelerated towards a positively biased grid. Collisions between electrons and gas molecules in this grid cause the molecules to become ionised, which are then accelerated towards a negatively biased collector. The drain current through this collector is then proportional to the pressure in the system. Mass spectrometers are also employed as residual gas analysers (RGA) in order to assess the contribution of partial pressures to a vacuum system as well as to perform leak detection.

B.2 Key to Vacuum Symbols

The key to the vacuum symbols[2] used in Figs. 2.2 and 2.3, is shown in Fig. B.5.

[2]From DIN 28401.

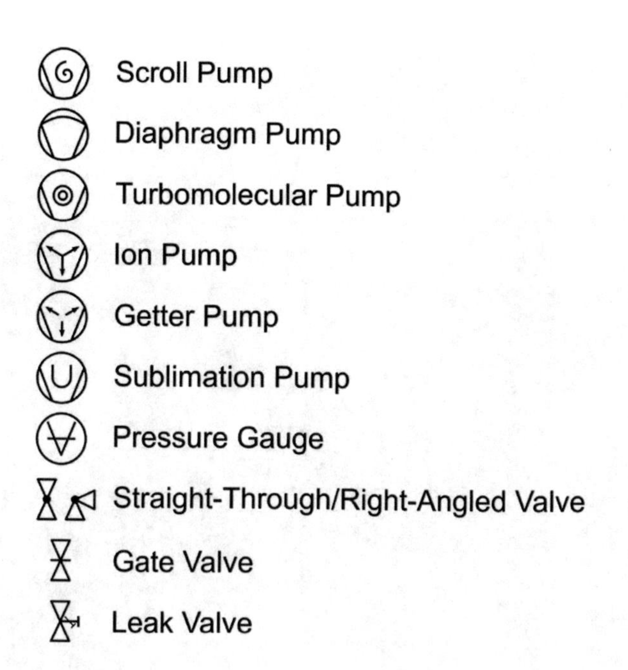

Fig. B.5 Vacuum symbols used in the UHV chamber schematics in Figs. 2.2 and 2.3

B.3 The Assembly of the Single-Chamber Electronic Characterisation UHV System

In Sect. 2.2.4, the motivation behind the design and commissioning of a new UHV system to perform the PES experiments required for this thesis was explained. Here, several photographs of this system are shown, presenting further aspects of this system.

After the initial commissioning of the system, one limiting factor was that only one sample at a time could be present within the system because there was no facility for sample storage. Figure B.6a shows this system at this stage of assembly. A sample was introduced through the load lock, directly onto the transfer arm, which was then pumped separately from the main chamber. This was improved upon by the addition to the system shown in Fig. B.6b. At the previous load lock entry point, an additional transfer arm was added which contained storage for four samples. This was then used as the load lock, allowing up to four samples to be entered simultaneously, reducing pumping time. Furthermore, as the original transfer arm could be pumped individually to both the main chamber and the new load lock, this area could be used as a small preparation chamber, by the addition of a sputter ion gun, as shown in Fig. B.6b.

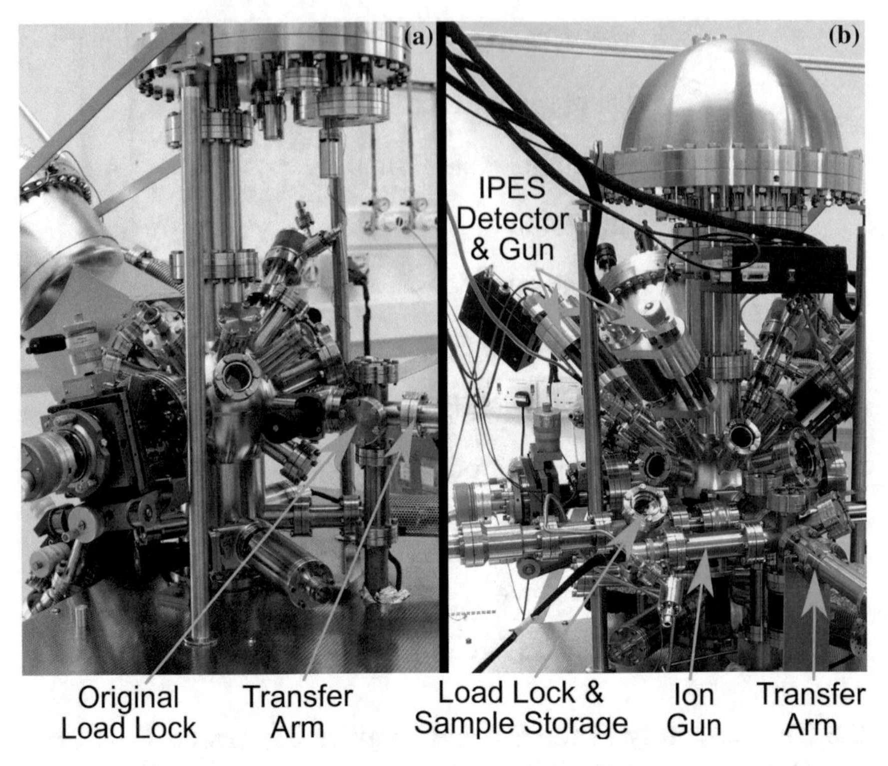

Fig. B.6 **a** Single chamber UHV system after initial commissioning showing the load lock which was able to contain only one sample at a time. **b** Single chamber UHV system after the addition of a sample storage area

In Fig. B.7, the operating principles of the monochromated x-ray source, as described in Sect. 2.3.2, are shown. In Fig. B.7a, the Rowland circle schematic from Fig. 2.12 is overlaid, showing the location of the different peripherals. Figure B.7b, c show the interior view of the crystals during assembly and how they are mounted to the 3–axis manipulator in order to optimise its position. Also shown in Fig. B.7a is a piece of tubing that connects the monochromator chamber to the bottom of the main chamber. This added parallel component increases the conductance to this area of the system, increasing the effective pumping speed.

During optimisation of the monochromated x-ray source, it is necessary to first direct the x-ray beam toward the centre of the chamber, and then to focus the beam in order to both maximise the intensity, and also to produce the best resolution, which is achieved by changing the positions of the x-ray source, the crystal, and also the sample stage. This is facilitated by using a sample that contains a phosphor in order to visibly locate the x-ray beam, and also the silver foil, used for spectrometer calibration in Sect. 2.3.3. This sample is shown within the UHV system during an optimisation procedure in Fig. B.8. The beam is trained on the phosphor,

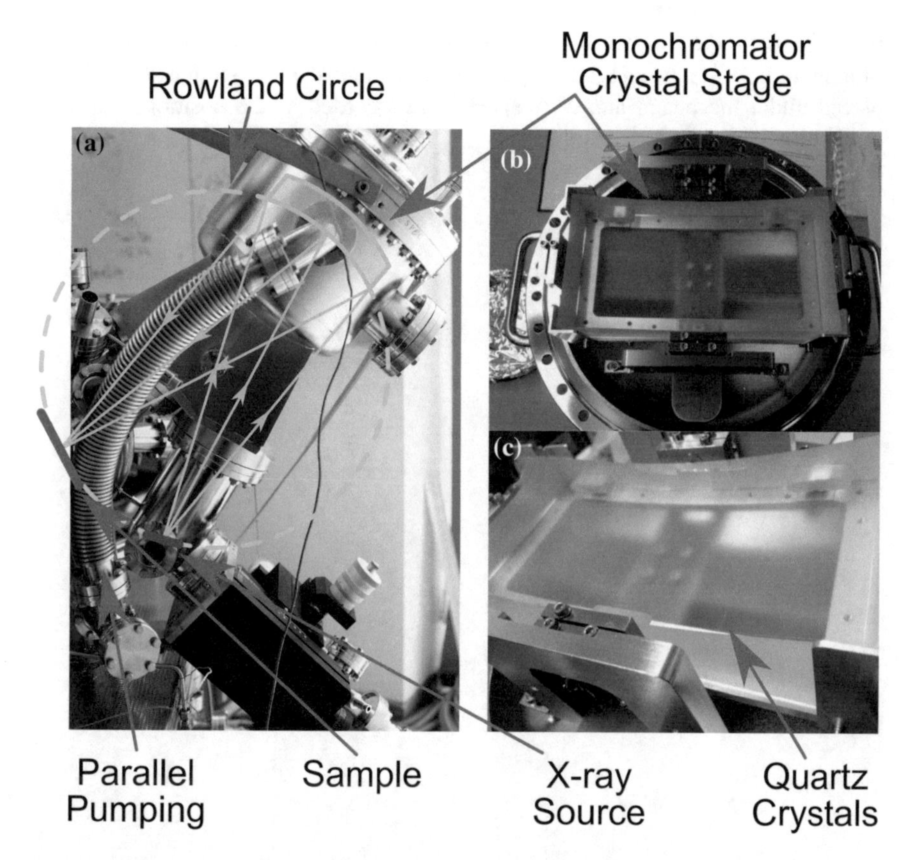

Fig. B.7 **a** Rowland circle overlaid onto the UHV system, showing the locations of the x-ray source, sample, and crystal. **b**, **c** Monochromator crystal and the supporting stage

Fig. B.8 Inside view of the UHV system during optimisation of the monochromator. The sample on the manipulator contains a piece of silver foil and a phosphor-coated foil. The green spot is the luminescence from the x-ray beam. The heating filament used for annealing samples in vacuo is located beneath the sample stage

resulting in the green spot caused by the luminescence of the phosphor under x-ray illumination. After locating the beam to the centre of the chamber, the silver is placed under the beam, and XPS spectra are recorded as the positions are finely tuned to achieve the best resolution.

Appendix C
In-depth Experimental Details

In Chap. 2, descriptions and discussions of the instrumentation and data extraction methods used in this thesis were given. Throughout the main results chapters, it was described how each technique was applied; however, for the sake of succinctness, certain details were omitted which were either beyond the scope of the study, or else offered little towards the interpretation of the data and the conclusions drawn in each chapter. Therefore, for the sake of completeness and reproducibility, these details shall be specified in this appendix, grouped by technique.

C.1 Raman Spectroscopy Details

C.1.1 Chapter 3: $CuSbS_2$ and Chap. 4: Cu_3BiS_3

The Horiba Scientific Jobin-Yvon LabRam HR system Raman spectrometer consisted of a confocal microscope coupled to a single grating spectrometer, equipped with a notch filter and a CCD camera detector. Spectra were measured using an incident wavelength of 514.5 nm from an argon ion laser in backscattering geometry with an exposure time of 30 s and 5 acquisitions, over a range of 100–1500 cm^{-1}. Before measurements, the spectrometer was calibrated to the laser wavelength and the Si 520 cm^{-1} mode.

C.1.2 Chapter 6: Single Crystal CZTS

The Renishaw inVia Raman spectrometer consisted of a confocal microscope coupled to a single grating spectrometer, equipped with a notch filter and a CCD camera detector. Spectra were measured using an incident wavelength of 532 nm from a Cobolt Samba CW DPSS laser in backscattering geometry with an exposure

© Springer International Publishing AG, part of Springer Nature 2018

T. J. Whittles, *Electronic Characterisation of Earth-Abundant Sulphides for Solar Photovoltaics*, Springer Theses, https://doi.org/10.1007/978-3-319-91665-1

time of 15 s and 30 acquisitions, over a range of 100–1400 cm^{-1}, or using an incident wavelength of 633 nm from a He–Ne laser in backscattering geometry with an exposure time of 30 s and 30 acquisitions, over a range of 100–1100 cm^{-1}. Before measurements, the spectrometer was calibrated to the laser wavelength and the Si 520 cm^{-1} mode.

C.2 Optical Absorption Details

C.2.1 Chapter 3: $CuSbS_2$ and Chap. 5: SnS and SnS_2

FTIR absorption spectroscopy was performed using the experimental system described in Sect. 2.5. The low temperatures were obtained using an Oxford Instruments CFV2 continuous-flow helium cryostat. The vacuum pressures were 2 mbar for the optical path in the spectrometer and 1×10^{-6} mbar in the sample environment of the cryostat.

C.3 Density Functional Theory Details

For comparison between the calculated pDoS and the XPS/UPS VB spectra, the pDoS curves were corrected using the standard photoionisation cross-sections [14], then convolved with a Gaussian function to account for thermal broadening and the spectrometer resolution, and then further convolved with a Lorentzian function to account for lifetime broadening. For the DoS before experimental corrections were applied, the pDoS curves were convolved with a Gaussian function, so that features could be more easily distinguished. The broadening parameters are given in Table C.1.

Table C.1 FWHM of the broadening functions applied to the pDoS curves for analysis

Material	XPS Comparison		UPS Comparison		Analysis
	Gaussian FWHM	Lorentzian FWHM	Gaussian FWHM	Lorentzian FWHM	Gaussian FWHM
CAS	0.38	0.25	—	—	0.30
CBS	0.38	0.30	—	—	0.30
Tin Sulphides	0.41	0.35	—	—	0.20
CZTS	0.43	0.30	0.16	0.38	0.30

C.3.1 Chapter 3: $CuSbS_2$, Chap. 4: Cu_3BiS_3 and Chap. 6: CZTS

The screened hybrid functional HSE06 [15] was used for geometry optimization and the DoS calculations, with the explicit inclusion of relativistic spin-orbit coupling for CBS due to the presence of the heavy element, Bi. HSE06 includes 25% Hartree–Fock exchange, which is screened with a parameter of $\omega = 0.11\,\text{bohr}^{-1}$, with 75% exchange and full correlation from the GGA functional, PBE [16]. This method has been used successfully in calculations of CAS [17–19], CBS [20–24], CZTS [25, 26], and related materials. The PAW method [27] was used to describe the interaction between valence and core electrons, scalar relativistic pseudopotentials were used, and Bi 5*d* states were treated as valence. A cutoff energy of 400 eV, increased to 560 eV for the optimisation to minimize the errors from Pulay stress [28], and a Γ-centred k-point mesh of $3 \times 2 \times 4$, $4 \times 7 \times 2$, and $6 \times 6 \times 6$ were used to give convergence to within 1 meV per atom for CAS, CBS, and CZTS, respectively. A convergence criterion of 0.01 eV Å^{-1} was used in the optimization for the forces on each atom.

C.3.2 Chapter 5: Tin Sulphides

The room temperature structures of each of the tin sulphides were relaxed using the PBE GGA functional [16] with the D3 dispersion correction applied to account for VdW forces [29]; using the Becke–Johnson damping variant [30]. A plane-wave basis set with a 550 eV cutoff was employed with PAW pseudopotentials [31, 32] treating the Sn 5*s*, 4*d* and 5*p* and S 3*s* and 3*p* states as valence electrons. The Brillouin zones of SnS, SnS_2 and Sn_2S_3 were sampled using Γ-centred Monkhorst–Pack k-point grids [33], with $8 \times 4 \times 8$, $8 \times 8 \times 6$, and $4 \times 8 \times 3$ subdivisions, respectively. The ion positions, cell shape, and volume were optimized to a tolerance of 10^{-7} eV on the electronic wave functions and 0.005 eV Å^{-1} on the forces.

Electronic-structure calculations were then carried out on these optimized structures using the HSE06 hybrid functional [15, 35, 36]. The DoS of the valence and conduction bands was simulated using denser Brillouin zone sampling meshes with $12 \times 6 \times 12$, $12 \times 12 \times 8$, and $6 \times 12 \times 5$ subdivisions for SnS, SnS_2 and Sn_2S_3, respectively, with interpolation being performed using the Blöchl-corrected tetrahedron method [37]. For these calculations, the electronic minimisation was performed to a tolerance of 10^{-6} eV.

C.4 Photoemission Spectroscopy Details

All PES measurements presented in this thesis were performed in the Electronic Characterisation UHV system described in Sect. 2.2.4. The base pressure of this system throughout the study was less than 2×10^{-10} mbar, with hydrogen as the

Table C.2 Experimental parameters for PES measurements in Chaps. 3/4/5/6

Technique & Region		Source Power (W)	E_P (eV)	Bias (V)	Step Size (eV)	Flux (μA)
XPS	Survey	250/250/200/300	50	0	0.5	—
	Core-Levels & VB	250/250/200/300	10	0	0.03	—
	Auger	—/—/—/300	10	0	0.1	—
	SEC	9/9/9/9	10	-10	0.03	—
UPS	SEC & VB	—/—/—/Maximum	2	-2	0.01	—
IPES	CBM	—/—/1.04A[a]/1.03A[a]	—	0	0.1	—/—/4/3.9–10.0

[a]Electron gun filament current

main residual gas. The x-ray source used for XPS measurements was monochromated Al Kα radiation.[3] The UV source used for UPS measurements was a helium discharge lamp and He(I)[4] was chosen as the excitation line. Details of the operating parameters for XPS, UPS, and IPES, are given in Table C.2.

C.5 In Vacuo Cleaning Details

Samples were sputtered and annealed in order to remove surface contaminants, including oxides and hydrocarbons, which form due to handling in ambient atmosphere. Low energy sputtering and a grazing angle of incidence[5] were used in order to minimise sample damage, except where stated. This cleaning facilitated the measurement of a more representative surface of the samples, and hence prevented the VB spectra from being overwhelmed by signal from the contamination.

C.5.1 Chapter 3: $CuSbS_2$

During cleaning, the no-TT and TT samples of CAS were monitored by XPS and were considered to be clean when the C 1*s* peak and peaks due to antimony oxide were no longer visible on the survey spectra. Such cleanliness was achieved by

[3]hν = 1486.6 eV.
[4]hν = 21.22 eV.
[5]20°.

sputtering with 500 eV Ar^+ ions[6] in 5 minute steps, followed by annealing at 200 ° C for 60 min. In total, the no-TT sample received 20 min of sputtering and the TT sample received 30 min of sputtering.

C.5.2 Chapter 4: Cu_3BiS_3

During cleaning, the sample of CBS was monitored by XPS and was considered to be clean when sputtering cycles no longer produced noticeable changes to the survey spectra following the removal of oxidised bismuth. Such cleanliness was achieved by sputtering with 500 eV Ar^+ ions[7] in 5 min steps for 30 min total, followed by annealing at 150 °C for ~20 h and then at 250 °C for ~24 h before allowing the sample to cool to ambient temperature.

C.5.3 Chapter 5: Tin Sulphides

Each of the crystals required differing active feedback cleaning procedures, with details given in Tables C.3, C.4 and C.5 for SnS, SnS_2, and Sn_2S_3 respectively, with the results comments determining the subsequent cleaning cycle.

SnS

Table C.3 In vacuo cleaning cycles for SnS

#	Cycle	Energy or Temperature	Time (min)	I_{flux} (μA)	Results
1	Anneal 1	200°C	60	—	Reduction in contaminants
2	Sputter 1	500 eV	5	7	Reduction in contaminants
3	Sputter 2	500 eV	5	6	Reduction in contaminants
4	Sputter 3	500 eV	5	7	Reduction in contaminants
5	Sputter 4	500 eV	5	6	Reduction in contaminants
6	Sputter 5	500 eV	5	8	Reduction in contaminants
7	Sputter 6	500 eV	10	7	Reduction in contaminants Development of metallic Sn
8	Anneal 2	200°C	20	—	No real change
9	Anneal 3	250°C	20	—	Reduction in metallic Sn
10	Anneal 4	230°C	65	—	Reduction in metallic Sn
11	Anneal 5 Sputter 6	250°C 500 eV	15 10	— 7	Reduction in metallic Sn
12	Anneal 6	250°C	20	—	No metallic Sn (Clean)

I_{flux} is the flux ion current at the sample

[6]Resulting in a flux ion current of 10 μA at the sample.
[7]Resulting in a flux ion current of 5 μA at the sample.

SnS_2

Table C.4 In vacuo cleaning cycles for SnS_2

#	Cycle	Energy or Temperature	Time (min)	I_{flux} (μA)	Results
1	Sputter 1	500 eV	5	8	Reduction in contaminants
2	Sputter 2	500 eV	5	8	Reduction in contaminants
3	Sputter 3	500 eV	5	8	Reduction in contaminants
4	Sputter 4	500 eV	5	8	Reduction in contaminants
5	Sputter 5	500 eV	5	8	Reduction in contaminants
6	Anneal 1	200°C 225°C 250°C	30 60 150	— — —	Increase in contaminants Development of Sn^{2+} species
7	Sputter 6	500 eV	10	6	Reduction in contaminants No change to Sn^{2+}
8	Sputter 7	500 eV	10	7	No real change
9	Sputter 8	500 eV	15	7	No real change
10	Sputter 9	500 eV	15	8	No real change
11	Anneal 2	300°C	60	—	Redevelopment of Sn^{2+} species (Cannot remove this by cleaning)
12	Cleave	—	—	—	Minimal contaminants (Clean)

I_{flux} is the flux ion current at the sample

Sn_2S_3

Table C.5 In vacuo cleaning cycles for Sn_2S_3

#	Cycle	Energy or Temperature	Time (min)	I_{flux} (μA)	Results
1	Sputter 1	500 eV	5	7	Reduction in contaminants
2	Anneal 1	250°C	100	—	Increase in contaminants
3	Sputter 2	500 eV	5	7	Reduction in contaminants
4	Anneal 2	250°C	400	—	Reduction in contaminants (Clean)

I_{flux} is the flux ion current at the sample

C.5.4 Chapter 6: CZTS

Cleaning of the CZTS Crystal

The CZTS crystal required an active feedback cleaning procedure, with the details shown in Table C.6, with the results comments determining the subsequent cleaning cycle.

Annealing of the CZTS Crystal to 300 °C

The annealing details involved for the crystal up to 300 °C in order to induce the order/disorder transition are detailed in Table C.7. These details are in relation to Sect. 6.9.2.

Table C.6 In vacuo cleaning cycles for the single crystal CZTS

#	Cycle	Energy or Temperature	Time (min)	I_{flux} (μA)	Result	Spectral Reference
1	Prior to cleaning	—	—	—	—	*As In*
2	Anneal 1	240°C	1330	—	SnO_2 reduced	—
3	Sputter 1 Anneal 2	100 eV 240°C	4 3060	< 1[a] —	SnO_2 reduced	*Cl. 1*
4	Sputter 2–7 Anneal 3–8	100 eV 230°C	5×6 15×6	< 1[a] —	SnO_2 eliminated	*Cl. 2*
5	Sputter 8–13 Anneal 9–14	100 eV 230°C	5×6 15×6	< 1[a] —	No real change	*Cl. 3*
6	Sputter 14–17 Anneal 15–18	150 eV 230°C	5×4 15×4	< 1[a] —	No real change	—
7	Anneal 19	250°C	190	—	No real change	*Clean*

Multiple numbers represent repeated cycles. I_{flux} is the flux ion current at the sample
[a]Too low to measure

Table C.7 In vacuo annealing details for the single crystal CZTS up to 300 °C

#	Cycle	Temperature	Time (min)	Spectral Reference	Comment
1	Prior to annealing	—	—	*Clean*	—
2	Anneal 1	220°C	45	—	—
3	Anneal 2	250°C	105	—	—
4	Anneal 3	300°C	250	*@300°C*	Spectra at 300°C
5	Cooling	300°C	1885	*Cooling*	Spectra during cooling to 25°C
6	—	—	—	*Cold*	Spectra after cooling to 25°C

Annealing of the CZTS Crystal above 300 °C and Aggressive Sputtering
The annealing and sputtering details involved for annealing the crystal above 300 °C and sputtering the crystal hard to induce structural changes are detailed in Tables C.8 and C.9, respectively. These details are in relation to Sect. 6.9.3.

Table C.8 In vacuo annealing details for the single crystal CZTS above 300 °C

#	Cycle	Temperature	Time (min)	Spectral Reference	Comment
1	Prior to annealing	—	—	*Clean*	—
2	Anneal 1	150°C	20	—	—
3	Anneal 2	180°C	20	—	—
4	Anneal 3	250°C	15	—	—
5	Anneal 4	300°C	230	—	—
6	Anneal 5	330°C	75	—	—
7	Anneal 6	360°C	1235	*@360°C*	Spectra at 360°C
8	Anneal 7	410°C	130	*@410°C*	Spectra at 410°C
9	Anneal 8	450°C	175	*@450°C*	Spectra at 450°C
10	Anneal 9	475°C	1145	*@475°C*	Spectra at 475°C
11	Anneal 10	450°C	4230	*@450°C*	Spectra at 450°C
12	—	—	—	*Cold*	Spectra after cooling to 25°C

Table C.9 In vacuo aggressive sputtering and annealing details of the single crystal CZTS

#	Cycle	Energy or Temperature	Time (min)	Iflux (μA)	Spectral Reference	Comment
1	Prior to annealing	—	—	—	Clean	—
2	Sputter 1	2000 eV	30	20	Sp.	—
3	Anneal 1	150°C	15	—	—	—
4	Anneal 2	160°C	15	—	—	—
5	Anneal 3	180°C	10	—	—	—
6	Anneal 4	215°C	40	—	—	—
7	Anneal 5	260°C	15	—	—	—
8	Anneal 6	300°C	15	—	—	—
9	Anneal 7	320°C	40	—	—	—
10	Anneal 8	350°C	35	—	—	—
11	Anneal 9	400°C	45	—	—	—
12	Anneal 10	430°C	20	—	—	—
13	Anneal 11	460°C	105	—	—	—
14	Anneal 12	—	—	—	Ann.	Spectra after cooling to 25°C

I_{flux} is the flux ion current at the sample

Appendix D
Extra Data

D.1 Extra Data for Chap. 3

*D.1.1 The Fitting of Two S 2*p *Doublets*

In Sect. 3.3.2, the S 2*p* region was fitted with two doublets, which were attributed to the two different coordination environments for sulphur in the CAS crystal structure. This fitting model would appear somewhat tenuous, given the small shift between these two doublets, and therefore reasons for this fitting is presented here.

First, when fitting this region with a single S 2*p* doublet, no satisfactory fit could be achieved. A comparison for the TT sample between the two-doublet fit shown in Fig. 3.6, and a single-doublet fit is shown in Fig. D.1. The quality of fit for the two-doublet model is superior to that of the single-doublet model. With two doublets, the peak envelope (black line) matches the data very well; however, when using only one doublet, the peak envelope (pink shaded area) does not fully represent the data (green and yellow shaded areas). Such an asymmetric lack of fitting[8] cannot be rectified by lineshape changes and hence, two doublets are required.

Also, in a study of other samples of CAS, which were grown by different techniques, the two-doublet model applied to the S 2*p* regions of these samples also produced a more satisfactory fit; the results of which are shown in Fig. D.2. This demonstrates that there is strong evidence for the accuracy of the application of this model, and that the extra doublet is a feature of the material itself, rather than remnant from the growth, or other changes.

Furthermore, the two-doublet model results in a ratio between the doublets of 1:1: expected because of the mixing ratio of the different coordination environments in the crystal structure.[9]

[8]Under- and over-fitting both on the same side of the peak features.

[9]See Fig. 3.1.

© Springer International Publishing AG, part of Springer Nature 2018

T. J. Whittles, *Electronic Characterisation of Earth-Abundant Sulphides for Solar Photovoltaics*, Springer Theses, https://doi.org/10.1007/978-3-319-91665-1

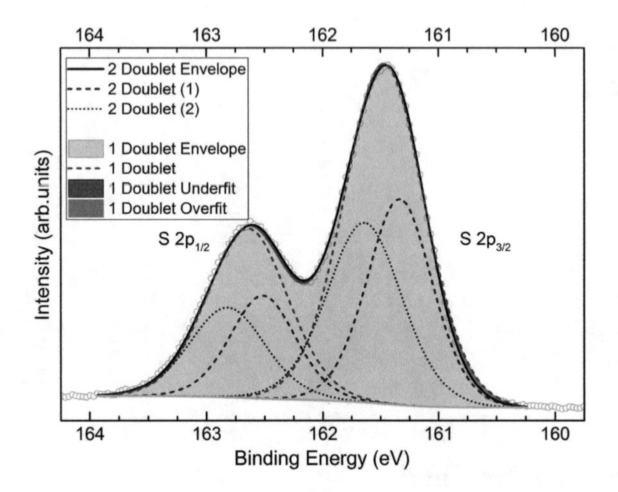

Fig. D.1 XPS S 2*p* region of the CAS sample measured in Chap. 3, demonstrating the quality of fit when using one or two S 2*p* doublets. Peak envelope of the two doublet model shown in black, peak envelope of the one doublet model shaded in pink, with the difference between this envelope and the experimental data shaded in green and yellow

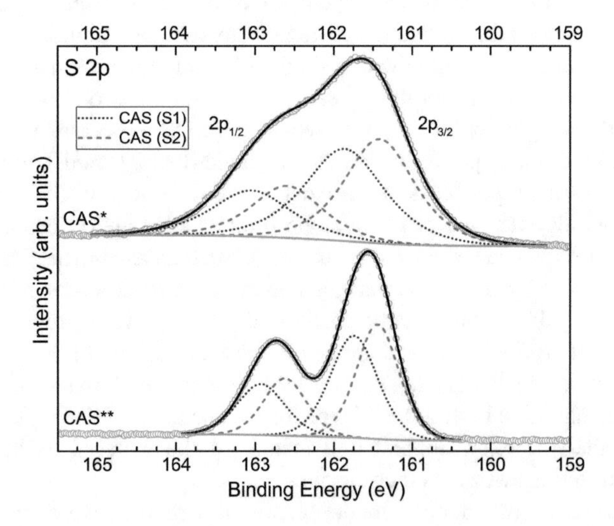

Fig. D.2 XPS S 2*p* region of independently measured samples of CAS (CAS* and CAS**), demonstrating the two doublet model. CAS* has noticeably wider peak FWHM because of the poorer crystallinity of that sample

D.2 Extra Data for Chap. 4

D.2.1 XPS CL Fitting Procedure for Bismuth and Sulphur

Because of the complexity of the overlapping region of the Bi 4*f* and S 2*p* doublets[10] for the CBS sample in Chap. 4, the S 2*s* and Bi 4*d* regions were also recorded in order to corroborate the fitting. These two regions should be representative of the main doublets, because, as they all lie within a similar KE regime for the ejected photoelectrons, their $IMFP_e$ [38] and photoelectron escape depths are similar, and therefore the ratios between the area of a bulk species[11] and surface species[12] should be the same.

In order to create a robust fit, some assumptions must be made in order to prevent an unmanageable number of variables. First, it was assumed that any peaks not arising from CBS will not shift BE due to cleaning, and will only change either in intensity, FWHM [39], or both. Therefore, these were fixed to the BE that was measured in the sample before surface cleaning, which was used to generate the fitting model because here the contaminant peaks are largest and this is where one has most confidence in the fitted BE.

Second, the number of species of an element is the same no matter which orbital is being used, and the ratios between the species within an orbital will be the same when considering different orbitals. This will be true for the areas of the peaks, because although the photoionisation cross sections will be different between orbitals, they are the same within an orbital. This will also be true for the FWHM of the peaks, because although the core-hole lifetimes will be different for the different species and different orbitals, it will be the same difference between the orbitals as the instrumental broadening remains the same throughout. Hence, for the other corroborative regions, these ratios were fixed.

Bearing these assumptions in mind, the fitting procedure used for the Bi and S peaks for CBS can be summarised as follows:

1. Determine the number, FWHM ratio and area ratio of S peaks for the sample before cleaning using the S 2*s* region.
2. Translate these peaks into S 2*p* doublets using appropriate constraints.
3. Determine the number, FWHM ratio and area ratio of Bi peaks for the sample before cleaning by modelling the remaining intensity with Bi 4*f* doublets.
4. Translate these doublets into Bi 4*d* doublets using appropriate constraints.
5. Translate all the above steps to the sample after cleaning to determine the effect of cleaning in terms of peak intensities and BE shifts for the peaks associated with CBS.

The sample before surface cleaning was chosen to be fitted first as these spectra contain the greater number of prominent peak crests, giving better confidence to the

[10]Shown in Fig. 4.6.
[11]Cu_3BiS_3.
[12]Bi_2O_3 or S–O.

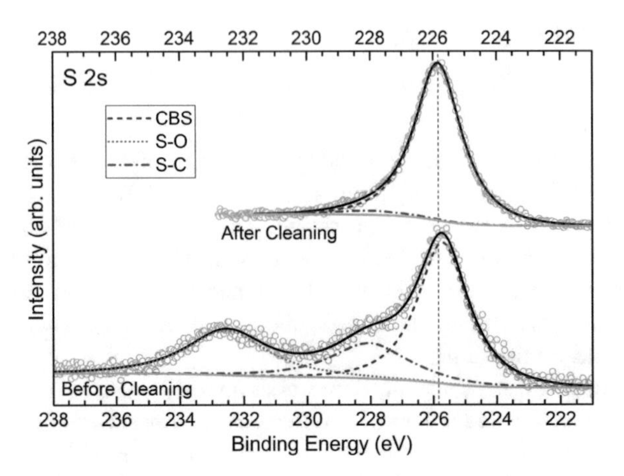

Fig. D.3 XPS S 2*s* region of the CBS sample from Chap. 4 before and after surface cleaning. Peak envelope shown in black

number of peaks chosen. The S 2*s* region was fitted first, as this would give the number of small S 2*p* peaks present in the overlap region, allowing more confidence in the number and positions of the larger Bi 4*f* peaks. Three peaks were determined in the S 2*s* region for the sample before cleaning, as shown in the bottom spectra in Fig. D.3. This then reduced to two peaks after cleaning, where one of the peaks is almost insignificant. The assignments for these peaks are in agreement with those for S 2*p* that are described in Sect. 4.3.2.

The S 2*s* peaks are then converted to S 2*p* doublets,[13] and are fixed with regards to area and FWHM ratios. The remaining area in the overlap region is then taken up by Bi 4*f* doublets. An initial model used two doublets for the sample before cleaning, one for Bi in CBS (red dash) and another at higher BE for oxidised bismuth (pink dot), these two doublets are evident from the peak crests seen in Fig. D.4.

Attempting to optimise the fit described above[14] for the Bi 4*f* and S 2*p* region of the sample before cleaning resulted in an unsatisfactory fit for several reasons, which are illustrated in the top spectra of Fig. D.4. There is always missing intensity at low BE around 157 eV (α), the S 2*p* doublet attributed to S–C (green shading) fits with a BE (β) that is too close to the BE of the S 2*p* doublet associated with CBS (red shading) to be in agreement with the literature [40], and several other features (γ) where the model either under- or overcompensates the experimental data. This suggests that another Bi doublet is present because of the missing intensity around 157 and 163 eV, which also corresponds to the doublet separation

[13]Using the parameters detailed in Sect. 4.2.2.
[14]Three S 2*p* doublets and two Bi 4*f* doublets.

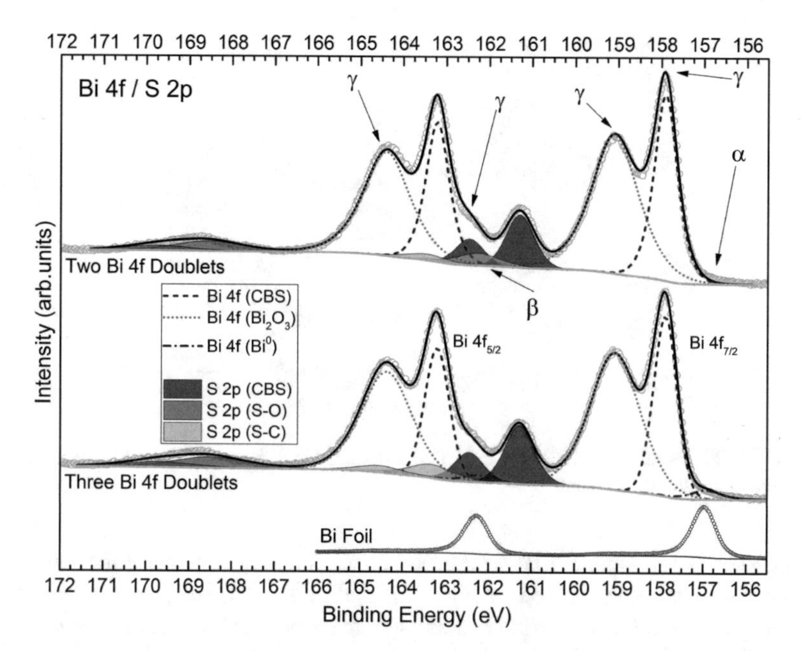

Fig. D.4 XPS Bi 4*f* and S 2*p* overlap region of the CBS sample from Chap. 4 with (middle spectrum) and without (top spectrum) including a contribution from metallic bismuth. Shown in the bottom spectrum is the Bi 4*f* region from an independently measured Bi foil. Peak envelopes shown in black

for Bi 4*f*. A doublet around these BE would be consistent with metallic bismuth,[15] and reasons for this assignment are discussed in Sect. 4.3.2. Confidence in this extra species is heightened by the fact that without it, line shape and background alterations would not give a better fit. With this doublet included (blue dash dot), a much more robust fit is achieved, as shown in the middle spectrum of Fig. D.4, and it is this model which is propagated to the clean sample, which also gives a satisfactory fit.[16] The area of the metallic bismuth peaks are small, and as such do not affect the BE of the other peaks. It is noted that the line shape of metallic species should be asymmetric [41]; however, because of the relative intensity of this species and the added complexity associated with asymmetrical line shapes, a symmetrical line shape models this species satisfactorily. The assignments for the peaks in this region can be found in Sect. 4.3.2.

The Bi 4*d* region was fitted only as corroboration to the peak model adopted in the overlap region and not as a dictatorial region like the S 2*s* region. This is because there are many reports on the S 2*s* region being used as an alternative [42–54], but few on using the Bi 4*d* region [55]. Hence, there is less confidence in the

[15]Also shown in Fig. D.4 is a metallic bismuth spectrum from an independently measured Bi foil (blue data).
[16]See Fig. 4.6.

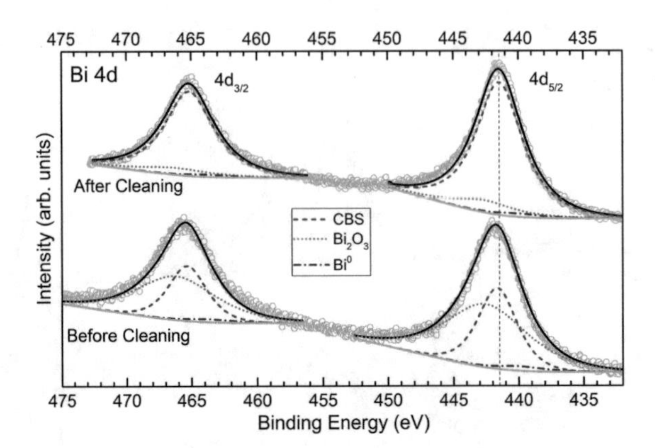

Fig. D.5 XPS Bi 4d region of the CBS sample from Chap. 4 before and after surface cleaning. Peak envelope shown in black

fitting used here. Also, the Bi 4*d* region has a poorer SNR because the intensity is lower and also, as evidenced in Fig. D.5, the width of these peaks is very large, meaning that peaks will overlap and as such, there will be less confidence using this region for an initial fit. Nevertheless, a corroborative fit was achieved, which is in agreement with the model used for the overlap region. Due to the size of the metallic bismuth peak (blue dash dot), the BE was constrained to the standard reference energy for metallic Bi $4d_{5/2}$ at 440 eV [56], in agreement with the Bi $4f_{7/2}$ peak located at the standard value for metallic bismuth as well. Background subtraction proved difficult here because the region lies toward the lower BE side of the Cu LMM Auger features, creating a steep background, as can be seen in Fig. D.5. As such, a Shirley background did not result in a good fit, so a Tougaard background was used [57]. It is believed that a modified Shirley background would provide the best fit; however, such rigour is unnecessary for this region. Using the appropriate constraints,[17] reasonable fits were achieved for the sample both before and after cleaning, strengthening confidence in the fit within the overlapping region. The largest peak in this region (red dash) corresponds to Bi in CBS, as it is in the overlap region, and the peak at higher BE (pink dot) is due to Bi_2O_3: in agreement the literature [58].

D.2.2 Binding Energies for the Secondary Regions

The binding energies of the fitted peaks in Sect. D.2.1 for the S 2*s* and Bi 4*d* regions for the CBS sample from Chap. 4 are detailed in Table D.1, with the binding energies for all of the fitted peaks in this study given in Tables 4.1, 4.2, 4.3 and 4.4.

[17]As detailed above and using the parameters detailed in Sect. 4.2.2.

Table D.1 XPS BE of the secondary regions of the CBS sample in Chap. 4 before and after surface cleaning

Sample	Cu_3BiS_3		Bi_2O_3	Contamination		
	Bi $4d_{5/2}$	S $2s$	Bi $4d_{5/2}$	Bi $4d_{5/2}$ (Bi^0)	S $2s$ (S-C)	S $2s$ (S-O)
Before Cleaning	441.63 (3.54)	225.71 (1.87)	442.52 (7.47)	440.00 (4.49)	228.04 (2.84)	232.55 (2.97)
After Cleaning	441.47 (4.22)	225.84 (1.75)	442.51 (5.61)	440.00 (4.51)	228.04 (2.84)	—

All values given in eV and the peak FWHM are given in parentheses

D.2.3 Calculated XRD Pattern

In Chap. 4, the calculated diffraction pattern for wittichenite Cu_3BiS_3 was presented, calculated using the method in Sect. 2.6.2. Here, the data generated from these calculations are given (Table D.2).

Table D.2 Calculated reflections and intensities for wittichenite [59] Cu_3BiS_3

2θ (°)	Intensity (Norm. %)	2θ (°)	Intensity (Norm. %)	2θ (°)	Intensity (Norm. %)
15.70	13.26	37.33	14.43	51.78	10.50
17.05	12.35	37.39	11.50	52.32	15.79
19.48	42.12	38.42	14.86	52.81	12.26
23.02	32.50	39.02	6.26	53.12	7.64
24.58	9.41	39.55	8.10	54.23	9.60
26.53	8.67	41.23	5.94	54.27	8.72
26.61	12.38	41.36	12.70	54.37	12.22
27.99	41.79	42.93	6.93	54.63	7.80
28.79	16.91	44.31	6.33	57.34	6.33
28.98	49.06	45.21	21.37	57.67	7.33
28.98	38.56	47.23	10.46	58.35	7.00
30.25	26.14	47.74	10.60	60.05	7.19
31.25	100.00	47.81	8.96	61.37	12.51
31.71	5.72	47.89	9.82	61.58	5.21
31.78	13.61	48.38	5.07	63.84	8.53
33.81	48.40	49.89	30.74	66.00	5.66
34.77	15.25	50.25	11.24	67.29	7.14
35.92	6.42	51.73	17.46		

All values given in eV and the peak FWHM are given in parentheses

D.3 Extra Data for Chap. 6

D.3.1 Calculated XRD Patterns

In Chap. 6, calculated diffraction patterns were presented, calculated using the method in Sect. 2.6.2. Here, the data generated from these calculations are given Tables D.3–D.10.

Sphalerite ZnS

Table D.3 Calculated reflections and intensities for sphalerite [60] ZnS

2θ (°)	Intensity (Norm. %)
28.57	100.00
33.10	11.88
47.52	60.82
56.38	40.85

Herzenbergite SnS

Table D.4 Calculated reflections and intensities for herzenbergite [61] SnS

2θ (°)	Intensity (Norm. %)	2θ (°)	Intensity (Norm. %)
22.00	57.13	45.56	38.43
27.50	100.00	51.15	11.83
31.55	70.15	51.37	20.87
31.66	35.08	54.33	18.60
32.00	95.16	56.54	13.16
39.08	28.67	56.76	35.11

Berndtite SnS_2

Table D.5 Calculated reflections and intensities for berndtite [62] SnS_2

2θ (°)	Intensity (Norm. %)
15.06	85.42
28.31	28.75
32.22	100.00
42.01	62.21
50.12	26.60
52.64	19.73
55.16	20.54
60.84	15.13

Ottemannite Sn_2S_3

Table D.6 Calculated reflections and intensities for ottemannite [60] Sn_2S_3

2θ (°)	Intensity (Norm. %)	2θ (°)	Intensity (Norm. %)	2θ (°)	Intensity (Norm. %)
12.62	15.28	31.20	40.97	42.71	14.88
16.10	60.77	31.86	100.00	46.89	68.38
21.48	71.20	32.53	26.50	48.05	14.59
23.71	27.52	33.50	24.69	48.51	28.78
26.56	29.78	33.78	39.45	49.43	21.06
27.70	22.09	37.93	16.12	53.63	12.51
28.79	44.71	39.87	14.53		
30.87	20.61	40.13	52.51		

Covellite CuS

Table D.7 Calculated reflections and intensities for covellite [64] CuS

2θ (°)	Intensity (Norm. %)
27.13	10.78
27.68	14.75
29.28	45.02
31.79	100.00
32.87	44.12
38.85	10.47
47.93	72.26
58.67	16.24
59.34	42.87

Chalcocite Cu_2S

Table D.8 Calculated reflections and intensities for chalcocite [65] Cu_2S

2θ (°)	Intensity (Norm. %)	2θ (°)	Intensity (Norm. %)	2θ (°)	Intensity (Norm. %)
23.81	12.94	32.74	10.51	37.42	33.89
23.83	14.38	32.86	38.32	37.48	35.90
24.75	10.46	33.58	10.76	38.61	14.07
26.88	10.53	33.59	13.93	38.64	18.56
26.89	11.16	34.21	15.16	38.64	12.10
27.20	13.90	35.00	10.10	40.21	14.88
27.22	12.87	35.41	13.24	40.81	33.06
27.98	17.34	35.44	12.15	45.95	100.00
28.00	17.13	35.52	10.05	46.53	20.53
28.24	14.55	36.23	24.18	48.37	44.41
28.26	14.57	36.27	20.83	48.40	45.98
30.25	16.90	36.37	14.83	48.47	17.21
30.28	18.26	37.37	37.15	53.63	18.83
32.37	10.74	37.40	32.87	53.81	10.52

CTS

Table D.9 Calculated reflections and intensities for CTS [66]

2θ (°)	Intensity (Norm. %)
28.46	100.00
32.97	16.48
47.33	58.10
56.15	41.22
58.88	4.21

Kesterite CZTS

Table D.10 Calculated reflections and intensities for kesterite [67, 68] CZTS

2θ (°)	Intensity (Norm. %)	hkl	2θ (°)	Intensity (Norm. %)	hkl
16.34	1.32	002	47.33	39.04	204
18.25	4.82	101	50.35	0.30	222
23.13	1.48	110	50.46	0.15	006
28.45	100.00	112	51.08	0.55	301
29.66	1.99	103	53.27	0.30	310
32.94	10.34	200	56.10	27.47	312
33.02	5.14	004	56.20	13.66	116
36.98	1.05	202	56.81	0.18	303
37.91	2.08	211	56.86	0.52	215
40.70	0.69	114	58.87	4.01	224
44.91	0.98	213	61.58	0.11	206
44.97	0.63	105	62.13	0.30	321/107
47.28	19.59	220	64.11	0.38	314

References

1. International Energy Agency (IEA). Key World Energy Statistics 2016; 2016.
2. Jean J, Brown PR, Jaffe RL, Buonassisi T, Bulović V. Pathways for solar photovoltaics. Energy Environ. Sci. 2015; 8(4):1200–19.
3. Green MA, Emery K, Hishikawa Y, Warta W, Dunlop ED, Levi DH, Ho-Baillie AWY. Solar cell efficiency tables (version 49). Prog. Photovoltaics Res. Appl. 2017;25(1):3–13.
4. Banu S, Ahn SJ, Ahn SK, Yoon K, Cho A. Fabrication and characterization of cost-Efficient $CuSbS_2$ thin film solar cells using hybrid inks. Sol. Energy Mater. Sol. Cells 2016;151:14–23.
5. Sinsermsuksakul P, Sun L, Lee SW, Park HH, Kim SB, Yang C, Gordon RG. Overcoming efficiency limitations of SnS-based solar cells. Adv. Energy Mater. 2014;4(15):1400496.
6. Sun K, Yan C, Liu F, Huang J, Zhou F, Stride JA, Green M, Hao X. Over 9% efficient kesterite Cu_2ZnSnS_4 solar cell fabricated by using $Zn_{1-X}Cd_XS$ buffer layer. Adv. Energy Mater. 2016;6(12):1600046.
7. U.S. Geological Survey. Mineral Commodity Summaries 2017; 2017.
8. ASTM International. Standard Tables for Reference Solar Spectral Irradiances: Direct Normal and Hemispherical on 37° Tilted Surface. ASTM G173-03(2012). West Conshohocken 2012.
9. Woodruff DP, Delchar TA. 1 - Introduction. In: Modern techniques of surface science. Cambridge: Cambridge University Press; 1994; p. 1–14.
10. Hofmann P. Surface physics: an introduction. Philip Hofmann; 2013.
11. Weston GF. Ultrahigh vacuum practice. Cambridge: Butterworth & Co. Ltd; 1985.
12. Audi M, de Simon M. Ion pumps. Vacuum 1987;37(8–9):629–36.

13. Chung Y-W. Fundamental concepts in ultrahigh vacuum, surface preparation, and electron spectroscopy. In: Practical guide to surface science and spectroscopy. Amsterdam: Elsevier; 2001; p. 1–22.
14. Yeh JJ, Lindau I. Atomic subshell photoionization cross sections and asymmetry parameters: $1 \leqslant Z \leqslant 103$. At. Data Nucl. Data Tables 1985;32(1):1–155.
15. Krukau AV, Vydrov OA, Izmaylov, AF, Scuseria GE. Influence of the exchange screening parameter on the performance of screened hybrid functionals. J. Chem. Phys. 2006;125 (22):224106.
16. Perdew JP, Burke K, Ernzerhof M. Generalized gradient approximation made simple. Phys. Rev. Lett. 1996;77(18):3865–8.
17. Kumar M, Persson C. $CuSbS_2$ and $CuBiS_2$ as potential absorber materials for thin-film solar cells. J. Renew. Sustain. Energy 2013;5(3):31616.
18. Temple DJ, Kehoe AB, Allen JP, Watson GW, Scanlon DO. Geometry, electronic structure, and bonding in $CuMCh_2$ (M = Sb, Bi; Ch = S, Se): alternative solar cell absorber materials? J. Phys. Chem. C 2012;116(13):7334–40.
19. Yang RX, Butler KT, Walsh A. Assessment of hybrid organic–inorganic antimony sulfides for earth-abundant photovoltaic applications. J. Phys. Chem. Lett. 2015;6(24):5009–14.
20. Kehoe AB, Temple DJ, Watson GW, Scanlon DO. Cu_3MCh_3 (M = Sb, Bi; Ch = S, Se) as candidate solar cell absorbers: insights from theory. Phys. Chem. Chem. Phys. 2013;15 (37):15477.
21. Kumar M, Persson C. Cu_3BiS_3 as a potential photovoltaic absorber with high optical efficiency. Appl. Phys. Lett. 2013;102(6):3–7.
22. Tumelero MA, Faccio R, Pasa AA. Unraveling the native conduction of trichalcogenides and its ideal band alignment for new photovoltaic interfaces. J. Phys. Chem. C 2016;120(3):1390–99.
23. Ganose AM, Butler KT, Walsh A, Scanlon DO. Relativistic electronic structure and band alignment of BiSI and BiSeI: candidate photovoltaic materials. J. Mater. Chem. A 2016;4:2060–8.
24. Savory CN, Ganose AM, Scanlon DO. Exploring the PbS–Bi_2S_3 series for next generation energy conversion materials. Chem. Mater. 2017;29(12):5156–7.
25. Paier J, Asahi R, Nagoya A, Kresse G. Cu_2ZnSnS_4 as a potential photovoltaic material: a hybrid Hartree-Fock density functional theory study. Phys. Rev. B 2009;79(11):115126.
26. Botti S, Kammerlander D, Marques MAL. Band structures of Cu_2ZnSnS_4 and $Cu_2ZnSnSe_4$ from many-body methods. Appl. Phys. Lett. 2011;98(24):241915.
27. Blöchl PE. Projector augmented-wave method. Phys. Rev. B 1994;50(24):17953–79.
28. Pulay P. Ab initio calculation of force constants and equilibrium geometries in polyatomic molecules. Mol. Phys. 1969;17(2):197–204.
29. Grimme S, Antony J, Ehrlich S, Krieg H. A consistent and accurate Ab initio parametrization of density functional dispersion correction (DFT-D) for the 94 elements H-Pu. J. Chem. Phys. 2010;132(15):154104.
30. Grimme S, Ehrlich S, Goerigk L. Effect of the damping function in dispersion corrected density functional theory. J. Comput. Chem. 2011;32(7):1456–65.
31. Blöchl PE. Projector augmented-wave method. Phys. Rev. B 1994;50(24):17953–79.
32. Kresse G, Joubert D. From ultrasoft pseudopotentials to the projector augmented-wave method. Phys. Rev. B 1999;59(3):1758–75.
33. Monkhorst HJ, Pack JD. Special points for brillouin-zone integrations. Phys. Rev. B 1976;13 (12):5188–92.
34. Heyd J, Scuseria GE, Ernzerhof M. Hybrid functionals based on a screened coulomb potential. J. Chem. Phys. 2003;118(18):8207–15.
35. Heyd J, Scuseria GE, Ernzerhof M. Erratum: "Hybrid functionals based on a screened coulomb potential" [J. Chem. Phys. 118, 8207 (2003)]. J. Chem. Phys. 2006;124(21):219906.
36. Blöchl PE, Jepsen O, Andersen OK. Improved tetrahedron method for brillouin-zone integrations. Phys. Rev. B 1994;49(23):16223–33.

37. Powell CJ, Jablonski A, Salvat F. NIST databases with electron elastic-scattering cross sections, inelastic mean free paths, and effective attenuation lengths. Surf. Interface Anal. 2005;37(11):1068–71.
38. Nelin CJ, Bagus PS, Brown MA, Sterrer M, Freund HJ. Analysis of the broadening of x-ray photoelectron spectroscopy peaks for ionic crystals. Angew. Chemie - Int. Ed. 2011;50 (43):10174–7.
39. Paraknowitsch JP, Thomas A, Schmidt J. Microporous sulfur-doped carbon from thienyl-based polymer network precursors. Chem. Commun. 2011;47(29):8283.
40. Doniach S, Sunjic M. Many-electron singularity in x-ray photoemission and X-ray line spectra from metals. J. Phys. C Solid State Phys. 1970;3(2):285–91.
41. Yin J, Jia J. Synthesis of Cu_3BiS_3 nanosheet films on TiO_2 nanorod arrays by a solvothermal route and their photoelectrochemical characteristics. CrystEngComm 2014;16(13):2795.
42. Grigas J, Talik E, Lazauskas V. X-ray photoelectron spectra and electronic structure of Bi_2S_3 crystals. Phys. status solidi 2002;232(2):220–30.
43. Debies TP, Rabalais JW. X-ray photoelectron spectra and electronic structure of Bi_2X_3 (X = O, S, Se, Te). Chem. Phys. 1977;20(2):277–83.
44. Chen R, So MH, Che C-M, Sun H. Controlled synthesis of high crystalline bismuth sulfide nanorods: using bismuth citrate as a precursor. J. Mater. Chem. 2005;15(42):4540.
45. Fang Z, Liu Y, Fan Y, Ni Y, Wei X, Tang K, Shen J, Chen Y. Epitaxial growth of CdS nanoparticle on Bi_2S_3 nanowire and photocatalytic application of the heterostructure. J. Phys. Chem. C 2011;115(29):13968–76.
46. Liao X-H, Wang H, Zhu J-J, Chen H-Y. Preparation of Bi_2S_3 nanorods by microwave irradiation. Mater. Res. Bull. 2001;36(13–14):2339–46.
47. Liufu S-C, Chen L-D, Yao Q, Wang C-F. Bismuth sulfide thin films with low resistivity on self-assembled monolayers. J. Phys. Chem. B 2006;110(47):24054–61.
48. Panigrahi PK, Pathak A. The growth of bismuth sulfide nanorods from spherical-shaped amorphous precursor particles under hydrothermal condition. J. Nanoparticles 2013;2013, 1–11.
49. Purkayastha A, Yan Q, Raghuveer MS, Gandhi DD, Li H, Liu ZW, Ramanujan RV, Borca-Tasciuc T, Ramanath G. Surfactant-directed synthesis of branched bismuth telluride/sulfide core/shell nanorods. Adv. Mater. 2008;20(14):2679–83.
50. Tamašauskaitė Tamašiūnaitė L, Šimkūnaitė-Stanynienė B, Naruškevičius L, Valiulienė G, Žielienė A, Sudavičius A. EQCM study of electrochemical modification of Bi_2S_3 films in the Zn^{2+}-containing electrolyte. J. Electroanal. Chem. 2009;633(2):347–53.
51. Tamašauskaitė-Tamašiūnaitė L, Valiulienė G, Žielienė A, Šimkūnaitė-Stanynienė B, Naruškevičius L, Sudavičius A. EQCM study on the oxidation/reduction of bismuth sulfide thin films. J. Electroanal. Chem. 2010;642(1):22–9.
52. Wang H, Zhu J-J, Zhu J-M, Chen H-Y. Sonochemical method for the preparation of bismuth sulfide nanorods. J. Phys. Chem. B 2002;106(15):3848–54.
53. Zhong J, Xiang W, Liu L, Yang X, Cai W, Zhang J, Liang X. Biomolecule-assisted solvothermal synthesis of bismuth sulfide nanorods. J. Mater. Sci. Technol. 2010;26(5):417–22.
54. Debies TP, Rabalais JW. X-ray photoelectron spectra and electronic structure of Bi_2X_3 (X = O, S, Se, Te). Chem. Phys. 1977;20(2):277–83.
55. Nyholm R, Berndtsson A, Martensson N. Core level binding energies for the elements Hf to Bi (Z = 72-83). J. Phys. C Solid State Phys. 1980;13(36):L1091–6.
56. Repoux M. Comparison of background removal methods for XPS. Surf. Interface Anal. 1992;18(7):567–70.
57. Debies TP, Rabalais JW. X-ray photoelectron spectra and electronic structure of Bi_2X_3 (X = O, S, Se, Te). Chem. Phys. 1977;20(2):277–83.
58. Kocman V, Nuffield EW. The crystal structure of wittichenite, Cu_3BiS_3. Acta Crystallogr. Sect. B Struct. Crystallogr. Cryst. Chem. 1973;29(11):2528–35.
59. Skinner BJ. Unit cell edges of natural and synthetic sphalerites. Am. Mineral. 1961;46 (11/12):1388–411.

60. Hofmann W. Ergebnisse Der Strukturbestimmung Komplexer Sulfide. Zeitschrift für Krist. - Cryst. Mater. 1935;92(1–6).
61. Hazen RM, Finger LW. The crystal structures and compressibilities of layer minerals at high pressure. I. SnS_2 Berndtite. Am. Mineral. 1978;63:289–92.
62. Kniep R, Mootz D, Severin U, Wunderlich H. Structure of tin(II) tin(IV) trisulphide, a redetermination. Acta Crystallogr. Sect. B Struct. Crystallogr. Cryst. Chem. 1982;38(7):2022–3.
63. Evans Jr, HT, Konnert JA. Crystal structure refinement of covellite. Am. Mineral. 1976;61:996–1000.
64. Evans HT. Crystal structure of low chalcocite. Nat. Phys. Sci. 1971;232(29):69–70.
65. Chen X, Wada H, Sato A, Mieno M. Synthesis, electrical conductivity, and crystal structure of $Cu_4Sn_7S_{16}$ and structure refinement of Cu_2SnS_3. J. Solid State Chem. 1998;139(1):144–51.
66. Hall SR, Szymanski JT, Stewart JM. Kesterite, $Cu_2(Zn,Fe)SnS_4$ and Stannite $Cu_2(Fe,Zn)SnS_4$, structurally similar but distinct minerals. Can. Mineral. 1978;16(2):131–7.
67. Lafond A, Choubrac L, Guillot-Deudon C, Fertey P, Evain M, Jobic S. X-ray resonant single-crystal diffraction technique, a powerful tool to investigate the kesterite structure of the photovoltaic Cu_2 $ZnSnS_4$ compound. Acta Crystallogr. Sect. B Struct. Sci. Cryst. Eng. Mater. 2014;70(2):390–4.
68. Fluck E. New notations in the periodic table. Pure Appl. Chem. 1988;60(3):431–6.

Glossary

The work presented throughout this thesis involves concepts pertinent to physics, chemistry and materials science, and as such, terminology can overlap to a certain extent and become confusing. Certain protocols have therefore been followed throughout this work to keep this confusion to a minimum. Where this differs from the standard, or where there is the possibility of confusion, clarifications and justifications are made here:

- Contemporary British English spelling and grammatical preference is used throughout, except in previous publications that required American English.
- The symbol '$\sim$' is used to mean 'of the order of', rather than 'approximately'. Where 'approximately' is required, the symbol '$\approx$' is used.
- All element and orbital symbols have their usual meanings.
- Simplified Kröger-Vink notation is used to represent defects, where the charge state is not given because it is not of interest here. So X_Y represents a defect where species X occupies the usual site of species Y. X and Y can be elements, or a vacancy, V.
- References to groups of the periodic table use the old notation with roman numerals rather than the IUPAC-approved notation [69]. As such, copper is group I, rather than group 11. This use is common within semiconductor research because it can be intuitively used to describe the electron configurations of certain elements.
- The solar cells studied throughout this thesis deal only with heterojunction devices consisting of a pn junction. The term 'absorber layer' refers to the p-type material unless explicitly stated. The term 'window layer' refers to the n-type material, being preferred over 'buffer layer', with regards to the materials studied here.
- The term 'bandgap' is used throughout, as opposed to 'band gap' or 'band-gap' in order to emphasise the fact that this term is a single concept, removing ambiguity from terms such as wide-bandgap.

© Springer International Publishing AG, part of Springer Nature 2018

T. J. Whittles, *Electronic Characterisation of Earth-Abundant Sulphides for Solar Photovoltaics*, Springer Theses, https://doi.org/10.1007/978-3-319-91665-1

- The term ‘x-ray’ is used throughout, as opposed to ‘xray’, ‘X-ray’, or ‘X ray’ because it is the best form of a clumsy term in my opinion. Better would be to adopt a descriptive term, such as the German ‘Röntgenstrahlung’, paying homage to the discoverer, and this term is not totally unfeasible given that the related ‘bremsstrahlung’ is a term that is already in use in English.
- The term ‘metallic’, as used here, refers not only to elements that are classified as metals, but any element that has an oxidation state of zero.
- Polycrystalline samples that have the possibility of contamination or impurities, are termed ‘technical’.
- The number of elements in a compound leads to the terms: ‘binary’, ‘ternary’, and ‘quaternary’ compounds. Here, these terms are applied only to the parent materials or structures, rather than any subsequently derived mixtures, alloys or substitutions. For example, $CuInSe_2$ and $CuGaSe_2$ are ternary compounds, but $Cu(In,Ga)Se_2$ (CIGS), an alloy of these two materials, is not called a quaternary compound as it isn’t a different structure like CZTS: a true quaternary compound.
- For band offsets of heterojunctions, the sign of the value of the offset is determined by the difference between the second material and the first material. For example, the valence band offset (VBO) of a heterojunction A/B, is the position of the valence band of material B minus the position of the valence band of material A. Consequently, a positive VBO means that valence band of material A is above that of material B on an energy band diagram, and vice versa. This matter is explained fully in Sect. 1.2.2, using Fig. 1.8.
- Problems involved with the pluralisation of terms that can be replaced with an acronym or initialism are removed by making no change to the acronym when used in its plural sense. For example, BE can stand for both ‘binding energy’ and ‘binding energies’, with context dictating which is intended. This removes the clumsiness of appending acronyms with an ‘s’, and avoids confusion of how to pluralise FWHM.
- Acronyms and initialisms are treated grammatically as though the full representation is voiced, rather than the shortened version. For example ‘a SEC value’ is correct and reads as ‘a secondary electron cutoff value’, rather than ‘an SEC value’.
- With regards to spectra obtained via a certain type of spectroscopy, for which there exists an acronym or initialism, the full name of the spectroscopic technique becomes the noun adjunct, rather than exchanging ‘spectroscopy’ for ‘spectrum’. As such, the term ‘photoemission spectroscopy (PES) spectrum’ is used rather than ‘photoemission spectrum’. This avoids the breaking of acronyms such as ‘PE spectrum’.
- For terms which are generally represented by a symbol, but have names consisting of multiple words, and which also appear frequently in the discussion text as prose, these are represented variously by their corresponding symbol in cases such as equations and diagrams, and by their acronym or initialism within prose. For example, ϕ and WF both representing the term ‘work function’.

Printed in the United States
By Bookmasters